REVIEWS OF REACTIVE INTERMEDIATE CHEMISTRY

BICENTENNIAL
1807
(W)WILEY
2007
BICENTENNIAL

THE WILEY BICENTENNIAL–KNOWLEDGE FOR GENERATIONS

*E*ach generation has its unique needs and aspirations. When Charles Wiley first opened his small printing shop in lower Manhattan in 1807, it was a generation of boundless potential searching for an identity. And we were there, helping to define a new American literary tradition. Over half a century later, in the midst of the Second Industrial Revolution, it was a generation focused on building the future. Once again, we were there, supplying the critical scientific, technical, and engineering knowledge that helped frame the world. Throughout the 20th Century, and into the new millennium, nations began to reach out beyond their own borders and a new international community was born. Wiley was there, expanding its operations around the world to enable a global exchange of ideas, opinions, and know-how.

For 200 years, Wiley has been an integral part of each generation's journey, enabling the flow of information and understanding necessary to meet their needs and fulfill their aspirations. Today, bold new technologies are changing the way we live and learn. Wiley will be there, providing you the must-have knowledge you need to imagine new worlds, new possibilities, and new opportunities.

Generations come and go, but you can always count on Wiley to provide you the knowledge you need, when and where you need it!

WILLIAM J. PESCE
PRESIDENT AND CHIEF EXECUTIVE OFFICER

PETER BOOTH WILEY
CHAIRMAN OF THE BOARD

REVIEWS OF REACTIVE INTERMEDIATE CHEMISTRY

Edited by

Matthew S. Platz
Department of Chemistry
Ohio State University
Columbus, OH

Robert A. Moss
Department of Chemistry
Rutgers University
New Brunswick, NJ

Maitland Jones, Jr.
Department of Chemistry
Princeton University
Princeton, NJ

WILEY-INTERSCIENCE

A John Wiley & Sons, Inc., Publication

Published by John Wiley & Sons, Inc., Hoboken, New Jersey
Published simultaneously in Canada

For general information on our other products and services or for technical support, please contact our
Customer Care Department within the United States at (800) 762-2974, outside the United States at
(317) 572-3993 or fax (317) 572-4002.

Wiley also publishes its books in a variety of electronic formats. Some content that appears in print
may not be available in electronic formats. For more information about Wiley products, visit our web
site at www.wiley.com.

Wiley Bicentennial Logo: Richard J. Pacifico

Library of Congress Cataloging in Publication Data is available.

ISBN: 978-0-471-73166-5

Printed in the United States of America

10 9 8 7 6 5 4 3 2 1

CONTENTS

■ PREFACE

In 2004, Moss, Platz and Jones edited Reactive Intermediate Chemistry. This book contained chapters written by leading experts on the chemistry of the reactive intermediates commonly encountered in mechanistic organic chemistry; carbocations, radicals, carbanions, singlet and triplet carbenes, nitrenes and nitrenium ions. A three-dimensional approach was offered integrating venerable methods of chemical analysis of reaction products, direct observational studies of reactive intermediates (RI's) and high accuracy calculations of the geometries, potential energy surfaces and spectra of RI's. The book was aimed at beginning graduate students and newcomers to a particular field to provide him or her with an introductory chapter that would rapidly allow them to pursue advanced work.

Such is the richness and intellectual vibrancy of the field of RI chemistry that an additional book was needed to cover silicon, germanium and tin centered RI's, as well as tetrahedral intermediates and topics of increasing importance such as quantum mechanical tunnelling, conical intersections, solid-state chemistry, and combustion chemistry. These topics are covered in this new book.

We hope *Reviews of Reactive Intermediate Chemistry* well captures the continuing evolution and breadth of Reactive Intermediate Chemistry, assists chemists to appreciate the state of the art and encourages new research in this area.

MATTHEW S. PLATZ
ROBERT A. MOSS
MAITLAND JONES, JR.

M. J. Bearpark
Department of Chemistry
Imperial College London
South Kensington campus
London SW7 2AZ UK
email: m.bearpark@imperial.ac.uk

L. M. Campos
University of California, Los Angeles
Department of Chemistry and
  Biochemistry
607 Charles E. Young Drive East
Los Angeles, CA 90095-1569
email: lcampos@chem.ucla.edu

M. Garcia-Garibay
University of California, Los Angeles
Department of Chemistry and
  Biochemistry
607 Charles E. Young Drive East
Los Angeles, CA 90095-1569
email: mgg@chem.ucla.edu

K. S. Gates
Department of Chemistry
125 Chem Bldg.
University of Missouri
Columbia, MO 65211
email: GatesK@missouri.edu

J. P. Guthrie
The University of Western Ontario
Department of Chemistry
London, Ontario, Canada
N6A 5B8
email: peter.guthrie@uwo.ca

C. M. Hadad
Department of Chemistry
Ohio State University
100 West 18th Avenue
Columbus, OH 43210
email: hadad.1@osu.edu

W. M. Kwok
Department of Chemistry
The University of Hong Kong
Pokfulam Road
Hong Kong
email: kwokwm@hkucc.hku.hk

V. Ya. Lee
Department of Chemistry
Graduate School of Pure and Applied
  Sciences
University of Tsukuba
Tsukuba, Ibaraki 305-8571, Japan
email: leevya@chem.tsukuba.ac.jp

C. Ma
Department of Chemistry
The University of Hong Kong
Pokfulam Road
Hong Kong
email: macs@hkucc.hku.hk

J. K. Merle
National Institute of Standards
   and Technology
100 Bureau Drive, Stop 8380
Gaithersburg, MD 20899-8380
email: john.merle@nist.gov

D. L. Phillips
Department of Chemistry
The University of Hong Kong
Pokfulam Road
Hong Kong
email: phillips@hkucc.hku.hk

M. A. Robb
Department of Chemistry
Imperial College London
South Kensington campus
London SW7 2AZ UK
email: mike.robb@imperial.ac.uk

A. Sekiguchi
Department of Chemistry
Graduate School of Pure and Applied
   Sciences
University of Tsukuba
Tsukuba, Ibaraki 305-8571, Japan
email: sekiguch@chem.tsukuba.ac.jp

R. S. Sheridan
Department of Chemistry/216
University of Nevada, Reno
Reno, NV 89557
email: rss@chem.unr.edu

J. P. Toscano
Department of Chemistry
Johns Hopkins University
3400 N. Charles Street
Baltimore, MD 21218
email: jtoscano@jhu.edu

P. G. Wenthold
Department of Chemistry
Purdue University
560 Oval Drive
West Lafayette, Indiana, 47907-2084
email: pgw@purdue.edu

PART 1

REACTIVE INTERMEDIATES

Tetrahedral Intermediates Derived from Carbonyl Compounds, Pentacoordinate Intermediates Derived from Phosphoryl and Sulfonyl Compounds, and Concerted Paths Which Avoid Them

J. PETER GUTHRIE

Department of Chemistry, University of Western Ontario, London, Ontario, Canada N6A 5B7

Reviews of Reactive Intermediate Chemistry. Edited by Matthew S. Platz, Robert A. Moss, Maitland Jones, Jr.
Copyright © 2007 John Wiley & Sons, Inc.

1.1. TETRAHEDRAL INTERMEDIATES

This chapter will deal mainly with tetrahedral intermediates from carbonyl derivatives, with some discussion on the much less-studied analogs for phosphorus and sulfur. It will also address the issue of concerted mechanisms which can sometimes bypass these intermediates.

   Carbonyl reactions are extremely important in chemistry and biochemistry, yet they are often given short shrift in textbooks on physical organic chemistry, partly because the subject was historically developed by the study of nucleophilic substitution at saturated carbon, and partly because carbonyl reactions are often more difficult to study. They are generally reversible under usual conditions and involve complicated multistep mechanisms and general acid/base catalysis. In thinking about carbonyl reactions, I find it helpful to consider the carbonyl group as a (very) stabilized carbenium ion, with an O^- substituent. Then one can immediately draw on everything one has learned about carbenium ion reactivity and see that the reactivity order for carbonyl compounds:

$$CH_2{=}O > CH_3CH{=}O > PhCH{=}O > (CH_3)_2C{=}O > CH_3COPh$$

corresponds almost perfectly to the order for carbenium ions (see Table 1.1).

$$CH_3CH_2^+ > (CH_3)_2CH^+ > Ph(CH_3)CH^+ \sim (CH_3)_3C^+ > (CH_3)_2(Ph)C^+$$

   The difference between carbonyl chemistry and (simple) carbocation chemistry is a result of much greater stability of the carbonyl group relative to a simple carbenium

TABLE 1.1. Reactivity of carbonyl compounds and carbenium ions.[a]

	$CH_3CH_2^+$	$(CH_3)_2CH^+$	$Ph(CH_3)CH^+$	$(CH_3)_3C^+$	$(CH_3)_2(Ph)C^+$
pK_R^+	-29.6[b]	-22.7[b]	-16.2[c]	-16.4[d]	-13.1[e]
	$CH_2{=}O$	$CH_3CH{=}O$	$PhCH{=}O$	$(CH_3)_2C{=}O$	CH_3COPh
$\log K_{H_2O}$[f]	1.61	-1.72	-3.82	-4.60	-6.92

[a]All in aqueous solution at 25°C; standard states are 1M ideal aqueous solution with an infinitely dilute reference state, and for water the pure liquid.
[b]Reference 1.
[c]Reference 2.
[d]Reference 3.
[e]Reference 4.
[f]Reference 5.

ion. This means that for many carbonyl group/nucleophile combinations the carbonyl compound is more stable than the adduct, which is not the case for what are traditionally considered carbenium ions until one gets to stabilized triaryl cations (e.g., crystal violet) or to very non-nucleophilic solvents such as magic acid.[6]

Thus carbonyl chemistry can be considered as analogous to S_N1 chemistry and is in fact inherently faster than S_N2 chemistry (not that S_N2 reactions cannot be fast, but this requires a strong thermodynamic driving force: for a comparable driving force the carbonyl reaction is faster).

The big difference is that for simple carbenium ions the cation is a transient intermediate and the covalent adduct is the normally encountered form, while for carbonyl compounds the "carbenium ion" is the stable form (with a few exceptions) and the covalent adduct is the transient intermediate. In fact, in many cases, the tetrahedral intermediate is too unstable to be detected (at least with current techniques) and yet the rate of overall reaction is strongly influenced by the height of this thermodynamic barrier. By Hammond's Postulate, a reaction leading to a high energy intermediate will have a transition state resembling this intermediate in structure and energy. If we can estimate the energy of the intermediate, then we have taken the first step toward estimating the rate of reaction.

For many carbonyl reactions, attempts have been made to prepare catalytic antibodies which accelerate the reaction. Such antibodies are normally obtained by challenging the immune system of a suitable animal with a compound resembling the tetrahedral intermediate in the reaction of interest. The idea is that if the antibody binds to and thus stabilizes the tetrahedral intermediate it will facilitate the reaction.[7,8] If the intermediate is a tetrahedral intermediate based on carbon then the analog is often a phosphate or phosphonate derivative, which is a stable tetrahedral species with a geometry and surface charge distribution resembling those of the intermediate in the reaction to be catalyzed.[9] A complimentary idea is that anything which resembles the transition state for an enzyme-catalyzed reaction, but is unreactive, will be a very strong inhibitor of that reaction.[10,11] Thus mimics of the tetrahedral intermediate can be strong inhibitors of enzymes catalyzing reactions which proceed by way of reactive tetrahedral intermediates.

1.1.1. Evidence for Tetrahedral Species as Reactive Intermediates

As early as 1899, Stieglitz[12] proposed a tetrahedral intermediate for the hydrolysis of an imino ether to an amide. Thus it was clear quite early that a complicated overall transformation, imino ether to amide, would make more sense as the result of a series of simple steps. The detailed mechanism proposed, although reasonable in terms of what was known and believed at the time, would no longer be accepted, but the idea of tetrahedral intermediates was clearly in the air. Stieglitz stated of the aminolysis of an ester that "it is now commonly supposed that the reaction takes place with the formation of an intermediate product as follows:" referring to work of Lossen.[13] (Note that the favored tautomer of a hydroxamic acid was as yet unknown.)

$$C_6H_5-C\overset{O}{\underset{OC_2H_5}{\diagup}} \quad + \quad NH_2OH \quad \rightleftharpoons \quad C_6H_5-\overset{OH}{\underset{OC_2H_5}{\overset{|}{C}}}-NHOH \quad \rightleftharpoons \quad C_6H_5-C\overset{OH}{\underset{NOH}{\diagup}} \quad + \quad C_2H_5OH$$

For many reactions of aldehydes or ketones with nucleophiles, the tetrahedral adduct is more or less readily detectable. Formaldehyde is overwhelmingly converted to methylenediol in water,[14] acetaldehyde is about 50% hydrated in water,[15] and acetone is only slightly converted to the hydrate, although the hydrate is readily detected by modern NMR instruments (the signal for the hydrate CH_3 is somewhat smaller than that for the ^{13}C satellite for the CH_3 of the keto form).[16] Thus, it is reasonable to assume that all carbonyl compounds can undergo nucleophilic addition, even when it is not directly detectable. For functional groups such as esters, the adduct with water or alcohol or even alkoxide is, for normal esters, at such low concentrations as to be undetectable. However, electron-withdrawing groups favor the addition of nucleophiles, so that CF_3COOMe will add $MeO^{-17,18}$ and the equilibrium constant in methanol can be determined by ^{19}F NMR titration; at high concentrations of methoxide the conversion is essentially complete.[19]

A more difficult challenge is to establish that a tetrahedral intermediate is on the reaction path for the transformation of a carbonyl containing functional group. Isotopic exchange occurring with rates and a rate law very similar to hydrolysis provides strong evidence that the tetrahedral intermediate is on the reaction path and is partitioning between proceeding on to product or reverting to starting material with the loss of isotope.[20] This simple interpretation assumes that proton transfers involving the tetrahedral intermediate are fast relative to breakdown, which need not always be true.[21]

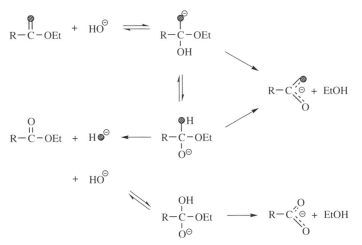

In other ester reactions, there may be concern that the reaction might be concerted, bypassing the tetrahedral intermediate. We will return to this question later. If the properties of Nu: or Lv: can be varied so that the relative leaving group abilities within the tetrahedral intermediate change from "Lv:" being poorer than "Nu:" to "Lv:" being better than "Nu:" (allowing where necessary for any other factors which influence relative leaving group ability), then there will be a change in rate determining step if the mechanism is stepwise by way of a tetrahedral intermediate. This will show up as a break in a linear free energy relation (whether Hammett, or Taft, or Brønsted plot) for the stepwise mechanism, but as a simple linear relationship for the concerted mechanism[22] (see below). This test requires that the two competing steps of the stepwise reaction (breakdown of the intermediate to starting material or to product) have sufficiently different slopes for the linear free energy relation to give a clear break. This need not be the case if both are fast; that is, if the intermediate is of relatively high energy, so that by Hammond's Postulate the two transition states are close to the structure of the intermediate (and necessarily also to each other) and thus respond similarly to changes in reactant structure.

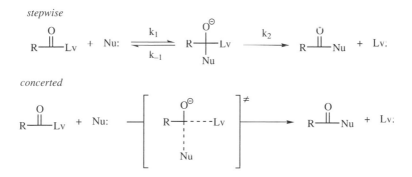

If the formation and breakdown steps of a mechanism involving a tetrahedral intermediate respond differently to changes in pH or catalyst concentration, then one can find evidence from plots of rate versus pH or rate versus catalyst concentration for a change in rate determining step and thus for a multistep mechanism. An example would be the maximum seen in the pH rate profile for the formation of an imine from a weakly basic amine (such as hydroxylamine).[23] On the alkaline side of the maximum, the rate determining step is the acid-catalyzed dehydration of the preformed carbinolamine; on the acid side of the maximum, the rate determining step is the uncatalyzed addition of the amine to form the carbinolamine. The rate decreases on the acid side of the maximum because more and more of the amine is protonated and unable to react.

If some change in reaction conditions leads to a change in the products of a reaction, without changing the observed rate, then there must be an intermediate which partitions in ways which respond to these changed reaction conditions, and formation of the intermediate must be rate determining. For instance, the products from the hydrolysis of the iminolactone shown below change with changing pH over a range where there is no change in the observed rate law.[24]

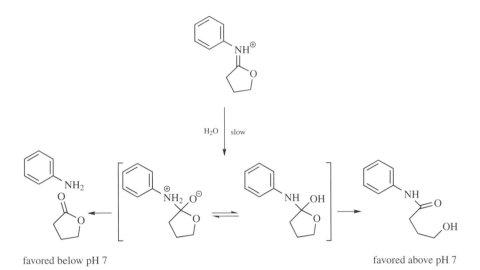

favored below pH 7   favored above pH 7

1.1.2. Stable Analogs

It is worth noting that the reactivity and short lifetime of most tetrahedral intermediates are a consequence of the presence of several electronegative atoms on a single center, with at least one of these atoms bearing a hydrogen. This means that an elimination pathway is accessible, which leads to a neutral product that is likely to be more stable than the tetrahedral intermediate. Without at least one electronegative atom bearing a hydrogen, any elimination must lead to a cationic species, which in most cases provides an additional barrier to reaction. Such analogs of tetrahedral intermediates are in fact well-known materials, acetals, aminals, orthoesters, and so forth and are relatively stable (compared with tetrahedral intermediates) because they do not have a facile elimination pathway. They are nonetheless reactive, especially to acid or, in some cases, simply exposure to polar solvents.[25] Mixed orthoacid derivatives [acetals of amides, $R\text{-}C(OR')_2(NR''_2)$, monothioorthoesters, $R\text{-}C(OR')_2(SR'')$, and even mixed orthoesters, $R\text{-}C(OR')_2(OR'')$] are also prone to disproportionation, especially in the presence of even traces of acid.[26] Thus $HC(OEt)_2(OR)$, R=cyclohexyl, becomes a mixture of $HC(OEt)_3$, $HC(OEt)_2(OR)$, $HC(OEt)(OR)_2$, and $HC(OR)_3$.[26] Monothioorthoesters have a distinct tendency to go to mixtures of orthoesters and trithioorthoesters: $HC(OEt)_2(SEt)$ goes to $HC(OEt)_3$ and $HC(SEt)_3$.[26]

1.1.3. Special Cases

There are some special cases where tetrahedral intermediates are unusually stable; there are three phenomena which lead to this stability enhancement. The first is an unusually reactive carbonyl (or imine) compound which is very prone to addition. An example of such a compound is trichoroacetaldehyde or chloral, for which the covalent hydrate can be isolated. A simple way to recognize such compounds is to think of the carbonyl group as a (very) stabilized carbocation, bearing an O^- substituent.

Groups which would destabilize a carbocation (H, or an electron-withdrawing group) will make the carbonyl more reactive to addition, both kinetically and thermodynamically. Formaldehyde is a peculiar case, because it is overwhelmingly converted to methylenediol in water, but upon evaporation it breaks down to gaseous formaldehyde rather than remaining as the liquid diol. It can (with acid catalysis) be trapped either as paraformaldehyde or trioxane. Similarly, hexafluoroacetone hydrate is a liquid, with useful solvent properties and little tendency to lose water.[27] CF_3 groups are even more destabilizing to an adjacent C^+ than H. The same reasoning explains why one can titrate methyl trifluoroacetate with methoxide in dry methanol, observing formation of the anionic tetrahedral species by ^{19}F NMR.[19]

The second special case is addition of a very good nucleophile; hydrogen cyanide and bisulfite are the most common examples, and cyanohydrins, α-cyanoamines and bisulfite adducts (α-hydroxy sulfonates) are commonly stable enough to isolate, at least for reactive carbonyl compounds. All these compounds are prone to fall apart under suitable conditions, regenerating the carbonyl compound.

The third phenomenon which favors tetrahedral intermediates is intramolecularity, and if a nucleophile is contained in the same molecule as a carbonyl group, it will show an enhanced tendency to add; the less entropy is lost in this addition (the fewer free rotations must be frozen out) the more the addition is favored. A famous example of this phenomenon is tetrodotoxin (**1**), the toxin of the puffer fish.[28] This molecule is a hemiorthoester in which there is an O^- on a carbon atom which also has two alkoxy groups, yet it does not break down to give a lactone. The explanation is that a secondary alcohol is held very close to the lactone carbonyl and thus there is an entropic advantage to the addition relative to a corresponding intermolecular reaction. In addition, there are numerous electron-withdrawing groups which enhance the reactivity of the lactone carbonyl toward addition.

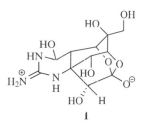

1

A more recent example is the twisted amide (**2**) devised by Kirby,[29] which despite the lack of electron-withdrawing groups (other than nitrogen) is completely hydrated upon protonation on nitrogen; here the "amide" is unable to delocalize the nitrogen electrons onto the carbonyl, which means there is none of the usual amide stabilization.

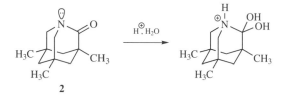

2

TABLE 1.2. Equilibrium constants for addition of nucleophiles to carbonyl compounds.[a]

Nucleophile/ Carbonyl compound	H_2O	HOMe	RSH[b]	RNH_2[c]	R_2NH[d]	NH_2OH	HCN	HSO_3^-
CH_2O	41[e]	1310[e]	1.4×10^6 [f,g]	3.4×10^6[h]	2.6×10^6[i]	–	9.18×10^8[e]	6.6×10^9[j]
CH_3CHO	0.019[e]	0.741[e]	36.[k,l]	–	61.[s]	–	3.7×10^4[e]	6.8×10^5[o]
$(CH_3)_2CHCHO$	0.011[e]	0.360[e]	16[p]	8.5[m]	1.7[n]	–	–	4.8×10^4[j]
$(CH_3)_3CCHO$	4.2×10^{-3e}	0.128[e]	4.8[p,q]					
PhCHO	1.5×10^{-4r}	3.6×10^{-3s} 0.09[t]			1.5[s]	11.3[u]	236[v]	6.4×10^3 [e]
4-pyridine-CHO	0.023[w]	0.50[w]	193[h, g]	87[w]		1500[h]		
4Cl-Ph-CHO	$4.0 \times 10^{-4\,r}$	0.24[t]	2.3[h, g]			24[h]	3.0×10^2 [h]	1.1×10^{4h}
4-NO_2-Ph-CHO	$3.1 \times 10^{-3\,r}$	3.0[t]				153[x]	1820[v]	
CH_3COCH_3	2.5×10^{-5e}	2.2×10^{-4e}				1.0[u]	14[e]	230[e]
Ph-CO-CH_3	$1.2 \times 10^{-7\,y}$						0.77[z,aa]	5.5[bb]
Ph-CO-CF_3	1.40[cc]			100[dd, ee]		1.5×10^{3dd}	760[dd]	2.3×10^{3dd}
$HCOOCH_3$	$3 \times 10^{-7\,ff}$ $8 \times 10^{-8\,s}$				$4 \times 10^{-5\,s}$			
CH_3COOCH_3	6×10^{-9ff} 9×10^{-11s}				$1 \times 10^{-8\,s}$			
CF_3COOCH_3	0.1[ff]							
$HCOSCH_2CH_3$	$3 \times 10^{-4\,ff}$							
$CH_3COSCH_2CH_3$	$6 \times 10^{-9\,ff}$ $8 \times 10^{-7\,s}$				$1 \times 10^{-4\,s}$			
$CF_3COSCH_2CH_3$	$2 \times 10^{-3\,ff}$							
$HCON(CH_3)_2$	$2 \times 10^{-14\,ff}$ $1 \times 10^{-12\,s}$	$8 \times 10^{-13\,s}$						
$CH_3CON(CH_3)_2$	$6 \times 10^{-15\,ff}$ 3×10^{-14s}	$2 \times 10^{-14\,s}$	$1.1 \times 10^{-12\,s}$					
$CF_3CON(CH_3)_2$	$5 \times 10^{-9\,s}$							
$PhCON(CH_3)_2$	$5 \times 10^{-14\,s}$	2×10^{-14s}						

[a]All in aqueous solution at 25°C unless otherwise noted; equilibrium constants have dimensions of M^{-1}. [b]Various alkane thiols, of similar equilibrium reactivity. [c]Methylamine or a primary alkyl amine of similar reactivity. [d]Dimethylamine or a secondary alkyl amine of similar reactivity [e]Reference 30. [f]Reference 14. [g]RSH is mercaptoethanol. [h]Reference 31. [i]Reference 32. [j]Reference 33. [k]Reference 21. [l]RSH is ethanethiol [m]Reference 34. [n]Reference 35. [o]Reference 36. [p]Reference 37. [q]RSH is 2-methoxyethanethiol. [r]Reference 38. [s]Reference 39. [t]Reference 40. [u]Reference 23. [v]Reference 41. [w]Reference 31. [x]Reference 42. [y]Reference 5. [z]Reference 43. [aa]In ethanol. [bb]Reference 44. [cc]Reference 45. [dd]Reference 46. [ee]RNH_2 is n-butylamine. [ff]Reference 47.

1.1.4. Equilibrium Constants

Table 1.2 gives a representative sampling of equilibrium constants for additions to various types of carbonyl compounds. Notice that there are numerous gaps in the table. This means that much remains to be done in the study of carbonyl addition reactions. In trying to devise schemes for predicting the equilibrium constants for such reactions, the scarcity of experimental data is a serious handicap. There are many fewer equilibrium constants for additions to imines, and even fewer cases where

TABLE 1.3 Equilibrium constants for addition to imines.[a]

Imine/ nucleophile	Ph-CH=N-CH$_2$Ph	Ph-CH=N-Ph	4-NO$_2$-Ph-CH=N-Ph	4-NO$_2$-Ph-CH=N-Ph-4-OCH$_3$
HCN	8.1 × 10^{3} [b,c]			
n-BuSH		5.21[d]	27.5[d]	
MeOH		1.5 × 10^{-3} [e]	7.1 × 10^{-3} [e]	0.6 × 10^{-3}[e]
		2.0 × 10^{-3} [f]	8.1 × 10^{-3}[f]	

[a]In methanol at 25°C unless otherwise noted; equilibrium constants have dimensions M^{-1}.
[b]Reference 48.
[c]In aqueous solution.
[d]Reference 49.
[e]Reference 50.
[f]Reference 51.

there is any kind of systematic set. Table 1.3 gives some representative values. There are also a few equilibrium constants for the addition of water to imines, but these do not overlap with the other additions.

1.1.5. Indirect Equilibrium Constants

For many addition reactions of carbonyl compounds, it is not possible to measure equilibrium constants directly because they are too unfavorable, and there is no selectively sensitive assay for the adduct. Two indirect methods allowing calculation of these equilibrium constants have been reported. The first takes advantage of the existence, for many unstable tetrahedral adducts, of orthoester analogs, which are stable because there are no OH (or NH or SH) groups in the analog, where they are present in the adduct of interest. If one can prepare and purify the analog, then its heat of hydrolysis can be measured, its solubility can be measured or estimated, and its entropy can be estimated by standard methods. This means that the free energy of formation of the analog can in principle be determined. Then one needs only to calculate the equilibrium constant for the hypothetical hydrolysis which converts the orthoester analog into the tetrahedral adduct,[30] to be able to calculate the free energy of formation of the adduct. From this, plus the free energies of formation of the carbonyl compound and the nucleophile, one can calculate the equilibrium constant for the addition reaction. The nice thing about the hypothetical hydrolysis is that one can say with confidence that its free energy change will be small. This must be so because in this hydrolysis the number of OH and CO bonds is conserved (so that by the bond energy additivity approximation ΔH will be zero), and the number of molecules is the same before and after the reaction (so that to a first approximation ΔS will be zero). The free energy change does depend on symmetry (the number of OR groups on the LHS), on steric interactions (OH is smaller than OR and thus will have smaller steric interactions), and on electronic effects (there is a small dependence on σ* for the R^1,R^2, R^3 groups). This method has been applied to esters,[52] amides,[53] and thioesters.[47] Various values determined by this approach are included in the tables in this chapter.

$$\begin{array}{c} R^1 \\ R^2 \end{array}\!\!\!\!\!\! \diagdown\!\!\!\!\!\! C\!-\!OR \;+\; H_2O \;\; \rightleftharpoons \;\; \begin{array}{c} R^1 \\ R^2 \end{array}\!\!\!\!\!\! \diagdown\!\!\!\!\!\! C\!-\!OH \;+\; HOR \\ R^3 R^3$$

A quite different and complimentary approach is to assume that addition of a nucleophile to an acyl derivative (RCOX) would follow the linear free energy relationship for addition of the nucleophile to the corresponding ketone (RCOR′, or aldehyde if R=H) if conjugation between X and the carbonyl could be turned off, while leaving its polar effects unchanged.[39] This can be done if one knows or can estimate the barrier to rotation about the CO–X bond, because the transition state for this rotation is expected to be in a conformation with X rotated by 90° relative to RCO. In this conformation X is no longer conjugated, so one can treat it as a pure polar substituent. Various values determined by this approach are included in the tables in this chapter.

1.1.6. Equations for the Effect of R, R′

For a given type of reaction (addition of a particular nucleophile to a particular functional group), one can get useful predictive equations on the basis of the Hammett $\rho\sigma$ or the Taft $\rho^*\sigma^*$ formalism. Unfortunately, there is some ambiguity in the literature about the definition and, consequently, the numerical values of Taft σ^* parameters. The values which some authors give to σ^* for a substituent X correspond to what other authors would say is the σ^* value for the related substituent CH_2-X. The problem arises because Taft used several different definitions of σ^* [54] which led to different and inconsistent values. These have then been quoted, not always consistently, by various textbook authors. For example, for OCH_3, Wiberg[55] gives 0.52, the value for CH_2OCH_3,[54] while Hine[56] gives no value for OCH_3 and 0.64[57] for CH_2OCH_3. Carroll[58] gives -0.22 which is the value for ortho substituted benzenes;[54] Perrin[59] gives 1.81.

In this chapter, the definitions used by Perrin in his book on pK_a prediction[59] (which also includes a very convenient compilation of σ^* values) will be used. One must be alert to the importance of the number of hydrogens directly attached to the carbonyl carbon; several groups have pointed out that aldehydes and ketones give separate but parallel lines, with formaldehyde displaced by the same amount again.[60] What this means is that given one equilibrium constant for an aldehyde (or ketone) one may estimate the equilibrium constant for other aldehydes (or ketones) from this value and ρ^* for the addition using a value from experiment, if available, or estimated if necessary. This assumes that there is no large difference in steric effects between the reference compound and the unknown of interest.

1.1.7. Equations for Effect of Nu

Sander and Jencks introduced a linear free energy relationship for nucleophilic addition to carbonyls. The equilibrium nucleophilicity of a species HNu is given by a parameter γ, defined as the logarithm of the equilibrium constant for addition of HNu to pyridine-4-carboxaldehyde relative to methylamine.

$$\gamma = \log\left(\frac{K_{Hnu}}{K_{CH_3NH_2}}\right)$$

What this implies is that given one equilibrium constant for addition of a nucleophile of known γ to a carbonyl compound, one could estimate the equilibrium constant for addition of another nucleophile to the same carbonyl compound. This requires knowing the slope of the plot of $\log K$ versus γ; this slope is not very sensitive to the nature of the carbonyl compound, but it is at least known that K_{H2O}/K_{MeOH} depends on the electron-withdrawing power of the groups bonded to the carbonyl,[30] and thus more information is needed to estimate an equilibrium constant for strongly electron-withdrawing substituents. From Ritchie's studies of nucleophile addition to trifluoroacetophenone,[46] we can derive a slope for $\log K$ versus γ of 0.42, distinctly less than the value of 1 for formaldehyde or simple benzaldehydes.

1.1.8. Anomeric Effect

Another effect which can influence the equilibrium constants for addition to carbonyl groups is the presence of lone pairs in the adducts. Given the fragment RO-C-X, one can have contributing structures $RO^+=C\ X$ (in valence bond language) or overlapping of the lone pair orbital on oxygen with the antibonding orbital of the C–X bond (in molecular orbital language), which acts to make conformations with such an interaction more stable than those without such interaction. The number of these interactions is likely to have an important effect on the equilibrium constant.

1.1.9. Estimation of Equilibrium Constants for Tetrahedral Intermediate Formation

Addition of water is the best studied reaction, and so there are numerous equations permitting one to estimate $\log K$ from σ or σ^*, provided a suitable reference compound has been studied. For most cases, except where strong short range inductive effects are important, the ρ value can be estimated. For nucleophiles other than water, one can either use the same sort of linear free energy relation, provided one has a suitable reference reaction where K is known, or use an orthogonal approach, going from the K for water addition to the K for the desired nucleophile using the γ method. Because the slope of a plot of $\log K$ versus γ depends on the nature of the substrate carbonyl compound, this requires some knowledge of the appropriate slope parameter for at least a closely related system. Fortunately, the slope is not a strong function of the electronic nature of the carbonyl compound; even for $PhCOCF_3$ the slope only falls from 1.0 to 0.42. One must also note that anomeric effects will have

TABLE 1.4. Linear free energy relationships for addition to carbonyl groups: variation in carbonyl group.[a]

Compound	Nucleophile	σ values used	ρ	Experimental data
RCOR'	H_2O	σ^*	1.70[b]	
ArCHO	H_2O	σ	1.71[c]	
ArCHO	HO^-	σ^+	2.37	[H, 4-Cl, 3-Cl, 4-CF_3, 3-NO_2, 3NO_2-4-Cl, 4-NO_2, 3,5-$(NO_2)_2$][c]
ArCHO	MeOH	σ^+	1.58	[4-OMe, 4-Me, H, 3-OMe, 4-Cl, 4-Br, 3-Cl, 3-Br, 3-NO_2, 4-NO_2][d]
ArCHO	MeO^-	σ^+	2.48	[4-OMe, 4-Me, H, 3-OMe, 4-Cl, 4-Br, 3-Cl, 3-Br, 3-NO_2, 4-NO_2][d]
RCOR'	MeOH	σ^*	1.82	[R,R' = Me_2, Me, $ClCH_2$, $(ClCH_2)_2$][e]
ArCHO	HCN	σ^+	1.01[f]	
ArCHO	CN^-	σ^+	1.49[f]	
ArCHO	HSO_3^-	σ^+	1.25[g]	
$ArCOCH_3$	HSO_3^-	σ^+	1.05	[4-OMe, 4-Me, H, 4-Cl, 4-Br, 3-Br, 4-NO_2][h]
ArCHO	NH_2OH	σ^+	1.18	[3-NO_2, 4-NO_2, 4-Cl, H, 4-NMe_3^+],[i] [4-OMe, 4-NMe_2][j,k]
$ArCOCH_3$	NH_2OH	σ^+	1.66	[3-NO_2, 4-Br, 4-F, 4-Me, 4-OMe][l]
RCOR'	HSO_3^-	σ^*	0.37	[CH_3,CH_3],[m] [CH_3OCH_2, CH_3OCH_2][n]
$RCOOCH_3$	H_2O	σ^*	3.08	[Me, Et, iPr, CF_3, $ClCH_2$, $NCCH_2$, $MeOCH_2$][o]
$RCOSCH_2CH_3$	H_2O	σ^*	2.06	[CH_3, CF_3][p]
$RCON(CH_3)_2$	H_2O	σ^*	1.99	[CH_3, CF_3][q]
PhCOR	H_2O	σ^*	2.06	[CH_3,[r] CF_3][s]
PhCOR	HSO_3^-	σ^*	1.00	[CH_3,[h] CF_3][s]
PhCOR	NH_2OH	σ^*	1.80	[CH_3,[t] CF_3][s]

[a]All in aqueous solution at 25°C unless otherwise noted.
[b]Reference 60.
[c]Reference 38.
[d]Reference 40.
[e]Reference 30.
[f]Reference 41.
[g]Reference 61.
[h]Reference 44 (reported ρ = 1.2, ρ = 0.95 based on an extrapolation of the correlation line for PhCOCH$_3$ + HNu).
[i]Reference 42.
[j]Reference 62.
[k]At 30°C.
[l]Reference 63.
[m]Reference 5 and references cited therein.
[n]Reference 64.
[o]Reference 65.
[p]Reference 47.
[q]Reference 39.
[r]Based on an extraplolation of the correlation line for PhCOCH$_3$ + HNu.
[s]Reference 46.
[t]Based on an extrapolation of the correlation line for ArCOCH$_3$ + NH$_2$OH.

an important influence on the observed log K, so this approach can be used only for aldehydes and ketones. For acyl derivatives, the anomeric effects must be different and the magnitude of this effect is not yet known.

There are anomalies for compounds with CF_3 directly attached to a carbonyl group. Equilibrium constants for addition to such a carbonyl group are higher than expected, relative to the CH_3 compound. However, the rate constants for hydroxide addition to esters do not show this phenomenon. This might indicate that when CF_3 is directly attached to carbonyl (which is formally treated as $C^+\text{-}O^-$), there is, in addition to the field effect measured by σ^*, an important inductive contribution which augments the field effect. Alternatively, it may just reflect the large uncertainties in free energy changes based on extended thermochemical calculations.

Despite many papers over many years, there is still a serious shortage of information that allows linear free energy relation treatment of these reactions. The available linear free energy relations, some of them calculated for this chapter, are collected in Tables 1.4 and 1.5. There are definite indications that ρ is

TABLE 1.5 Linear free energy relationships for addition to carbonyl groups; variation in nucleophile.[a]

Substrate	Slope	Experimental data
CH_2O	1.0[b]	
Py 4-CHO	1.0[b]	
4-ClPhCHO	1.0[b]	
CH_3COCH_3	0.92[c]	
CH_3CHO	1.06[c]	
i-PrCHO	0.909	$[H_2O, MeOH]$,[d] $[MeOCH_2CH_2SH]$,[e] $[HSO_3^-]$[f]
PhCHO	1.02[c]	
$MeOCH_2COCH_2OMe$	0.67	$[H_2O, HSO_3^-]$[g]
$PhCOCF_3$	0.42	$[H_2O, H_2O_2, HCN, HSO_3^-]$[h]
$PhCOCH_3$	1.04	$[NH_2OH,$[i] $HCN,$[j,k] $HSO_3^-]$[l]
$p\text{-}NO_2PhCHO$	0.96	$[H_2O,$[m] $HCN,$[n] $NH_2OH,$[o] HSO_3^- [p]$]$

[a]All in aqueous solution at 25°C unless otherwise noted.
[b]Reference 31.
[c]Reference 5.
[d]Reference 30.
[e]Reference 37.
[f]Reference 33.
[g]Reference 64.
[h]Reference 46.
[i]Extrapolated from a correlation based on data of Lamaty.[63]
[j]Reference 43.
[k]In ethanol.
[l]Reference 44.
[m]Reference 38.
[n]Reference 41.
[o]Reference 42.
[p]Extrapolated from a correlation based on data in reference 66.

TABLE 1.6 Cross terms.

System	First correlation	Second correlation	Cross term
RCOR'	σ^*	γ	-0.19
PhCOR'	σ^*	γ	-0.13
ArCHO	σ	γ	-0.12
ArCOMe	σ	γ	-0.22
RCOR'	γ	σ^*	-0.19
ArCHO	γ	σ	-0.06

different for different nucleophiles and that Δ is different for different carbonyl compounds, though in neither case is the sensitivity very large. There are insufficient data to tell how elaborate a model must be. The simplest model for the observations is:

$$\log K = \rho_0^* \, \Sigma\sigma^* + \Delta_0\gamma + a_{\gamma\sigma} \, \Sigma\sigma^* \, \gamma + \text{const}$$

However, the data do not permit a proper test, although they indicate (see Table 1.6) that $a_{\gamma\sigma}$ is between -0.12 and -0.22 (with the exception of aromatic aldehydes where one sequence gives -0.12 and the other gives -0.06).

One reason why the necessary measurements have not been done is that it is not easy to get a set of compounds that would give clean reactions and have a strong electron withdrawing group. Cyanide can act as a nucleophile in the S_N2 sense as well as at a carbonyl group, so that alternative modes of reactions are possible for $ClCH_2COCH_3$ (including a Darzens-like reaction of the cyanohydrin anion). FCH_2COCH_3 might serve, but it is unpleasantly toxic. $CF_3CH_2COCH_3$ would be good but it is not commercially available and it might slowly eliminate HF by an Elcb mechanism. Many polar substituents will also form enolates by ionization and thus lead to complications. However, despite all of these difficulties, it would be very desirable to have more data to unscramble the linear free energy relations controlling these important reactions.

Another sign of complexity which has largely been ignored is that CH_2O, but not simple aldehydes or ketones, shows a dispersion of $\log K - \gamma$ plots with nitrogen nucleophiles falling on a line parallel to but higher than the line for other nucleophiles. The same phenomenon is seen for $PhCOCF_3$! The common feature is that both have carbonyl groups with destabilizing substituents (two Hs or one CF_3). It is not obvious why this should be, but the phenomenon seems real.

In principle, it should be possible to use computational thermochemistry to calculate free energies of formation for unknown tetrahedral intermediates. In practice this remains difficult because of the problem of estimating solvation energies. There is no doubt that computational methods will become increasingly important in this as in other areas.

1.1.10. Mechanisms of Tetrahedral Intermediate Formation and Breakdown

Uncatalyzed mechanisms for the breakdown of a tetrahedral intermediate are relatively rare because they require generation of a cation and an anion:

It is for this reason that orthoesters and acetals are (comparatively) stable in the absence of an acid. Alternatively, one can have an uncatalyzed mechanism involving preliminary tautomerization to a zwitterion, but the thermodynamic cost of this imposes a considerable barrier to reaction.

By contrast base-catalyzed mechanisms are generally fast, provided, of course, that one of the heteroatoms defining the tetrahedral intermediate has an ionizable proton.

Finally, acid-catalyzed mechanisms are generally fast but must overcome the relatively low basicity of the tetrahedral intermediate (with electron-withdrawing substituents necessarily present, the basicity is low).

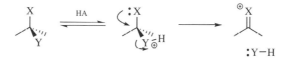

It is helpful to think of these as displacement reactions: if the leaving group Y is poor, then a good "nucleophile" X is needed; whereas if Y is a good leaving group, then a poor "nucleophile" will suffice. Thus rapid reaction will often require enhancing either X (by base catalysis) or Y (by acid catalysis). For example, the dehydration of carbinolamines derived from strongly basic amines can proceed by an uncatalyzed path,[34,67] but carbinolamines derived from weakly basic amines require acid catalysis.[23] The breakdown of cyanohydrins requires base catalysis,[68] and does not occur in acid; the cyano group is not very basic, and with strong acid one gets hydrolysis to the amide or acid instead.

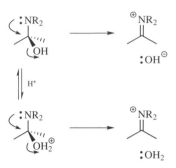

Much of the complication in the chemistry of acyl transfer reactions can be understood in terms of the relative leaving group abilities of the possible leaving groups. Thus, it is reasonable that oxygen exchange should accompany the hydrolysis of esters either in acid or in base,[20] because in each case the competing leaving groups are very similar.

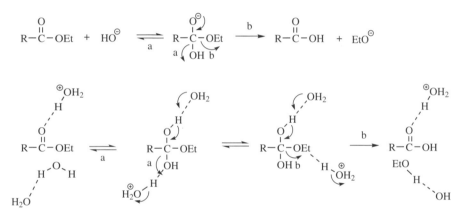

For amide hydrolysis in base, the initial adduct would revert to starting materials (without remarkable stabilization, an amide ion is a hopeless leaving group, so that path b does not compete with path a), but a not very difficult proton transfer gives an intermediate in which the amine is the better leaving group and path b′ can compete with path a.[69]

For amide hydrolysis in acid, proton transfer to give a cationic intermediate is easy, and breakdown to products is favored over reversion to starting material;[70] process b is hopelessly bad, but process b' is better than a.

$$
\begin{array}{ccc}
\overset{\oplus}{O}{-}H & \overset{H}{O} & \overset{\oplus}{O}{-}H \\
\| & | & \| \\
R{-}C{-}NR_2 & R{-}C{-}NR_2 \xrightarrow{\ b\ } & R{-}C{-}OH \ + \ {}^{\ominus}NR_2 \\
& a\ OH\ b &
\end{array}
$$

$$
\begin{array}{cc}
H_2O & \overset{\oplus}{H_2O}{-}H \\
\end{array}
$$

$$
\begin{array}{ccc}
\overset{H}{O} & & \overset{\oplus}{O}{-}H \\
| & \xrightarrow{\ b'\ } & \| \\
R{-}C{-}\overset{\oplus}{N}HR_2 & & R{-}C{-}OH \ + \ NHR_2 \\
OH\ b' & &
\end{array}
$$

Aminolysis of simple esters is surprisingly difficult, despite the greater thermodynamic stability of amides than esters; the problem is that the initial tetrahedral intermediate preferentially reverts to starting material (not only is the amine the better leaving group, but loss of alkoxide would lead to an *N*-protonated amide), and only trapping of this intermediate by proton transfer allows the reaction to proceed.[53,71]

$$
\begin{array}{ccc}
O & O^{\ominus} & O \\
\| & | & \| \\
R{-}C{-}OEt \ + \ NH_2R \xrightleftharpoons[a]{} & R{-}C{-}OEt \xrightarrow{\ b\ } & R{-}C{-}\overset{\oplus}{N}H_2R \ + \ EtO^{\ominus} \\
& \overset{\oplus}{N}H_2R\ b &
\end{array}
$$

$$
-H^+
$$

$$
\begin{array}{ccc}
O^{\ominus} & & O \\
| & \xrightarrow{\ b'\ } & \| \\
R{-}C{-}OEt & & R{-}C{-}NHR \ + \ EtO^{\ominus} \\
NHR\ b' & &
\end{array}
$$

1.1.11. Rates of Breakdown of Tetrahedral Intermediates

Rates of addition to carbonyls (or expulsion to regenerate a carbonyl) can be estimated by appropriate forms of Marcus Theory.[72–75] These reactions are often subject to general acid/base catalysis, so that it is commonly necessary to use Multidimensional Marcus Theory (MMT)[76,77] to allow for the variable importance of different proton transfer modes. This approach treats a concerted reaction as the result of several orthogonal processes, each of which has its own reaction coordinate and its own intrinsic barrier independent of the other coordinates. If an intrinsic barrier for the simple addition process is available then this is a satisfactory procedure. Intrinsic barriers are generally insensitive to the reactivity of the species, although for very reactive carbonyl compounds one finds that the intrinsic barrier becomes variable.[77]

Alternatively one can make use of No Barrier Theory[78–81] (NBT), which allows calculation of the free energy of activation for such reactions with no need for an empirical intrinsic barrier. This approach treats a real chemical reaction as a result of several simple processes for each of which the energy would be a quadratic function of a suitable reaction coordinate. This allows interpolation of the reaction hypersurface; a search for the lowest saddle point gives the free energy of activation. This method has been applied to enolate formation,[82] ketene hydration,[83] carbonyl hydration,[80] decarboxylation,[84] and the addition of water to carbocations.[79]

Both these methods require equilibrium constants for the microscopic rate determining step, and a detailed mechanism for the reaction. The approaches can be illustrated by base and acid-catalyzed carbonyl hydration. For the base-catalyzed process, the most general mechanism is written as general base catalysis by hydroxide; in the case of a relatively unreactive carbonyl compound, the proton transfer is probably complete at the transition state so that the reaction is in effect a simple addition of hydroxide. By MMT this is treated as a two-dimensional reaction: proton transfer and C–O bond formation, and requires two intrinsic barriers, for proton transfer and for C–O bond formation. By NBT this is a three-dimensional reaction: proton transfer, C–O bond formation, and geometry change at carbon, and all three are taken as having no barrier.

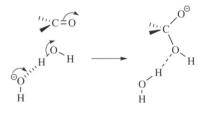

For acid catalyzed hydration, the general mechanism is:

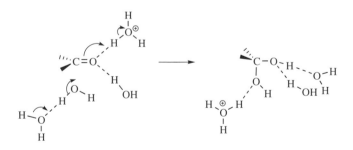

and is written as a general acid-catalyzed process; with the more basic carbonyl compounds the proton transfer from hydronium ion may be complete at the transition state. A second water molecule acts as a general base to deprotonate the nucleophilic water because the product of simple attack, a cationic tetrahedral intermediate, would be significantly more acidic than water and thus would lose a proton to the solvating water. By MMT this is treated as a three-dimensional reaction: proton transfer from hydronium ion, proton transfer from water to water, and C–O bond formation, and requires intrinsic barriers for proton transfer and C–O bond formation. By NBT this is treated as a four-dimensional reaction: proton transfer from hydronium ion, proton transfer from water to water, C–O bond formation, and geometry change at carbon. Treating proton transfer as a no barrier process is clearly only an approximation because there is a small intrinsic barrier to proton transfer between electronegative atoms[76b] but this seems to be a workable approximation as long as there are also heavy atom bond changes in the overall reaction.

1.2. PENTACOORDINATE INTERMEDIATES INVOLVING P

Phosphate esters have a variety of mechanistic paths for hydrolysis.[85] Both C–O and P–O cleavage are possible depending on the situation. A phosphate monoanion is a reasonable leaving group for nucleophilic substitution at carbon and so S_N2 or S_N1 reactions of neutral phosphate esters are well known. PO cleavage can occur by associative (by way of a pentacoordinate intermediate), dissociative (by way of a metaphosphate species), or concerted (avoiding both of these intermediates) mechanisms.

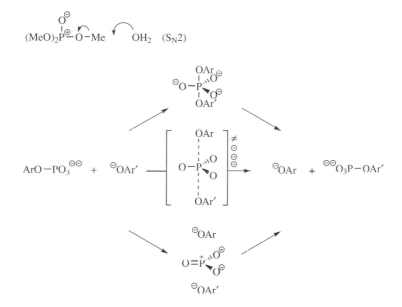

The pentacoordinate intermediate is the analog of the tetrahedral intermediate, and stable phosphoranes are the analogs of ortho esters and related species in carbon chemistry. $Ph_3P(OPh)_2$[86] and $P(OPh)_5$[87] were reported in 1959, and in 1958 a general synthesis of pentaalkoxy phosphoranes containing an unsaturated five-membered ring was reported.[88,89] In 1964 a synthesis of pentaethoxyphosphorane was devised[90] which led to the preparation of a number of saturated and unsaturated pentaalkoxy phosphoranes. A less hazardous route using an alkyl benzene sulfenate as an oxidizing agent makes these compounds more accessible.[91] Thus, analogs of the putative intermediate in the associative mechanisms are known, but these compounds are very sensitive to water; much more so than simple orthoesters.

For a number of reactions of cyclic di- and triesters of phosphoric acid, there are exchange data which can be rationalized on the assumption of trigonal bipyramidal intermediates which readily interconvert by pseudorotation.[92] This constitutes a strong argument that at least these cyclic esters react by an associative mechanism and is suggestive evidence that simple trialkyl phosphates also react by this mechanism. The pH dependence of exocyclic versus endocyclic cleavage of methyl ethylene phosphate is readily interpreted in terms of the effect of ionization of the intermediate on the pseudorotation of these pentacoordinate intermediates.[93]

Analogous to tetrodotoxin are phosphoranoxides **3**,[94] **4**,[95,96] and **5**,[97] where chelation, steric bulk, and proper arrangement of electron-withdrawing and electron-donating substituents make them stable enough to isolate.

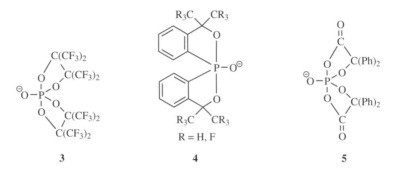

3   4   5

1.2.1. Bonding in Pentacoordinate Phosphorus and Sulfur Compounds

These compounds are often referred to as hypervalent. The apical bonds in a trigonal bipyramid are described by molecular orbitals constructed from a p-orbital on the central atom and σ-bonding orbitals (p- or sp^n hybrid) on the apical ligands. The molecular orbitals can be drawn as:

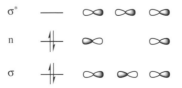

For a discussion of hypervalent bonding see reference 98. This picture indicates that there is an accumulation of partial negative charge on the apical ligands and thus a partial positive charge on the central atom.[99] Thus the apical ligands should be electronegative.

TABLE 1.7. Equilibrium constants for addition or elimination from phosphoric acid esters.[a]

Starting compound	log K for elimination	log K for addition of water[b]	log K for addition of hydroxide[c]
H_3PO_4	-23[d]	-12	—
$H_2PO_4^-$	-20[e]	-19	—
$HPO_4^=$	-26[f]	-25	—
$(EtO)_3PO$	-46[g]	-10	-3
$(EtO)_2PO_2H$	-21[h]	-12	—
$(EtO)_2PO_2^-$	-35[i]	-18	-16
$EtOPO_3H_2$	-21[j]	-12	—
$EtOPO_3H^-$	-18[k]	-18	—
$EtOPO_3^=$	-27[l]	-24	-27

[a]All in aqueous solution at 25°C; standard states are 1 M ideal solution with an infinitely dilute reference state, and the pure liquid for water; equilibrium constants from reference 100, except as noted.
[b]$K = $ [adduct]/[orthophosphate]
[c]$K = $ [adduct]/[orthophosphate][HO$^-$]
[d]$K = $ [HPO$_3$]/[H$_3$PO$_4$]
[e]$K = $ [PO$_3^-$]/[H$_2$PO$_4^-$]
[f]$K = $ [PO$_3^-$][HO$^-$]/[HPO$_4^=$]
[g]$K = $ [(EtO)$_2$PO$^+$][EtO$^-$]/[(EtO)$_3$PO]; estimated as described in Section 1.4.3.
[h]$K = $ [EtOPO$_2$][EtOH]/[(EtO)$_2$PO$_2$H]
[i]$K = $ [EtOPO$_2$][EtO$^-$]/[(EtO)$_2$PO$_2^-$]
[j]$K = $ [HPO$_3$][EtOH]/[EtOPO$_3$H$_2$]
[k]$K = $ [PO$_3^-$][EtOH]/[EtOPO$_3$H$^-$]
[l]$K = $ [PO$_3^-$][EtO$^-$]/[EtOPO$_3^=$]

1.2.2. Indirect Equilibrium Constants

By methods analogous to those used for the tetrahedral intermediates related to carboxylic acid derivatives, Guthrie proceeded from the heat of formation of pentaethoxyphosphorane to free energies of the $P(OEt)_n(OH)_{5-n}$ species.[100] This allowed the calculation of the equilibrium constants for addition of water or hydroxide to simple alkyl esters of phosphoric acid; see Table 1.7.

1.3. PENTACOORDINATE INTERMEDIATES INVOLVING S

Sulfate monoesters can react by dissociative paths, and this is the favored path.[101] Whether such reactions are concerted or involve a very short-lived sulfur trioxide intermediate has been the subject of debate.[102,103] Benkovic and Benkovic reported evidence suggesting that the nucleophile is present (though there is little bond formation) in the transition state for the reaction of amines with *p*-nitrophenyl sulfate.[104]

Alkyl esters of sulfuric or sulfonic acids normally react with C–O cleavage; only when this is disfavored, as in aryl esters, does one see S–O cleavage. Sulfate diester and sulfonate ester reactions (with S-O cleavage) have been discussed in terms of concerted or stepwise (addition elimination) mechanisms,[105,106] but recent authors[22] have favored concerted mechanisms. In suitable sulfonates, with an ionizable hydrogen next to the sulfur, there are also stepwise elimination addition pathways by way of sulfenes[107] or analogs.

The simplest sulfur analogs of tetrahedral intermediates are the sulfuranes

none of which are known experimentally, although computational results suggest

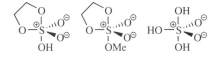

that they are at least energy minima.[108,109–111] Other than various halogen derivatives, the sulfuranes closest to those of interest here which have actually been prepared are the chelated derivatives **6**[112] and **7**[105] prepared by Martin et al. The former is an analog of the adduct of a sulfone, whereas the latter is an analog of a sulfonate ester adduct.

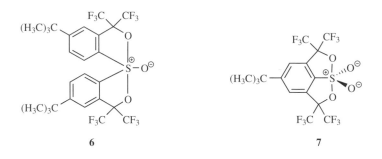

6   **7**

1.4. CONCERTED REACTIONS

1.4.1. General Principles

A useful general rule when considering concerted reactions is that a concerted reaction path is followed in order to avoid unstable intermediates.[113] A concerted path has more things happening (more partial bonds, more atoms undergoing geometry change) so it is to be expected that such a path will be slower (will have a higher intrinsic barrier) than an alternative stepwise path,[76,114–116] unless the stepwise path is disfavored by leading to a high-energy species. The classic examples of this principle are the S_N2 and E2 reactions. The S_N2 is observed when the S_N1 alternative is disfavored because of the instability of the carbocation which would have to form. The other stepwise alternative is the pentacoordinate species with five full bonds to carbon, and this is almost invariably too high energy to be a viable reaction intermediate.

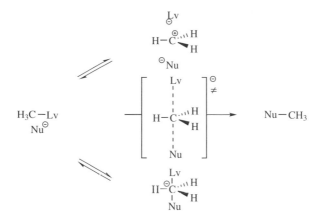

However an example of a [10-C-5] species has been reported.[117,118]

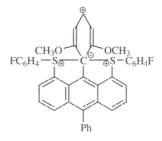

Similarly, the E2 is observed when both the E1 and E1cb alternatives are disfavored. Thus we expect that a concerted acyl transfer will be most likely when the intermediates in both the stepwise alternatives (tetrahedral intermediate and acylium ion) are of high energy.[119,120] Similarly, a phosphoryl transfer is expected to be concerted when both the pentacoordinate intermediate and the metaphosphate species are of high energy, and to shift to a stepwise path when one or the other is accessible. Sulfonyl transfer will be concerted when the corresponding stepwise alternatives (pentacoordinate sulfurane or sulfur trioxide for a sulfate monoester; sulfurane and O-alkylated sulfur trioxide for a sulfate diester; and sulfurane and a sulfonylium ion, RSO_2^+, for a sulfonate ester) are high energy species. Because so few sulfuranes have been prepared, the sulfurane species seem to be more inaccessible than is the case for phosphate esters, and this, in isolation, would suggest that there is a greater likelihood of concerted pathways for sulfate or sulfonate derivatives. However, the dissociative intermediates are all unlikely and very high energy species, which would suggest that stepwise reaction by an addition–elimination mechanism is more likely. Kice reviewed the evidence[106] and concluded that sulfonylium ions are much more difficult to form than the corresponding acylium ions.

1.4.2. Evidence for Concerted Reactions

For acyl transfer, oxygen exchange has been observed in various reactions,[20,69,70] providing evidence supporting a stepwise addition–elimination mechanism. It is of course now generally accepted that most acyl transfer reactions occur by stepwise mechanisms, although in some cases concerted mechanisms are believed to be preferred. For various simple phosphate esters, oxygen exchange into the unreacted ester has been observed accompanying hydrolysis.[121] This suggests that at least some phosphate ester reactions occur by stepwise mechanisms, although there are also situations where concerted mechanisms have been proposed.

Oae found that for both base- and acid-catalyzed hydrolysis of phenyl benzenesulfonate, there was no incorporation of ^{18}O from solvent into the sulfonate ester after partial hydrolysis.[122,123] This was interpreted as ruling out a stepwise mechanism, but in fact it could be stepwise with slow pseudorotation. In fact this nonexchange can be explained by Westheimer's rules[92] for pseudorotation, assuming the same rules apply to pentacoordinate sulfur. For the acid-catalyzed reaction, the likely intermediate would be **8** for which pseudorotation would be disfavored because it would put a carbon at an apical position. Further protonation to the cationic intermediate is unlikely even in 10 M HCl (the medium for Oae's experiments) because of the high acidity of this species: a Branch and Calvin calculation[124] (See Appendix), supplemented by allowance for the effect of the phenyl groups (taken as the difference in pK_a between sulfuric acid and benzenesulfonic acid[125]), leads to a pK_a of -7 for the first pK_a of this cation; about -2 for the second pK_a, and about 3 for the third pK_a. Thus, protonation by aqueous HCl to give the neutral intermediate is likely but further protonation to give cation **9** would be very unlikely.

For the intermediates in base-catalyzed hydrolysis of a sulfate ester (**10**), pseudorotation about any of the equatorial bonds will necessarily put at least one O^- in an apical position, which is strongly disfavored.[126]

Okuyama et al. have presented evidence that at least some sulfenates and sulfinate derivatives react by way of hypervalent intermediates.[127–129]

For acyl transfer, phosphoryl transfer, and sulfonyl transfer, the primary kind of evidence in favor of concerted mechanisms for some reactions is a linear Bronsted plot of log k versus pK_a^{nuc} for a range of nucleophiles, spanning $pK_a^{nuc} - pK_a^{lg} = 0$, coupled with an assumption that the β values for addition and elimination would be quite different. Much of this argument is based on the large β_{eq} values which are interpreted as meaning that the oxygen in an aryl ester bears a substantial δ^+ charge which will be markedly diminished only by cleavage of the bond from oxygen to the carbonyl carbon (or phosphoryl phosphorus or sulfonyl sulfur). If the carbonyl carbon is regarded as C^+-O^- (for which there is considerable support[130–132]), then the β_{eq} values reflect the interaction of the alcoholic or phenolic group with this (+) charge, and formation of a tetrahedral intermediate, with cancellation, will drastically change the interaction without significant C–O bond cleavage.

The problem is that "proving" concerted reaction requires negative evidence: a Bronsted plot with a clear break is strong evidence for a stepwise reaction; absence of a break could mean a concerted reaction or similar β values for both modes of breakdown of the intermediate, which is likely if the intermediate is high in energy relative to starting materials and products. The situations where concerted reactions are proposed are in fact ones where the tetrahedral intermediate is indeed likely to be of high energy, because while a good leaving group (electron deficient phenol) will favor addition to form a tetrahedral intermediate, the same phenol is a poor nucleophile which makes addition unfavorable.

From the *a priori* point of view, when would one expect concerted reactions? On the basis of the model presented earlier, concerted reactions occur when both the stepwise alternatives require high energy intermediates. Then a concerted path, avoiding both bad intermediates, can have a transition state lower in free energy than either. If one stepwise intermediate is much higher in energy than the other, then any change in structure from the lower energy intermediate toward the higher energy one is likely to raise, not lower the free energy, and thus a concerted path becomes unlikely. For the reaction of aryloxides with aryl acetates, an analysis of the energetics[120] suggested that the energies of the acylium ion with two phenoxides and of the tetrahedral intermediates were comparable, which predisposes this system to becoming concerted. Unfortunately, the only equilibrium data for acylium ions are for acetylium ion,[133] a few alkanecarboxylium ions, and benzoylium[134] ion and nothing is known about substituent effects. For phosphate esters, nucleophilic substitution of monoester dianions is likely to be concerted because both stepwise intermediates are bad, with dissociative reaction by a nearly free monomeric metaphosphate intermediate being an alternative absent a good nucleophile, while for diesters or triesters, the dissociative intermediate is high in energy relative to the associative intermediate making concerted reaction unlikely. For sulfate diesters and sulfonate esters, the high energy of the dissociative intermediates make concerted reactions with S–O cleavage unlikely, while for sulfate monoesters the dissociative stepwise reaction or concerted reaction (with a very open transition state) look feasible.

By linear free energy relation arguments, Williams et al. concluded that in the case of a five-membered ring sultone the reaction with a phenoxide was either stepwise or, if concerted, had a transition state close to the pentacoordinated intermediate.[135]

Thus, there is suggestive evidence that both stepwise intermediates for sulfonyl transfer reactions may be relatively high-energy species. Now we will try to estimate the energetics for such species; first for the simplest parent cases, even though they react by other mechanisms, and then for the aryl esters which do react with S–O cleavage. The goal is not to get estimates good enough to estimate the rate but to see if what is now known is enough to rule out some mechanistic paths. We will see that this is, in fact, the case.

First we will look at hydroxide attack on sulfate diesters, and estimate the free energy changes for the two stepwise limiting cases corresponding to concerted displacement.

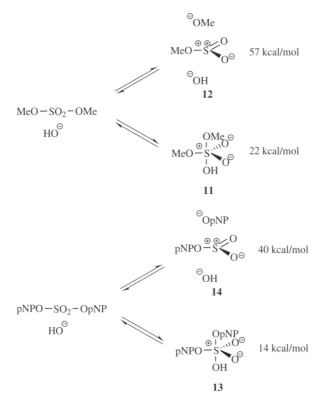

For species **11** we will use the intrinsic barrier for hydroxide addition to trimethyl phosphate, $\widetilde{G} = 19$ (calculated using rate and equilibrium data from reference 100) and assume the same value for the attack of hydroxide at sulfur on dimethyl sulfate. This (nonobservable) rate will be estimated using a Brønsted type plot from the rate constants for diaryl sulfates (diphenyl sulfate,[136] and bis *p*-nitrophenyl sulfate), estimated from the rate for phenyl dinitrophenyl sulfate,[137] assuming equal contributions for the two nitro groups. This gives $\beta_{lg} = -0.95$, and thus for dimethyl sulfate log $k = -11.3$ and $\Delta G^{\neq} = 33$, which affords $\Delta G° = 22$ kcal/mol for the formation of **11**.

For the same reaction, Thatcher and Cameron calculated (MP2/6-31+G*//HF/3-21+G* with continuum solvation[111]) $\Delta G° = 21$ kcal/mol.

For species **12**, we first estimate the pK_a of HO-SO$_2^+$ using the method of Branch and Calvin[124] (see Appendix), knowing full well that this will not be accurate because the central atom is highly charged, the estimate for sulfuric acid with S^{++} is too acidic, and resonance will also make a contribution, so that the crude estimate will not be acidic enough. However, the errors may partly cancel.

$$\log K_a = -16 + 13.2*2 - 13.2/2.8 + 2*4/2.8 + 3.4 + \log(1/3) = 11.5$$

From this and the relation between the equilibrium for ester formation and the pK_a of the acid,[100] we estimate the free energy change for the reaction

$$CH_3OH + HO\text{-}SO_2^+ \rightleftharpoons CH_3O\text{-}SO_2^+ + H_2O$$

as $\Delta G° = 8.6$ kcal/mol. Then from the thermodynamic cycle:

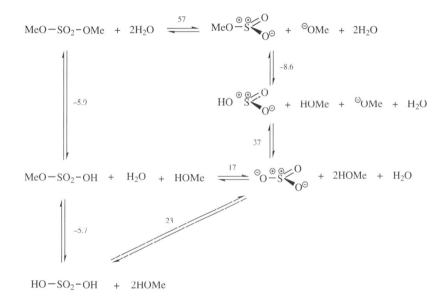

we estimate $\Delta G° = 57$ kcal/mol for formation of **12** and methoxide from (McO)$_2$SO$_2$. In this and all following thermodynamic cycles, the numbers are free energies, in kcal/mol, in the direction indicated by the arrow next to the number. (Numbers used in the cycle: $\Delta G°$ for hydrolysis of dimethyl sulfate;[125] $\Delta G°$ for hydrolysis of monomethyl sulfate—calculated from the pK_a;[125] $\Delta G°$ for dissociation of sulfuric acid to SO$_3$.[138]) Despite all the uncertainties in this calculation, it looks like sulfate diester hydrolysis should be stepwise or very close to it, because the dissociative intermediate **12** is 35 kcal/mol higher in energy than the associative intermediate **11**.

To estimate the effects of changing from methyl to p-nitrophenyl ester on the addition reaction, we use the change in equilibrium constant for addition of hydroxide to acetate esters, which was estimated[120] as $\Delta\Delta G° = 4.3$ kcal/mol. We assume the same change applies to sulfates and phosphates. Then, from the free energy change for addition of hydroxide to dimethyl sulfate, we get $\Delta G° = 14$ kcal/mol for the reaction given below.

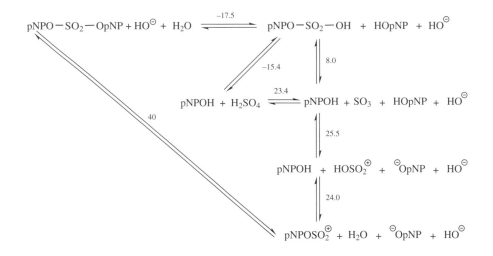

The starting points are the free energies of hydrolysis for pNPOSO$_3$H[138] and pNPOPO$_3$H$_2$.[100] From these we may deduce an equation relating ΔG_{hydrol} to pK_a of HOX for

$$pNPOX + H_2O \rightleftharpoons pNPOH + HOX$$

as $\Delta G_{hydrol} = -12.69 + 0.98pK_a$. From this we may calculate ΔG_{hydrol} for (pNPO)$_2$SO$_2$ as -17.5, and for pNPOSO$_2^+$ as -24.0 kcal/mol, respectively. Then the energy of the dissociative corner can be calculated following the thermodynamic cycle:

(Numbers used in this cycle: $\Delta G°$ for dissociation of sulfuric acid to sulfur trioxide;[138] $\Delta G°$ for hydrolysis of bis-p-nitrophenyl sulfate, estimated as described above; $\Delta G°$ for hydrolysis of mono-p-nitrophenyl sulfate;[138] $\Delta G°$ for esterification to give pNPOSO$_2^+$, estimated as described above; $\Delta G°$ for ionization of protonated SO$_3$, estimated as described above; $\Delta G°$ for ionization of p-nitrophenol[139].)

The reaction of hydroxide with dimethyl sulfate clearly should not be concerted: the dissociative corner, **12**, is far too high in energy, yet the reaction would not show

[18]O exchange because pseudorotation is strongly inhibited. Similarly, the reaction of hydroxide with bis-*p*-nitrophenyl sulfate should not be concerted because **14** is far too high in energy. The free energy of activation for reaction (estimated from data of Hengge[137]) is 22 kcal/mol, enough higher than the equilibrium free energy change for intermediate formation to be reasonable.

Now we turn to the reaction of water with the sulfate monoester monoanion, which, in the case of aryl esters, is believed to react either by a dissociative path, or a concerted path with a transition state resembling the dissociative limit. There is a problem for this reaction in the case of an alkyl ester. Simple attack of water leads to a very acidic species with H_2O^+ bonded to S^+; loss of a proton to solvent water would be extremely fast, occurring before the O–S bond was fully formed. For an alkyl ester, microscopic reversibility would require that the very similar leaving groups, MeO^- and HO^-, depart by the same mechanism. For the intermediate acting as limiting case in a concerted reaction, this would require a complex with two H^+ ions, in fact an acid catalyzed path. There is no problem with a fully stepwise reaction, since H^+ could diffuse from one position to another to allow MeO^- to depart as MeOH. For an aryl ester, with a much better leaving group, the mechanism of loss of ArO^- is not required to be the same as that for loss of HO^-.

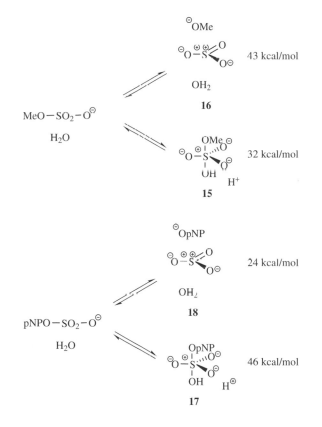

For species **15**, we estimate as follows:

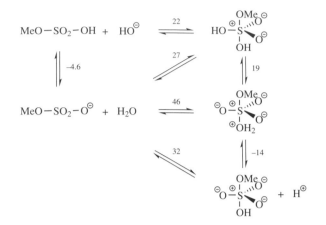

(Numbers used in this cycle: $\Delta G°$ for hydroxide plus monomethyl sulfate, assumed to be the same as for dimethyl sulfate estimated above; $\Delta G°$ for ionization of monomethyl sulfate;[138] $\Delta G°$ for tautomerization of the anionic adduct, based on pK_a values estimated by the method of Branch and Calvin; $\Delta G°$ for ionization of the apically protonated adduct, based on a pK_a estimated by the method of Branch and Calvin.)

For species **16**, we estimate the dissociative process as follows.

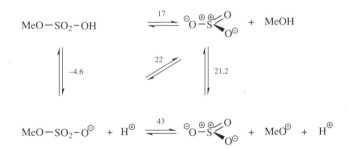

(Numbers used in this cycle: $\Delta G°$ for dissociation of monomethyl sulfate to give sulfur trioxide, estimated above; $\Delta G°$ for ionization of monomethyl sulfate;[138] $\Delta G°$ for ionization of methanol, calculated from the pK_a of methanol, 15.54.[139])

The energies for the stepwise intermediates for the two paths are within 11 kcal, suggesting a concerted mechanism is possible for S–O cleavage, but that the reaction would be very slow. In fact, of course, C–O cleavage predominates for alkyl esters.

Next, we do the estimations for *p*-nitrophenyl sulfate.

$$pNPO-SO_2-O^{\ominus} + H_2O + H^{\oplus} \xrightleftharpoons{24} pNPO^{\ominus} + SO_3 + H^{\oplus} + H_2O$$

$$\Big\updownarrow {\scriptstyle -12.6} \qquad\qquad\qquad\qquad\qquad \Big\updownarrow {\scriptstyle -9.4}$$

$$pNPOH + HSO_4^{\ominus} + H^{\oplus} \xrightleftharpoons{46} pNPOH + SO_3 + H^{\oplus} + OH^{\ominus}$$

$$\Big\updownarrow {\scriptstyle -3.8} \qquad\qquad\qquad\qquad\qquad \Big\updownarrow {\scriptstyle 19}$$

$$pNPOH + H_2SO_4 \xrightleftharpoons{23} pNPOH + SO_3 + H_2O$$

(Numbers used in this cycle: ΔG° for dissociation of H_2SO_4 to give SO_3;[138] ΔG° for acid dissociation of H_2SO_4;[125] ΔG° for hydrolysis of *p*-nitrophenyl sulfate mono-anion;[138] ΔG° for ionization of *p*-nitrophenol.[139])

Then from the free energy change for addition of hydroxide to monomethyl sulfate and the correction from methyl to *p*-nitrophenyl used above we get $\Delta G^{\circ} = 18 \, kcal/mol$ for

$$pNPO-SO_2-OH + HO^{\ominus} \xrightleftharpoons{} \underset{\substack{| \\ OH}}{\overset{\substack{OpNP \\ |}}{{}^{\ominus}O-\overset{\oplus}{S}{\cdots}^{{}^{\cdots}}\!\!O^{\ominus}}}{\overset{}{}}\!\!{}^{\blacktriangledown}OH$$

Allowing for proton transfer equilibria leads to:

$$pNPO-SO_2-O^{\ominus} + H_2O \xrightleftharpoons{46} \underset{\substack{| \\ OH}}{\overset{\substack{OpNP \\ |}}{{}^{\ominus}O-\overset{\oplus}{S}{\cdots}^{{}^{\cdots}}\!\!O^{\ominus}}}{}^{\blacktriangledown}O^{\ominus} + H^{\oplus}$$

$$\Big\updownarrow {\scriptstyle 25.8} \qquad\qquad\qquad\qquad\qquad \Big\updownarrow {\scriptstyle 2.7}$$

$$pNPO-SO_2\cdot OH + HO^{\ominus} \xrightleftharpoons{18} \underset{\substack{| \\ OH}}{\overset{\substack{OpNP \\ |}}{{}^{\ominus}O-\overset{\oplus}{S}{\cdots}^{{}^{\cdots}}\!\!O^{\ominus}}}{}^{\blacktriangledown}OH$$

(Numbers used in the cycle: ΔG° for addition of hydroxide to *p*-nitrophenyl sulfate, see above; ΔG° for proton transfer from *p*-nitrophenyl sulfate to hydroxide, based on pK_a values;[125] ΔG° for ionization of the monoanionic adduct of *p*-nitrophenyl sulfate, estimated by the method of Branch and Calvin, supplemented by the difference in pK_a between sulfuric acid and *p*-nitrophenyl sulfate.)

It is clear that the water reaction of *p*-nitrophenyl sulfate monoanion should occur by a dissociative mechanism, because **17** is too high in energy. (This was in fact assumed in the derivation of the numbers for SO$_3$ formation from aryl sulfates,[138] but the independently calculated value for the associative intermediate shows that the reaction is clearly expected to be dissociative via **18** or very close to it.)

Now we turn to the reactions of esters of sulfonic acids. Here we have less basis for estimation because the structural changes are more serious. It seems likely that **20** (or **22**) is, if anything, less stable than **12**, because **12** has at least the possibility of p-electron release from the RO group, which **20** does not, and the ferociously electron deficient S^{++} will need all the stabilization it can get. On the contrary, carbon is less electronegative than oxygen. For lack of anything better, we will assume similar energetics for dissociation. Moreover, a trigonal bipyramidal intermediate with a C in place of a neutral O should be favored, so that **19** or **21** should be easier to form than **1** or **3**. Benzenesulfonylium ion is not likely to be stabilized significantly by π-overlap because the charges on the sulfur cannot be delocalized onto the benzene ring, in contrast to the benzoylium ion. Benzenesulfonylium ion may then be less stable than methanesulfonylium ion because of the greater electronegativity of sp^2 than sp^3 carbons.

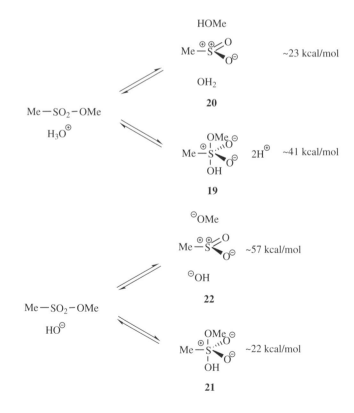

For acid-catalyzed hydrolysis of a sulfonate we use the following cycles.

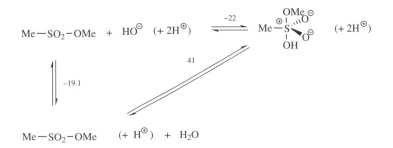

(Numbers used in the cycle: $\Delta G°$ for addition of hydroxide, assumed to be the same as for dimethyl sulfate; $\Delta G°$ for ionization of the neutral adduct, based on a pK_a estimated by the method of Branch and Calvin; $\Delta G°$ for ionization of the cationic adduct, based on a pK_a estimated by the method of Branch and Calvin.)

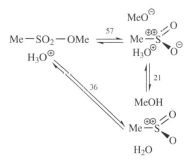

(Numbers used in this cycle: $\Delta G°$ for dissociation of methyl methanesulfonate to methanesulfonylium ion and methoxide, assumed to be the same as for dimethyl sulfate, estimated above; $\Delta G°$ for ionization of methanol.[130])

In this case, the dissociative path via **20** looks slightly favored though a concerted path looks possible. For this reaction the likely mechanism would be as follows:

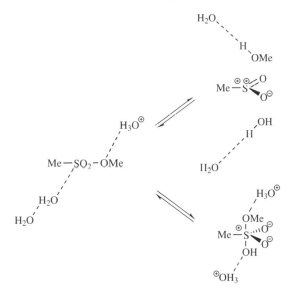

35

For the alkaline hydrolysis of a simple sulfonate ester, the associative mechanism via **21** looks better. The assumption that $\Delta G°$ for the additions of hydroxide to a sulfate or an analogous sulfonate ester are similar is supported by similar rate constants for analogous phosphate, phosphonate, and phosphinate esters. The rate constants for alkaline hydrolyses of $(MeO)_3PO$ ($k = 1.6 \times 10^{-4}$),[100] $MePO_3Me_2$ ($k = 2.5 \times 10^{-3}$),[140] and Et_2PO_2Et ($k = 1.2 \times 10^{-4}$)[141] are all similar, suggesting that the free energies of addition are also similar. This certainly suggests that the free energies of addition to dimethyl sulfate and methyl methanesulfonate will also be similar.

1.4.3. Possible Concerted Reactions of Phosphate Esters

The free energy of dissociation for triethyl phosphate to give the diethoxymetaphosphylium ion can be calculated as follows. The pK_a of HPO_3 is taken as -1.4, the value for HNO_3. The pK_a for $(HO)_2PO^+$ is then estimated as -6.4, using the increment of $5\, pK_a$ units per step from Pauling's rules.[142] Then the pK_a for $(EtO)PO_2H^+$ is assumed to be -7.1 (increment of 0.66 per ethoxy[100]). This allows a calculation of the ΔG for replacement of OH in $(EtO)PO_2H^+$ by OEt, using eq. (4) in reference 100, as:

$$(HO)(EtO)PO^+ + EtOH \rightleftharpoons (EtO)_2PO^+ + H_2O \qquad \Delta G = 7.17$$

The free energy of esterification for diethyl phosphate will be taken as $+3.2\,kcal/mol$ (average value per step of hydrolysis[100]); the free energy of dissociation of diethyl phosphate to ethyl metaphosphate and ethanol is $+28\,kcal/mol$ (Table 1.7); the free energy change for proton transfer from ethanol to ethyl metaphosphate is $+30.9\,kcal/mol$ (based on the pK_a value above and 15.5 for ethanol). This leads to the cycle and $\log K = -46$ for the dissociation of $(EtO)_3PO$ to $(EtO)_2PO^+$; this is the origin of the value in Table 1.7.

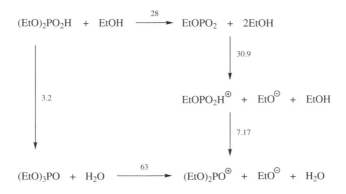

We can now calculate the energies of the two stepwise intermediates relative to the starting materials for mono-, di-, and triesters of phosphoric acid. Starting with the triester we get:

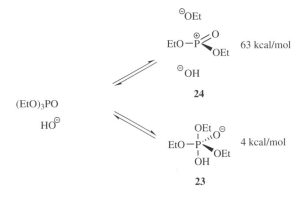

Here the thermodynamics strongly favor stepwise reaction by way of pentacoordinate intermediate **23**.

Turning now to the diester monoanion we get:

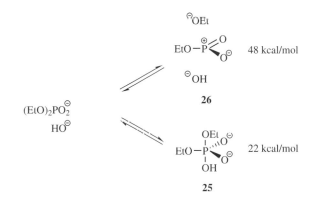

Here the thermodynamics still favor stepwise reaction by way of pentacoordinate intermediate, **25**, but the preference is weaker than in the triester case above. A concerted path might be barely possible here but would be expected to be close in structure and transition state energy to the pentacoordinate intermediate.

Finally we consider the monoester dianion:

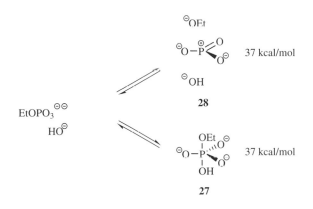

Here the thermodynamics seem set up for a concerted process, since both intermediates are bad. The reaction is, however, likely to be very slow for an alkyl ester.

For comparison one could look at the well known case of the monoester monoanion, which is known to have a transition state close to the metaphosphate anion.

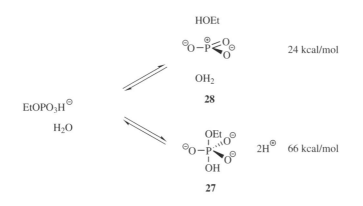

The associative energy is calculated from the cycle given below.

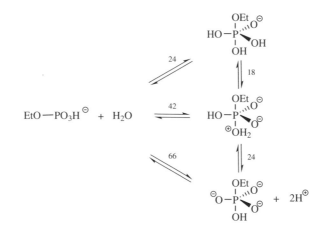

(Numbers used in this cycle: $\Delta G°$ for addition of water to give a monoanionic adduct, Table 1.7; $\Delta G°$ for proton transfer reactions, based on pK_a values estimated by the method of Branch and Calvin.) The dissociative energy is taken from Table 1.7.

There is an extra complication for the associative limit of this reaction. Addition of water to the monoanion would give a species with very acidic hydrogens, so that dissociation must be expected to be concerted. By microscopic reversibility, the very similar leaving group ethanol must depart by an analogous path. Thus the mechanism becomes:

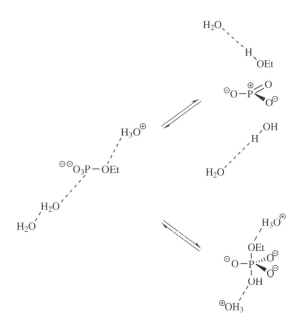

The zwitterionic form of monoethyl phosphate is unlikely to have a significant life-time because loss of a proton from cationic oxygen would be very fast.

Set against this argument based on energetics is the work of Williams, who has presented evidence for concerted reactions involving aryl diphenyl phosphates.[145] The key assumption here is that a linear Brønsted type plot requires a single transition state with no change in rate-determining step. This might be consistent with a stepwise reaction where breakdown in either direction is fast because the leaving groups are good. Hengge has reviewed the literature[143] and concluded that phosphate monoesters undergo hydrolysis by loose transition states close to the dissociative limit; that diesters and triesters with good leaving groups react by concerted mechanisms; and that triesters react by stepwise associative mechanisms. This analysis did not include consideration of the energetics of the dissociative species (alkyl metaphosphate ester or dialkyl metaphosphate cation). The analysis presented here suggests that concerted mechanisms will be strongly disfavored. More weight needs to be given to these simple energetic considerations.

1.5. CONCLUSION AND OUTLOOK

Tetrahedral intermediates vary enormously in stability relative to the corresponding carbonyl compounds, from extremes like hexafluoroacetone hydrate where it is difficult to remove the nucleophile from the adduct, to amide hydrates where the obligatory intermediate in acyl transfer is present at undetectably low concentrations. Linear free-energy relations provide a route to calculating the equilibrium constant for tetrahedral intermediate formation from carbonyl compounds, although there is still a shortage of experimental information on which to base these methods. There are several indirect ways to calculate these equilibrium constants if experiment is not feasible. Direct computational methods are coming along, but there remains a problem in calculating solvation energies. In the not very distant future, computational methods will become an important source of equilibrium information.

Calculating rate constants for the formation and breakdown of tetrahedral intermediates is possible provided the corresponding equilibrium constant is known. The mechanism must be known or postulated; these mechanisms often involve proton transfer steps.

Much less is known about the thermodynamics of the pentacoordinate intermediates in phosphoryl and sulfonyl chemistry, although such species clearly exist and are intermediates in at least some of the reactions of these classes of compounds.

Concerted mechanisms are possible for acyl, phosphoryl, and sulfonyl transfer. By thinking about the stepwise limits for each possible concerted process, one can, even with quite crude calculations, make some judgment about the feasibility of the concerted process. This area of chemistry is as yet unsettled and will see some changes in what is now generally accepted.

SUGGESTED READING

J. P. Guthrie, "Thermodynamics of Metastable Intermediates in Solution," *Acc. Chem. Res.* **1983**, *16*, 122.

J. Fastrez, "Estimation of the Free Energies of Addition of Nucleophiles to Conjugated Carbonyl Compounds and to Acyl Derivatives," *J. Am. Chem. Soc.* **1977**, *99*, 7004.

E. G. Sander and W. P. Jencks, "Equilibria for Addition to the Carbonyl Group," *J. Am. Chem. Soc.* **1968**, *90*, 6154.

J. P. Guthrie, "Hydration and Dehydration of Phosphoric Acid Derivatives: Free Energies of Formation of the Pentacoordinate Intermediates for Phosphate Ester Hydrolysis and of Monomeric Metaphosphate," *J. Am. Chem. Soc.* **1977**, *99*, 3391.

J. P. Guthrie, "Multidimensional Marcus Theory: an Analysis of Concerted Reactions," *J. Am. Chem. Soc.* **1996**, *118*, 12878.

J. P. Guthrie, "No Barrier Theory: Calculating Rates of Chemical Reactions from Equilibrium Constants and Distortion Energies," *ChemPhysChem* **2003**, *4*, 809.

A. Williams, *Concerted Organic and Bioorganic Mechanisms*, CRC Press: Boca Raton, FL, 2000.

M. I. Page and A. Williams, *Organic and Bioorganic Mechanisms*, Longman: London, 1997.

Appendix. The method of Branch and Calvin.[124]

This method estimates pK_a values for oxygen acids from that of the reference molecule water (corrected for the number of acidic hydrogens) using terms for electrostatic effects (based on formal charges) and inductive effects. Their equation is:

$$\log K_a = -16 + \sum I_{atom}\alpha_i + \sum I_{charge}\alpha_i + \log\left(\frac{n}{m}\right)$$

where -16 is the acid dissociation constant for water per hydrogen, I_{atom} is an inductive effect parameter for a particular atom, α_i is the fall-off factor, I_{charge} is the electrostatic parameter for unit charge, n is the number of equivalent acidic hydrogens in the acid, and m is the number of equivalent basic sites in the conjugate base. The parameters given by Branch and Calvin are:

Inductive constants for elements and formal charge	
$\alpha_i = 1/2.8$	$I_S = +3.4$
$I_{charge} = \pm 12.3$	$I_{Se} = +2.7$
$I_{Cl} = +8.5$	$I_{Te} = +2.4$
$I_{Br} = +7.5$	$I_N = +1.3$
$I_I = +6$	$I_P = +1.1$
$I_O = +4$	$I_{As} = +1.0$
	$I_C = -0.4$

By analogy with the proposal of Branch and Calvin, we can calculate the pK_a of an oxonium ion $X-OH_2^+$ by the related equation

$$\log K_a = 1.3 + \sum I_{atom}\alpha_i + \sum I_{charge}\alpha_i + \log\left(\frac{n}{m}\right)$$

where 1.3 is the $\log K_a$ value for hydronium ion, per acidic hydrogen.

For an O methylated oxonium ion, $X-O(Me)H^+$, we use a related equation based on the pK_{BH+} for dimethyl ether.[144] Per acidic hydrogen

$$\log K_a = 2.5 + \sum I_{atom}\alpha_i + \sum I_{charge}\alpha_i + \log\left(\frac{n}{m}\right)$$

Thus, for example, the K_a for $HOSO_2^+$ is calculated as

$$\log K_a = -16 + 13.2*2 - 13.2/2.8 + 2*4/2.8 + 3.4 + \log(1/3) = 11.5.$$

and K_a for $MeOSO_3(^{3-})OH_2^+$ is calculated as

$$\log K_a = +1.3 + 13.2 - 13.2*3/2.8 + 4*4/2.8 + 3.4 + \log(2/1) = 9.8.$$

Every time a hydroxyl substitutent is replaced by an alkoxyl, a correction of 0.66 is added;[100] this is based on the pK_a increments observed for H_3PO_4 and its mono and dialkyl esters.

REFERENCES

1. A. C. McCormack, C. A. McDonnell, R. A. More O'Ferrall, A. C. O'Donoghue, and S. N. Rao, *J. Am. Chem. Soc.* **2002**, *124*, 8575.

2. T. L. Amyes, J. P. Richard, and M. Novak, *J. Am. Chem. Soc.* **1992**, *114*, 8032.

3. M. M. Toteva and J. P. Richard, *J. Am. Chem. Soc.* **1996**, *118*, 11434.

4. J. P. Richard, V. Jagannadham, T. L. Amyes, M. Mishima, and Y. Tsuno, *J. Am. Chem. Soc.* **1994**, *116*, 6706.

5. J. P. Guthrie, *Can. J. Chem.* **1978**, *56*, 962.

6. G. Olah, *Angew. Chem. Int. Ed. Eng.* **1995**, *34*, 1393.

7. L. Pauling, *Chem. Eng. News* **1946**, *24*, 1375.

8. L. Pauling, *Am. Scientist* **1948**, *36*, 51.

9. E. M. Driggers and P. G. Schultz, *Adv. Protein Chem.* **1996**, *49*, 261.

10. G. Lienhard, *Science* **1973**, *180*, 149.

11. R. Wolfenden, *Acc. Chem. Res.* **1972**, *5*, 10.

12. J. Stieglitz, *Am. Chem. J.* **1899**, *21*, 101.

13. W. Lossen, *Justus Liebigs Annalen der Chemie* **1889**, *252*, 170.

14. P. Le Henaff, *Bull. Soc. Chim. France* **1968**, *11*, 4687.

15. J. Kurz, *J. Am. Chem. Soc.* **1967**, *89*, 3524.

16. J. Hine and R. Redding, *J. Org. Chem.* **1970**, *35*, 2769.

17. F. Swarts, *Bull. Soc. Chim. Belg.* **1926**, *35*, 412.

18. M. L. Bender, *J. Am. Chem. Soc.* **1953**, *75*, 5986.

19. J. P. Guthrie, *Can. J. Chem.* **1976**, *54*, 202.

20. M. Bender, *J. Am. Chem. Soc.* **1951**, *73*, 1626.

21. H. F. Gilbert and W. P. Jencks, *J. Am. Chem. Soc.* **1977**, *99*, 7931.

22. A. Williams, *Concerted Organic and Bioorganic Mechanisms*, CRC Press, Boca Raton, FL, **2000**.

23. W. P. Jencks, *J. Am. Chem. Soc.* **1959**, *81*, 475.

24. G. L. Schmir and B. Cunningham, *J. Am. Chem. Soc.* **1965**, *87*, 5692.

25. R. A. McClelland, *J. Am. Chem. Soc.* **1978**, *100*, 1844.

26. J. W. Scheeren and W. Stevens, *Rec. Trav. Chim. Pays-Bas* **1969**, *88*, 897.

27. A. L. Henne, J. W. Shepard, and E. J. Young, *J. Am. Chem. Soc.* **1950**, *72*, 3577.

28. R. B. Woodward, *Pure Appl. Chem.* **1964**, *9*, 49.

29. A. Kirby, I. Komarov, and N. Feeder, *J. Chem. Soc. Perkin Trans. II* **2001**, 522.

30. J. P. Guthrie, *Can. J. Chem.* **1975**, *53*, 898.

31. E. G. Sander and W. P. Jencks, *J. Am. Chem. Soc.* **1968**, *90*, 6154.

32. R. G. Kallen and W. P. Jencks, *J. Bio. Chem.* **1966**, *241*, 5864.

33. T. M. Olson and M. R. Hoffmann, *Atm. Env.* **1989**, *23*, 985.

34. J. Hine, F. A. Via, J. K. Gorkis, and J. C. Craig, *J. Am. Chem. Soc.* **1970**, *92*, 5186.

35. J. Hine and J. Mulders, *J. Org. Chem.* **1967**, *32*, 2200.

36. E. Betterton, Y. Erei, and M. Hoffmann, *Env. Sci. Tech.* **1988**, *22*, 92.

37. G. E. Lienhard and W. P. Jencks, *J. Am. Chem. Soc.* **1966**, *88*, 3982.

38. R. A. McClelland and M. Coe, *J. Am. Chem. Soc.* **1983**, *105*, 2718.

39. J. Fastrez, *J. Am. Chem. Soc.* **1977**, *99*, 7004.

40. M. R. Crampton, *J. Chem. Soc, Perkin Trans. 2* **1975**, 185.

41. W. M. Ching and R. G. Kallen, *J. Am. Chem. Soc.* **1978**, *100*, 6119.

42. A. Malpica and M. Calzadilla, *J. Phys. Org. Chem.* **2003**, *16*, 202.

43. A. Lapworth and R. H. F. Manske, *J. Chem. Soc.* **1930**, 1976.

44. P. R. Young and W. P. Jencks, *J. Am. Chem. Soc.* **1979**, *101*, 3268.

45. R. Stewart and J. D. Van Dyke, *Can. J. Chem.* **1970**, *48*, 3961.

46. C. D. Ritchie, *J. Am. Chem. Soc.* **1984**, *106*, 7087.

47. J. P. Guthrie, *J. Am. Chem. Soc.* **1978**, *100*, 5892.

48. J. Atherton, J. Blacker, M. Crampton, and C. Grosjean, *Org. Biomolec. Chem.* **2004**, *2*, 2567.

49. T. Oakes and G. Stacy, *J. Am. Chem. Soc.* **1972**, *94*, 1594.

50. Y. Ogata and A. Kawasaki, *J. Org. Chem.* **1974**, *39*, 1058.

51. J. Toullec and S. Bennour, *J. Org. Chem.* **1994**, *59*, 2831.

52. J. P. Guthrie, *J. Am. Chem. Soc.* **1973**, *95*, 6999.

53. J. P. Guthrie, *J. Am. Chem. Soc.* **1974**, *96*, 3608.

54. R. W. Taft, "Separation of Polar, Steric, and Resonance Effects in Reactivity," in *Steric Effects in Organic Chemistry* (Ed. M. S. Newman), Wiley, New York, **1956**, pp. 556 f.

55. K. B. Wiberg, *Physical Organic Chemistry*, Wiley, New York, **1964**, p. 595.

56. J. Hine, *Physical Organic Chemistry*, 2nd edition; McGraw-Hill, New York, **1962**.

57. P. Ballinger and F. A. Long, *J. Am. Chem. Soc.* **1960**, *82*, 795.

58. F. A. Carroll, *Perspectives on Structure and Mechanism in Organic Chemistry*, Brooks/Cole, Pacific Grove, CA, **1998**.

59. D. D. Perrin, B. Dempsey, and E. P. Serjeant, *pK_a Prediction for Organic Acids and Bases*, Chapman and Hall, London, **1981**.

60. P. Greenzaid, Z. Luz, and D. Samuel, *J. Am. Chem. Soc.* **1967**, *89*, 749.

61. P. Geneste, G. Lamaty, and J. P. Roque, *Tetrahedron* **1971**, *27*, 5561.

62. M. Calzadila, A. Malpica and T. Cordova, *J. Phys. Org. Chem.* **1999**, *12*, 708.

63. G. Lamaty, J. Roque, A. Natat, and T. Silou, *Tetrahedron* **1986**, *42*, 2657.

64. J. Hine, L. R. Green, and P. C. Meng, *J. Org. Chem.* **1976**, *41*, 3343.

65. J. P. Guthrie and P. A. Cullimore, *Can. J. Chem.* **1980**, *58*, 1281.

66. P. Geneste, G. Lamaty and J. P. Roque, *Rec. Trav. Chim. Pays-Bas* **1972**, *91*, 188.

67. J. Hine, J. C. Craig, J. G. Underwood, and F. A. Via, *J. Am. Chem. Soc.* **1970**, *92*, 5194.

68. A. Lapworth, *J. Chem. Soc.* **1903**, 995.

69. H. Slebocka-Tilk, A. J. Bennet, J. W. Keillor, R. S. Brown, J. P. Guthrie, and A. Jodhan, *J. Am. Chem. Soc.* **1990**, *112*, 8507.

70. A. J. Bennet, H. Slebocka-Tilk, R. S. Brown, J. P. Guthrie, and A. Jodhan, *J. Am. Chem. Soc.* **1990**, *112*, 8497.

71. W. P. Jencks, *Chem. Rev.* **1972**, *72*, 705.

72. A. O. Cohen and R. A. Marcus, *J. Phys. Chem.* **1968**, *72*, 4249.

73. R. A. Marcus, *J. Phys. Chem.* **1968**, *72*, 891.

74. R. A. Marcus, *J. Am. Chem. Soc.* **1969**, *91*, 7224.

75. R. A. Marcus, *Annu. Rev. Phys. Chem.* **1964**, *15*, 155.

76. a. J. P. Guthrie, *J. Am. Chem. Soc.* **1996**, *118*, 12878. b. J. P. Guthrie, *J. Am. Chem. Soc.* **1996**, *118*, 12886.

77. J. P. Guthrie, *J. Am. Chem. Soc.* **2000**, *122*, 5529.

78. J. P. Guthrie, *ChemPhysChem.* **2003**, *4*, 809.

79. J. P. Guthrie and V. Pitchko, *J. Phys. Org. Chem.* **2004**, *17*, 548.

80. J. P. Guthrie and V. Pitchko, *J. Am. Chem. Soc.* **2000**, *122*, 5520.

81. J. P. Guthrie, *Can. J. Chem.* **2005**, *83*, 1.

82. J. P. Guthrie, *J. Phys. Org. Chem.* **1998**, *11*, 632.

83. J. P. Guthrie, *Can. J. Chem.* **1999**, *77*, 934.

84. J. Guthrie, *Bioorganic Chem.* **2002**, *30*, 32.

85. G. Thatcher and R. Kluger, *Adv. Phys. Org. Chem.* **1989**, *25*, 99.

86. L. Horner, H. Oediger, and H. Hoffmann, *Justus Liebigs Annalen der Chemie* **1959**, *626*, 26.

87. I. N. Zhmurova and A. V. Kirsanov, *Zh. Obsh. Khim.* **1959**, *29*, 1687.

88. G. H. Birum and J. L. Dever, *134th National Meeting of the American Chemical Society, Chicago, Ill., Sept 1958, Abstracts, 101P* **1958**.

89. V. A. Kukhtin, *Dokl. Akad. Nauk SSSR* **1958**, *121*, 466; *Chem. Abst.* **1959**, *53*, 11056.

90. D. B. Denney and H. M. Relles, *J. Am. Chem. Soc.* **1964**, *86*, 3897.

91. L. Chang, D. B. Denney, and D. Z. Denney, *J. Am. Chem. Soc.* **1977**, *99*, 2293.

92. F. H. Westheimer, *Acc. Chem. Res.* **1968**, *1*, 70.

93. R. Kluger, F. Covitz, E. Dennis, L. D. Williams, and F. Westheimer, *J. Am. Chem. Soc.* **1969**, *91*, 6066.

94. G. Roschenthaler and W. Storzer, *Ang. Chem. Int. Ed. Eng.* **1982**, *21*, 208.

95. I. Granoth and J. C. Martin, *J. Am. Chem. Soc.* **1978**, *100*, 5229.

96. I. Granoth and J. C. Martin, *J. Am. Chem. Soc.* **1979**, *101*, 4618.

97. A. Dubourg, R. Roques, G. Germain, J. P. Declercq, B. Garrigues, D. Boyer, A. Munoz, A. Klaebe, and M. Comtat, *J. Chem. Res.* **1982**, 180. See, however, ref. 96; this may be a hydrogen bonded complex rather than a salt.

98. O. Curnow, *J. Chem. Educ.* **1998**, *75*, 910.

99. J. C. Martin, *Science* **1983**, *221*, 509.

100. J. P. Guthrie, *J. Am. Chem. Soc.* **1977**, *99*, 3391.

101. S. J. Benkovic, "Hydrolytic Reactions of Inorganic Esters", in *Comprehensive Chemical Kinetics, Vol. 10* (Eds. C. H. Bamford and C. F. H. Tipper), Elsevier, Amsterdam, **1972**.

102. W. Jencks, *Chem. Soc. Rev.* **1981**, *10*, 345.

103. W. P. Jencks, *Acc. Chem. Res.* **1980**, *13*, 161.

104. S. J. Benkovic and P. A. Benkovic, *J. Am. Chem. Soc.* **1966**, *88*, 5504.

105. C. W. Perkins, S. R. Wilson, and J. C. Martin, *J. Am. Chem. Soc.* **1985**, *107*, 3209.

106. J. Kice, *Adv. Phys. Org. Chem.* **1980**, *17*, 65.

107. J. F. King, *Acc. Chem. Res.* **1975**, *8*, 10.

108. X. Lopez, A. Dejaegere, and M. Karplus, *J. Am. Chem. Soc.* **2001**, *123*, 11755.

109. D. Cameron and G. Thatcher, *in The Anomeric Effect and Associated Stereoelectronic Effects* Thatcher, G. R. J., Ed. *ACS Symposium Series, American Chemical Society, Washington, DC, Vol. 539*, **1993**, pp. 256.

110. D. Cameron and G. Thatcher, *J. Org. Chem.* **1996**, *61*, 5986.

111. G. R. J. Thatcher and D. R. Cameron, *J. Chem. Soc. Perkin Trans. II* **1996**, 767.

112. E. F. Perozzi, J. C. Martin, and I. C. Paul, *J. Am. Chem. Soc.* **1974**, *96*, 6735.

113. A. Williams, *Acc. Chem. Res.* **1989**, *22*, 387.

114. F. G. Bordwell, *Acc. Chem. Res.* **1970**, *3*, 281.

115. F. G. Bordwell, *Acc. Chem. Res.* **1972**, *5*, 374.

116. M. J. S. Dewar, *J. Am. Chem. Soc.* **1984**, *106*, 209.

117. T. R. Forbus, and J. C. Martin, *J. Am. Chem. Soc.* **1979**, *101*, 5057.

118. T. Forbus, and J. C. Martin, *Heteroatom Chem.* **1993**, *4*, 113.

119. J. P. Guthrie and D. C. Pike, *Can. J. Chem.* **1987**, *65*, 1951.

120. J. P. Guthrie, *J. Am. Chem. Soc.* **1991**, *113*, 3941.

121. P. C. Haake and F. J. Westheimer, *J. Am. Chem. Soc.* **1961**, *83*, 1102.

122. D. Christman and S. Oae, *Chem. Ind. (London)* **1959**, 1251.

123. S. Oae, T. Fukumoto, and R. Kiritani, *Bull. Chem. Soc. Jpn* **1963**, *36*, 346.

124. G. Branch and M. Calvin, *The Theory of Organic Chemistry: An Advanced Couse*, Prentice-Hall, New York, **1941**.

125. J. P. Guthrie, *Can. J. Chem.* **1978**, *56*, 2342.

126. E. T. Kaiser and F. J. Kezdy, *Prog. Bioorg. Chem.* **1976**, *4*, 239.

127. T. Okuyama, J. Lee, and K. Ohnishi, *J. Am. Chem. Soc.* **1994**, *116*, 6480.

128. T. Okuyama, T. Nakamura, and T. Fueno, *J. Am. Chem. Soc.* **1990**, *112*, 9345.

129. T. Okuyama, T. Nakamura, and T. Fueno, *Chem. Lett.* **1990**, 1133.

130. M. R. Siggel and T. D. Thomas, *J. Am. Chem. Soc.* **1986**, *108*, 4360.

131. M. Siggel, A. Streitwieser, and T. Thomas, *J. Am. Chem. Soc.* **1988**, *110*, 8022.

132. K. Wiberg, J. Ochterski, and A. Streitwieser, *J. Am. Chem. Soc.* **1996**, *118*, 8291.

133. N. C. Deno, C. U. Pittman, and M. J. Wisotsky, *J. Am. Chem. Soc.* **1964**, *86*, 4370.

134. N. C. Deno, R. W. Gaugler, and M. J. Wisotsky, *J. Org. Chem.* **1966**, *31*, 1967.

135. T. Deacon, C. R. Farrar, and A. Williams, *J. Am. Chem. Soc.* **1978**, *100*, 2525.

136. E. T. Kaiser, I. R. Katz, and T. F. Wulfers, *J. Am. Chem. Soc.* **1965**, *87*, 3781.

137. J. Younker, and A. Hengge, *J. Org. Chem.* **2004**, *69*, 9043.

138. J. P. Guthrie, *J. Am. Chem. Soc.* **1980**, *102*, 5177.

139. W. P. Jencks and J. Regenstein, "Physical and Chemical Data", in *Handbook of Biochemistry and Molecular Biology, Vol. 1*, 3rd Edition (Ed. G. Fassman), Chemical Rubber Company, Cleveland, **1976**, p 305.

140. R. F. Hudson, and L. Keay, *J. Chem. Soc.* **1956**, 2463.

141. V. E. Belskii, M. V. Efrmova, I. M. Shermergorn, and A. N. Pudovik, *Izvest. Akad. Nauk SSSR, Ser. Khim.* **1969**, 307.

142. L. Pauling, *General Chemistry*, W.H. Freeman, San Francisco, 1947.

143. J. Purcell and A. C. Hengge, *J. Org. Chem.* **2005**, *70*, 8437.

144. P. Bonvicini, G. Modena, and G. Scorrano, *J. Am. Chem. Soc.* **1973**, *95*, 5960.

145. S. A. Ba-Saif, M. A. Waring, and A. Williams, *J. Am. Chem. Soc.* **1990**, *112*, 8115.

CHAPTER 2

Silicon-, Germanium-, and Tin-Centered Cations, Radicals, and Anions*

VLADIMIR Ya. LEE

Department of Chemistry, Graduate School of Pure and Applied Sciences, University of Tsukuba, Tsukuba, Ibaraki 305-8571, Japan

AKIRA SEKIGUCHI

Department of Chemistry, Graduate School of Pure and Applied Sciences, University of Tsukuba, Tsukuba, Ibaraki 305-8571, Japan

*In this contribution we will deal only with the low-coordinated (tricoordinated) cations, radicals, and anions. The vast number of hypercoordinated (penta-, hexacoordinated) species is outside the scope of the present review and will not be considered.

Reviews of Reactive Intermediate Chemistry. Edited by Matthew S. Platz, Robert A. Moss, Maitland Jones, Jr.
Copyright © 2007 John Wiley & Sons, Inc.

2.1. Si-, Ge-, AND Sn-CENTERED CATIONS

2.1.1. Introduction

The carbocations, or carbenium ions according to the latest IUPAC recommendations, are commonly described as the trivalent species R_3C^+ bearing a positive charge. The carbocations are typically sp^2-hybridized, thus featuring the diagnostic planar structure, bond angles of $120°$, and the vacant unhybridized $2p_z$-orbital of the central C atom lying perpendicular to the R_3C plane. The classical example of such carbocations is the simplest methyl cation CH_3^+ isoelectronic to BH_3. Because carbocations are electron-deficient species, they are highly electrophilic, participating in a vast number of reactions with nucleophiles, which brings about their great synthetic importance.

The chemistry of the heavy analogs of carbocations of the type $RR'R''E^+$ (E = Si, Ge, Sn) was developed much later than that of carbocations, and it is still a field of very active investigation. At the early stages of $RR'R''E^+$ chemistry, it very quickly became apparent that there is a great difference between the carbocations and their heavy analogs, caused by the sharply different properties of C and other elements of group 14 (principally electronegativity, size, and polarizability). Thus, the synthetic approaches, which successfully provide the stable carbocations in organic chemistry, appeared to be rather inefficient in the synthesis of silylium ions because of the high electrophilicity of the latter species leading to their intrinsic kinetic instability. Another major problem associated with $RR'R''E^+$ chemistry is the degree of real "freedom" of cations from external nucleophiles, such as counterions and solvents; in other words there was a question about the *ionic* versus *covalent* bonding nature of such cationic species. Therefore, the successful synthesis of silylium, germylium, and stannylium ions has required the creation of new synthetic strategies based upon the utilization of counterions and solvents of particularly low nucleophilicity to prevent their reaction or coordination to the cationic part. The first structurally characterized silylium ions were reported in the early 1990s; however, their true silylium ion nature has been severely criticized (see Section 2.1.4). Meanwhile, the use of the almost nonnucleophilic borate and carborane anions and benzene and toluene as the solvents of choice finally made possible the synthesis of the true $RR'R''E^+$ cations free from any covalent interactions with either counterions or solvents. Some of these cations are intramolecularly stabilized by conjugation with cyclic π-systems, whereas the acyclic cations are almost entirely electronically unperturbed, being the genuine heavy analogs of tricoordinate carbenium ions.

The chemistry of the heavy carbenium ion analogs has been the subject of several preceding reviews,[1] of which the latest one by Müller[1p] is the most comprehensive. In this section we will deal briefly with the most essential achievements made in the 1990s, and will emphasize the latest progress in the field, particularly concerning the synthesis and structural studies of the stable noncoordinated ("free") cations of the type $RR'R''E^+$.

2.1.2. Synthesis of $RR'R''E^+$ Cations (E = Si, Ge, Sn)

As mentioned in the Introduction, Si is substantially more electropositive, larger in size, and more polarizable than C. This would suggest that silylium ions R_3Si^+, for example, could be easily prepared and will be even more thermodynamically stable than the corresponding carbon analogs R_3C^+. This indeed was the case in the gas phase, where numerous silylium ions were detected and identified by means of mass spectrometry.[1a] However, the situation was entirely different in the condensed phase, that is, in solution, where the inherent high electrophilicity of the silylium ions resulted in their great instability, which was undoubtedly kinetic in origin. As a result, silylium ions exhibited an extremely high reactivity toward a variety of σ- and π-donors, even including low-nucleophilic counterions and solvents, which results in the easy formation of donor–acceptor complexes or (in extreme cases) covalently bonded tetravalent compounds. The utilization of the relatively inert arene solvents (benzene, toluene) and counterions of particularly low nucleophilicity (borate $B(C_6F_5)_4{}^-$, carborane $CB_{11}H_6Br_6{}^-$ anions) finally resolved this synthetic problem, and led to the successful preparation of the cationic species $RR'R''E^+$, in which the positive charge actually resides on the element atom E (see Section 2.1.4). This synthetic achievement along with the right choice of bulky substituents, resulted in the isolation of the first "free" (noncoordinated) cations, progress that allowed the direct study of the cations structures both in the crystal and in solution (see Section 2.1.5). However, despite the evident progress made in the past decade, the number of efficient synthetic approaches for the preparation of heavier group 14 element-centered cations of the type $RR'R''E^+$ (E = Si, Ge, Sn) is still very limited. Generally, they can be classified into the following groups depending on the starting material used.

2.1.2.1. From Halides RR′R″EX

In contrast to the common S_N1 reaction $R_3C-X \rightarrow R_3C^+ + X^-$ in organic chemistry as a general method for the generation of carbocations R_3C^+, the same synthetic approach is definitely not the best one for the preparation of the stable heavy analogs of carbenium ions $RR'R''E^+$, because of the strong E−X bonds (much higher halophilicity of Si, Ge, and Sn compared with that of C) and enhanced reactivity of the halides toward the developing cationic species $RR'R''E^+$. For these reasons, the cations generated by the above method were characterized as being not "true" silylium ions, but rather strongly polarized donor–acceptor complexes having only a partial positive charge localized on the Si atom (Scheme 2.1).[2,3] This was evidenced by their ^{29}Si NMR chemical shift values (62.7 ppm (Scheme 2.1, A), 76.6 ppm (Scheme 2.1, B)).

2.1.2.2. From Hydrides RR′R″EH

This is perhaps one of the most effective and straightforward method for the preparation of $RR'R''E^+$ cations. The driving force of this so-called "hydride transfer reaction," involving the oxidation of $RR'R''EH$ hydrides with powerful Lewis acids such as Ph_3C^+, is the relative strength of the breaking and forming bonds: stronger C−H versus weaker E−H bonds. A number of the heavy analogs of carbenium ions, intra- or intermolecularly stabilized with coordination to π-bonds, counterions, or solvent molecules, can be readily accessed by this synthetic route (Scheme 2.2, A–E).[4–8] The free cations can also be synthesized employing this synthetic approach (see Section 2.1.5.2.1).

$$Me_3Si\text{-}Br + AlBr_3 \xrightarrow{CH_2Br_2} Me_3Si^{\delta+}\text{-}Br-AlBr_3{}^{\delta-} \quad \textbf{(A)}$$

$$Et_3Si\text{-}OTf + BCl_3 \longrightarrow Et_3Si^{\delta+}\text{-}OTf-BCl_3{}^{\delta-} \quad \textbf{(B)}$$

Scheme 2.1

$$Et_3SiH + Ph_3C^+ \cdot TPFPB^- \xrightarrow[C_7H_8]{C_6H_6} [Et_3Si(C_7H_8)]^+ \cdot TPFPB^- + Ph_3CH \quad \textbf{(A)}$$

$$i\text{-}Pr_3SiH + Ph_3C^+ \cdot [CB_{11}H_6X_6]^- \xrightarrow{C_7H_8} i\text{-}Pr_3Si^+ \cdot [CB_{11}H_6Br_6]^- + Ph_3CH \quad \textbf{(B)}$$

$$Ph_3GeH + Ph_3C^+ \cdot ClO_4^- \xrightarrow{CH_2Cl_2} Ph_3Ge(OClO_3) + Ph_3CH \quad \textbf{(C)}$$

$$Bu_3SnH + B(C_6F_5)_3 \xrightarrow{C_6H_6} [Bu_3Sn(C_6H_6)]^+ \cdot HB(C_6F_5)_4^- \quad \textbf{(D)}$$

$$Bu_3SnH + Ph_3C^+ \cdot TFPB^- \xrightarrow{CD_2Cl_2} Bu_3Sn^+ \cdot TFPB^- + Ph_3CH \quad \textbf{(E)}$$

Scheme 2.2

2.1.2.3. From RR′R″E–Alkyl and RR′R″E–ERR′R″

One of the most impressive examples of R_3E C bond cleavage, generating the corresponding R_3E^+ cations, was recently demonstrated by Lambert's group by the reaction of allylic derivatives $Mes_3E-CH_2-CH=CH_2$ (E = Si, Ge, Sn) with $[Et_3Si(C_6H_6)]^+ \cdot TPFPB^-$ forming the intermediate carbenium ion $Mes_3E-CH_2-CH^+-CH_2SiEt_3$, which finally undergoes fragmentation to produce the more stable Mes_3E^+ cations (Scheme 2.3).[9]

$$Mes_3E\text{-}CH_2\text{-}CH=CH_2 + [Et_3Si(C_6H_6)]^+ \cdot TPFPB \xrightarrow{C_6H_6} Mes_3E\text{-}CH_2\text{-}CH^+ CH_2\text{-}SiEt_3 \cdot TPFPB^-$$

$$\xrightarrow{\hspace{1cm}} Mes_3E^+ \cdot TPFPB^- + H_2C=CH\text{-}CH_2\text{-}SiEt_3$$

Scheme 2.3

The R_3E^+ cations can be effectively generated by the oxidative cleavage of E–E bonds of R_3E-ER_3 by the action of a strong Lewis acid. Thus, $t\text{-}Bu_3E-Et\text{-}Bu_3$ readily reacted with $Ph_3C^+ \cdot TFPB^-$ in the presence of nitriles to form the nitrile complexes of the $t\text{-}Bu_3E^+$ cations[10] (Scheme 2.4, A), whereas $n\text{-}Bu_3Sn-Snn\text{-}Bu_3$ was elegantly oxidized with the $CB_{11}Me_{12}\bullet$ free radical to produce the $n\text{-}Bu_3Sn^+$ cation weakly coordinated to Me groups of the $CB_{11}Me_{12}^-$ anions[11] (Scheme 2.4, B).

$$t\text{-}Bu_3E\text{-}Et\text{-}Bu_3 + 2 Ph_3C^+ \cdot TFPB^- \xrightarrow{R\text{-}C\equiv N} 2 [t\text{-}Bu_3E(:N\equiv C\text{-}R)]^+ \cdot TFPB^- \quad \textbf{(A)}$$

$$[E = Si, Ge, Sn; R = Me, t\text{-}Bu]$$

$$n\text{-}Bu_3Sn\text{-}Snn\text{-}Bu_3 + 2 CB_{11}Me_{12}\bullet \xrightarrow{pentane} 2 n\text{-}Bu_3Sn^+ \cdot CB_{11}Me_{12}^- \quad \textbf{(B)}$$

Scheme 2.4

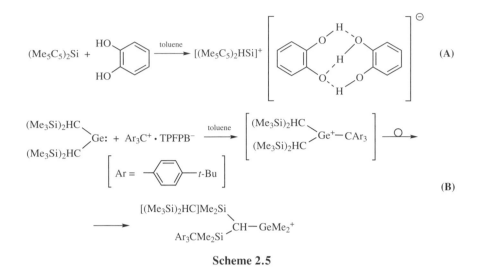

Scheme 2.5

2.1.2.4. From Heavy Carbene Analogs RR′E:

The addition of cationic species to heavy analogs of carbenes expands the coordination number of the central element from 2 to 3, resulting in the formation of element-centered cations strongly stabilized by the intramolecular electron donation. However, there are very few examples of such a synthetically rather attractive approach: the unusual reaction of decamethylsilicocene with catechol forming a silyl cation in the form of protonated decamethylsilicocene[12] (Scheme 2.5, A) and reaction of the stable Lappert's germylene [(Me$_3$Si)$_2$CH]$_2$Ge: with [(4-t-Bu−C$_6$H$_4$)]$_3$C$^+$•TPFPB$^-$ yielding the unexpected intramolecularly stabilized germyl cation after a series of consecutive rearrangements[13] (Scheme 2.5, B).

2.1.2.5. From Free Radicals RR′R″E•

This last synthetic route, involving the one-electron oxidation of the free radicals RR′R″E• with an appropriate Lewis acid such as Ph$_3$C$^+$, is one of the best methods for the extremely fast and clean formation of the element-centered cations RR′R″E$^+$. Although this approach requires the presence of the radical species as readily available starting materials, the recent synthesis of stable silyl-substituted radicals of the type (t-Bu$_2$MeSi)$_3$E• (E = Si, Ge, Sn) (see Section 2.2.4.1.2) made such an approach a rather attractive and easily accessible synthetic route to the stable and free (t-Bu$_2$MeSi)$_3$E$^+$ cations (Scheme 2.6).[14]

$$(t\text{-Bu}_2\text{MeSi})_3\text{E} \bullet \; + \; \text{Ph}_3\text{C}^+ \bullet \text{TPFPB}^- \quad \xrightarrow[\text{[E = Ge, Sn]}]{\text{C}_6\text{H}_6} \quad (t\text{-Bu}_2\text{MeSi})_3\text{E}^{+\bullet}\, \text{TPFPB}^-$$

Scheme 2.6

2.1.3. Reactions and Synthetic Applications

The reactivity studies and synthetic utilization of the heavy analogs of carbenium ions is still not sufficiently developed; however, even now it is clear that the major synthetic interest of silylium, germylium, and stannylium ions $RR'R''E^+$ (E = Si, Ge, Sn) lies in a direction quite parallel to that of the classical carbocations. Thus, the typical carbocation reactions in organic chemistry are: (1) reaction with nucleophiles to form a substitution product (novel C−C σ-bond) by the S_N1 mechanism; (2) elimination of a proton to form an elimination product (novel C=C π-bond) by the E1 mechanism; and (3) electrophilic addition to alkenes to form new cationic adducts (cationic polymerization). In respect to the $RR'R''E^+$ cations, whose enhanced electrophilicity (compared with their carbon counterparts) was exploited as a major synthetic advantage, the reaction paths (1) and (3) were mainly applied, both resulting in the formation of novel cationic species. Thus, the silylium ion smoothly adds to the >C=C< double bond to produce stable β-silyl carbocations,[15] and to the −C≡C− triple bond to form persistent silyl-substituted vinyl cations.[16] The $[Et_3Si(arene)]^+$ cation is, perhaps, one of the most synthetically useful silylium ion reagents, which was recently widely employed for the generation of a variety of carbenium and silylium ions. The most important contribution to this field has been made by the group of Reed, who, for example, generated $H^+\bullet[CHB_{11}R_5X_6]^-$ (R = H, Me, Cl; X = Cl, Br, I), the strongest currently known Brønsted superacid, by the simple treatment of $[Et_3Si(arene)]^+\bullet[CHB_{11}R_5X_6]^-$ with HCl.[17] The Brønsted acidity power of this superacid is extremely high, so it can readily protonate at ambient temperature such stable aromatic systems as fullerene C_{60} and Me-substituted benzenes $C_6Me_nH_{6-n}$ (n = 0, 1, 2, 3, 5, 6) to generate the fullerene cation $[HC_{60}]^{+17b}$ and benzenium ions $[HC_6Me_nH_{6-n}]^{+17a-c}$ respectively. The treatment of $[Et_3Si(arene)]^+\bullet[CHB_{11}Me_5X_6]^-$ (X = Cl, Br) with alkyl triflates ROTf (R = Me, Et) cleanly resulted in the formation of $R^+\bullet[CHB_{11}Me_5X_6]^-$, extremely electrophilic alkylating reagents that are stronger than alkyl triflates.[18] Thus, the high electrophilic power of $Me^+\bullet[CHB_{11}Me_5Br_6]^-$ was spectacularly demonstrated by its reactions with benzene C_6H_6 and alkanes R−H (R = C_4H_9, C_5H_{11}, C_6H_{13}), forming the corresponding toluenium ion $[Me(C_6H_6)]^+$ and tertiary carbenium ions R^+, respectively.[18] Certainly, such extremely high reactivity of $R^+\bullet[CHB_{11}Me_5X_6]^-$ exceeds that of the conventional alkyl triflates. Other examples of the practical application of heavy analogs of carbenium ions are listed below. As a start, one can mention the utilization of stannyl cations R_3Sn^+ (R = Me, Bu) as superior leaving groups in electrophilic aromatic *ipso*-substitution reactions widening the scope of the Friedel–Crafts acylation, Vilsmeier formylation, sulfinations, and sulfonations.[19] The chiral silyl cation complexes with acetonitrile were claimed to be novel Lewis acid catalysts for Diels–Alder cycloaddition reactions.[20] Expectedly, heavy carbenium analogs may promote the cationic polymerization of simple alkenes. Thus, the stable *sec*-alkyl β-stannylcarbocation, which was believed to be formed through the addition of intermediate Me_3Sn^+ cation to the C=C double bond, effectively polymerizes a number of alkenes, for example, isobutene, to produce high-molecular weight polymers.[21] The silanorbornyl cations were shown to be the key intermediates in the metal-free catalytic intramolecular hydrosilylation of C=C double bonds under mild conditions.[22] The stannylium ion $[n\text{-Bu}_3\text{Sn}]^+\cdot\text{TPFPB}^-$, generated in situ from $n\text{-Bu}_3\text{SnH}$ and $[\text{Ph}_3\text{C}]^+\cdot\text{TPFPB}^-$, serves as an effective catalyst for allylation of *ortho*-anisaldehyde with $n\text{-Bu}_3\text{Sn}-\text{CH}_2-\text{CH}=\text{CH}_2$, providing an excellent *ortho–para* regioselectivity.[23]

The reactivity of the stable free cations of heavier group 14 elements is still largely unknown. We can mention only the enhanced electrophilicity of the free germylium ion $(t\text{-Bu}_2\text{MeSi})_3\text{Ge}^+\cdot\text{TPFPB}^-$, which readily reacts with acetonitrile to give the corresponding nitrile complex $[(t\text{-Bu}_2\text{MeSi})_3\text{Ge} :\text{N}{\equiv}\text{C}-\text{CH}_3]^+\cdot\text{TPFPB}^-$, can be reduced with LiAlH_4 to form the hydride $(t\text{-Bu}_2\text{MeSi})_3\text{GeH}$, undergoes one-electron reduction with $t\text{-BuLi}$ to produce the free radical $(t\text{-Bu}_2\text{MeSi})_3\text{Ge}\bullet$, and quickly causes the ring-opening polymerization of THF.[14a] However, a comprehensive study of the reactivity of the stable free cations of heavier group 14 elements, and particularly the search for their wide synthetic applications in organometallic and organic chemistry, is a topic for future investigations.

2.1.4. Early Studies of RR′R″E⁺ Cations: Free or Complexed?

As discussed above, the silyl cations (and other cations of heavier group 14 elements) are more stable than the corresponding carbocations due to the more electropositive nature of Si, Ge, and Sn compared to that of C. Indeed, the recent investigation of the relative hydride affinities for silylium and carbenium ions and measurement of the equilibrium constants of hydride transfer reactions by FT ion cyclotron resonance spectroscopy demonstrated that the silylium ions in the gas phase are significantly more thermodynamically stable than the corresponding carbenium ions, and the positive charge of the silylium ions is mostly localized at the Si atom.[24] Numerous observations of silyl cations in the mass spectra of organosilicon compounds,[1a] reliably supported by theoretical calculations,[1j] provided solid evidence for the real existence of silicon-centered cations in the gas phase. In sharp contrast, synthesis of silylium ions in the condensed media has appeared to be a long-standing problem, whose story is full of controversial reports and hot debates concerning the real nature of the species synthesized. Thus, in 1970–80s, several groups claimed to have prepared silylium ions, either as reactive intermediates or as detectable species.[25–34] One such famous claim was the synthesis of triphenylsilyl perchlorate by Lambert et al.,[30] stating the existence of the stable Ph_3Si^+ cation as a silicon analog of the trityl cation Ph_3C^+. However, the following investigation by Olah and co-workers[35] has disproved this kind of statement, clearly demonstrating the covalent, rather than ionic, nature of the bonding between the Ph_3Si- and OClO_3-parts on the basis of NMR spectral and X-ray crystallographic data. Thus, until the early 1990s, the synthesis of real silylium ions, bearing a positive charge on Si atoms, had not been achieved.[36] Among the major problems associated with the synthetic difficulties in producing silylium ions, one should mention the very high oxo- and halophilicity of silylium ions compared with that of the corresponding carbenium analogs. This definitely prevented the use of the traditional leaving groups (tosylates, halides), widely applied to the generation of carbocations in organic chemistry, for the preparation of silylium ions due to the high Si–X and Si–O bond dissociation energies. Another problem originates from the difference in the size of the Si and C atoms: the bonds to Si are longer than those to C, which results in the overall decrease in the hyperconjugative stabilization on going from C to Si. It therefore became clear that the successful synthesis of true silylium ions required nonclassical approaches greatly different from the traditional organic chemistry methods. There are three factors of primary importance that might bring about success or failure in the synthesis of silylium ions: counterion, solvent, and substituents. The right choice of counterions and solvents was the major problem to overcome in the 1990s. The major requirement of counterions was their low nucleophilicity to prevent the formation of tight contact ion pairs with silylium ions or, in the extreme case, formation of covalently bonded compounds (triphenylsilyl perchlorate[30,35]). The solvent should also correspond to the same requirement of low nucleophilicity to avoid its possible coordination to the silylium ion electrophilic center. In either the case of counterion or solvent coordination to the silylium ion, there is a significant electronic perturbation around the cationic center and appreciable (or complete) transfer of the positive charge onto the nucleophilic counterpart. In other words, in such cases the silylium ion character is greatly diminished or completely lost. The great progress in resolving the silylium ion problem was achieved after the successful introduction of the extremely low nucleophilic anions $B(C_6F_5)_4^-$ and $CB_{11}H_6Br_6^-$ as the counterions of choice, and employment of nonpolar aromatic solvents (benzene, toluene) as the solvents of choice. Thus, the first major breakthrough came in 1993, when the groups of Lambert[4a] and Reed[37] published the crystal structures of their Et_3Si^+ and $i\text{-}Pr_3Si^+$ silylium ions. Lambert's $[Et_3Si(toluene)]^+\text{•}TPFPB^-$ (**1**$^+$**•TPFPB**$^-$) was prepared by the hydride transfer reaction between Et_3SiH and $Ph_3C^+\text{•}TPFPB^-$ in benzene (Scheme 2.7).[4a]

The crystal structure analysis of **1**$^+$**•TPFPB**$^-$ revealed no cation-anion interaction; however, there was a "distant" coordination of cationic center to a toluene molecule with a long Si$-$C bond distance of 2.18 Å. The toluene geometry was almost undistorted and essentially planar, which was interpreted by the authors as an indication of a very small degree of bonding interaction with the Si atom and, consequently, no or little charge transfer from the Si to the C atom. Finally, the authors concluded that their silylium ion **1**$^+$ represents a stable silyl cation lacking coordination to counterion and only very weakly coordinated to toluene solvent. However, two experimental points were in sharp contradiction to such a proposition: (1) the geometry around the silylium ion center was not planar (342° instead of the 360° required by the ideal trigonal-planar geometry expected for the true silylium ion species); and (2) the ^{29}Si NMR chemical shift of **1**$^+$**•TPFPB**$^-$ was observed at 92.3 ppm, far more upfield than the several hundreds ppm expected for the planar sp^2-structure of silylium ion. Such greatly problematic issues gave rise to the very lively discussion

$$Et_3SiH + Ph_3C^+\text{•}TPFPB^- \xrightarrow{C_6H_6} [Et_3Si(benzene)]^+\text{•}TPFPB^- + Ph_3CH$$

1$^+$**•TPFPB**$^-$

Scheme 2.7

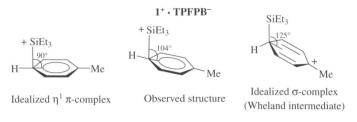

Scheme 2.8

about the real nature of **1⁺•TPFPB⁻**, during which Lambert's original claim of the nearly free silylium ion was severely criticized and objected by both theoreticians and experimentalists (Scheme 2.8).

Pauling calculated the bond order between the Si and *para*-C atom of a coordinated toluene molecule in **1⁺•TPFPB⁻** as 0.35, a value that cannot be neglected.[38] Olah and co-workers calculated that the ^{29}Si NMR chemical shift of the uncomplexed planar Et₃Si⁺ cation should appear at a very low field, 354.6 ppm[39a] or even at 371.3 ppm,[39b] whereas the observed value of 92.3 ppm[4a] corresponds well to the covalently bonded compound. Consequently, they concluded that the real structure of **1⁺•TPFPB⁻** is best regarded as a Wheland σ-complex (Scheme 2.8).[39] The formation of this σ-complex was also confirmed in the gas phase by radiolytic experiments and FT ion cyclotron resonance mass spectrometry.[40] Theoretical calculations by Cremer et al.[41] corroborated Olah's conclusion[39] concerning the degree of deshielding of the cationic Si atom: the ^{29}Si NMR chemical shifts of the free silylium ions R₃Si⁺ (R = Me, Et) in the gas phase were calculated at about 400 ppm, in noncoordinating solvents at 370–400 ppm, and in weakly coordinating solvents at 200–370 ppm, whereas in nucleophilic solvents, the silylium ion character is lost by the covalent bonding between the silylium ion and solvent molecules. Schleyer et al. also concluded that the pyramidalization at the silyl cation center and relatively high-field ^{29}Si NMR chemical shift of **1⁺•TPFPB⁻** were the evidences for its σ-complex structure.[42] Reed and co-workers suggested that the real structure of [Et₃Si(toluene)]⁺ is a hybrid of an η¹ π-complex and a σ-complex (Scheme 2.8) with a predominant contribution of a silylium ion–toluene π-complex;[43] Lambert has finally concurred with this idea.[4c,44]

Another milestone discovery in the field of silyl cations chemistry was achieved by Reed and co-workers in the same year, 1993, when Lambert published his Et₃Si⁺ study. Reed synthesized his *i*-Pr₃Si⁺• (CB₁₁H₆Br₆)⁻, (**2⁺•(CB₁₁H₆Br₆)**⁻, by the hydride transfer reaction of *i*-Pr₃SiH and Ph₃C⁺•(CB₁₁H₆Br₆)⁻ in toluene,[37] taking advantage of the very low nucleophilicity of the carborane anion[45] (Scheme 2.9).

$$i\text{-Pr}_3\text{SiH} + [\text{Ph}_3\text{C}]^+ \cdot [\text{CB}_{11}\text{H}_6\text{Br}_6]^- \xrightarrow{\text{toluene}} [i\text{-Pr}_3\text{Si}]^+ \cdot [\text{CB}_{11}\text{H}_6\text{Br}_6]^- + \text{Ph}_3\text{CH}$$

$$\mathbf{2^+ \cdot [CB_{11}H_6Br_6]^-}$$

Scheme 2.9

$2^{+} \cdot [CB_{11}H_6Br_6]^{-}$

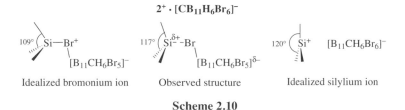

| Idealized bromonium ion | Observed structure | Idealized silylium ion |

Scheme 2.10

The i-Pr$_3$Si$^+$ cation exhibited no interaction with toluene solvent; however, the Si-cationic center was weakly bound to a counterion through its Br atoms with the long Si—Br bond of 2.479(9) Å. Reed's i-Pr$_3$Si$^+$ cation[37] was more planar (351° versus 342°) and more deshielded (109.8 ppm versus 92.3 ppm) than Lambert's Et$_3$Si$^+$ cation,[4a] points that allowed Reed to claim that $2^+ \cdot (CB_{11}H_6Br_6)^-$ represented a closer approach to the free silylium ion and possessed the highest degree of silylium ion character yet observed. However, this claim was rejected by Olah et al.,[39] who assigned the structure of $2^+ \cdot (CB_{11}H_6Br_6)^-$ to a polarized silylbromonium zwitterion rather than the true silylium ion. Reed finally concluded that his $2^+ \cdot (CB_{11}H_6Br_6)^-$ can be actually viewed as "…lying on a continuum between a bromonium ion and a silylium ion" with the major contribution of silylium rather than bromonium ion character (Scheme 2.10).[43]

Several other silylium ion derivatives R$_3$Si$^+ \cdot (CB_{11}H_6Br_6)^-$ (R$_3$Si$^+$ = Et$_3$Si$^+$, t-Bu$_2$MeSi$^+$, and t-Bu$_3$Si$^+$) have also been prepared by Reed and co-workers using the same synthetic protocol: hydride abstraction from R$_3$SiH by the Ph$_3$C$^+ \cdot (CB_{11}H_6Br_6)^-$.[46] All these compounds were structurally characterized showing features very similar to $2^+ \cdot (CB_{11}H_6Br_6)^-$: a long Si-Br bond distance (2.43–2.48 Å), a tendency to planarization around the Si-cationic center (345–351°), and downfield shifted ^{29}Si NMR resonances (105–115 ppm). The high degree of silylium ion character of these compounds was demonstrated by their smooth reactions with organic halides to form silyl halides R$_3$SiX and with water to give protonated silanols R$_3$Si(OH$_2$)$^+$.[46] In the following studies, Reed et al. have expanded the range of their least-coordinating anions by synthesizing hexachloro- and hexaiodocarboranes CB$_{11}$H$_6$X$_6^-$ (X = Cl, I) in addition to the original hexabromocarborane (X=Br).[5] The utilization of such weakly nucleophilic anions for the stabilization of silylium ions resulted in the preparation of novel derivatives of the i-Pr$_3$Si$^+$ cation, i-Pr$_3$Si$^+ \cdot (CB_{11}H_6X_6)^-$ (X = Cl, Br, I), of which the hexachloro derivative exhibited the highest degree of silylium ion character, whereas the hexaiodo derivative possesses the most halonium ion character and the most covalent Si–X bond.[5] Thus, i-Pr$_3$Si$^+ \cdot (CB_{11}H_6Cl_6)^-$ represented the closest approach to trialkylsilylium ions known at that time.[1i]

Investigations in the field of germylium and stannylium ions met with little success in the 1990s. Lambert et al. reported the synthesis of the protonated digermyl and distannyl ethers (R$_3$E)$_2$OH$^+ \cdot$TPFPB$^-$ (R = Me, Et; E = Ge, Sn), which possess significant amounts of germylium and stannylium ions character.[47]

$$R_3E-O^+(H)-ER_3 \leftrightarrow R_3E^+ \cdot HO-ER_3$$

In the field of stannylium ions, n-$Bu_3Sn^+ \cdot B(C_6F_5)_3H^-$, synthesized by the oxidation of stannyl hydride with $B(C_6F_5)_3$ (Section 2.1.2.2, Scheme 2.2, D), was characterized by Lambert et al. as a tricoordinate stannyl cation because of its low-field ^{119}Sn NMR resonance of 360 ppm.[7] The same n-Bu_3Sn^+ cation, having a TPFPB$^-$ counterion, exhibited a less deshielded value of +263 ppm (at room temperature),[15] which, however, was corrected in subsequent studies to +434 ppm (at $-60°C$).[23,48] n-$Bu_3Sn^+ \cdot TFPB^-$, prepared by the traditional hydride transfer method by Kira and co-workers (Section 2.1.2.2, Scheme 2.2, E), also exhibited a low-field resonance of the central Sn atom at 356 ppm.[8] However, in the following investigation Edlund et al.[49] pointed out that the above values of 360[7] and 356[8] ppm for the cationic Sn atom are best described as covalently bound arene complexes, quite similar to the above-discussed situation with the silylium ions–Wheland σ-complexes problem. On the basis of an empirical correlation between the ^{29}Si and ^{119}Sn NMR chemical shifts, the resonance of the truly trigonal-planar free Me_3Sn^+ was expected to be observed in the range 1500–2000 ppm, which was also supported by theoretical caluclations.[49]

2.1.5. Recent Developments

2.1.5.1. *Intramolecularly Coordinated Cations*   As cations, intra- or intermolecularly stabilized by coordination to either n- or π-donors, are not truly tricoordinated but rather tetra- or pentacoordinated compounds, we will just briefly overview the latest achievements in this field.

A series of nitrile complexes of t-Bu_3E^+ cations of the type [t-$Bu_3E^+(N{\equiv}C{-}R)]^+ \cdot TFPB^-$ (E = Si, Ge, Sn; R = Me, t-Bu) was synthesized according to Scheme 2.4, A (Section 2.1.2.3). These complexes uniformly displayed the tetracoordinated central atom E featuring a greatly distorted tetrahedral geometry.[10] An unusual reaction of bromosilirene with $[Et_3Si(C_6H_6)]^+ \cdot TPFPB^-$ resulted in the formation of the cationic derivative $3^+ \cdot TPFPB^-$, exhibiting the properties of delocalized halogen-bridged bissilylium ion (Scheme 2.11).[50] However, the ^{29}Si NMR resonance of $3^+ \cdot TPFPB^-$ (X = Br) was observed at 90.8 ppm (far upfield of the free silylium ions range), which gives evidence for the small extent of its silylium ion character and appreciable degree of the bromonium ion character.

In the field of germyl cations, several remarkable examples were recently reported. Thus, the transient germyl cation, prepared as shown in Scheme 2.5, B (Section

Scheme 2.11

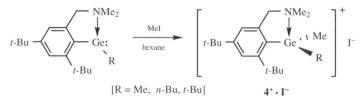

[R = Me, *n*-Bu, *t*-Bu]   **4⁺ · I⁻**

Scheme 2.12

2.1.2.4), exhibited a weak interaction to one of the aryl substituents, resulting in the partial transfer of the positive charge onto this Ar group, which acquired the properties of the cyclohexadienylium ion.[13] The germyl cations **4⁺•I⁻**, intramolecularly stabilized by nitrogen ligand coordination, were synthesized by the alkylation of heteroleptic germylenes with MeI (Scheme 2.12).[51] X-ray analysis of **4⁺•I⁻** (R = *t*-Bu) revealed no bonding interaction between the cationic and anionic portions of the molecule; the nitrogen coordination to the Ge cationic center does not greatly influence its geometry, which was more planar than tetrahedral (351.5°).

An interesting example of an intramolecularly π-coordinated germyl cation **5⁺•TPFPB⁻** was recently prepared by the dehalogenation of a bicyclic bromide with [Et₃Si(C₆H₆)]⁺•TPFPB⁻ in benzene (Scheme 2.13).[52] The cation **5⁺•TPFPB⁻** was free from any interactions to either solvent or counterion; however, there was a significant donor-acceptor interaction between the empty 4p orbital on the Ge cationic center and the C=C double bond. Such an interaction was clearly evidenced by the appreciable lengthening of the C=C bond (1.411(9) Å) and shortening of the Ge−C$_{double bond}$ interatomic distances (2.415(7) and 2.254(7) Å), as well as by the strong bending of the C−C unit toward the cationic Ge atom. The formulation of such three-center two-electron bonding was supported by NBO calculations: the C=C bond was definitely electron deficient (1.28 e), whereas the 4p-orbital on the cationic Ge atom was significantly occupied (0.42 e). Such effective through-space orbital interaction resulted in the formation of the bishomoaromatic system, whose negative NICS(1) value of −11.0 indicated the presence of an aromatic ring current in **5⁺**.[52]

The crystal structure of the *n*-Bu₃Sn⁺ cation **6⁺•CB₁₁Me₁₂⁻**, prepared according to Scheme 2.4, B (Section 2.1.2.3), was recently reported by Michl et al.[11] Although *n*-Bu₃Sn⁺ in **6⁺•CB₁₁Me₁₂⁻** lacked solvent coordination, it appeared to be weakly

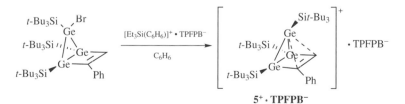

5⁺ · TPFPB⁻

Scheme 2.13

$$\text{Me}_3\text{E-EMe}_3 \text{ (or Me}_4\text{Pb)} + \text{CB}_{11}\text{Me}_{12}\cdot \xrightarrow{\text{pentane}} 2 \text{ Me}_3\text{E}^+ \cdot \text{CB}_{11}\text{Me}_{12}^-$$

$$[\text{E = Ge, Sn}] \qquad\qquad\qquad\qquad \mathbf{7^+} \cdot \mathbf{CB_{11}Me_{12}^-}$$

$$[\text{E = Ge, Sn, Pb}]$$

Scheme 2.14

coordinated to the Me groups of the $\text{CB}_{11}\text{Me}_{12}^-$ counterion with long Sn–C(Me) bond distances of 2.81 Å. Consequently, the geometry around the Sn atom was not perfectly planar (353.1°). The ^{119}Sn NMR resonance of $\mathbf{6^+}\cdot\mathbf{CB_{11}Me_{12}^-}$ was observed at 454.3 ppm,[11] far upfield from the 1700 ppm estimated for trialkylstannylium ions in the gas phase,[49] indicating that the same ion aggregation through Me coordination also occurred in solution. Employing the same synthetic approach, Michl and co-workers synthesized a series of novel stannylium ions $\mathbf{7^+}\cdot\mathbf{CB_{11}Me_{12}^-}$ (Scheme 2.14).[53]

Quite similar to the case of $\mathbf{6^+}\cdot\mathbf{CB_{11}Me_{12}^-}$, all cations $\mathbf{7^+}$ exhibited a significant interaction with the Me groups of the $\text{CB}_{11}\text{Me}_{12}^-$ counterion with long E–C(Me) bond distances of 2.5–3.0 Å (EXAFS). Such an interaction moves the NMR chemical shifts of the central nuclei in $\mathbf{7^+}\cdot\mathbf{CB_{11}Me_{12}^-}$ to a higher field (^{119}Sn NMR signal for E = Sn: 335.9 ppm in CD_2Cl_2 at −60 °C; ^{207}Pb NMR signal for E = Pb: 1007.4 ppm in CD_2Cl_2 at room temperature).[53] The cation–anion interaction was estimated to be predominantly ionic with some contribution of the covalent bonding (from 1/4 to 1/3), and the strength of this interaction was found to increase in the order $\text{Me}_3\text{Pb}^+ < \text{Me}_3\text{Sn}^+ \ll \text{Me}_3\text{Ge}^+$.

An important contribution to silylium ion chemistry has been made by the group of Müller, who very recently published a series of papers describing the synthesis of intramolecularly stabilized silylium ions as well as silyl-substituted vinyl cations and arenium ions by the classical hydride transfer reactions with $\text{Ph}_3\text{C}^+\cdot\text{TPFPB}^-$ in benzene.[54] Thus, the transient 7-silanorbornadien-7-ylium ion $\mathbf{8^+}$ was stabilized and isolated in the form of its nitrile complex $[\mathbf{8}(\text{N}\equiv\text{C}-\text{CD}_3)]^+\cdot\text{TPFPB}^-$ (Scheme 2.15), whereas the free $\mathbf{8^+}$ was unstable and possibly rearranged at room temperature into the highly reactive [PhSi$^+$/tetraphenylnaphthalene] complex.[54a,i]

In contrast, 2-silanorbornyl cation $\mathbf{9^+}\cdot\text{TPFPB}^-$ (Scheme 2.15) was quite stable due to its effective internal complexation to the C=C double bond and showed no coordination to either counterion or solvent molecules.[54b,h,i] The intramolecular π-complexation was evident by the ^{29}Si NMR chemical shift of $\mathbf{9^+}\cdot\text{TPFPB}^-$ observed at 87.7 ppm, which is in the range typical for the silyl cation–π arene complexes. Thus, $\mathbf{9^+}\cdot\text{TPFPB}^-$ possesses some degree of silylium ion character, although alternatively it can be viewed as a bridged β-silyl carbocation. The scope of such a synthetic approach was later expanded involving other heavier group 14 elements (Ge, Sn, Pb) to produce the novel stable norbornyl cations $\mathbf{10^+}\cdot\text{TPFPB}^-$ free from aromatic solvent interaction (Scheme 2.15).[54c,i] The element-centered cationic nature of all $\mathbf{10^+}\cdot\text{TPFPB}^-$ was manifested in their low-field shifted signals: ^{29}Si NMR resonance at 80.2–87.2 ppm for E = Si, ^{119}Sn NMR resonance at 334.0 ppm for E = Sn, and ^{207}Pb NMR resonance at 1049 ppm for E = Pb;

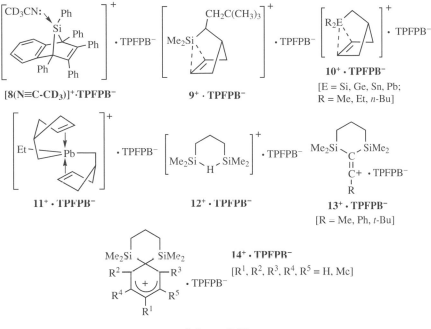

Scheme 2.15

however, these values were markedly smaller than those for the free tricoordinate ions. Interestingly, cations of the type described above can be stabilized by the π-interaction with the two C=C double bonds from a pair of cyclopentenyl ligands (Scheme 2.15).[54d] Such a plumbylium ion **11⁺·TPFPB⁻** was stable at room temperature for several weeks, exhibiting the low-field shifted ²⁰⁷Pb NMR resonance at 807 ppm. The crystal structure analysis of **11⁺·TPFPB⁻** revealed well-separated cation and anion parts, and a trigonal-bipyramidal coordination geometry around the Pb atom.[55d] The silyl cation **12⁺·TPFPB⁻** with a three-center two-electron Si–H–Si bond, involving hydrogen as a bridging atom, displayed a low-field ²⁹Si NMR resonance at 76.7 ppm as a doublet due to the ²⁹Si–¹H coupling with a coupling constant $^1J = 39$ Hz (Scheme 2.15).[54e,f] The stable β-disilacyclohexenylidene-substituted vinyl cations **13⁺·TPFPB⁻**, formed upon the intramolecular addition of the transient silylium ion across the C≡C triple bond, are free from the solvent interaction (Scheme 2.15).[16a,54f] The high stability of **13⁺·TPFPB⁻** is definitely due to hyperconjugation with the two β-silyl substituents, as well as to electron donation from the α-phenyl group. The crystal structure of **13⁺** (R = t-Bu) with the hexabromocarborane counterion, **13⁺·[CB₁₁H₆Br₆]⁻**, revealed a free vinyl cation with a markedly short C=C double bond of 1.221 Å, which approaches the length of a normal C≡C triple bond.[16b,54f] On the contrary, the =C–Si bonds are rather long (1.984 and 1.946 Å), which provides solid evidence for the β-silyl hyperconjugation giving to **13⁺** cation some silylium ion character. A series of bissilylated arenium ions **14⁺•TPFPB⁻** was prepared by the intramolecular complexation of transient silylium ions with aromatic rings (Scheme 2.15).[54f,g] The unusual thermodynamic stability of **14⁺•TPFPB⁻** was ascribed to the hyperconjugative effects of the two silyl substituents.

2.1.5.2. Free (Noncoordinated) Cations

2.1.5.2.1. Cyclic Conjugated Cations
The existence of the heavy analogs of the famous cyclopropenylium ion, comprising Si, Ge, Sn, and Pb atoms, was theoretically predicted 10 years ago by Schleyer et al.[55] However, their calculations on the model $E_3H_3^+$ (E = C, Si, Ge, Sn, Pb) species revealed a preference for the nonclassical hydrogen-bridged structure over the classical cyclopropenylium-type compound for the heaviest representatives $E_3H_3^+$ (E = Ge, Sn, Pb). In marked contrast to such theoretical expectations, the first heavy analog of $C_3R_3^+$, cyclotrigermenylium ion **15⁺•TPB⁻**, synthesized by the reaction of cyclotrigermene **16** with Ph_3C^+•TPB⁻ in benzene (Scheme 2.16), was isolated and structurally characterized as a full analog of the cyclopropenylium ion featuring the same equilateral triangle geometry.[56a,b] The skeletal Ge–Ge bond distances in **15⁺•TPB⁻** were intermediate between the Ge=Ge double and Ge–Ge single bonds of the precursor **16**. All these structural peculiarities imply that **15⁺•TPB⁻** actually possesses a 2π-electron aromaticity, thus representing a heavy homolog of the simplest Hückel aromatic system. The extra stabilization of the **15⁺** cation, gained through such aromaticity, allowed for its existence without a stabilizing interaction with external or internal nucleophiles. And indeed, **15⁺•TPB⁻** lacked any bonding interaction to either counterion or solvent molecules, being truly a free germyl cation.

The discrepancies between the calculated[55] hydrogen-bridged nonplanar C_{3v} structure for the model $Ge_3H_3^+$ and experimental[56a,b] cyclopropenylium-type planar D_{3h} structure for the real $Ge_3(Sit-Bu_3)_3^+$ should definitely be ascribed to the substituent effect: small H atoms possessing very high migratory and bridging aptitude versus very bulky electropositive t-Bu₃Si substituents rather reluctant to migrate. In the following studies, the same authors improved the method for the preparation of **15⁺** utilizing other low nucleophilic anions, such as TPFPB⁻, TFPB⁻, and TSFPB⁻.[56c–e] The novel cyclotrigermenylium derivatives of the type $[(t-Bu_3E)_3Ge_3]^+$•Ar_4B^- (E = Si, Ge; Ar_4B^- = TPFPB⁻, TFPB⁻, TSFPB⁻) were synthesized, which appeared to be significantly more thermally stable than the original **15⁺•TPB⁻**. The structural features of all these cations $[(t-Bu_3E)_3Ge_3]^+$•Ar_4B^- were nearly identical to each

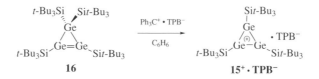

Scheme 2.16

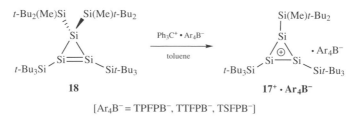

$$[Ar_4B^- = TPFPB^-, TTFPB^-, TSFPB^-]$$

Scheme 2.17

other and to those of the original **15⁺•TPB⁻**, which implied the absence of any observable influence of counterions on the cyclotrigermenylium framework.[56e] In other words, in the crystalline state the cyclotrigermenylium ion is entirely free from any nucleophilic interactions. The same is true for the solution structure of cyclotrigermenylium ion: its ^{29}Si NMR resonance is practically solvent and counterion independent indicating its free status in the condensed phase.[56e]

The silicon version of the cyclotrigermenylium ion, cyclotrisilenylium ion **17⁺•Ar₄B⁻** (Ar₄B⁻ = TPFPB⁻, TTFPB⁻, TSFPB⁻), was prepared very recently by Sekiguchi and co-workers by the reaction between the cyclotrisilene **18** and Ph₃C⁺•Ar₄B⁻ in toluene (Scheme 2.17).[57] **17⁺•Ar₄B⁻** was also characterized as a 2π-electron aromatic compound and a free cyclotrisilenylium ion. The preference for the planar cyclopropenylium-type D_{3h} structure for the parent Si₃H₃⁺ was manifested in the preceding calculations, showing that such a structure is the global minimum on the Si₃H₃⁺ potential energy surface.[58]

In the crystal, **17⁺** exhibited no interaction with either counterion or solvent molecules; its Si₃ skeleton represents a nearly regular triangle with Si–Si bond distances intermediate between the Si=Si and Si–Si bonds of the precursor **18**. All NMR chemical shifts of **17⁺** were practically solvent and counterion independent, which proved its "freedom" in the solution. The ^{29}Si NMR resonances of the skeletal Si atoms in **17⁺** are expectedly strongly deshielded due to the distributed positive charge on them: 284.6 and 288.1 ppm.[57]

Oxidation of the less bulky substituted cyclotrisilene **19** with [Et₃Si(C₆H₆)]⁺•TPFPB⁻ in benzene surprisingly resulted in the formation of the cyclotetrasilenylium ion **20⁺•TPFPB⁻**, which was also free from any counterion or solvent interaction in both the solid state and solution (Scheme 2.18).[59] In the cationic portion of the molecule **20⁺**, the Si1, Si2, and Si3 atoms exhibited a planar geometry whereas

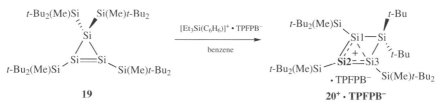

Scheme 2.18

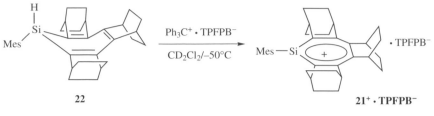

Scheme 2.19

Si4 revealed a distorted tetrahedral configuration, and the Si1–Si2 and Si2–Si3 bond distances of 2.240(2) and 2.244(2) Å were intermediate between the Si=Si and Si–Si bond lengths of the precursor **19**. This was explained by the delocalization of the positive charge over the Si1, Si2, and Si3 atoms, accompanied by the Si1–Si3 through-space orbital interaction, resulting in the overall homoaromaticity of **20⁺**. The hypothesis of homoaromaticity was further supported by the observation of an extremely low-field shifted signal of Si2, the central atom of the Si_3 homoaromatic system, at 315.7 ppm.

An interesting example of the heavy analog of a 6π-electron system, silatropylium ion **21⁺•TPFPB⁻**, was synthesized by Komatsu et al. by the hydride transfer reaction between the precursor silepin **22** and Ph₃C⁺•TPFPB⁻ in CD₂Cl₂ at −50°C (Scheme 2.19).[60] **21⁺•TPFPB⁻** was stable below −50°C; at higher temperatures it reacted with the solvent CD₂Cl₂ through the chlorine abstraction reaction. The ²⁹Si NMR resonance of **21⁺•TPFPB⁻** was observed at a rather low field, 142.9 ppm, on which basis the authors concluded that **21⁺** represents a nearly free silylium ion. The considerable deshielding of all ring C atoms in **21⁺** (compared with those of the neutral precursor **22**) indicated a significant delocalization of the positive charge over the entire seven-membered ring. Such effective delocalization led to the overall aromaticity of **21⁺**, which was supported by the NICS calculations for the parent compound C_6H_7Si: NICS(1) = −7.3.

The same group of Komatsu very recently reported another example of a cationic silaaromatic compound, 2-silaimidazolium ion **23⁺•TPFPB⁻**, generated by the treatment of silaheterocycle **24** with [Et₃Si(C₆H₆)]⁺•TPFPB⁻ in CD₂Cl₂ or C₇D₈ (Scheme 2.20).[61] **23⁺•TPFPB⁻** was stable below −10°C; at higher temperatures it unavoidably decomposed. The aromaticity of **23⁺** was deduced on the basis of its NMR spectral data and further supported by calculations: the NICS(1) and ASE for the model compound were estimated as −6.5 and 5.6 kcal/mol, respectively.

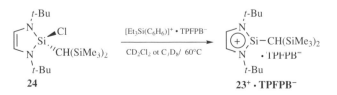

Scheme 2.20

$$(Me_5C_5)_2Si: + (Me_5C_5H_2)^+ \cdot TPFPB^- \xrightarrow{CH_2Cl_2} (\eta^5\text{-}Me_5C_5)Si{:}^+ \cdot TPFPB^- + 2Me_5C_5H$$

$$\mathbf{26} \qquad\qquad\qquad\qquad\qquad\qquad \mathbf{25^+ \cdot TPFPB^-}$$

Scheme 2.21

Interestingly, when the same compound **24** was oxidized with $[Ag(C_6H_6)_3]^{+\bullet}$ TPFPB$^-$ in CH_2Cl_2 at room temperature, only the SET process was observed to give the corresponding cation radical **24$^{+\bullet}$•TPFPB$^-$** as a stable compound.[62] The ESR spectrum of **24$^{+\bullet}$•TPFPB$^-$** revealed a multiplet signal ($g = 2.0029$) due to the coupling with two olefinic protons, two nitrogen atoms, and one chlorine atom. Both ESR and X-ray crystallography displayed the free state of the cation radical **24$^{+\bullet}$** in solution and in the crystal. Calculations on the model compound disclosed notable $\sigma_{Si-X}^*-\pi$ (X = Cl, C) orbitals mixing, resulting in the total stabilization of the cation radical **24$^{+\bullet}$**.

The sharply increasing interest in the chemistry of monovalent silyliumylidene ion derivatives of the type RSi$^+$ stimulated a great number of experimental studies in this field, which recently culminated in the synthesis of the very remarkable $(\eta^5\text{-}Me_5C_5)Si^+$ cation **25$^+$•TPFPB$^-$** (Scheme 2.21).[63] A stable **25$^+$•TPFPB$^-$** was obtained by the oxidation of decamethylsilicocene **26** with $(Me_5C_5H_2)^{+\bullet}$TPFPB$^-$ in CH_2Cl_2 (Scheme 2.21).

In the solid state, the $(\eta^5\text{-}Me_5C_5)Si^+$ cation in **25$^+$•TPFPB$^-$** displayed only a weak interaction with the anionic part, featuring an almost ideal pentagonal-pyramidal geometry (Figure 2.1).

In the ^{1}H NMR spectrum of **25$^+$•TPFPB$^-$**, only one singlet for the five Me groups of the Me_5C_5 unit was observed, even at low temperature, at 2.23 ppm, which indicated the retention of the pentagonal-pyramidal cationic structure in solution. The ^{29}Si NMR resonance of **25$^+$•TPFPB$^-$** was measured at the very high field region of -400.2 ppm, which is diagnostic for π–complexes of a divalent Si atom. Thus, both X-ray diffraction and NMR spectroscopy disclosed the structure of **25$^+$** as a cationic π-complex having an $\eta^5\text{-}Me_5C_5$ ligand bound to a "naked" Si center. Alternatively, the $(\eta^5\text{-}Me_5C_5)Si^+$ cation can be viewed as a pentahaptocoordinated analog of the desired monohaptocoordinated silyliumylidene ion $(\eta^1\text{-}Me_5C_5)Si^+$. The reactivity of **25$^+$• TPFPB$^-$** was particularly interesting. Thus, it reacts with $(Me_3Si)_2NLi$ to produce, in the first step, the transient silylene $[(Me_3Si)_2N](Me_5C_5)Si:$. It then dimerizes to finally yield *trans*-1,2-diaminodisilene $(\eta^1\text{-}Me_5C_5)[(Me_3Si)_2N]Si=Si[N(SiMe_3)](\eta^1\text{-}Me_5C_5)$, which has a Me_5C_5 group σ-bonded to the Si atom.[63]

Figure 2.1

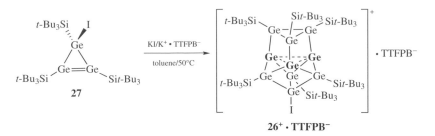

Scheme 2.22

The trishomoaromatic free germyl cation **26⁺•TTFPB⁻** was recently described by Sekiguchi and co-workers.[64] This rather unusual pergermacationic cluster was obtained by heating iodocyclotrigermene **27** at 50°C in toluene in the presence of KI and K⁺•TTFPB⁻ (Scheme 2.22). The crystal structure analysis of **26⁺•TTFPB⁻** revealed the absence of any observable interaction of the cationic part to either counterion or solvent, implying the free status of **26⁺** in the crystalline form. The three central Ge atoms of the cluster structure are formally tricoordinated, carrying no substituents, and the interatomic distances between them are very long (3.2542–3.2642 Å). Theoretical calculations on the model cluster compound convincingly demonstrated the even delocalization of the positive charge over the three Ge atoms to form a three-center two-electron bond, thus giving rise to the trishomocyclotrigermenylium system. The trishomoaromaticity of **26⁺** was further supported by ASE and NICS calculations, which were estimated as −19.2 kcal/mol and −26.4.[64]

2.1.5.2.2. Acyclic Tricoordinate Cations

The most challenging stable tricoordinate cations of the type R_3E^+ (R = alkyl, aryl, silyl; E = Si–Pb), featuring an ideal trigonal-planar geometry and independence of both counterions and solvents, became a focus of particular attention and the subject of a long-standing search by organometallic chemists. The first major breakthrough in the synthesis of such highly desirable compounds was achieved by the group of Lambert, who reported the preparation of the free tricoordinate trimesitylsilylium ion Mes₃Si⁺•TPFPB⁻ (**28⁺•TPFPB⁻**) in 1997.[65] As the classical hydride transfer reaction of Mes₃SiH with Ph₃C⁺•TPFPB⁻ did not proceed due to the large steric bulkiness of the Mes groups, a novel synthetic approach involving an allyl substituent as a leaving group ("allyl leaving group approach")[9] was developed (see also Section 2.1.2.3, Scheme 2.3). Thus, the reaction of allyltrimesitylsilane with β-silylcarbocation (Et₃SiCH₂CPh₂⁺)•TPFPB⁻ cleanly produced **28⁺•TPFPB⁻**, which was stable in solution for several weeks (Scheme 2.23). **28⁺•TPFPB⁻** exhibited a

$$Mes_3Si\text{-}CH_2\text{-}CH=CH_2 + R^+ \cdot TPFPB^- \xrightarrow{C_6H_6} Mes_3Si^+ \cdot TPFPB^- + H_2C=CH\text{-}CH_2\text{-}R$$

$$[R = Et_3SiCH_2CPh_2] \qquad \mathbf{28^+ \cdot TPFPB^-}$$

Scheme 2.23

single ^{29}Si NMR resonance at 225.5 ppm, which was the same in different aromatic solvents[9,65] and with different counterions,[9,44] but greatly shifted to the high-field region in the presence of nucleophilic reagents (CD$_3$CN, Et$_3$N).

The following computations supported the existence of Mes$_3$Si$^+$ as a free cation lacking observable coordination to solvent and having a calculated value of the ^{29}Si NMR chemical shift very close to the experimental value: 230.1 (GIAO/HF) and 243.9 (GIAO/DFT) ppm versus 225.5 ppm.[66]

However, TPFPB$^-$ derivatives have an evident drawback: they often form oils or clathrates precluding crystallization, therefore **28$^+$•TPFPB$^-$** has failed to crystallize. To have crystals suitable for X-ray analysis, the TPFPB$^-$ counterion was substituted by the carborane CB$_{11}$HMe$_5$Br$_6$$^-$ anion, and accordingly **28$^+$•(CB$_{11}$HMe$_5$Br$_6$)$^-$** was synthesized by the reaction of trimesitylallylsilane Mes$_3$Si$-$CH$_2$$-CH=CH_2$ with Et$_3$Si$^+$•(CB$_{11}$HMe$_5$Br$_6$)$^-$ in benzene.[67,68] The crystal structure analysis of **28$^+$•(CB$_{11}$HMe$_5$Br$_6$)$^-$** revealed that this is indeed free from counterion and solvent interaction, sp^2-silylium ion, featuring a trigonal-planar geometry.[67] The solid state ^{29}Si NMR resonance of **28$^+$•(CB$_{11}$HMe$_5$Br$_6$)$^-$** was practically the same as that of **28$^+$•TPFPB$^-$** in solution: 226.7 ppm versus 225.5 ppm.

The range of "allyl leaving group approach" utilization was later expanded to other heavier group 14 elements. Thus, the trimesitylstannylium ion derivative Mes$_3$Sn$^+$•TPFPB$^-$ **29$^+$•TPFPB$^-$** was generated by the treatment of Mes$_3$Sn$-$CH$_2$$-CH=CH_2$ with either [Et$_3$Si(C$_6$H$_6$)]$^+$•TPFPB$^-$ or [Et$_3$SiCH$_2$CPh$_2$$^+$]•TPFPB$^-$ in benzene (Scheme 2.3).[9] The Mes$_3$Sn$^+$ cation in **29$^+$•TPFPB$^-$** exhibited a very low field ^{119}Sn signal at 806 ppm, which was independent of the solvent, thus demonstrating its free status. The trimesitylgermylium ion Mes$_3$Ge$^+$ was also prepared by the same synthetic route, and the degree of its cationic character was estimated to be comparable with those of the analogous silylium and stannylium ions.[9]

The tridurylsilylium ion Dur$_3$Si$^+$, prepared according to the allyl leaving group experimental procedure described above, exhibited a single ^{29}Si NMR resonance at 226.8 ppm, which was very close to that of the Mes$_3$Si$^+$ cation of 225.5 ppm.[69] This fact was interpreted as the manifestation of the free tricoordinate nature of the Dur$_3$Si$^+$ cation. However, the similar tridurylstannylium ion Dur$_3$Sn$^+$ revealed a ^{119}Sn NMR chemical shift at 720 ppm, which was shifted to lower frequency

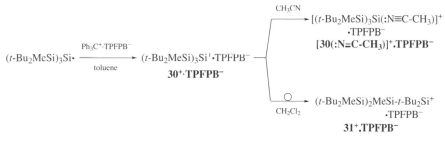

Scheme 2.24

compared with the Mes_3Sn^+ cation (806 ppm), thus possessing ca. 70% of the true stannylium ion character.[69]

The one-electron oxidation of the stable silyl radical $(t\text{-}Bu_2MeSi)_3Si\bullet$ with $Ph_3C^+\bullet TPFPB^-$ in toluene resulted in the formation of the transient silylium ion $(t\text{-}Bu_2MeSi)_3Si^+$ as a TPFPB$^-$ derivative **30$^+$•TPFPB$^-$**, stabilized in the form of its nitrile complex **[30(:N≡C-CH$_3$)]$^+$•TPFPB$^-$** (Scheme 2.24).[70] Without stabilizing complexation, **30$^+$•TPFPB$^-$** underwent an isomerization involving a fast 1,2-methyl shift from the peripheral Si atom to the central cationic Si atom to yield a new silyl cation **31$^+$•TPFPB$^-$** (Scheme 2.24). The driving force for such a methyl migration seems to be an additional stabilization, which **31$^+$** can gain through the hyperconjugative interaction of the silylium ion center with the neighboring Si–Si σ-bonds. The cationic Si atom in **31$^+$•TPFPB$^-$** resonated at a very low field: 303 ppm at $-50°C$ in CD_2Cl_2, implying its existence in solution as a free silylium ion.

In contrast to the above case, the one-electron oxidation of the stable germyl and stannyl radicals, having the same $t\text{-}Bu_2MeSi$ substituents, opened the door to the easiest and most straightforward access to the free tricoordinate germyl and stannyl cations. These species, germylium ion $(t\text{-}Bu_2MeSi)_3Ge^+$ and stannylium ion $(t\text{-}Bu_2MeSi)_3Sn^+$, isolated as stable TPFPB$^-$ derivatives, **32$^+$•TPFPB$^-$** and **33$^+$•TPFPB$^-$**, were prepared by Sekiguchi et al. by the oxidation of $(t\text{-}Bu_2MeSi)_3Ge\bullet$ and $(t\text{-}Bu_2MeSi)_3Sn\bullet$ radicals in benzene (see Section 2.1.2.5, Scheme 2.6).[14] The crystal structure analysis of **32$^+$•TPFPB$^-$** undoubtedly confirmed that the cationic portion of the molecule **32$^+$** displayed no interaction with either TPFPB$^-$ counterion or benzene solvent molecules, that is, **32$^+$** is indeed a free cation in the solid state.[14a] Despite the enormous electrophilicity, **32$^+$** maintained its "freedom" in solution as well, which was evidenced by its solvent-independent ^{29}Si NMR resonance: 49.9 ppm in CD_2Cl_2, 49.9 ppm in $CDCl_3$, and 50.3 ppm in C_6D_6. Interestingly, the Si–Ge bonds in cation **32$^+$** were substantially longer than those in the starting radical $(t\text{-}Bu_2MeSi)_3Ge\bullet$ (av. 2.5195(10) Å versus av. 2.4535(4) Å).[14a] This was realized in terms of the different extent of $4p_z(Ge)$–σ*(Si–C(t-Bu)) hyperconjugation: it is more important for the $(t\text{-}Bu_2MeSi)_3Ge\bullet$ radical, whose $4p_z$-orbital is singly occupied, and less important for cation **32$^+$**, whose $4p_z$-orbital is vacant. The enhanced electrophilicity of **32$^+$**, discussed in Section 2.1.3, which greatly exceeds that of the cyclotrigermenylium ion $(t\text{-}Bu_3Si)_3Ge^+$ **15$^+$**, can be understood, because acyclic **32$^+$** lacks the highly stabilizing π-conjugation effects diagnostic for three-membered cyclic **15$^+$**.

The X-ray diffraction analysis of **33$^+$•TPFPB$^-$** also disclosed the free status of the stannylium ion $(t\text{-}Bu_2MeSi)_3Sn^+$, featuring a perfectly trigonal-planar geometry around the cationic sp^2-hybridized Sn atom.[14b] The $5p_z(Sn)$–σ*(Si–C(t-Bu)) hyperconjugative effects were also operative in this case, being responsible for the shortening of the Si–Sn bonds in cation **33$^+$** compared with those in the $(t\text{-}Bu_2MeSi)_3Sn\bullet$ radical. The free status of **33$^+$** in solution was impressively manifested by its extremely low-field ^{119}Sn NMR resonance of 2653 ppm, representing the most deshielded Sn nucleus of all low-coordinated stannyl cations. This value greatly exceeded the value expected on the basis of the ^{29}Si–^{119}Sn empirical correlation (1500–2000 ppm)[49] and the other value (ca. 1000 ppm) calculated for a free triorganostannylium ion,[71] and reasonably agreed with the 2841 ppm calculated for the model $(H_3Si)_3Sn^+$ cation at the GIAO-B3LYP/6-311G(d) level.[14b]

The most recent accomplishment in the field of stable free cations of heavier group 14 elements was recently reported by Lambert and co-workers, who utilized their allyl leaving group approach to synthesize Tip₃Sn⁺•TPFPB⁻ (**34⁺•TPFPB⁻**), by the reaction of $Tip_3Sn-CH_2-CH=CH_2$ with Ph₃C⁺•TPFPB⁻.[72] The crystal structure analysis of **34⁺•TPFPB⁻** revealed no bonding interaction of cation **34⁺** with counterion, solvent, and methinyl hydrogen atoms of the *i*-Pr groups, implying that **34⁺** is a truly free tricoordinate planar stannylium ion. The ^{119}Sn NMR chemical shift of **34⁺•TPFPB⁻** was measured at 714 ppm, which was taken as evidence for the true stannylium ion nature of **34⁺**. This was further supported by the GIAO calculations at the MPW1PW91 level providing a value of 763 ppm for the stannylium ion.

2.2. Si-, Ge-, AND Sn-CENTERED FREE RADICALS

2.2.1. Introduction

Free radicals are usually defined as atoms or as a group of atoms with an unpaired electron, that is, as open shell species. One of the most fundamental issues in free radical chemistry is the question of their configurations, which is to a large extent determined by the nature of the SOMO: whether it is of the p-type or sp³-hybrid type. Accordingly, the corresponding radicals are to be considered as either π-radicals with the trigonal-planar geometry or σ-radicals featuring the pyramidal configuration. Typically, the free radicals can adopt one of the three most widely widespread configurations: rigid pyramidal (**A**), flexible pyramidal (**B**), and flexible planar (**C**) (Figure 2.2).

The A-type radicals should be assigned to the class of the σ-radicals, C-type radicals belong to the class of the π-radicals, and B-type radicals are intermediate between them. The geometry of the free radicals is primarily governed by the effects of the substituents, both electronic and steric. The electronic effect of substituents can be clearly rationalized in terms of the frontier orbital interactions. Thus, it is well known that the electronegative σ-accepting and π-donating groups (halogens, amino groups, etc.) greatly increase the degree of the pyramidalization at the radical center. This can be explained by the attractive SOMO (unpaired electron)–LUMO (antibonding σ*-orbital of the central element–electronegative substituent bond) interaction, which becomes progressively more favorable with the pyramidalization at the radical center. The electronegative substituents raise the SOMO energy level through their π-donation and lower the LUMO level by their σ-acceptance, thus decreasing the SOMO–LUMO energy gap and making

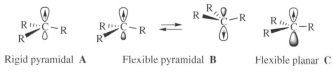

Rigid pyramidal **A**   Flexible pyramidal **B**   Flexible planar **C**

Figure 2.2. Geometry of organic free radicals.

their interaction more favorable. Accordingly, the electropositive σ-donating and π-accepting groups (silyl, germyl, etc.) have the opposite effect on the configuration of the radical center, greatly favoring the flattening of the radical structures.

On the contrary, the steric effect of the bulky groups may also play an important role in determining the radical configuration. Thus, the huge groups at the radical center may prefer to be removed from each other as far as possible to decrease their steric repulsion. Definitely, such a remote arrangement of the bulky substituents can be achieved in the planar geometry of the radical center, that is, the bulkiness of the substituents is in direct relation to the degree of planarization of the molecule.

A great deal of information on the electronic structure and geometry of radicals in solution can be extracted from their ESR spectra, as it is well established that the values of hyperfine coupling constants (hfcc), arising from the spin density of the ns-orbitals, markedly increase with increasing of the SOMO s-character. The pyramidalization of the radicals is manifested in higher values of their hfccs (σ-radicals), whereas smaller values of the hfccs are indicative of the more planar radicals (π-radicals).

Another powerful tool for the structural elucidation of free radicals is X-ray diffraction analysis, which provides unique and reliable information about the structures of the radical species in the crystalline form. Needless to say, the use of such an informative method is limited to the case of stable (isolable) radicals only.

The chemistry of the heavy analogs of organic free radicals, that is, radicals centered on the Si, Ge, Sn, and Pb atoms, has been thoroughly reviewed several times, particularly that of the silyl radicals.[73] Therefore, in the present review we will briefly bring together the most important discoveries in this field with particular attention paid to the most recent developments and progress, especially in the synthesis of the stable representatives of heavier group 14 elements centered radicals.[74]

2.2.2. Transient Species

2.2.2.1. Generation   In the vast majority of cases, the simple tricoordinate radicals of the heavier group 14 elements of the type $RR'R''E\bullet$ [R, R', R'' = H, alkyl (Me, Et, n-Pr, n-Bu, t-Bu, $Ph_nMe_{3-n}C-CH_2$, cyclo-C_3H_5, etc.), aryl (Ph, $Me_nC_6H_{5-n}$, 2,4,6-Me_3-C_6H_2, 2,4,6-Et_3-C_6H_2, 2,4,6-i-Pr_3-C_6H_2, etc.); E = Si, Ge, Sn, Pb] can be generated in one of the following ways: (1) by the photolysis or thermolysis of the corresponding hydrides R_3EH in the presence of the radical initiators $(t$-$BuO)_2$[75–87] or AIBN[85] (Scheme 2.25, A); (2) by the γ-irradiation of R_4E[88–93, 95] or photolysis of $R_3E-Hg-ER_3$[80,94] (Scheme 2.25, B and C); or (3) by the thermolysis or photolysis of either R_3E-ER_3[80,82,84,96–101] or $R_3E-ER_2-ER_3$[102] (Scheme 2.25, D and E).

While the first two methods have been widely utilized for the generation of a whole range of the heavier group 14 elements centered free radicals, the last method has been mainly employed for the generation of germyl and stannyl radicals, being rarely used for the generation of silyl radicals because of the significant Si–Si bond strength. Other remarkable methods employed for the generation of radicals are: the reduction of the chlorides R_3ECl with metallic sodium[103] (Scheme 2.25, F); photolysis of anionic derivatives R_3ELi[104] and $(R_3E)_2Ca$[105] (Scheme 2.25, G and H);

$$R_3EH \xrightarrow[\text{or } \Delta/\text{AIBN}]{h\nu/(t\text{-BuO})_2} R_3E \cdot \qquad \text{(A)}$$

$$R_4E \xrightarrow{\gamma\text{-irradiation}} R_3E \cdot \qquad \text{(B)}$$

$$R_3E\text{-Hg-}ER_3 \xrightarrow{h\nu} R_3E \cdot \qquad \text{(C)}$$

$$R_3E\text{-}ER_3 \xrightarrow[h\nu/(t\text{-BuO})_2]{\Delta \quad \text{or}} R_3E \cdot \qquad \text{(D)}$$

$$R_3E\text{-}ER_2\text{-}ER_3 \xrightarrow{h\nu} R_3E \cdot + \cdot ER_2\text{-}ER_3 \qquad \text{(E)}$$

$$R_3ECl \xrightarrow[-\text{NaCl}]{\text{Na}} R_3E \cdot \qquad \text{(F)}$$

$$R_3ELi \xrightarrow{h\nu} R_3E \cdot \qquad \text{(G)}$$

$$(R_3E)_2Ca \xrightarrow{h\nu} R_3E \cdot + \cdot Ca(ER_3) \qquad \text{(H)}$$

$$R_2E: \ [\text{or } (R_2N)_2E:] \xrightarrow{h\nu} R_3E \cdot \ [\text{or } (R_2N)_3E \cdot] \qquad \text{(I)}$$

$$Ph_3GeX \xrightarrow{+ e^- \text{ (cathode)}} Ph_3Ge \cdot + X^- \qquad \text{(J)}$$

$$HGeCl_3 \xrightarrow{- e^- \text{ (anode)}} Cl_3Ge \cdot + H^+ \qquad \text{(K)}$$

Scheme 2.25

photolysis of heavy carbene analogs $R_2E:$[105] or $(R_2N)_2E:$[106,107] (Scheme 2.25, I); and electrochemical generation of $Ph_3Ge\bullet$ radicals by the one-electron reduction of halides Ph_3GeX (X = Cl, Br, I)[108] (Scheme 2.25, J) and $Cl_3Ge\bullet$ radicals by the one-electron oxidation of $HGeCl_3$ (Scheme 2.25, K).[109]

2.2.2.2. Structure

2.2.2.2.1. Electronic Spectroscopy   Initially, the main attention of researchers was focused on the simplest, albeit most fundamental, parent species $H_3E\bullet$ (E = Si, Ge, Sn). Thus, the electronic spectrum of the $H_3Si\bullet$ radical, generated by the reaction of $F\bullet$ (or $Cl\bullet$) upon the excimer-laser photolysis of F_2 (or Cl_2) with H_4Si, revealed strong vibrational progression between 365 and 410 nm, observed by resonance-enhanced multiphoton ionization spectroscopy (REMPI).[110] Similarly, the resonance of the $H_3Ge\bullet$ radical, generated by the reaction of either $F\bullet$ or $Cl\bullet$ radicals with H_4Ge, was observed in the region of 370–430 nm by REMPI spectroscopy.[111] The transient UV absorption band of $H_3Si\bullet$, generated by the flash photolysis of a $CCl_4/H_4Si/N_2$ mixture, was observed in the region of 205–250 nm (absorption maximum at ~215 nm), which corresponds to its pyramidal structure.[112] The recent microwave spectroscopy study of $F_3Si\bullet$, generated by the glow discharge of F_3Si–SiF_3, has proved its pyramidal geometry deduced from the analysis of the rotational spectrum.[113]

2.2.2.2.2. ESR Spectroscopy The first low-temperature ESR spectra of the transient radicals $H_3E•$ (E = Si, Ge, Sn) were recorded in the matrices of inert gases.[114] The following ESR parameters were determined at 4.2 K for the above radicals (g-factor; hfcc (in mT) for the central nuclei (^{29}Si, ^{73}Ge, and ^{119}Sn); matrix inert gas): $H_3Si•$ (2.0030; 19; Xe); $H_3Ge•$ (2.0073; 7.5; Xe); and $H_3Sn•$ (2.0170; 38; Kr).[114] The large hfcc values indicate the pronounced pyramidal nature of those radicals, whose structural parameters were estimated as follows (s-character of bond orbitals; bond angles (degrees)): $H_3Si•$ (0.285; 113.5); $H_3Ge•$ (0.295; 115); and $H_3Sn•$ (0.310; 117). Such nonplanar geometry is markedly different from the planar geometry of $H_3C•$, which exhibited an ESR resonance at $g = 2.0020$ in a Xe matrix, corresponding to calculated values of s-character of bond orbitals = 0.333 and bond angles = 120°.[114] The subsequent reinvestigation of the ESR parameters of the $H_3Si•$ radical, generated by either γ-irradiation of H_4Si adsorbed on a silica gel surface at 77 K[115] or by the reaction of H_4Si with H atoms in a Kr matrix,[116] gave essentially the same values of the ^{29}Si hfccs of 18.2 mT[115] and 19.0 mT.[116] Such large hyperfine splitting, and consequently large s-character of the SOMO, definitely confirm the pyramidal configuration of the $H_3Si•$ radical, in agreement with Pauling's hypothesis that an $AX_3•$-type radical is expected to be nonplanar if X is more electronegative than A.[117] The ESR spectrum of the $H_3Ge•$ radical, generated in a matrix of nonmagnetic isotopes of Xe, has also been reported.[118] The ESR spectra of the perhalogenated silyl radicals $F_3Si•$[119] and $Cl_3Si•$,[120] generated by the radiolysis of F_3SiH in an SF_6 matrix and irradiation of $SiCl_4$, exhibited very large ^{29}Si hfccs of 49.8 mT ($F_3Si•$),[119] 41.6 mT ($Cl_3Si•$ in Cl_3SiCH_3 matrix),[120a] and 44.0 mT ($Cl_3Si•$ in Cl_4Si matrix),[120b] indicating that these radicals are strongly pyramidal due to the great difference in electronegativities of the Si and F atoms. The pyramidality of silyl radicals progressively increased along the series $(CH_3)_3Si• < (CH_3)_2ClSi• < (CH_3)Cl_2Si• < Cl_3Si•$, as evidenced by the increasing values of their hfccs: 129 < 215 < 308 < 440.[120b] The simple trialkylsilyl radicals were found to be pyramidal; however, the degree of pyramidality decreased from $Me_3Si•$ to $t-Bu_3Si•$ because of the increasing bulk of the substituents (^{29}Si hfccs in mT): $Me_3Si•$ (18.1, 18.3),[121] $Et_3Si•$ (17.0),[122] $t-Bu_3Si•$ (16.3).[123] The ^{29}Si hfcc values dramatically decreased upon the increasing silyl substitution of the Si radicals, indicating their remarkable planarity (^{29}Si hfccs in mT): $Me_2(Me_3Si)Si•$ (13.7), $Me(Me_3Si)_2Si•$ (7.1), $(Me_3Si)_3Si•$ (6.5).[124] An unusual example of the bicyclic bridgehead silyl radical $HC(CH_2)_3Si•$ was also reported as being pyramidal, which was deduced from its small β-hydrogen coupling constant of 0.15 mT.[125] Phenyl-substituted Ge radicals $Ph_3Ge•$, $Ph_2MeGe•$, and $PhMe_2Ge•$ were also estimated to be pyramidal judging from their α-CH_3 hfcc.[126] The ESR conformational analysis of the series of neopentyl $Me_{3-n}(t-BuCH_2)_nGe•$ (n = 1–3) and trimethylsilyl substituted $Me_{3-n}(Me_3SiCH_2)_nGe•$ (n = 1–3) germyl radicals revealed the existence of two kinds of methylene protons due to the nonplanar geometry and slow inversion of these radicals.[127] Similarly, the tris(2-phenyl-2-methylpropyl)germyl radical $R_3Ge•$ (R = $PhMe_2CCH_2$), generated by the reaction of $t-BuO•$ radical with R_3GeH, exhibited nonequivalence of the two methylene protons at −120°C due to hindered rotation around the C–Ge bonds.[128] Such nonequivalence implies a rigid pyramidal configuration of the $R_3Ge•$ radical on the ESR time scale at low temperature. The

ESR spectrum of the $R_3Sn\bullet$ radical (R = 2,4,6-triisopropylphenyl), spontaneously generated upon dissolving distannane $R_3Sn–SnR_3$ in deoxygenated toluene, revealed at $-140°C$ in the solid state the ^{119}Sn hfcc of 163 mT.[27] Such a radical is more planar than the $Ph_3Sn\bullet$ radical (^{119}Sn hfcc = 186.6 mT)[129] but less planar than the $Me_3Sn\bullet$ radical (^{119}Sn hfcc = 161.1 mT).[103]

2.2.2.2.3. Theoretical Calculations   The most intensively theoretically studied free radicals, centered on the heavier group 14 elements, are the parent $H_3E\bullet$ (E = Si, Ge, Sn) species. The halogen- and methyl-substituted derivatives of the types $X_3E\bullet$ and $Me_3E\bullet$ (X = F, Cl; E = Si, Ge, Sn) have also been the focus of recent theoretical investigations. In complete accord with the experimental results, a number of sophisticated calculations at different levels of theory have reliably established that the parent radicals $H_3E\bullet$ have pyramidal geometries in their ground states, in sharp contrast to the methyl radical $H_3C\bullet$, which is well known to possess a planar geometry. Thus, a systematic DFT study at the NL-SCF/TZ2P level revealed the following trend in the changes of the H–E–H bond angles in the $H_3E\bullet$ (E = C, Si, Ge, Sn) series: 120.00 ($H_3C\bullet$), 112.66° ($H_3Si\bullet$), 112.44° ($H_3Ge\bullet$), and 110.56° ($H_3Sn\bullet$), that is, a steady increase in the pyramidalization degree was observed upon descending group 14.[130] Accordingly, the inversion barriers of the pyramidal C_{3v} form of the $H_3E\bullet$ species through its planar D_{3h} form also monotonically increase going down group 14: 0.0 kcal/mol ($H_3C\bullet$), 3.7 kcal/mol ($H_3Si\bullet$), 3.8 kcal/mol ($H_3Ge\bullet$), 7.0 kcal/mol ($H_3Sn\bullet$).[130] Such a phenomenon had been typically explained in the framework of the second-order Jahn–Teller effect, namely, by the stabilizing mixing between the nonbonding SOMO and E–H antibonding LUMO, which results in pyramidalization around the E-radical center. This pyramidalization effect is stronger for the bigger, and more electropositive, heavier E atoms, because the SOMO–LUMO energy gap becomes smaller due to the raising of the SOMO levels and lowering of the LUMO levels. However, strong evidence for the significant role of steric effects on the geometry of $H_3E\bullet$ has recently been provided.[130] The authors have concluded that $H_3C\bullet$ is planar, predominantly because of the great steric repulsion between the hydrogen atoms. This steric effect overcomes, in the particular case of $H_3C\bullet$, the electronic effect, which always favors a pyramidal structure. The heavier analogs of the methyl radical, $H_3E\bullet$ (E = Si, Ge, Sn), have elongated E–H bonds, thus decreasing the degree of steric repulsion between the hydrogen atoms and, accordingly, increasing the degree of pyramidalization around the E-radical center.

The influence of the nature of the substituents on the geometry of free radicals $H_3E\bullet$ was also studied computationally. Thus, $Me_3Si\bullet$ was predicted at the UHF/6-21G level to be pyramidal (inversion barrier 13.3 kcal/mol)[131] in contrast to $Me_3C\bullet$, which is essentially planar at the UHF/4-31G level (inversion barrier 1.2 kcal/mol).[132] There are two main reasons for such a difference: the first is the decrease in the HOMO–LUMO energy gap for $Me_3Si\bullet$ stabilizing the SOMO upon pyramidalization (electronic effect), whereas the second reason is the decrease of the van der Waals repulsion of the Me substituents on going from $Me_3C\bullet$ to $Me_3Si\bullet$ due to the longer Si–C bonds (steric effect).[131] When the hydrogen atoms in $H_3Si\bullet$ were successively changed to a Me group and F atom, the pyramidalization degree and the inversion barriers of the resulting radicals progressively increased at the UHF/3-21G level: 5.1 kcal/mol ($H_3Si\bullet$), 6.3 kcal/mol ($H_2MeSi\bullet$), 13.4 kcal/mol ($H_2FSi\bullet$).[131] Further increase in the degree of halogen substitution at the radical center of $R_3E\bullet$ results in a more pronounced pyramidalization and sharp rise of the inversion barriers. Thus, in the row of the trihalogen substituted radicals of the type $X_3E\bullet$ (X = F, Cl; E = C, Si, Ge), both the degree of pyramidalization and the inversion barrier markedly increase on going from C to Ge derivatives: $F_3C\bullet$ (27.4 kcal/mol),[133] $F_3Si\bullet$ (106.44°, 68.1 kcal/mol, at UHF/3-21G level),[131] $Cl_3C\bullet$ (116.1°, 2.97 kcal/mol, at UHF/STO-3G level),[134] $Cl_3Si\bullet$ (108.0°, 39.40 kcal/mol, at UHF/STO-3G level),[135] $Cl_3Ge\bullet$ (107.6°, 56.13 kcal/mol, at UHF/STO-3G level).[135] However, later calculations by the same authors gave different estimates of the degree of pyramidalization (calculated at the UHF/3-21G* level) and inversion barriers (calculated at the UMP2/3-21G*//UHF/6-31G*(3-21G*) levels) in the series of fluorides $F_3E\bullet$ and chlorides $Cl_3E\bullet$ (E = Si, Ge, Sn): $F_3Si\bullet$ (107.2°, 38.5 kcal/mol), $F_3Ge\bullet$ (106.1°, 28.7 kcal/mol), $F_3Sn\bullet$ (104.6°, 20.7 kcal/mol), $Cl_3Si\bullet$ (109.6°, 36.5 kcal/mol), $Cl_3Ge\bullet$ (108.3°, 25.6 kcal/mol), $Cl_3Sn\bullet$ (107.1°, 19.3 kcal/mol).[136] Such a general tendency of the increase in pyramidalization at the heavier group 14 elements radical centers, upon the increasing electronegativity of the substituents, was explained in terms of the stabilizing SOMO–LUMO interaction, as was discussed above (Section 2.2.1). In contrast, substitution with the electropositive H_3Si substituents causes a significant planarization of the silyl radical: the Si–Si–Si bond angle in the $(H_3Si)_3Si\bullet$ radical was estimated at the UMP2/DZP level as 114.80° (in contrast to 108.14° for $F_3Si\bullet$ and 109.51° for $Cl_3Si\bullet$ radicals).[137]

2.2.2.3. Synthetic Applications

Although many group 14 element centered free radicals were established to be important reactive intermediates, participating in a number of organometallic reactions, only two of them have received particular attention from synthetic chemists. They are n-$Bu_3Sn\bullet$ and $(Me_3Si)_3Si\bullet$, generated from tributyltin hydride n-Bu_3SnH and tris(trimethylsilyl)silane $(Me_3Si)_3SiH$, promoting a majority of the radical chain reactions. Tributyltin hydride is by far the most widely used mediator for the highly selective reduction of functional groups and formation of C–C bonds by either inter- or intramolecular cyclization, proceeding usually in high yields.[138] However, there are some drawbacks for the use of this radical reagent, the main one being the incomplete removal of toxic tin by-products from the final products. Therefore, tris(trimethylsilyl)silane was recently suggested as a valuable synthetic alternative to tributyltin hydride in many radical chain reactions.[73c,d,f,h] Both n-Bu_3SnH and $(Me_3Si)_3SiH$ were found to be very efficient reducing reagents for a variety of functional groups: halides, chalcogen groups, thiono esters, isocyanides, etc. The synthetic application aspects of n-Bu_3SnH and $(Me_3Si)_3SiH$ have been thoroughly discussed with a number of practical examples in the recent reviews.[73c,d,f,h,138] In this contribution we will just refer to some recent examples of synthetic utilization of both compounds: n-Bu_3SnH, as a reagent of intramolecular cyclization of allylic propiolates,[139] intramolecular cyclization of enynes,[140] stannylformylation of 1,6-dienes accompanied by ring closure,[141] desulfonylation of heterocyclic α-sulfones,[142] diastereoselective 1,4-addition to α,β-unsaturated (−)-8-phenylmenthyl ester,[143] addition to ynals to produce 2-iodomethylene cyclopentanols and cyclohexanols,[144] radical cyclization of oxime ethers;[145] (Me₃Si)₃SiH as a reagent for reduction of carbonyl groups,[146] Z–E isomerization of some alkenes through the addition–elimination path,[147] and addition to β-alkenyloxyenones to form tetrahydrofuranyl systems.[148]

2.2.3. Persistent Radicals

The term "persistent" is typically applied to long-lived radicals,[74a] which in a majority of cases can be characterized by ESR spectroscopy.[149] However, the lifetime of those radicals can vary rather broadly from several minutes to months, depending on the environment around the radical center. In this contribution, we will use the terminology of "persistent" radicals for those radical species with a *relatively* long lifetime; however, those persistent radicals that can be isolated as individual room temperature stable compounds and in many cases characterized by X-ray crystallography will be specifically named as "stable" radicals (for the chemistry of stable radicals, see Section 2.2.4).

Many persistent radicals centered on the heavier group 14 elements are currently known.[74a] In this section we will emphasize only several of the most remarkable of them: the first representatives synthesized by Lappert and co-workers, recent examples of the persilyl-substituted silyl radicals prepared by Matsumoto's group, and the latest representatives reported by Apeloig's group.

The first landmark syntheses of persistent radicals of the heavier group 14 elements were reported nearly 30 years ago by Lappert's group.[150] The first radical of this series, [(Me₃Si)₂CH]₃Sn•, was prepared by photolysis of the stable stannylene [(Me₃Si)₂CH]₂Sn: with visible light in benzene (Scheme 2.26).[150a,c]

The overall process seems to involve the initial disproportionation of the stannylene into the pair of radicals (Me₃Si)₂CH• and [(Me₃Si)₂CH]Sn•, followed by the valency-increasing reaction of the starting stannylene [(Me₃Si)₂CH]₂Sn: with the radical (Me₃Si)₂CH•. The ESR spectrum of [(Me₃Si)₂CH]₃Sn• revealed a central quartet ($g = 2.0094$) due to the coupling of the unpaired electron with the three equivalent protons of the (Me₃Si)₂CH substituents. Most importantly, a pair of satellites from ^{117,119}Sn nuclei with the hfccs of 169.8 and 177.6 mT was clearly observed at higher gains. The magnitude of these values, which are comparable with those of the transient Sn-centered radicals (*vide supra*), definitely indicates the appreciable pyramidality of [(Me₃Si)₂CH]₃Sn•. This radical was remarkably stable in benzene solution in the absence of air: no decrease in the intensity of its ESR signal was observed for more than 1 month at room temperature. Employing the same approach, a series of Ge- and Sn-centered radicals R₃E• has been prepared by photolysis of the stable heavy carbene analogs R₂E: [R = CH(SiMe₃)₂ or N(SiMe₃)₂; E = Si or Ge] in benzene or hexane solutions at room temperature (Scheme 2.27, A).[150b,c]

$$R_2Sn: \xrightarrow{h\nu} RSn\bullet \; + \; R\bullet \xrightarrow{R_2Sn:} R_3Sn\bullet$$
$$R = CH(SiMe_3)_2$$

Scheme 2.26

$$2 R_2E: \xrightarrow{h\nu} R_3E \bullet + RE \bullet \qquad\qquad\qquad (A)$$

$$(R = CH(SiMe_3)_2, N(SiMe_3)_2; E = Si, Ge)$$

$$Si_2Cl_6 + 6 RLi \xrightarrow{h\nu} R_3Si \bullet + (RSiCl_2)_2 + SiCl_4 + RSi \bullet \quad (B)$$

Scheme 2.27

The persistent silyl radicals $R_3Si \bullet$ were synthesized by the reaction of Si_2Cl_6 with 6 equivalents of RLi in benzene followed by photolysis of the reaction mixture (Scheme 2.27, B).[150b,c]

The ESR spectra of the radicals $R_3E \bullet$ exhibited the anticipated splitting patterns, for example, $[(Me_3Si)_2CH]_3Ge \bullet$ gives a decet (coupling with ^{73}Ge nucleus, $I = 9/2$) of septets (coupling with the 3 equivalent ^{14}N nuclei, $I = 1$).[150b]

All radicals were remarkably stable in solution at room temperature (g-value, hfcc in mT, half-life): $[(Me_3Si)_2CH]_3Si \bullet (2.0027, a(^{29}Si) = 19.3, t_{1/2} \sim 10\,min)$, $[(Me_3Si)_2CH]_3Ge \bullet$ $(2.0078, a(^{73}Ge) = 9.2$, unchanged after 4 months), $[(Me_3Si)_2CH]_3Sn \bullet (2.0094, a(^{117}Sn)$ $= 169.8$ and $a(^{119}Sn) = 177.6, t_{1/2} \sim 1$ year), $[(Me_3Si)_2N]_3Ge \bullet (1.9991, a(^{73}Ge) = 17.1$, $t_{1/2} > 5$ months), $[(Me_3Si)_2N]_3Sn \bullet (1.9912, a(^{117}Sn) = 317.6$ and $a(^{119}Sn) = 342.6, t_{1/2}$ ~ 3 months).[150b] The remarkable persistence of these radicals was mainly attributed to the great steric hindrance to dimerization due to the bulky $(Me_3Si)_2CH$ or $(Me_3Si)_2N$ groups.[150b] Another factor contributing to the enhanced stability of the radicals is the low strength of the E–H bonds (C−H: 104 kcal/mol; Si–H: 81 kcal/mol; Ge–H: 73 kcal/ mol; Sn–H: 70 kcal/mol),[151] which do not favor the quenching of radicals by hydrogen abstraction from the solvent.[150c] The third stabilizing factor is the disadvantage of the disproportionation pathway to form doubly bonded species, which is typical for carbon-centered radicals.[150a] The large values of the hfccs undoubtedly provide evidence for the pyramidal structures of all of the above-persistent radicals; the increased pyramidality of amino-substituted radicals $[(Me_3Si)_2N]_3E \bullet$ (E = Ge, Sn) compared with that of the alkyl-substituted radicals $[(Me_3Si)_2CH]_3E \bullet$ (E = Si, Ge, Sn) is in accord with the greater electronegativity of the N atom versus the C atom. However, the extent of pyramidality of amino-substituted radicals $[(Me_3Si)_2N]_3E \bullet$ (E = Ge, Sn) is less than that of $Cl_3Ge \bullet$, which has $a(^{73}Ge) = 22.9\,mT$.[152] Thus, all persistent radicals $R_3E \bullet$ (R = $CH(SiMe_3)_2$ or $N(SiMe_3)_2$; E = Si, Ge, Sn) possess a highly pronounced pyramidal geometry, being intermediate between the planar $CH_3 \bullet$ and nearly tetrahedral $X_3E \bullet$ (X = F, Cl; E = Si, Ge, Sn), in full accord with the increasing electronegativity difference between the central atom and substituents.

Several remarkable Si-centered radicals were recently synthesized by Matsumoto's group. Thus, the highly persistent tris(trialkylsilyl)silyl radicals $(Et_nMe_{3-n}Si)_3Si \bullet$ ($n = 1-3$) **35** were generated by three methods: (1) hydrogen abstraction from hydrides $(Et_nMe_{3-n}Si)_3SiH$ by t-BuO$\bullet$ radical upon photolysis in the presence of $(t$-BuO$)_2$; (2) Si–Si bond breaking upon photolysis of tetrakis(trialkylsilyl)silanes $(Et_nMe_{3-n}Si)_4Si$; and (3) central Si–Si bond breaking upon photolysis of hexakis(trialkylsilyl)silanes $(Et_nMe_{3-n}Si)_3Si$-Si$(SiMe_{3-n}Et_n)_3$ (Scheme 2.28).[153]

$$(Et_nMe_{3-n}Si)_3SiH \xrightarrow[\text{(t-BuO)}_2]{\text{hv}} (Et_nMe_{3-n}Si)_3Si \cdot \xleftarrow{\text{hv}} (Et_nMe_{3-n}Si)_4Si$$

35

$(n = 1\text{-}3)$

hv $\uparrow$

$$(Et_nMe_{3-n})_3Si - Si(SiMe_{3-n}Et_n)_3$$

Scheme 2.28

All these radicals exhibited similar ESR spectra with the two pairs of satellites resulting from the coupling of the unpaired electron with the α-^{29}Si nucleus (larger hfcc) and β-^{29}Si nuclei (smaller hfcc). The ESR characteristics of the above radicals, in comparison with the previously described $(Me_3Si)_3Si\bullet$ radical, are given below (T in °C, g-factor, $a(\alpha$-^{29}Si) in mT, $a(\beta$-^{29}Si) in mT): $(Me_3Si)_3Si\bullet$ (-25, 2.0053, 6.38, 0.71);[154] $(EtMe_2Si)_3Si\bullet$ (15, 2.0060, 6.28, 0.71); $(Et_2MeSi)_3Si\bullet$ (15, 2.0060, 6.03, 0.73); $(Et_3Si)_3Si\bullet$ (15, 2.0063, 5.72, 0.79).[153] As one can notice, the $a(\alpha$-^{29}Si) values progressively decreased upon the increase of the steric bulk around the radical center with the increasing number of Et groups, which in turn means an increasing degree of planarity on going from $(Me_3Si)_3Si\bullet$ to $(Et_3Si)_3Si\bullet$ radicals. This also implies that the delocalization of the radical spin density over the σ^*-orbitals of the Si–C bonds of silyl substituents becomes more effective in the same direction. In contrast to $(Me_3Si)_3Si\bullet$, whose ESR spectrum was observed at -25°C,[154] all $(Et_nMe_{3-n}Si)_3Si\bullet$ radicals **35** exhibited remarkable stability in solution at room temperature due to the significant shielding effect of the Et groups. The half-life time of these radicals depends on the generation method, the most long-lived silyl radicals were generated from the hexakis(trialkylsilyl)disilanes: 3 h for $(EtMe_2Si)_3Si\bullet$, one day for $(Et_2MeSi)_3Si\bullet$, and 1.5 months for $(Et_3Si)_3Si\bullet$ at 15°C.[153] Further increasing the degree of the substituents' bulkiness, the same authors prepared $(i\text{-}Pr_3Si)_3Si\bullet$ **36** by hydrogen abstraction from the corresponding hydride $(i\text{-}Pr_3Si)_3SiH$ under the action of t-BuO$\bullet$ radical, which was photochemically generated from $(t\text{-}BuO)_2$ (Scheme 2.29).[155]

The ESR parameters of the radical **36**, g-factor $= 2.0061$ and $a(\alpha$-^{29}Si) $= 5.56$ mT, provided convincing evidence for the significant planarity of the Si-radical center. Radical **36** was remarkably persistent with a half-life of 5 days at 15°C, which was much longer than that of the $(Et_nMe_{3-n}Si)_3Si\bullet$ ($n = 1$-3) radicals **35** generated by the same method.[153] A new representative of the persistent persilyl radicals, tris(tert-butyldimethylsilyl)silyl radical $(t\text{-}BuMe_2Si)_3Si\bullet$ **37**, was independently generated by one of the following methods: (1) one-electron oxidation of $(t\text{-}BuMe_2Si)_3SiNa$ with $NO^+BF_4^-$ or $Ph_3C^+BPh_4^-$; (2) one-electron reduction of $(t\text{-}BuMe_2Si)_3SiBr$ with an equimolar amount of Na; or (3) hydrogen abstraction from hydrosilane

$$(i\text{-}Pr_3Si)_3SiH \xrightarrow[\text{(t-BuO)}_2]{\text{hv}} (i\text{-}Pr_3Si)_3Si \cdot + t\text{-}BuOH$$

36

Scheme 2.29

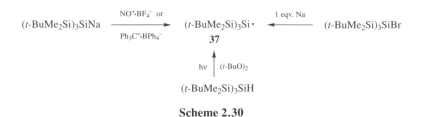

Scheme 2.30

(t-BuMe$_2$Si)$_3$SiH by t-BuO• radical generated photochemically from (t-BuO)$_2$ (Scheme 2.30).[156]

This new radical **37** exhibited an ESR resonance with g-factor = 2.0055 and a(α-^{29}Si) = 5.71 mT. Judging from the hfcc values, the degree of planarity of (Et$_3$Si)$_3$Si• (a(α-^{29}Si) = 5.72 mT),[153] (t-BuMe$_2$Si)$_3$Si• (a(α-^{29}Si) = 5.71 mT),[156] and (i-Pr$_3$Si)$_3$Si• (a(α-^{29}Si) = 5.56 mT)[155] were estimated to be almost the same, although the last species should be recognized as the most sterically crowded and, consequently, most planar. The half-life of radical **37** was dependent on its generation method: ~ one day at 15°C (generated by either oxidation or reduction reactions) and 10 h at 15°C (generated by the hydrogen abstraction reaction).[156]

A family of persistent silyl-substituted silyl radicals was very recently reported by the group of Apeloig. Tris(pentamethyldisilyl)silyl radical (Me$_3$SiMe$_2$Si)$_3$Si• **38** was generated from a variety of precursors: hydrogen abstraction from the corresponding hydride (Me$_3$SiMe$_2$Si)$_3$SiH by t-Bu$_2$Hg, thermolysis or photolysis of the dimercurial compound (Me$_3$SiMe$_2$Si)$_3$Si-Hg-Hg-Si(SiMe$_2$SiMe$_3$)$_3$, and photolysis of hexakis(pentamethyldisilyl)disilane (Me$_3$SiMe$_2$Si)$_3$Si-Si(SiMe$_2$SiMe$_3$)$_3$ (Scheme 2.31).[157]

The ESR spectrum of radical **38** revealed a central signal (g-factor = 2.0065) accompanied by two pairs of satellites from the α- and β-^{29}Si atoms. The value of a(α-^{29}Si) of 5.99 mT gives evidence for the essentially planar configuration of this radical, which was also theoretically supported. Thus, calculations at the UB3LYP//6-31G*//UB3LYP/6-31G* level revealed the Si–Si–Si bond angle in **38** of 118.0° (120° for the ideal trigonal-planar sp^2-hybridized radical), whereas the same Si–Si–Si bond angle for (H$_3$Si)$_3$Si• radical (model for (Me$_3$Si)$_3$Si• radical) was estimated as 115.9°. Thus, the radical **5** (a(α-^{29}Si) = 5.99 mT)[157] is more planar than the (Me$_3$Si)$_3$Si• radical (a(α-^{29}Si) = 6.38 mT), but less planar than the (i-Pr$_3$Si)$_3$Si• radical **36** (a(α-^{29}Si) = 5.56 mT).[155] Quite parallel, the kinetic stability greatly increases in the order (Me$_3$Si)$_3$Si• < (Me$_3$SiMe$_2$Si)$_3$Si• < (i-Pr$_3$Si)$_3$Si•: the first one can be

$$\text{(Me}_3\text{SiMe}_2\text{Si)}_3\text{SiH} \xrightarrow[t\text{-Bu}_2\text{Hg}]{\Delta} \underset{\textbf{38}}{\text{(Me}_3\text{SiMe}_2\text{Si)}_3\text{Si•}} \xleftarrow{h\nu} \text{(Me}_3\text{SiMe}_2\text{Si)}_3\text{Si}-\text{Si(SiMe}_2\text{SiMe}_3)_3$$

$$\Delta \uparrow \text{ or } h\nu$$

$$\text{(Me}_3\text{SiMe}_2\text{Si)}_3\text{Si}-\text{Hg}-\text{Hg}-\text{Si(SiMe}_2\text{SiMe}_3)_3$$

Scheme 2.31

observed only at $-25°C$,[154] whereas the second one has a half-life of 6 min at room temperature,[157] and the last one has a half-life of 5 days at $15°C$.[155] A series of rather unusual metal- (Li, Hg) substituted silyl radicals was generated quite recently by Apeloig and co-workers (g-factor, $a(\alpha\text{-}^{29}Si)$ in mT): $(i\text{-}Pr_3Si)_2SiLi\bullet$ (2.0073, 3.2);[158] (thf)Li($t\text{-}Bu_2MeSi)_2Si\bullet$ (2.0067, 3.33);[159] $t\text{-}BuHg\text{-}Si(SiMet\text{-}Bu_2)_2\text{-}Hg\text{-}Si(SiMet\text{-}Bu_2)_2\bullet$ (1.984, 5.60);[160] $HSi(SiMet\text{-}Bu_2)_2\text{-}Si(SiMet\text{-}Bu_2)_2\bullet$ (1.983, 5.60);[160] (thf)Li$\text{-}Si(SiMet\text{-}Bu_2)_2\text{-}Hg\text{-}Si(SiMet\text{-}Bu_2)_2\bullet$ (1.984, 5.60).[160]

2.2.4. Stable Radicals

The development of sophisticated new experimental techniques during the last decade has made possible the isolation of stable representatives of the free radical species featuring an unpaired electron on the heavier group 14 elements, that is, silyl, germyl, and stannyl radicals. This great progress in the isolation of the stable radicals opens unprecedented possibilities for their structural characterization in the crystalline form, which in turn enables the direct comparison of the fundamental differences and similarities between the solution and solid state structures of the free radical species.[74]

All stable group 14 element-centered free radicals can be grouped into two large classes: neutral radicals and charged radicals. Each of these two classes can be further divided into the two subclasses of cyclic and acyclic radicals with greatly different physical and chemical properties. Thus, if the cyclic radicals could primarily benefit from the delocalization of their unpaired electron over the cyclic π-bond system, the acyclic trivalent radicals are able to take advantage of the hyperconjugative stabilization through the delocalization of the unpaired electron over the σ*-orbitals of the neighboring bonds.[74b] All these classes of the above-mentioned stable radicals will be briefly discussed below.

2.2.4.1. Neutral Radicals

2.2.4.1.1. Cyclic Radicals   The first compound of this type was reported by Power et al. in 1996. The cyclotrigermenyl radical **39**, highly protected by very large 2,6-$Mes_2\text{-}C_6H_3$ groups, was prepared by the reduction of chlorogermylene :Ge(Cl)(2,6-$Mes_2 C_6H_3$) with KC_8 in THF (Scheme 2.32).[161] The structure of this compound, featuring an odd electron on one of the Ge atoms and a double bond between the two other Ge atoms, is highly reminiscent of that of the organic cyclopropenyl radical.

The structure of **39** in solution was deduced on the basis of its ESR spectrum, which displayed a single resonance (g = 2.0069) with the hfcc a (^{73}Ge) = 1.6 mT. The small a value is suggestive of the localization of an unpaired electron in an orbital with mostly p-character implying the sp^2-hybridization of the Ge-radical centers and, hence, their planarity.

The all-silicon version of the cyclobutenyl radical, cyclotetrasilenyl radical **40**, was recently reported by Sekiguchi et al. Its synthesis was accomplished by the one-electron reduction of the precursor cationic species, cyclotetrasilenylium ion **20**$^+$•**TPFPB**$^{-59}$ with either $t\text{-}Bu_3SiNa$ or KC_8 in Et_2O (Scheme 2.33).[162] The four membered Si_4 ring of **40** is nearly planar with the Si1–Si2 and Si2–Si3 bonds

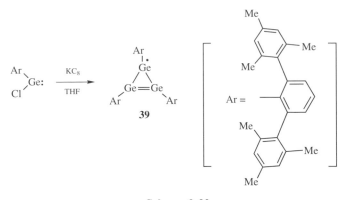

Scheme 2.32

intermediate between the typical Si–Si single and Si=Si double bonds, which provides evidence for the allylic-type structure of the radical **40**. The ESR spectrum of **40** revealed a central signal ($g = 2.0058$) with the three pairs of satellites ($a = 4.07$, 3.74, and 1.55 mT) resulting from the coupling of the unpaired electron with the two terminal (Si1, Si3) and the central (Si2) nuclei of the delocalized allylic Si1–Si2–Si3 unit of radical **40**. These small values of the hfccs are consistent with the suggestion of the planarity of radical **40** in solution.

A rather unusual bicyclic Ge-centered radical, 1,6,7-trigermabicyclo[4.1.0]hept-3-en-7-yl **41**, was recently synthesized by Sekiguchi and co-workers by the one-electron oxidation of the parent bicyclic anion **42⁻•K⁺** with B(C₆F₅)₃ in THF (Scheme 2.34).[163] The specific accommodation of the odd electron on the Ge3 atom was clearly observed in the crystal structure of **41**, which displayed nearly planar geometry around the tricoordinate Ge atom. The peculiar *endo*-conformation of **41** allowed the through-space intramolecular interaction between the SOMO(Ge3) and π(C=C) orbitals caused by their relative proximity: the distance between Ge3 and the C=C bond of 3.63 Å is in the range of van der Waals interactions. However, the *endo*-conformer of **41** easily undergoes partial isomerization to the *exo*-isomer in solution during ESR measurements, which resulted in the observation of two independent signals ($g = 2.0210$ and 2.0223) with almost identical intensities.[163] The planarity of the Ge-radical centers was manifested in the diagnostically small values of the hfcc (3.4 and 2.6 mT), which clearly indicates the sp²-hybridization of the

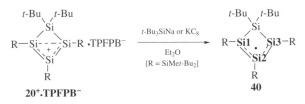

Scheme 2.33

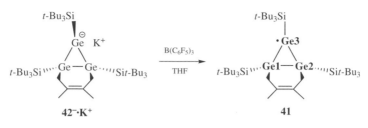

Scheme 2.34

radical centers. Apparently, in the crystalline form the single *endo*-conformation is insignificantly stabilized by the Ge•--C=C bond interaction (calculation provided a small preference for the *endo*- over the *exo*-form of 0.59 kcal/mol), whereas in solution the *endo–exo* radical inversion smoothly takes place to produce a mixture of both conformers.

Notably, the bicyclic radical **41**[163] is localized, in sharp contrast to the allylic-type delocalization of cyclotrigermenyl radical **39**[161] and cyclotetrasilenyl radical **40**.[162] The obvious reason for such a distinction is the absence of the neighboring to the Ge-radical center π-bond necessary for the effective through-bond delocalization of the unpaired electron in the radical **41**, whereas the through-space radical–C=C bond interaction is not sufficiently strong to induce the effective delocalization of the unpaired electron.

2.2.4.1.2. Acyclic Tricoordinate Radicals

As one can easily note, the above examples of the cyclic radicals of Si and Ge atoms have rather particular structures featuring the cyclic delocalization of the odd electrons. The simple tricoordinate acyclic radicals of the type $R_3E\bullet$ (E = Si, Ge, Sn, Pb), lacking the stabilization effects of the cyclic π-delocalization, constitute another, more general and even more challenging, class of stable organometallic radicals.[74b] Consequently, the search for such highly symmetrical species appeared to be of primary importance for synthetic chemists.

The major breakthrough in the development of such $R_3E\bullet$ species was achieved several years ago by Sekiguchi et al., who reported the isolation and full identification of a homologous series of $(t\text{-Bu}_2\text{MeSi})_3E\bullet$ (E = Si, Ge, Sn) radicals without π-bond conjugation.[74b] All of these radicals, kinetically and thermodynamically stabilized by the bulky electropositive silyl substituents, were prepared by the same very simple and straightforward method, which clearly demonstrated the generality of this synthetic approach.

The first member of the new class of stable radicals, $(t\text{-Bu}_2\text{MeSi})_3\text{Si}\bullet$**43**, was synthesized by the one-electron oxidation of the silylsodium derivative $(t\text{-Bu}_2\text{MeSi})_3\text{SiNa}$ with dichlorogermylene–dioxane complex $GeCl_2\bullet$diox in Et_2O (Scheme 2.35).[164]

Crystal structure analysis of **43** revealed that the radical is perfectly trigonal planar, implying the sp^2-hybridization of the central Si atom and, hence, the localization of the unpaired electron on its $3p_z$-orbital. The remarkable planarity of the radical **43** was explained by the great steric bulk of the voluminous $t\text{-Bu}_2\text{MeSi}$ groups, which prefered to move away from each other as far as possible to avoid the steric repulsion between them. On the contrary, the significant electron donation of the positive silyl substituents

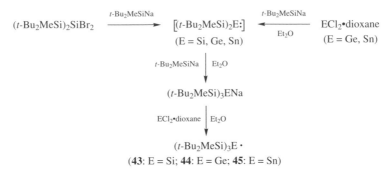

Scheme 2.35

results in the lowering of the inversion barrier at the Si radical center, also promoting its planarization. The spatial arrangement of t-Bu$_2$MeSi substituents in radical **43** favors the hyperconjugative delocalization of the unpaired electron over the antibonding σ^*-orbitals of Si–C(t-Bu) bonds, which notably contributes to the overall stabilization of the radical species. The structure of radical **43** in solution was deduced from its ESR spectrum, which revealed an intense signal ($g = 2.0056$) accompanied by a pair of satellites resulting from the coupling of the unpaired electron with the paramagnetic ^{29}Si nuclei and featuring the following values of the hfccs: $a = 5.80$ mT (α-Si) and 0.79 mT (β-Si). The small former value is undoubtedly indicative of the predominantly p-character of the SOMO, thus manifesting the planar geometry of the silyl radical **43** in solution. In other words, the persilyl radical **43** is indeed truly a π-radical.

The Ge analog of **43**, (t-Bu$_2$MeSi)$_3$Ge• **44**, was prepared in a quite similar, even simpler, way (Scheme 2.35).[164] The direct reaction of t-Bu$_2$MeSiNa with GeCl$_2$•diox in Et$_2$O resulted in the formation of intermediate germylsodium derivative (t-Bu$_2$MeSi)$_3$GeNa, which was subsequently oxidized with a second equivalent of GeCl$_2$•diox to produce the final product **44**. This germyl radical is totally isostructural to the above-described silyl radical **43** also featuring a trigonal-planar geometry around the radical center with the same spatial arrangement of the silyl substituents. The ESR spectrum of **44** displayed a central signal ($g = 2.0229$) along with the very characteristic decet of ^{73}Ge satellites exhibiting very small values of the hfcc of 2.00 mT, which should be definitely recognized as a direct indication of the planarity of germyl radical **44**.

The utilization of the same synthetic procedure brought about the successful preparation of the first and still the only representative of the stable stannyl radicals, (t-Bu$_2$MeSi)$_3$Sn• **45**, by the reaction of t-Bu$_2$MeSiNa with SnCl$_2$•diox in Et$_2$O (Scheme 2.35).[14b] Quite similar to the aforementioned silyl radical **43** and germyl radical **44**, stannyl radical **45** is trigonal planar in the crystalline form, and its planarity is also retained in solution, which was evidenced from the ESR spectrum displaying the central signal ($g = 2.0482$) with the pair of ^{119,117}Sn satellites featuring the diagnostically small value of the hfcc of 32.9 mT. Such a value is one order of magnitude less than those of all previously reported persistent stannyl radicals.

Undoubtedly, the uniform structural and spectral behavior of the trigonal-planar π-radicals (*t*-Bu₂MeSi)E• (E = Si, Ge, Sn) **43–45** both in the solid state and in solution should be ascribed to the immediate impact of the bulky electropositive silyl substituents. In contrast, it is well known that simple alkyl and aryl substituents cause a highly pronounced pyramidalization at the radical centers, where the unpaired electron typically occupies the orbital with a high s-contribution (σ-radicals).[73d,e,h]

2.2.4.2. Charged Radicals (Ion Radicals)

The appearance of the stable charged radical species, that is anion and cation radicals, as an independent class of highly reactive organometallic compounds, became possible only in the past few years after the preparation of the necessary precursors.[74b] Reviewing the recent accomplishments in this field, we will subdivide this class of stable charged radicals into the two groups of cyclic and acyclic representatives.

2.2.4.2.1. Cyclic Anion Radicals

The sole representative of this class of stable radicals, cyclotetrasilane anion radical **46**, was recently reported by Lappert's group.[165] The compound was prepared by the reduction of the dichloride **47** with metallic potassium in THF along with the doubly reduced species, cyclotetrasilane dianion **48** (Scheme 2.36). The anion radical **46** was isolated as a solvent-separated ion-pair, in which the Si₄ ring is square planar with the skeletal Si–Si bond lengths of 2.347(2) Å, slightly shorter than the typical bond lengths of cyclopolysilanes. The formulation of the delocalization of the unpaired electron over the cyclotetrasilane skeleton was clearly manifested by the ESR spectrum of **46**, which revealed a multiplet (g = 2.0025) of 15 lines (of total 17) with the hfcc of 0.35 mT due to the coupling of the unpaired electron with the eight ¹⁴N nuclei (I = 1).

2.2.4.2.2. Acyclic Anion Radicals

2.2.4.2.2.1. HEAVY ALKENE ANION RADICALS   The notably low-lying LUMOs of the heavy analogs of alkenes >E=E< (E = Si, Ge, Sn, Pb), compared with those of alkenes >C–C<, is a well-known and widely recognized inherent feature of the heavy alkene analogs. This is why the latter have been considered as a convenient and effective source for the generation of the corresponding anion radicals by one-electron reduction. The first report on the generation of persistent disilene anion

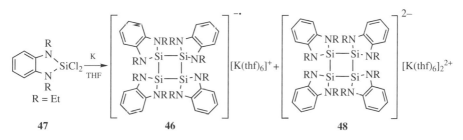

Scheme 2.36

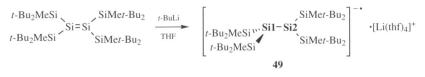

Scheme 2.37

radicals $[R_2Si=SiR_2]^{\bullet-}$ (R = i-Pr, t-Bu) by the reduction of 1,2-dihalides $R_2(X)Si-Si(X)R_2$ with alkali metals in THF was published by Weidenbruch in 1985.[166] More recently, Kira mentioned the generation of the tetrasilyldisilene anion radicals by the reduction of the disilenes with potassium in DME.[167] However, the isolation of the virtually stable anion radicals of disilenes, accompanied with their structural identification, was finally accomplished only a couple of years ago by Sekiguchi's group.

Thus, the first virtually stable disilene anion radical **49** was prepared by the reduction of the highly crowded disilene $(t$-Bu$_2$MeSi$)_2$Si=Si(SiMet-Bu$_2)_2$ with t-BuLi in THF (Scheme 2.37).[168] Upon reduction, the Si–Si bond of **49** became highly twisted (88°) and long (2.341(5) Å), which definitely should be ascribed to a decrease in the bond order on going from the starting disilene to a resulting disilene anion radical. One of the Si atoms (Si2) features a planar geometry while the other one (Si1) is pyramidal, which can be rationalized in the framework of the unpaired electron–negative charge separation: preferential accommodation of a single electron on the Si2 (radical center) and electron pair on the Si1 (anionic center) atoms. **49** exhibited a strong ESR resonance in the region typical for silyl-substituted silyl radicals (g = 2.0061) together with a pair of satellites ($a(^{29}$Si) = 2.45 mT). This hfcc value is less than half that of the structurally similar silyl radical $(t$-Bu$_2$MeSi$)_3$Si• **43** (5.80 mT),[164] which was explained by the delocalization of the unpaired electron over both Si1 and Si2 atoms.

The tin homolog of **49**, distannene anion radical **50**, was successfully prepared employing the same synthetic protocol: the direct reduction of distannene $(t$-Bu$_2$MeSi$)_2$Sn=Sn(SiMet-Bu$_2)_2$ with potassium mirror in THF in the presence of [2.2.2]cryptand (Scheme 2.38).[169]

The crystal structure of **50** is highly reminiscent of that of its silicon congener **49**: strongly twisted (74°) Sn–Sn bond with a bond length of 2.8978(3) Å, which is almost 9% longer than that of its precursor distannene. The separation of the unpaired electron and negative charge was also clearly manifested by the different geometry of both Sn atoms: one of them (Sn2) being essentially planar and the other (Sn1) distinctly pyramidal. Interestingly, such single electron–electron pair separation was preserved

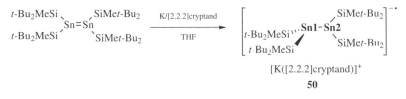

[K([2.2.2]cryptand)]$^+$

50

Scheme 2.38

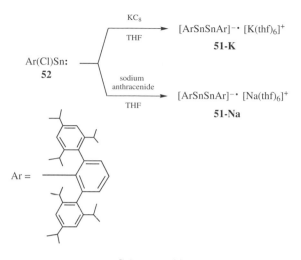

Scheme 2.39

in the solution of **50**, which was clearly demonstrated by its ESR spectrum showing a central resonance ($g = 2.0517$) with the two pairs of satellites resulting from the coupling with ^{119,117}Sn nuclei ($a(^{119,117}Sn_\alpha) = 34.0$ and $a(^{119,117}Sn_\beta) = 18.7$ mT). This was reasonably explained by the localization of the odd electron on one of the two Sn atoms, in contrast to the above case of the disilene anion radical **49**.[168]

2.2.4.2.2.2. HEAVY ALKYNE ANION RADICALS   The only example of compounds of this type, the anion radical of the valence isomer of distannyne **51–K**, was recently synthesized by Power by the reduction of chlorostannylene **52** with potassium graphite in THF (Scheme 2.39).[170]

Both substituents at the Sn atoms are highly *trans*-bent, and the Sn–Sn bond length of 2.8123(9) Å is quite normal for the Sn–Sn single bond. **51–K** features an ESR signal ($g = 2.0069$) together with the ^{119,117}Sn satellites, hfccs of which were simulated as 0.83 mT [$a(^{117}Sn)$] and 0.85 mT [$a(^{119}Sn)$]. Such small values of the hfccs provide evidence that the unpaired electron in anion radical **51–K** resides in an orbital of π-symmetry. All of the experimental data are consistent with the formulation of the resonance structure **51–K**, in which both tin atoms bear a lone pair and the unpaired electron is accommodated over their $5p_\pi$-orbitals (Figure 2.3).

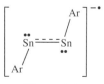

Figure 2.3

Such representation with a formal Sn–Sn bond order of 1.5 and highly *trans*-bent substituents is suggestive of the diminished Sn–Sn hybridization resulting from the poor $5p_\pi$–$5p_\pi$-orbitals overlap to form weak π-bonds. The anion radical sodium salt **51–Na**, prepared similarly by the reaction of chlorostannylene **52** with sodium anthracenide in THF, exhibited structural and spectral features identical to those of the above-described potassium derivative **51–K** (Scheme 2.39).[171]

2.2.4.3. Stable Biradicals of the Heavier Group 14 Elements

Biradicals are typically defined as the non-Kekulé structures that have an even number of electrons and have at least one bond less than the number required by the normal valency rules.[172] Among them the two classes of singlet and triplet biradicals can be distinguished, which are differentiated by the orientation of their spins: triplet biradicals feature the two electrons with parallel spins ($\uparrow\uparrow$), whereas in the singlet biradicals these electrons are spin-paired ($\uparrow\downarrow$). The chemistry of both singlet and triplet biradicals has been greatly developed in recent years with a number of direct observations of such reactive intermediates, and even their isolation as stable compounds in a limited number of cases.[173] Referring to the chemistry of the analogs of biradicals of the heavier group 14 elements, one can say that this field is only in its initial stage. To date, no examples of stable triplet biradicals of heavier group 14 elements have been reported in the literature, and only a couple of stable singlet biradicals (otherwise known as biradicaloids, that is, the closed-shell derivatives of the singlet biradicals exhibiting a weak partial coupling between the radical centers[174]) have been synthesized in recent years.[74b] Both these singlet biradicaloids represent a very interesting heteroatomic modification of the famous organic cyclobutane-1,3-diyl.[175] One can be referred also to the closely related papers describing the synthesis of the stable carbon- and boron-centered biradicaloids of the cyclobutane-1,3-diyl type.[176]

The first biradicaloid, 1,3-diaza-2,4-digermacyclobutane-1,3-diyl, **53**, was simply prepared by the reaction of the digermyne ArGeGeAr with trimethylsilyl azide in hexane (Scheme 2.40).[177]

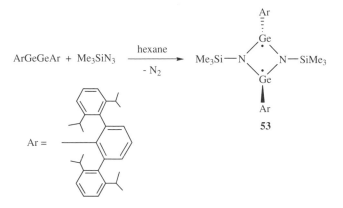

Scheme 2.40

The Ge_2N_2 ring of **53** is planar with both N atoms having a trigonal-planar geometry (359.97°) and in-plane arrangement of substituents. In contrast, both Ge atoms are distinctly pyramidal (322.10°) with a markedly out-of-plane *trans*-arrangement of the aryl substituents.

The formulation of **53** as a singlet biradical with odd electrons accommodated on the Ge atoms was evidenced from the trivalent state of both Ge atoms, sufficiently removed from each other by 2.755 Å, clearly outside the distance of the covalent bonding interaction between them. A theoretical calculation also revealed the absence of Ge–Ge bonding, exhibiting the nonbonding combination centered on Ge atoms as the HOMO of the molecule. Typically for the singlet biradicals, the ESR spectrum of **53** revealed no signals in the temperature range 77–300 K, whereas the normal NMR spectra were observed.

The second stable biradicaloid, 1,3-diaza-2,4-distannacyclobutane-1,3-diyl, **54**, was unexpectedly obtained by the reaction of chloro(amino)stannylene dimer $[Sn\{N(SiMe_3)_2\}(\mu\text{-}Cl)]_2$ and AgOCN in diethyl ether (Scheme 2.41).[178]

Similar to the above case of Power's biradicaloid **53**, the Sn_2N_2 ring of **54** is planar with the trigonal-planar geometry of N atoms and pyramidal configuration of Sn atoms. The orientation of substituents in **54** is also highly reminiscent of that in **53**, that is, a nearly in-plane arrangement of Me_3Si groups at the N atoms and almost perpendicular disposition of the Cl substituents at the Sn atoms. Such pyramidalization and tricoordination of Sn atoms definitely pointed to the radical character of both Sn atoms, which are separated from each other by 3.398 Å, well outside the range of the normal Sn–Sn covalent interaction indicating the absence of transannular 2,4-bonding. In other words, **54** should be considered to be a singlet biradicaloid possessing two Sn radical centers. Accordingly, **54** is diamagnetic exhibiting no ESR signals. The electron counting in **54**, two lone pairs from the N atoms and two odd electrons from the two Sn atoms, gives rise to the total of six, implying the six π-electron four-center system as the most appropriate bonding description of **54**. Theoretical calculations revealed the preference for the singlet state over the triplet one for **54** of 13.6 kcal/mol, which was ascribed to the σ-donating influence of the electropositive silyl substituents at the N atoms which increased the singlet–triplet energy gap $\Delta(S\text{-}T)$ in favor to the singlet ground state. Comparing germanium- and tin-containing biradicaloids **53** and **54**, one can note that they feature similar structural and bonding properties, which allow their reliable attribution to the new class of singlet biradicaloids of the heavier group 14 elements.

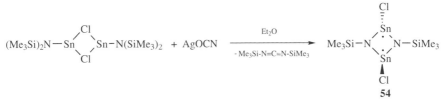

Scheme 2.41

2.3. Si-, Ge-, AND Sn-CENTERED ANIONS

2.3.1. Introduction

The common textbook description of the carbanion is the negatively charged species R_3C^- with a trivalent central carbon atom, in which the three orbitals are used for the formation of C–R bonds and the fourth one bears a lone pair. As this lone pair occupies one of the tetrahedral valencies, the carbanions are typically sp^3-hybridized, thus featuring a trigonal-pyramidal geometry. In this sense, carbanions are isoelectronic with amines, which also have a tetrahedral configuration. The pyramidalization degree markedly increases going down group 14, thus making the heavier analogs of carbanions, that is, silyl, germyl, stannyl, and plumbyl anions, more and more pyramidal. However, it should be noted that the degree of such pyramidalization is primarily determined by the nature of substituents. Quite similar to the case described in the above Radical section (see Section 2.2.1), the π-donating and σ-accepting substituents raise the inversion barrier of the heavier group 14 element-centered anions, resulting in their overall pyramidalization. In contrast, the π-accepting and σ-donating substituents cause a significant decrease in the inversion barriers manifested in the appreciable planarization of the anion species. One should also take into account the importance of the steric factor: the bigger the substituents at the anionic center, the more planar this anion tends to be. In the extreme case of very bulky electropositive substituents R, the favorable combination of both electronic and steric factors provides an unusual example of the planar geometry around the R_3E^- anionic center (E = Si, Ge) (vide infra). Except for the geometry of anionic species, there are a couple of other points of primary importance: first of all, the nature of interaction between the anionic and cationic parts of the molecule (covalent or ionic), and the aggregation state of the anionic molecule (monomeric, dimeric, or oligomeric). The major tools for the structural elucidation of the heavier group 14 element-centered anions are X-ray crystallography and NMR spectroscopy. The structures of the particular anionic compound in the crystalline form and in solution may greatly differ from each other, especially in view of the counter cation solvation effects. Below, in the most representative examples, we will demonstrate the major structural conclusions that can be drawn from the X-ray and NMR data. The chemistry of the heavy analogs of carbanions, particularly that of the silyl anions, has been repeatedly and comprehensively reviewed in recent years.[179] In the present section we will briefly summarize the most important achievements in this area made before 2000, and particular attention will be paid to the most recent discoveries since 2000.

2.3.2. Synthesis

The methods of generation of Si-centered anionic species are the most extensively studied and well developed.[179a,b,f,g] The choice of the particular synthetic method depends on the substituents on the Si atom.

$$\text{Me}_3\text{Si-SiMe}_3 + \text{MeLi} \xrightarrow{\text{Et}_2\text{O/HMPA}} \text{Me}_3\text{SiLi} + \text{Me}_4\text{Si} \qquad \textbf{(A)}$$

$$\text{Me}_3\text{Si-SiMe}_3 + \text{MeONa} \xrightarrow{\text{HMPA}} \text{Me}_3\text{SiNa} + \text{Me}_3\text{SiOMe} \qquad \textbf{(B)}$$

$$\text{Me}_3\text{Si-SiMe}_3 + 2\text{MH} \xrightarrow{\text{HMPA}} 2\text{Me}_3\text{SiM} + \text{H}_2 \qquad [\text{M = Na, K}] \qquad \textbf{(C)}$$

$$(\text{Me}_3\text{Si})_2\text{Hg} + 2\text{Li} \xrightarrow{\text{THF}} 2\text{Me}_3\text{SiLi} + \text{Hg} \qquad \textbf{(D)}$$

$$(\text{Me}_3\text{Si})_2\text{Hg} + \text{Mg} \xrightarrow{\text{DME}} (\text{Me}_3\text{Si})_2\text{Mg} + \text{Hg} \qquad \textbf{(E)}$$

$$\text{Et}_3\text{SiH} + \text{KH} \xrightarrow{\text{DME}} \text{Et}_3\text{SiK} + \text{H}_2 \qquad \textbf{(F)}$$

$$\textit{t}\text{-Bu}_3\text{SiX} + 2\text{M} \xrightarrow{\text{THF or heptane}} \textit{t}\text{-Bu}_3\text{SiM} + \text{MX} \quad [\text{M = Na, K; X = Br, I}] \qquad \textbf{(G)}$$

Scheme 2.42

2.3.2.1. (Alkyl)silyl Anions

1. Reduction of the Si–Si bond with RM, ROM, MH (M = alkali metal) (Scheme 2.42, **A–C**).[180–182]
2. Transmetalation of silylmercury compounds with Li or Mg (Scheme 2.42, **D, E**).[183,184]
3. Reduction of hydrosilanes with KH (Scheme 2.42, **F**).[182]
4. Reduction of the very bulky halosilanes with alkali metals (Scheme 2.42, **G**).[185]

2.3.2.2. (Aryl)silyl Anions

1. Reduction of chlorosilanes with Li or K (the most useful method) (Scheme 2.43, **A, B**).[186, 187]
2. Reduction of the Si–Si bond of disilanes with alkali metals (Scheme 2.43, **C**).[188]
3. Reduction of hydrosilanes with KH (Scheme 2.43, **D**).[182]

$$\text{Ar}_3\text{SiCl} + 2\text{Li} \xrightarrow{\text{THF}} \text{Ar}_3\text{SiLi} + \text{LiCl} \quad [\text{Ar = Ph, }\textit{o}\text{-tolyl}] \qquad \textbf{(A)}$$

$$\text{Ar}_n(\text{Alkyl})_{3-n}\text{SiCl} + 2\text{Li} \xrightarrow{\text{THF}} \text{Ar}_n(\text{Alkyl})_{3-n}\text{SiLi} + \text{LiCl} \qquad \textbf{(B)}$$

$$\text{Ph}_3\text{Si-SiPh}_3 + \text{M} \xrightarrow{\text{DME}} 2\text{Ph}_3\text{SiM} \quad [\text{M = Li, Na, K}] \qquad \textbf{(C)}$$

$$\text{Ph}_3\text{SiH} + \text{KH} \xrightarrow{\text{DME/HMPA}} \text{Ph}_3\text{SiK} + \text{H}_2 \qquad \textbf{(D)}$$

Scheme 2.43

$$\text{Me}_3\text{Si-SiAr}_2\text{-SiMe}_3 + \text{MeLi} \xrightarrow{\text{THF/Et}_2\text{O}} \text{Me}_3\text{Si-SiAr}_2\text{Li} + \text{Me}_4\text{Si} \qquad \textbf{(A)}$$

$$(\text{Me}_3\text{Si})_4\text{Si} + \text{MeLi} \xrightarrow{\text{THF/Et}_2\text{O}} (\text{Me}_3\text{Si})_3\text{SiLi} + \text{Me}_4\text{Si} \qquad \textbf{(B)}$$

$$[(\text{PhMe}_2\text{Si})_2\text{MeSi}]_2\text{Hg} + 2\text{Li} \xrightarrow{\text{toluene}} 2(\text{PhMe}_2\text{Si})_2\text{MeSiLi} + \text{Hg} \qquad \textbf{(C)}$$

$$(\text{Me}_3\text{Si})_3\text{Si-Si}(\text{SiMe}_3)_3 + 2\text{Li} \xrightarrow{\text{THF}} 2(\text{Me}_3\text{Si})_3\text{SiLi} \qquad \textbf{(D)}$$

$$cyclo\text{-}(\text{Ph}_2\text{Si})_4 + 2\text{Li} \xrightarrow{\text{THF}} \text{Li}(\text{Ph}_2\text{Si})_4\text{Li} \qquad \textbf{(E)}$$

$$cyclo\text{-}(\text{Ar}_2\text{Si})_3 + 2\text{Li} \xrightarrow{\text{dioxane}} \text{LiAr}_2\text{Si-Ar}_2\text{Si-SiAr}_2\text{Li} \qquad \textbf{(F)}$$
$$[\text{Ar} = 2\text{-}(\text{Me}_2\text{NCH}_2)\text{C}_6\text{H}_4]$$

$$\text{Ph}_3\text{Si-Ph}_2\text{SiCl} + 2\text{Li} \xrightarrow{\text{THF}} \text{Ph}_3\text{Si-Ph}_2\text{SiLi} + \text{LiCl} \qquad \textbf{(G)}$$

Scheme 2.44

2.3.2.3. (Silyl)silyl Anions

1. Reduction of the Si–Si bond of oligosilanes with organo- or silyllithium reagents (Scheme 2.44, **A, B**).[189, 190]
2. Transmetalation of silylmercury compounds with Li (Scheme 2.44, **C**).[191]
3. Reduction of the Si–Si bond of oligosilanes with Li (Scheme 2.44, **D–F**).[192–194]
4. Reduction of chlorooligosilanes with Li (Scheme 2.44, **G**).[195]

2.3.2.4. Silyl Anions with Functional Groups
(Hydrido)silyllithiums:

1. Reduction of chlorosilanes with Li (Scheme 2.45, **A**).[196]
2. Reduction of the Si–Si bond of oligosilanes with Li (Scheme 2.45, **B**).[197]

$$\text{HMes}_2\text{SiCl} + 2\text{Li} \xrightarrow{\text{THF}} \text{HMes}_2\text{SiLi} + \text{LiCl} \qquad \textbf{(A)}$$

$$\text{Ph}_2\text{HSi-SiHPh}_2 + 2\text{Li} \xrightarrow{\text{THF}} 2\text{Ph}_2\text{HSiLi} \qquad \textbf{(B)}$$

Scheme 2.45

(Halo)silyllithiums (silylenoids):
Reduction of 1,1-dihalosilanes with Li (Scheme 2.46).[198]

Silylenoids $\text{R}_2\text{Si}(\text{Li})\text{Cl}$ are unstable and easily undergo α-elimination of LiCl to form $\text{R}_2\text{Si:}$ or self-condensation to form $\text{R}_2\text{Si}=\text{SiR}_2$, then oligomeric substances.

$$R_2SiCl_2 + 2Li \xrightarrow{THF} [R_2Si(Li)Cl] + LiCl$$

Scheme 2.46

(Amino)silyllithiums:

1. Reduction of (amino)chlorosilanes with Li (Scheme 2.47, **A**).[199]
2. Transmetalation of (amino)silylstannanes with *n*-BuLi or *t*-BuLi (Scheme 2.47, **B**).[179a]

$$(Et_2N)_nPh_{3-n}SiCl + 2Li \xrightarrow{THF} (Et_2N)_nPh_{3-n}SiLi + LiCl \quad [n = 1,2] \qquad \textbf{(A)}$$

$$(Et_2N)_nPh_{3-n}Si\text{-}SnMe_3 + n\text{-}BuLi \xrightarrow{THF} (Et_2N)_nPh_{3-n}SiLi + Me_3Sn\text{-}n\text{-}Bu \quad [n = 1,2] \qquad \textbf{(B)}$$

Scheme 2.47

(Alkoxy)silyllithiums:

1. Reduction of (alkoxy)chlorosilanes with LDMAN (LDMAN = lithium 1-(dimethylamino)naphthalenide) (Scheme 2.48, **A**).[200]
2. Transmetalation of (alkoxy)silylstannanes with *n*-BuLi (Scheme 2.48, **B**).[201]

$$(RO)_nPh_{3-n}SiCl + 2LDMAN \xrightarrow{THF} (RO)_nPh_{3-n}SiLi + LiCl \qquad \textbf{(A)}$$

$$(t\text{-}BuO)Ph_2Si\text{-}SnMe_3 + n\text{-}BuLi \xrightarrow{THF} (t\text{-}BuO)Ph_2SiLi + Me_3Sin\text{-}Bu \qquad \textbf{(B)}$$

Scheme 2.48

2.3.2.5. Cyclic Anions (Silole and Germole Anions and Dianions)

Silole Monoanions:

Reductive cleavage of the Si–Si bond of bis(siloles) with alkali metals resulted in the formation of silole monoanions (Scheme 2.49).[202] Silole monoanions were found to be aromatic on the basis of NMR spectral data and calculations.

Silole and Germole Dianions:

Reduction of 1,1-dihalosiloles or -germoles with Li or K (Scheme 2.50, **A, B**).[203–206]

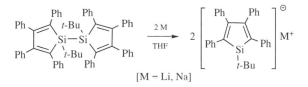

[M = Li, Na]

Scheme 2.49

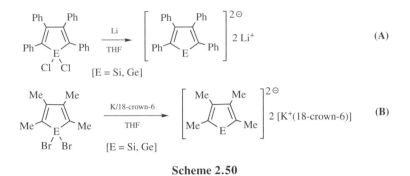

Scheme 2.50

The tetraphenylsilole dianion (Scheme 2.50, **A**, E = Si) has a unique crystal structure, in which one of the Li atoms is η^5-coordinated to the ring, whereas the other Li atom is η^1-bonded to the Si atom.[203] The corresponding germole dianion (Scheme 2.50, **A**, E = Ge) crystallized in two different forms depending on the crystallization temperature. The compound obtained upon crystallization from dioxane at −20°C exhibited a structure with both Li atoms being η^5-coordinated to the germole ring, whereas the compound crystallized at room temperature has one Li atom η^1-coordinated to the Ge atom and the other Li atom η^5-coordinated to the germole ring.[204] The tetramethylsilole[205] and -germole[206] dianions (Scheme 2.50, **B**) possess the theoretically predicted η^5–η^5-structure with the two K atoms situated above and below the five-membered ring. All these silole and germole dianion species are aromatic according to the X-ray and NMR spectral data and are supported by theoretical calculations.

2.3.2.6. Germyl, Stannyl, and Plumbyl Anions   The preparative methods for the synthesis of the germyl, stannyl, and plumbyl anions are essentially the same as those mentioned above for the silyl anions. The most widely used methods are: (1) reduction of halides R_3EX (R = alkyl, aryl; E = Ge, Sn, Pb; X = Cl, Br) with alkali metals; and (2) reductive cleavage of the E–E bond of $R_3E\text{-}ER_3$ (R = alkyl, aryl; E = Ge, Sn, Pb) with alkali metals or organolithium reagents.[179h] Due to the favorable polarization of the $E^{\delta-}-H^{\delta+}$ (E = Ge, Sn, Pb) bond, the direct metalation of hydrides R_3EH is possible with a variety of reducing reagents (*n*-BuLi, *t*-BuLi, *i*-Pr$_2$NLi, NaH, PhCH$_2$K, (Me$_3$Si)$_2$NK, etc.) (Scheme 2.51, **A**−**C**).[207–209]

$$\text{Mes}_3\text{GeH} + t\text{-BuLi} \xrightarrow{\text{THF}} \text{Mes}_3\text{GeLi} \qquad \textbf{(A)}$$

$$\text{Ph}_2\text{HGe-GeHPh}_2 + 2\,t\text{-BuLi} \xrightarrow{\text{THF}} \text{Ph}_2\text{LiGe-GePh}_2\text{Li} \qquad \textbf{(B)}$$

$$\text{Me}_3\text{SnH} + i\text{-Pr}_2\text{NLi} \xrightarrow{\text{THF}} \text{Me}_3\text{SnLi} \qquad \textbf{(C)}$$

Scheme 2.51

2.3.3. Structure

The most fundamental issues of the structures of heavier group 14 element-centered anionic derivatives R_3EM (R = alkyl, aryl, silyl; E = Si, Ge, Sn, Pb; M = alkali or alkaline earth metals) turned out to be the questions of their aggregation states (monomeric, dimeric, or oligomeric), nature of the E–M bond (covalent or ionic), and configuration of the anionic centers E (tetrahedral, pyramidal, or planar).[179a,b,f,g,h] The most important experimental techniques that are widely used to clarify these questions are NMR spectroscopy and X-ray diffraction analysis.

2.3.3.1. NMR Spectroscopy   NMR spectroscopy provides a major amount of the structural information of anionic species in solution, playing the analogous key role to ESR spectroscopy in the structural investigations of radical species. The observation of the scalar $^{29}Si-^{7}Li(^{6}Li)$ coupling in both ^{29}Si and $^{7}Li(^{6}Li)$ NMR spectra is typically interpreted in terms of the significant covalent contribution to the Si–Li bonding. Thus, the $^{29}Si-^{7}Li$ coupling in the ^{29}Si NMR spectrum of the solution of $(Me_3Si)_3SiLi(thf)$ in $C_7H_8-C_6D_6$ resulted in the room temperature observation of the well-resolved quartet for the central anionic Si atom at –189.4 ppm with the coupling constant ^{1}J ($^{29}Si-^{7}Li$) = 38.6 Hz.[210] This indicates that $(Me_3Si)_3SiLi$ is monomeric in solution, existing as a CIP in which the Si–Li bond is partially covalent. Other alkali metal derivatives $(Me_3Si)_3SiM$ (M = K, Rb, Cs) exhibited the similar highly upfield shifted values of the ^{29}Si NMR resonances for the central Si atom: -185.7, -184.4, and -179.4 ppm, respectively.[211] $Ph_3Si^{6}Li$ also exists in 2-Me–THF solution as a CIP, which was evident from its ^{29}Si NMR spectrum at 173 K exhibiting a triplet at -9 ppm with ^{1}J ($^{29}Si-^{6}Li$) = 17 Hz.[212] In other words, triphenylsilyllithium retains its Si–Li bonding in solution and is monomeric. Similarly, disilanyllithium $Me_3Si-Me_2Si^{6}Li$ in THF-d_8 solution at 180 K revealed a triplet resonance for the anionic Si atom at -74.9 ppm due to the coupling with one ^{6}Li nucleus ($I = 1$) with a coupling constant of 18.8 Hz.[213] This implies the monomeric aggregation state of $Me_3Si-Me_2Si^{6}Li$ in THF solution and partial covalent character of the Si–Li bond. 1,2–Dilithiodisilane $Li[2-(Me_2NCH_2)C_6H_4]_2Si-Si[(2-(Me_2NCH_2)C_6H_4]_2Li$ is also monomeric in THF at room temperature, retaining the Si–Li bonding in solution, as was demonstrated by the observation of the ^{29}Si NMR resonance of the anionic Si atoms as a quartet at -32.8 ppm (^{1}J ($^{29}Si-^{7}Li$) = 36 Hz).[194] For the group 14 elements heavier than Si, the degree of covalency of the E–Li bonds (E = group 14 element) was discussed to decrease in the order Ge > Sn > Pb with a considerable amount of covalent character for the Ge–Li bond.[214] In some cases, the Sn–Li bonds were also characterized to be predominantly covalent. Thus, for example, n-Bu_3SnLi exhibited a quartet resonance in the ^{119}Sn NMR spectrum in diethyl ether at 154 K due to coupling with the ^{7}Li nucleus (^{1}J ($^{119}Sn-^{7}Li$) = 402.5 Hz), implying a significant covalent contribution to the Sn–Li bond.[215]

2.3.3.2. X-Ray Crystallography   The crystal structure information on the heavier group 14 elements centered anions available in the scientific literature at the present time is vast. Therefore, in this section we will deal only with the most representative examples of such species. The major part of these structural data concerns the derivatives of silyllithium, for which the most common bonding situation was described as the one with a pyramidal Si anionic center and Si–Li bonding distance close to the sum of the covalent radii of Si and Li atoms (2.69 Å). The first ever structurally characterized silyllithium derivative was Me_3SiLi, which formed a hexamer $(Me_3SiLi)_6$ in the crystalline form.[216] The structure of the compound is represented by an Li_6 six-membered ring in the severely folded chair conformation, which otherwise would be better regarded as a strongly distorted octahedron (Figure 2.4).

The Si–Li bonds have an average value of 2.68 Å. Interestingly, upon complexation with TMEDA the hexameric structure was destroyed to form a dimer $(Me_3SiLi)_2 \cdot (TMEDA)_3$ instead with Si–Li bond lengths of 2.70 Å.[217] The crystal structure of one of the most readily available and commonly used silyllithium derivatives, Ph_3SiLi, was determined only in 1993. The solvated species $Ph_3SiLi \cdot (thf)_3$ exhibited a strong pyramidality at the Si atom and Si–Li distance of 2.672(9) Å.[218] The particular structures of some linear and branched oligosilyllithium derivatives also deserve the following considerations. The 1,4-dilithiooctaphenyltetrasilane (as a THF solvate) $(thf)_3 \cdot Li(Ph_2Si)_4Li \cdot (thf)_3$ was found to be monomeric in a planar zigzag-shaped conformation with Si–Li bond lengths of 2.714(10) Å.[219] The solvent-free pentamethyldisilanyllithium $(Me_3SiMe_2SiLi)_4$ is tetrameric in the crystalline form with the four Li atoms occupying the vertices of the Li_4 tetrahedron and Si–Li bonds of 2.683 Å.[220] The crystal structure of $(Me_3Si)_3SiLi \cdot (thf)_3$ was determined as monomeric with the normal Si–Li bond distance of 2.669(13) Å and distorted tetrahedral configuration of the Si atom.[210,218] In contrast, the structures of both Rb and Cs trimethylsilyl derivatives (as toluene solvates) are dimeric with the alkali metals playing the bridge role: $[(Me_3Si)_3SiRb]_2 \cdot (C_7H_8)$ and $[(Me_3Si)_3SiCs]_2 \cdot (C_7H_8)_3$.[211]

The first structurally characterized germyllithium derivative was $(Me_3Si)_3GeLi \cdot (donor)$ (donor = THF or PMDETA), exhibiting Ge–Li bond distances of 2.665(6) Å (THF donor) or 2.651(9) Å (PMDETA donor), which are slightly longer than the sum of the covalent radii of Ge and Li atoms (2.56 Å).[221] The first crystal structure of a stannyllithium derivative was reported for $Ph_3SnLi \cdot (PMDETA)$ with an Sn–Li bond of 2.817(7) Å, which is a bit longer than the sum of the covalent radii of Sn and Li atoms (2.74 Å).[222] The central Sn atom exhibited a pyramidal geometry (288°), which became even more pronounced in the corresponding Pb derivative $Ph_3PbLi \cdot (PMDETA)$ (283°).[223] The Pb–Li bond length in the latter compound is

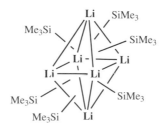

Figure 2.4

also insignificantly longer than the sum of the covalent radii of Pb and Li atoms: 2.858 Å versus 2.81 Å.

The structural information on the alkaline earth metal derivatives of the heavier group 14 elements is markedly limited. The first silylmagnesium compound to be structurally characterized was $(Me_3Si)_2Mg \cdot (DME)$, which is monomeric with distorted tetrahedral geometry of the Mg atom and Si–Mg bond length of 2.630(2) Å.[224] The "silicon Grignard reagent" $[Me_3SiMgBr \cdot (TMEDA)]_2$ forms a centrosymmetric dimer with two bridging Br atoms. One of the two Mg–Br distances (2.534 Å) is short, within the normal range for Grignard reagents, whereas the other one (3.220 Å) is longer than the sum of the ionic radii of Mg^{2+} and Br^-, indicating that the dimeric interaction is weak.[225] The first symmetrical organogermylmagnesium derivative for which its crystal structure was reported was $(Me_3Ge)_2Mg \cdot (DME)_2$ with Ge–Mg bond length of 2.7 Å.[226] The tin derivative of alkaline earth metals, $(Me_3Sn)_2Ca \cdot (thf)_4$, has a distorted octahedral configuration of a Ca atom bonded to the two Sn atoms and four O atoms of THF ligands with an Sn–Ca bond length of 3.2721(3) Å.[227]

2.3.4. Reactions and Synthetic Applications

The reactions of the heavier group 14 elements alkali and alkaline earth metal derivatives are to a large extent similar to those of the classical organolithium reagents. Thus, one of the most important synthetic applications of silyllithium derivatives is the substitution reaction involving the exchange of halogens (or other good leaving groups, for example, triflates) with silyl groups to form new Si–C, Si–Si, and other bonds:

$$R_3SiLi + R'X \rightarrow R_3SiR' + LiX$$

The most commonly used reagents in synthetic organosilicon chemistry are Ph_3SiLi and $(Me_3Si)_3SiLi$. These compounds and many other silyllithium reagents have been widely utilized in a variety of useful transformations: (1) S_N2 reaction with organic halides to give the corresponding substitution products; (2) 1,2-addition to C=C double bonds to form the silyl-substituted alkanes; (3) 1,2-addition to C≡C triple bonds to produce the silyl-substituted alkenes; and (4) 1,2-addition to carbonyl groups >C=O to form, after aqueous workup, the silyl-substituted alcohols. The reactivity of silylmagnesium derivatives has been far less studied; however, in a majority of cases it is the same as that of silyllithium derivatives resulting in the formation of identical products. The alkali metal derivatives of the group 14 elements that are heavier than Si (Ge, Sn, Pb) react in essentially the same way as silyllithium compounds. They also undergo nucleophilic substitution reactions with a variety of alkyl halides to form effectively group 14 element–carbon bonds. The substitution of aromatic acyl chlorides with germyl anions provides a straightforward route to α-germylketones, useful organometallic reagents. Reaction of germyllithium derivatives with aldehydes and ketones resulted in the formation of germyl-substituted alcohols, whereas their reaction with nitriles produced 3-germa-β-diketiminates. The anionic derivatives of the heavier group 14 elements are prone to SET (single

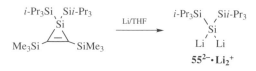

Scheme 2.52

electron transfer) reactions involving intermediate formation of the ion radical or radical species, due to their high-lying HOMOs from which one electron could be easily removed. All the above-mentioned types of reactions of the heavier group 14 element-centered anions have been described in detail in previous reviews.[179a,b,h]

2.3.5. Recent Developments

2.3.5.1. sp³-Anions
A great variety of acyclic anionic compounds of the type RR′R″EM or (RR′R″E)$_2$M (R, R′, R″ = alkyl, aryl, silyl; E = Si, Ge, Sn, Pb; M = Li, Na, K, Rb, Cs, Mg, Ca, Sr, Ba) have been prepared over past years. Due to the limited space of the present review we are unable to cover them all, so we will focus on the most interesting and instructive examples.

One such achievement that should be definitely recognized is the synthesis of the stable 1,1-dilithiosilane and 1,1-dilithiogermane derivatives R$_2$ELi$_2$ (R = t-Bu$_2$MeSi, i-Pr$_3$Si; E = Si, Ge). The dilithiosilane $\mathbf{55^{2-} \cdot Li_2^+}$ derivatives were prepared by a rather unusual reaction of the corresponding silacyclopropenes with metallic lithium in THF, which involves the breaking of the two endocyclic Si–C bonds of the starting silacyclopropene (Scheme 2.52).[228]

The anionic Si atom is nearly tetrahedral, being bonded to the two Si atoms of substituents and two Li atoms. The resonance of this Si anionic center in the ^{29}Si NMR spectrum of $\mathbf{55^{2-} \cdot ^6Li_2^+}$ was observed at a very high field (-292.0 ppm) as a quintet due to coupling with the 2 equivalent ^{6}Li atoms ($I = 1$) with a coupling constant of 15.0 Hz. This chemical shift is largely shifted upfield compared with that of (Me$_3$Si)$_3$SiLi (-189.4 ppm),[210] because of the sharply increased electron density

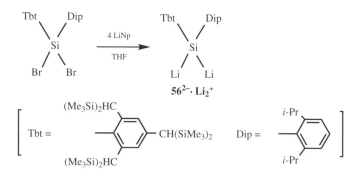

Scheme 2.53

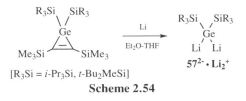

$[R_3Si = i\text{-}Pr_3Si, t\text{-}Bu_2MeSi]$

Scheme 2.54

on the central Si atom. Another derivative of 1,1-dilithiosilane $56^{2-} \cdot Li_2^+$ was prepared by Tokitoh and co-workers in the same year, 1999, by the reduction of the very crowded 1,1-dibromosilane Tbt(Dip)SiBr$_2$ (Tbt = 2,4,6-tris[bis(trimethylsilyl)methyl]phenyl, Dip = 2,6-diisopropylphenyl) with excess lithium naphthalenide (Scheme 2.53).[229] The formation of $56^{2-} \cdot Li_2^+$ was undoubtedly proved by trapping reactions with a variety of electrophiles; however, its isolation was precluded due to unavoidable intramolecular Li–H exchange reaction leading to the isomerization product.

A heavier analog of the 1,1-dilithiosilanes, 1,1-dilithiogermane derivative $57^{2-} \cdot Li_2^+$, was synthesized recently employing the same synthetic approach: reduction of germacyclopropenes with metallic lithium (Scheme 2.54).[230] In contrast to the monomeric dilithiosilane $55^{2-} \cdot [Li^+(thf)]_2$,[228] dilithiogermane has an unusual dimeric structure in the crystalline form, $\{57^{2-} \cdot [Li^+(thf)_2]\}_2$, in which both Ge atoms are not tetrahedral, but pentacoordinated, being bonded to the two Si atoms of substituents and three Li atoms.

Apart from their attractive and unique structures, both dilithiosilane $55^{2-} \cdot Li_2^+$ and dilithiogermane $57^{2-} \cdot Li_2^+$ were found to be extremely useful synthons for the straightforward preparation of a wide range of new multiply bonded as well as cyclic and polycyclic compounds comprising heavier group 14 elements. Thus, they have been successfully utilized for the synthesis of the novel heavy analogs of alkenes: disilenes and germasilenes $(t\text{-}Bu_2MeSi)_2Si=EAr_2$ (E = Si, Ge; Ar = Mes, Tip),[231] silastannene $(t\text{-}Bu_2MeSi)_2Si=SnTip_2$,[232] germastannene $(t\text{-}Bu_2MeSi)_2Ge=SnTip_2$,[233] 1,3-disila-2-gallata- and -indataallenic anionic compounds $[(t\text{-}Bu_2MeSi)_2Si=M=Si(SiMet\text{-}Bu_2)_2]^- \cdot [Li^+(thf)_4]$ (M = Ga, In),[234] sila- and germaborenes $(t\text{-}Bu_2MeSi)_2Si=BMes$[235] and $(t\text{-}Bu_2MeSi)_2Ge=BMes$,[235] and heavy analogs of cyclopropenes tetrakis(di-*tert*-butylmethylsilyl)cyclotrisilene,[236] tetrakis(di-*tert*-butylmethylsilyl)cyclotrigermene,[237] and tetrakis(di-*tert*-butylmethylsilyl)-1*H*-siladigermirene.[237]

A family of tricoordinated persilyl substituted anions of the heavier group 14 elements $(R_3Si)_3E^-$ (R = alkyl; E = Si, Ge, Sn) was synthesized in recent years. Sekiguchi and co-workers employed the straightforward one-electron reduction of the corresponding radicals $(t\text{-}Bu_2MeSi)_3E\bullet$ **43–45** with metallic Li or K to form anionic derivatives $(t\text{-}Bu_2MeSi)_3E^- \cdot M^+$ (E = Si, Ge, Sn; M = Li, K) **58**$^- \cdot M^+$ (Scheme 2.55).[238]

It is interesting that the structural diversity of **58**$^- \cdot M^+$ depends on the polarity of the solvating medium. Thus, in nonpolar hexane both silyl- and germyllithium derivatives **58a**$^- \cdot Li^+$ (E = Si) and **58b**$^- \cdot Li^+$ (E = Ge) adopt the nonsolvated monomeric structures, in which the central anionic atom (Si or Ge) was nearly planar and E–Si bonds were appreciably shorter than those of the neutral compounds $(t\text{-}Bu_2MeSi)_3EH$

$$(t\text{-}Bu_2MeSi)_3E \cdot \xrightarrow[\text{solvent}]{M} (t\text{-}Bu_2MeSi)_3E^- \cdot M^+$$

$$\textbf{58}^- . \textbf{M}^+$$

[E = Si, Ge, Sn; M = Li, K; solvent = hexane, heptane, benzene, THF]

Scheme 2.55

due to hyperconjugation between the $p_z(E) - \sigma^*$(adjacent Si–C bonds) orbitals.[238a] It should be noted that typically silyl- and germyllithiums have tetrahedral rather than planar geometry, and they are usually solvated by the coordinating solvents (THF).[179] The characteristic planarity of **58a,b**$^-$**•Li**$^+$ is to be attributed to the combined influence of both steric and electronic factors: bulkiness of the t-Bu$_2$MeSi substituents and intramolecular Li•••CH$_3$(t-Bu groups) agostic interaction.[238a]

The reduction of the stannyl radical $(t\text{-}Bu_2MeSi)_3Sn\bullet$ with alkali metals produces a variety of structural modifications depending on the solvent used (Scheme 2.55).[238b] Thus, in nonpolar heptane, a dimeric stannyllithium species [**58c**$^-$**•Li**$^+$]$_2$ (E = Sn) was formed, whereas in more polar benzene, the monomeric pyramidal structure **58c**$^-$**•[η^6-Li**$^+$**(C$_6$H$_6$)]** was produced. In the latter compound the Li$^+$ ion was covalently bonded to the anionic Sn atom being at the same time η^6-coordinated to the benzene ring. A similar monomeric pyramidal CIP **58c**$^-$**•[Li**$^+$**(thf)$_2$]** was prepared by reduction in polar THF; the addition of [2.2.2]cryptand to this compound resulted in the isolation of the "free" stannyl anion **58c**$^-$**•{K**$^+$**([2.2.2]cryptand)}**, in which the K$^+$ ion lacked its bonding to the Sn atom.[238b]

Among other interesting persilyl substituted anions of the type (R$_3$Si)$_3$E$^-$ (R = Me, Ph; E = Si, Ge, Pb) one can mention tris(trimethylsilyl)silanide (Me$_3$Si)$_3$Si$^-$ and -germanide (Me$_3$Si)$_3$Ge$^-$ as well as tris(triphenylplumbyl)plumbide (Ph$_3$Pb)$_3$Pb$^-$. The chemistry of the first two species was extensively studied by the group of Ruhland–Senge.[239] Thus, the alkali metal derivatives of (Me$_3$Si)$_3$Si$^-$ anion **59**$^-$**•M**$^+$ (**M = K, Rb, Cs**) were synthesized by the intermetallic exchange reaction between the readily available (Me$_3$Si)$_3$SiLi and alkali metal $tert$-butoxides t-BuOM (Scheme 2.56).[239a]

Similarly, the alkaline earth metal derivatives of (Me$_3$Si)$_3$Si$^-$ anion [**59**$^-$]$_2$**•M**$^{2+}$ (**M = Ca, Sr, Ba**)[239b] and (Me$_3$Si)$_3$Ge$^-$ anion [**60**$^-$]$_2$**•M**$^{2+}$ (**M = Ca, Sr, Ba**)[239c] were prepared by the metal exchange reaction between (Me$_3$Si)$_3$SiK (or (Me$_3$Si)$_3$GeK) and alkaline earth metal iodides MI$_2$ (Scheme 2.57). All of the above-mentioned compounds of the types **59**$^-$**•M**$^+$, [**59**$^-$]$_2$**•M**$^{2+}$, and [**60**$^-$]$_2$**•M**$^{2+}$ exhibited a highly pronounced pyramidal geometry around the anionic Si- and Ge-centers.

The anionic lead derivative [(Ph$_3$Pb)$_3$Pb]$^-$•[MgBr]$^+$ (**61**$^-$**•[MgBr]**$^+$) was surprisingly formed upon the reaction of PbBr$_2$ with PhMgBr (Scheme 2.58).[240] The crystal

$$(Me_3Si)_3SiLi + t\text{-}BuOM \xrightarrow{THF} (Me_3Si)_3SiM$$

$$[M = K, Rb, Cs] \qquad\qquad \textbf{59}^- . \textbf{M}^+$$

Scheme 2.56

$$(Me_3Si)_3EK \ + \ MI_2 \ \xrightarrow{\text{THF}} \ [(Me_3Si)_3E]_2M$$

$$[E = Si, Ge; M = Ca, Sr, Ba] \quad [\mathbf{59^-}]_2 \cdot \mathbf{M^{2+}} \ (\text{for } E = Si)$$
$$[\mathbf{60^-}]_2 \cdot \mathbf{M^{2+}} \ (\text{for } E = Ge)$$

Scheme 2.57

structure analysis of $\mathbf{61^-} \cdot [\mathbf{MgBr(thf)_5}]^+$ revealed a Pb_4 trigonal-pyramidal configuration of the molecule with the electron pair occupying the apex position, resulting in the large pyramidalization at the central anionic Pb atom (279°).

$$PbBr_2 \ + \ PhMgBr \ \xrightarrow{\text{THF}} \ (Ph_3Pb)_3Pb \ \cdot MgBr^+$$
$$\mathbf{61^-} \cdot \mathbf{MgBr^+}$$

Scheme 2.58

Remarkable progress has been achieved in the field of oligosilyllithium and - potassium derivatives in recent years by the efforts of the research groups of Apeloig and Marschner. Apeloig's group typically employed for the generation of the highly branched oligosilyllithiums by the reduction of silylmercury derivatives with metallic Li in either THF or hexane[241] to produce a variety of anionic products: $(Me_3Si)_3Si-SiMe_2-SiMe_2-Si(SiMe_3)_2Li(thf)_3$,[241a] $(Me_3SiMe_2Si)_3SiLi(thf)_3$ and its dimer $[(Me_3SiMe_2Si)_3SiLi]_2$,[241b] $[(thf)_2Li(i-Pr_3Si)_2Si]_2Hg$,[241c] unsolvated 1,1-dilithiosilane aggregate $\{[(t-Bu_2MeSi)_2SiLi_2][(t-Bu_2MeSi)_2HSiLi_2]\}$, and hydridosilyllithium dimer $[(t-Bu_2MeSi)_2HSiLi]_2$.[241d]

Marschner utilized a different synthetic approach: generation of the oligosilylpotassium derivatives by the cleavage of Si–Si bonds of oligosilanes with t-BuOK.[242] This allowed the preparation of the linear and cyclic mono- and dipotassium derivatives of oligosilanes: $(Me_3Si)_3Si-Si(SiMe_3)_2K(C_7H_8)$,[242a,b] $(Me_3Si)_2PhSi-Si(SiMe_3)_2K[(18-crown-6)(C_6H_6)]$,[242b] $(18\text{-crown-}6)K(Me_3Si)_2Si-Si(SiMe_3)_2K(18\text{-crown-}6)$,[242c] $(18\text{-crown-}6)K(Me_3Si)PhSi-Si(SiMe_3)_2K(18\text{-crown-}6)$,[242c] $cyclo\text{-}[(18\text{-crown-}6)K(Me_3Si)Si-Si(SiMe_3)_2\text{-}Si(SiMe_3)K(18\text{-crown-}6)\text{-}Si(SiMe_3)_2]$,[242c] $cyclo\text{-}[(18\text{-crown-}6)K\text{-}(Me_3Si)Si-SiMe_2-SiMe_2-Si(SiMe_3)K(18\text{-crown-}6)\text{-}SiMe_2-SiMe_2]$,[242d] $(18\text{-crown-}6)K-Si(SiMe_3)_2-(SiMe_2)_n-Si(SiMe_3)_2K(18\text{-crown-}6)$ ($n = 1–3$),[242e,f] and an interesting β-fluorosilyl anion, disilenoid $F(Me_3Si)_2Si-Si(SiMe_3)_2K(18\text{-crown-}6)$.[242g]

2.3.5.2. sp²-Anions The first stable silicon analogs of vinyllithium were reported only a couple of years ago. The first species, $Tip_2Si=Si(Tip)Li$ $\mathbf{62^-} \cdot \mathbf{Li^+}$, was originally suggested as the key reactive intermediate in the reduction of $Tip_2Si=SiTip_2$ with metallic lithium by Weidenbruch et al. in 1997.[243] The isolation and structural characterization of this compound was accomplished by the direct reaction of Tip_2SiCl_2 with metallic Li by Scheschkewitz in 2004 (Scheme 2.59).[244]

$$2 \ Tip_2SiCl_2 \ + \ 6 \ Li \ \xrightarrow{DME} \ Tip_2Si{=}Si(Tip)Li$$

$$\mathbf{62^-\cdot Li^+}$$

Scheme 2.59

The doubly bonded Si atoms resonated at 94.5 and 100.5 ppm, being shifted to a low field compared with those of disilene $Tip_2Si{=}SiTip_2$ (53.4 ppm). In the solid state $\mathbf{62^-\cdot[Li^+(dme)_2]}$ exists as a CIP with the Si–Li bonding distance of 2.853(3) Å. The Si=Si double bond length in $\mathbf{62^-\cdot[Li^+(dme)_2]}$ is 2.192(1) Å, longer than that of $Tip_2Si{=}SiTip_2$ (2.144 Å).

Also in 2004, the group of Sekiguchi reported the disilenyllithium derivative (t-$Bu_2MeSi)_2Si{=}Si(Mes)Li$ $\mathbf{63^-\cdot Li^+}$, synthesized by the reduction of tetrasila-1,3-butadiene with t-BuLi (Scheme 2.60).[245]

$$(t\text{-}Bu_2MeSi)_2Si{=}Si(Mes)\text{-}Si(Mes){=}Si(SiMet\text{-}Bu_2)_2 \ + \ t\text{-}BuLi \ \xrightarrow{THF} \ 2 \ (t\text{-}Bu_2MeSi)_2Si{=}Si(Mes)Li$$

$$\mathbf{63^-\cdot Li^+}$$

Scheme 2.60

The chemical shifts of the sp^2-Si atoms were observed at 63.1 (Si$_2Si{=}$) and 277.6 (Li(Mes)$Si{=}$) ppm. The Si=Si double bond in $\mathbf{63^-\cdot[Li^+(thf)_3]}$ is slightly twisted (twisting angle 15.2°), and its length is close to that in $\mathbf{62^-\cdot[Li^+(dme)_2]}$: 2.2092(7) Å versus 2.192(1) Å, respectively.

The one-electron reduction of disilene $(t\text{-}Bu_2MeSi)_2Si{=}Si(SiMet\text{-}Bu_2)_2$ with lithium, sodium, and potassium naphthalenides MNp (M = Li, Na, K) also resulted in the formation of disilenide derivatives, $(t\text{-}Bu_2MeSi)_2Si{=}Si(SiMet\text{-}Bu_2)M$ $\mathbf{64a\text{-}c^-\cdot M^+}$ (a: M = Li; b: M = Na; c: M = K) (Scheme 2.61).[246] Similar to the above case,[245] all metal-substituted sp^2-Si atoms diagnostically resonate at a very low field: 328.4 (for $\mathbf{64a^-}$), 325.6 (for $\mathbf{64b^-}$), and 323.1 (for $\mathbf{64c^-}$) ppm. Both sp^2–Si atoms in $\mathbf{64a^-\cdot[Li^+(thf)_2]}$ revealed planar geometries around them, and the Si=Si double bond length was determined to be 2.1983(18) Å. This value is similar to those of the above-described disilenyllithiums $\mathbf{62^-\cdot[Li^+(dme)_2]}$ (2.192(1) Å) and $\mathbf{63^-\cdot[Li^+(thf)_3]}$ (2.2092(7) Å), and is ca. 0.06 Å shorter than that of the starting (t-$Bu_2MeSi)_2Si{=}Si(SiMet\text{-}Bu_2)_2$. The disilenide derivatives $\mathbf{64a\text{-}c^-\cdot M^+}$ were believed

$$(t\text{-}Bu_2MeSi)_2Si{=}Si(SiMet\text{-}Bu_2)_2 \ + \ MNp \ \xrightarrow{THF} \ M(t\text{-}Bu_2MeSi)_2Si\text{-}Si(SiMet\text{-}Bu_2)_2M$$

$$[M = Li, Na, K]$$

$$\xrightarrow[\text{-}t\text{-}Bu_2MeSiM]{benzene} \ (t\text{-}Bu_2MeSi)_2Si{=}Si(SiMet\text{-}Bu_2)M$$

$$\mathbf{64a\text{-}c^-\cdot M^+}$$

Scheme 2.61

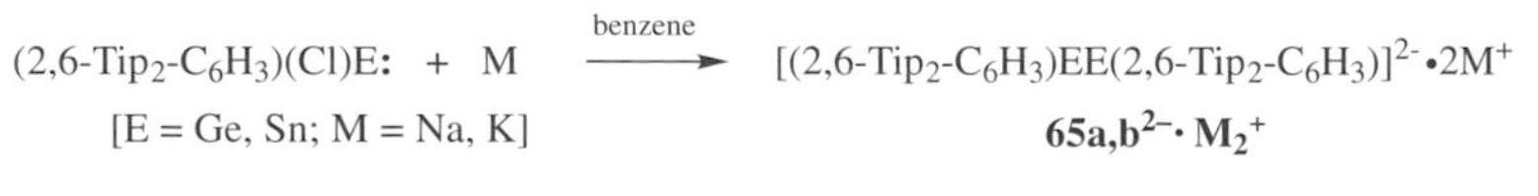

$$(2,6\text{-Tip}_2\text{-C}_6\text{H}_3)(\text{Cl})\text{E:} \; + \; \text{M} \; \xrightarrow{\;\text{benzene}\;} \; [(2,6\text{-Tip}_2\text{-C}_6\text{H}_3)\text{EE}(2,6\text{-Tip}_2\text{-C}_6\text{H}_3)]^{2-}\cdot 2\text{M}^+$$

$$[\text{E = Ge, Sn; M = Na, K}] \qquad\qquad \textbf{65a,b}^{2-}\cdot \textbf{M}_2^+$$

Scheme 2.62

to be produced through the intermediate formation of 1,2-dianions $(t\text{-Bu}_2\text{MeSi})_2\text{MSi}$–$\text{SiM}(\text{SiMe}t\text{-Bu}_2)_2$ followed by the β-elimination of $t\text{-Bu}_2\text{MeSiM}$ (Scheme 2.61).[246]

Power recently reported a rather unusual reduction of halides [2,6-Tip$_2$-C$_6$H$_3$](Cl)E: (E = Ge, Sn) with alkali metals M (M = Na or K), resulting in the formation of the doubly reduced species [2,6-Tip$_2$-C$_6$H$_3$]EE[2,6-Tip$_2$-C$_6$H$_3$]$^{2-}\cdot$2M$^+$ **65a,b**$^{2-}\cdot$**M**$_2^+$ (a: E = Ge, M = Na; b: E = Sn, M = K), in which the two heavier group 14 elements E are formally doubly bonded (Scheme 2.62).[247]

Both **65a,b**$^{2-}\cdot$**M**$_2^+$ have a *trans*-configuration of the aryl substituents. The Ge–Ge distance of 2.3943(13) Å in **65a**$^{2-}\cdot$**Na**$_2^+$ is comparable with those of the typical Ge=Ge double bonds, and the Sn–Sn distance of 2.7663(9) Å in **65b**$^{2-}\cdot$**K**$_2^+$ is also in the range of the normal Sn=Sn double bonds. Accordingly, the Ge–Na distance of ca. 3.1 Å is longer than the sum of the covalent radii of Ge and Na (2.86 Å), and the Sn–K distances of 3.579(2) and 3.591(2) Å are longer than the sum of the covalent radii of Sn and K (3.36 Å).

2.3.5.3. Cyclic and Polycyclic Anions

Several remarkable cyclic anionic compounds consisting of heavier group 14 elements have been reported in the scientific literature during the past five years, since the publication of a series of comprehensive reviews on silyl anions chemistry.[179] The first of them, the cyclotetragermanide ion **66**$^-\cdot$**Li**$^+$, was prepared unexpectedly by Weidenbruch et al. during the reduction of tetraaryldigermene Tip$_2$Ge=GeTip$_2$ (Tip = 2,4,6-triisopropylphenyl) with an excess of metallic lithium (Scheme 2.63).[248]

The four-membered ring of **66**$^-\cdot$**[Li**$^+$**(dme)$_3$]** was perfectly planar with two pairs of distinctly different skeletal Ge–Ge bonds of 2.5116(6) and 2.3679(6) Å: the former is typical for the Ge–Ge single bonds, whereas the latter is close to the values of Ge=Ge double bonds. Such crystallographic features led to the formulation of the allylic-type structure of **66**$^-$, in which the negative charge is delocalized over the Ge$_3$ fragment of the four-membered ring.

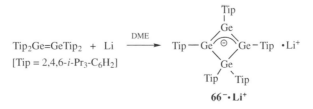

$$\text{Tip}_2\text{Ge}=\text{GeTip}_2 \; + \; \text{Li} \; \xrightarrow{\;\text{DME}\;}$$

$$[\text{Tip = 2,4,6-}i\text{-Pr}_3\text{-C}_6\text{H}_2]$$

66$^-\cdot$**Li**$^+$

Scheme 2.63

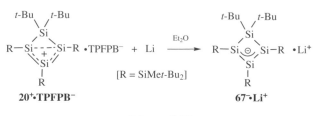

Scheme 2.64

A silicon version of the above compound, cyclotetrasilenide ion **67⁻•Li⁺**, was synthesized by Sekiguchi et al. by the two-electron reduction of cyclotetrasilenylium ion **20⁺•TPFPB⁻** with Li (Scheme 2.64).[249]

The compound **67⁻•[Li(thf)]⁺** has a cyclotetrasilene structure with the Li⁺ ion being directly bonded to one of the skeletal sp³-Si atoms and also coordinated to both doubly bonded Si atoms (η³-coordination). Such coordination causes a characteristic stretching of the Si=Si double bond, whose length (2.2245(7) Å) became intermediate between those of the typical Si—Si single (av. 2.34 Å) and Si=Si double (av. 2.20 Å) bonds. The trihaptocoordination of the Li⁺ ion is retained in the nonpolar solvents (toluene), in which **67⁻•Li⁺** exists as a CIP. However, in polar solvents (THF) the Li⁺ ion coordination to the Si₄ ring is destroyed, resulting in the formation of the SSIP due to the effective solvation of the countercation. The allylic anion structure for **67⁻•Li⁺** was supported by the observation of an extremely deshielded central Si atom of the Si₃ unit (δ = 273.0 ppm).

The latest syntheses of novel four- and five-membered cyclic anions of the heavier group 14 elements (Si, Ge, Sn) provided a fresh look at the question of aromaticity (or nonaromaticity) of the heavier analogs of 6π-electron aromatic systems. Thus, the story of the heavy analogs of cyclopentadienyl anion (Section 2.3.2.5) was continued by the preparation of the stannole dianion by Saito and co-workers.[250] The stannole dianion **68²⁻•Li₂⁺** was prepared by the metalation of the bis(1,1-stannole) with an excess of Li in THF at elevated temperatures;[250a] however, it can be better prepared by the lithiation of hexaphenylstannole with Li in THF under reflux conditions[250b,c] (Scheme 2.65).

In contrast to the case of the silole dianion[203b] and similar to the case of germole dianion[204] (Section 2.3.2.5), stannole dianion **68²⁻•[Li⁺(Et₂O)]₂** has both Li ions above and below the plane being pentahaptocoordinated to the SnC₄ five-membered ring.[250c] The stannole ring is nearly planar and the endocyclic C—C bonds are

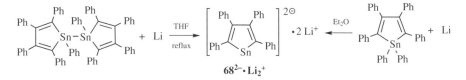

Scheme 2.65

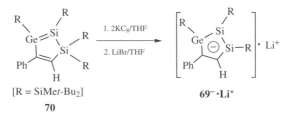

Scheme 2.66

almost equivalent $(1.422(6)-1.446(6)$ Å), thus suggesting some aromatic character for the compound. This was also supported by the observation of the ^{7}Li NMR resonance of $68^{2-} \cdot Li_2^+$ at a high field (-4.36 ppm) as well as by the calculated value of NICS(1) $= -5.96.^{250c}$

The story of the heavy analogs of 6π-electron cyclopentadienyl anions has culminated in the latest synthesis of a compound containing three heavier group 14 elements (two Si and one Ge) in the ring. This anionic species $69^- \cdot Li^+$ was prepared by the reduction of the disilagermacyclopentadiene precursor 70^{251} with potassium graphite KC_8 followed by the exchange of countercation from K^+ to Li^+ by treatment with LiBr (Scheme 2.66).^{252}

The Li^+ ion was found to be pentahaptocoordinated to the five-membered ring. All double bonds of the precursor **70** (Si=Ge and C=C) were stretched upon its reduction to produce $69^- \cdot [Li^+(thf)]$, whereas all single bonds of **70** (Si–Si, Si–C, and Ge–C) were shortened. Such structural trends definitely provide evidence for the delocalization of the negative charge over the five-membered ring of $69^- \cdot [Li^+(thf)]$, which in turn demonstrates its aromaticity. The NICS(1) value calculated at 1 Å above the ring center was typical for aromatic systems (-12.0). The anion $69^- \cdot Li^+$ exhibited an interesting solvent-dependent behavior: it preserves its aromaticity in the nonpolar solvents (toluene) (Figure 2.5, A) and is unable to enjoy aromatic delocalization in the polar solvents (THF) (Figure 2.5, B).^{252} Thus, in THF the coordination mode has been dramatically changed from the η^5- to η^1-mode, in which the anionic part acquires the properties of the localized cyclopentadienide derivative featuring Si=Si and C=C double bonds and having a negative charge on the Ge atom (Figure 2.5, B). Particularly important was the observation of the ^{7}Li NMR resonances of $69^- \cdot Li^+$ in different solvents: a highly upfield shifted value of -5.4 ppm due to the aromatic ring current effects (diagnostic for the classical aromatic cyclopentadienide

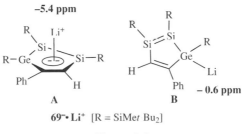

Figure 2.5

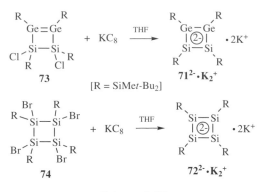

[R = SiMe*t*-Bu₂]

Scheme 2.67

derivatives) in toluene-d_8 (Figure 2.5, A) and normal value of -0.6 ppm (typical for the η^1-germyllithium derivatives) in THF-d_8 (Figure 2.5, B).

In contrast to **69⁻•Li⁺**, which was formulated above as the 6π-electron aromatic compound, other representatives of this class of cyclic compounds, the heavy analogs of cyclobutadiene dianion, were found to be nonaromatic. The two compounds of this type, disiladigermacyclobutadiene dianion **71²⁻•K₂⁺** and tetrasilacyclobutadiene dianion **72²⁻•K₂⁺**, were synthesized by the reductive dehalogenation of the corresponding precursors **73**[253] and **74**[254] with KC₈ (Scheme 2.67).[254]

In both compounds **71²⁻•[K⁺(thf)₂]₂** and **72²⁻•[K⁺(thf)₂]₂** the four-membered rings were nonplanar with potassium cations accommodated above and below the planes, being η^2-coordinated to the 1,3-positions of the ring. The skeletal Si–Si bonds in **72²⁻•[K⁺(thf)₂]₂** were not equivalent to each other. Certainly, such structural features of both **71²⁻•[K⁺(thf)₂]₂** and **72²⁻•[K⁺(thf)₂]₂** do not meet the classical criteria of aromaticity (ring planarity, cyclic bonds equalization), which was also supported by the NICS(1) calculations: $+4.3$ and $+6.1$ indicating the absence of the diatropic ring current.[254] Thus, the heavy analogs of cyclobutadiene dianion **71²⁻•K₂⁺** and **72²⁻•K₂⁺** should be recognized as nonaromatic compounds, in marked contrast to the well-established aromatic nature of the organic cyclobutadiene dianion derivative $(Me_3Si)_4C_4^{2-}•[Li^+(dme)]_2$ (dme = 1,2-dimethoxyethane).[255] The degree of partial delocalization of the negative charges over the four-membered rings of **71²⁻•K₂⁺** and **72²⁻•K₂⁺** in solution depends on the nature of the skeletal atoms. Thus, the solution structure of **71²⁻•K₂⁺** is the one in which the two negative charges preferentially accommodate on the more electronegative Ge atoms, making the two skeletal Si atoms doubly bonded (Figure 2.6, A).[254]

In contrast, the degree of delocalization of the two negative charges in **72²⁻•K₂⁺** is higher than that in **71²⁻•K₂⁺** (Figure 2.6, B).[254] Thus, the structural behavior of the heavy cyclobutadiene dianions is primarily driven by the relative electronegativities of the skeletal elements: the electronegativity difference favors electron localization (Figure 2.6, A versus B).

The newly synthesized heavy analogs of the 6π-electron compounds have found a very promising application as ligands for a new generation of transition metal

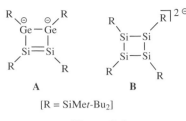

[R = SiMet-Bu₂]

Figure 2.6

complexes. Thus, the tetrasilacyclobutadiene (*t*-Bu₂MeSi)₄Si₄ was utilized as the ligand for the stable 18-electron anionic Co complex **75⁻•K⁺**, prepared by the reaction of **72²⁻•K₂⁺** with CpCo(CO)₂ (Scheme 2.68).[256]

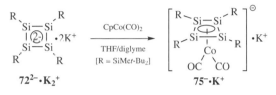

Scheme 2.68

Surprisingly, when the same disiladigermetene **73**[253] was reduced with alkaline earth metals (Mg, Ca) instead of alkali metals, strikingly different products were obtained: Mg or Ca derivatives of 1,3-disila-2,4-digermabicyclo[1.1.0]butane-2,4-dianions **76²⁻•Mg²⁺** and **76²⁻•Ca²⁺** (Scheme 2.69).[257]

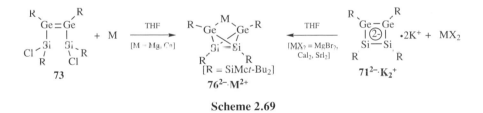

Scheme 2.69

Alternatively, such compounds (and additionally **76²⁻•Sr²⁺**) can be prepared from the disiladigermacyclobutadiene dianion **71²⁻•K₂⁺**[254] by its treatment with the salts MgBr₂, CaI₂, and SrI₂ (Scheme 2.69).[257] The bridgehead Si atoms of all dianions **76²⁻•M²⁺** (M = Mg, Ca, Sr) are very characteristically highly shielded: from –220.2 to –231.9 ppm. Interestingly, **76²⁻•M²⁺**, formally belonging to the class of bicyclo[1.1.0]butanes, actually acquires the structural properties of the much more rigid tricyclo[2.1.0.0²,⁵]pentanes.[257]

An example of a cyclotetrasilane dianion was recently reported by Matsumoto's group. This 1,3-dianion **77²⁻•K₂⁺** was prepared by reduction of the corresponding 1,3-dibromo precursor with potassium (Scheme 2.70).[258] Although the Si₄ ring of

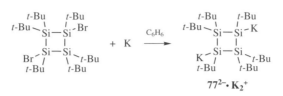

$$77^{2-} \cdot K_2^{+}$$

Scheme 2.70

$77^{2-} \cdot K_2^{+}$ was absolutely planar, the silyl anionic centers at the 1,3 positions were expectedly pyramidal.

An interesting bicyclic germyl anion $42^{-} \cdot K^{+}$ was readily prepared by the reduction of bicyclic iodide **78** with KC_8 (Scheme 2.71).[163] The Ge anionic center in $42^{-} \cdot [K^{+}(dme)_2]$ was highly pyramidalized with the normal value of the Ge–K bond distance (3.4324(10) Å). However, in polar solvents (THF) the Ge–K bond dissociates to produce $42^{-} \cdot K^{+}$ as an SSIP featuring the "free" anion 42^{-}.

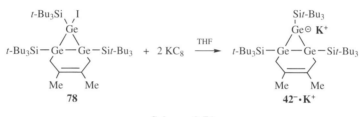

Scheme 2.71

2.4. CONCLUSION AND OUTLOOK

In contrast to organic chemistry, in which the study of carbocations, free radicals, and carbanions has successfully passed its centenary, the field of cations, radicals, and anions based on the heavier group 14 elements (Si, Ge, Sn) is still rather young and very actively developing. Being considered just several decades ago as phantom species, silyl, germyl, and stannyl cations, radicals, and anions quickly became a class of observable, spectroscopically detectable compounds. The chemistry of such amazing organometallics has reached its apogee in the past decade, when the experimental efforts of many research groups culminated in the synthesis of the stable and fully characterizable (in the majority of cases by X-ray diffraction analysis) compounds. Now, one can say that such highly desirable radical and ionic species have moved from the class of elusive reactive intermediates to the class of readily available and extremely useful organometallic reagents. Undoubtedly, all three classes of organometallic compounds, cations–radicals–anions, are tightly connected to each other, and their interplay is very important for the future development of organometallic chemistry. Thus, the interconversion between cations, radicals, and anions, when radicals can serve as very effective precursors for both cations and anions and

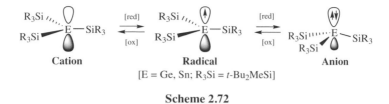

Scheme 2.72

vice versa, is a fundamental accomplishment which is greatly contributing to synthetic organometallic chemistry (Scheme 2.72).[14,164,238]

Looking to the future, one can expect further development of synthetic approaches to the stable derivatives of Si-, Ge-, and Sn-centered cations–radicals–anions based on modern sophisticated experimental techniques and advanced theories of bonding and reactivity of organometallic compounds. Apart from synthesis, the wide utilization and synthetic application of such derivatives as the heavy analogs of extremely synthetically useful carbocations, free radicals, and carbanions is highly desirable. One of such promising applications was recently suggested by theoreticians: using η^3-cyclotrisilenylium and η^3-cyclotrigermenylium ions as novel ligands for π-complexes with either main group or transition metal elements.[259] Now, it is the experimentalists turn to prepare and characterize these molecules.

2.5. LIST OF ABBREVIATIONS

THF	tetrahydrofuran
DME	1,2-dimethoxyethane
diox	1,4-dioxane
TMEDA	*N,N,N',N'*-tetramethylethylenediamine
PMDETA	*N,N,N',N'',N''*-pentamethyldiethylenetriamine
AIBN	azobis(isobutyronitrile)
LDMAN	lithium 1-(dimethylamino)naphthalenide
Np	naphthalenide
Mes	2,4,6-trimethylphenyl
Dur	2,3,5,6-tetramethylphenyl
Tip	2,4,6-triisopropylphenyl
Tbt	2,4,6-tris[bis(trimethylsilyl)methyl]phenyl
Dip	2,6-diisopropylphenyl
TPB⁻	tetraphenylborate
TFPB⁻	tetrakis[3,5-bis(trifluoromethyl)phenyl]borate
TPFPB⁻	tetrakis(pentafluorophenyl)borate
TTFPB⁻	tetrakis(2,3,5,6-tetrafluorophenyl)borate
TSFPB⁻	tetrakis{4-[*tert*-butyl(dimethyl)silyl]-2,3,5,6-tetrafluorophenyl}borate
HOMO	highest occupied molecular orbital

LUMO	lowest unoccupied molecular orbital
SOMO	singly occupied molecular orbital
DFT	density functional theory
GIAO	gauge independent atomic orbital
NICS	nucleus independent chemical shift
ASE	aromatic stabilization energy
SET	single electron transfer
hfcc	hyperfine coupling constant
CIP	contact ion pair
SSIP	solvent-separated ion pair
FT	Fourier transformation
EXAFS	extended X-ray absorption fine structure
REMPI	resonance-enhanced multiphoton ionization spectroscopy

2.6. SUGGESTED READING

Cations

J. B. Lambert, L. Kania, and S. Zhang, *Chem. Rev.* **1995**, *95*, 1191.

C. A. Reed, *Acc. Chem. Res.* **1998**, *31*, 325.

C. Maerker and P. v. R. Schleyer, in *The Chemistry of Organic Silicon Compounds, Vol. 2* (Eds. Z. Rappoport and Y. Apeloig), Wiley, Chichester, **1998**, Chapter 10.

P. Lickiss, in *The Chemistry of Organic Silicon Compounds, Vol. 2* (Eds. Z. Rappoport and Y. Apeloig), Wiley, Chichester, **1998**, Chapter 11.

I. Zharov and J. Michl, in *The Chemistry of Organic Germanium, Tin and Lead Compounds, Vol. 2, Part 1* (Ed. Z. Rappoport), Wiley, Chichester, **2002**, Chapter 10.

R. A. McClelland, in *Reactive Intermediate Chemistry* (Eds. R. A. Moss, M. S. Platz, and M. Jones Jr.), Wiley, Hoboken, **2004**, Chapter 1.

T. Müller, *Adv. Organomet. Chem.* **2005**, *53*, 155.

Radicals

C. Chatgilialoglu, *Chem. Rev.* **1995**, *95*, 1229.

J. Iley, in *The Chemistry of Organic Germanium, Tin and Lead Compounds* (Ed. S. Patai), Wiley, Chichester, **1995**, Chapter 5.

C. Chatgilialoglu and C. H. Schiesser, in *The Chemistry of Organic Silicon Compounds, Vol. 3* (Eds. Z. Rappoport and Y. Apeloig), Wiley, Chichester, **2001**, Chapter 4.

P. P. Power, *Chem. Rev.* **2003**, *103*, 789.

M. Newcomb, in *Reactive Intermediate Chemistry* (Eds. R. A. Moss, M. S. Platz, and M. Jones Jr.), Wiley, Hoboken, **2004**, Chapter 4.

V. Ya. Lee and A. Sekiguchi, *Eur. J. Inorg. Chem.* **2005**, 1209.

Anions

K. Tamao and A. Kawachi, *Adv. Organomet. Chem.* **1995**, *38*, 1.

P. D. Lickiss and C. M. Smith, *Coord. Chem. Rev.* **1995**, *145*, 75.

J. Belzner and U. Dehnert, in *The Chemistry of Organic Silicon Compounds, Vol. 2* (Eds. Z. Rappoport and Y. Apeloig), Wiley, Chichester, **1998**, Chapter 14.

A. Sekiguchi, V. Ya. Lee, and M. Nanjo, *Coord. Chem. Rev.* **2000**, *210*, 11.

P. Riviere, A. Castel, and M. Riviere-Baudet, in *The Chemistry of Organic Germanium, Tin and Lead Compounds, Vol. 2, Part 1* (Ed. Z. Rappoport), Wiley, Chichester, **2002**, Chapter 11.

S. Gronert, in *Reactive Intermediate Chemistry* (Eds. R. A. Moss, M. S. Platz, and M. Jones Jr.), Wiley, Hoboken, **2004**, Chapter 3.

2.7. REFERENCES

1. Reviews on the chemistry of the heavier group 14 elements centered cations: (a) H. Schwarz, in *The Chemistry of Organic Silicon Compounds* (Eds. S. Patai and Z. Rappoport), Wiley, Chichester, **1989**, Chapter 7. (b) J. B. Lambert and W. J. Schulz, in *The Chemistry of Organic Silicon Compounds* (Eds. S. Patai and Z. Rappoport), Wiley, Chichester, **1989**, Chapter 16. (c) S. H. Strauss, *Chemtracts: Inorganic Chemistry* **1993**, *5*, 119. (d) K. N. Houk, *Chemtracts: Organic Chemistry* **1993**, *6*, 360. (e) P. Riviere, M. Riviere-Baudet, and A. Castel, *Main Group Met. Chem.* **1994**, *17*, 679. (f) J. B. Lambert, L. Kania, and S. Zhang, *Chem. Rev.* **1995**, *95*, 1191. (g) P. v. R. Schleyer, *Science* **1997**, *275*, 39. (h) J. Belzner, *Angew. Chem. Int. Ed. Engl.* **1997**, *36*, 1277. (i) C. A. Reed, *Acc. Chem. Res.* **1998**, *31*, 325. (j) C. Maerker and P. v. R. Schleyer, in *The Chemistry of Organic Silicon Compounds, Vol. 2* (Eds. Z. Rappoport and Y. Apeloig), Wiley, Chichester, **1998**, Chapter 10. (k) P. Lickiss, in *The Chemistry of Organic Silicon Compounds, Vol. 2* (Eds. Z. Rappoport and Y. Apeloig), Wiley, Chichester, **1998**, Chapter 11. (l) J. B. Lambert, Y. Zhao, and S. M. Zhang, *J. Phys. Org. Chem.* **2001**, *14*, 370. (m) I. Zharov and J. Michl, in *The Chemistry of Organic Germanium, Tin and Lead Compounds, Vol. 2, Part 1* (Ed. Z. Rappoport), Wiley, Chichester, **2002**, Chapter 10. (n) V. Ya. Lee and A. Sekiguchi, in *The Chemistry of Organic Germanium, Tin and Lead Compounds, Vol. 2, Part 1* (Ed. Z. Rappoport), Wiley, Chichester, **2002**, Chapter 14. (o) A. Sekiguchi and V. Ya. Lee, *Chem. Rev.* **2003**, *103*, 1429. (p) T. Müller, *Adv. Organomet. Chem.* **2005**, *53*, 155.

2. G. A. Olah and L. D. Field, *Organometallics* **1982**, *1*, 1485.

3. G. A. Olah, K. Laali, and O. Farooq, *Organometallics* **1984**, *3*, 1337.

4. (a) J. B. Lambert, S. Zhang, C. L. Stern, and J. C. Huffman, *Science* **1993**, *260*, 1917. (b) J. B. Lambert and S. Zhang, *J. Chem. Soc., Chem. Commun.* **1993**, 383. (c) J. B. Lambert, S. Zhang, and S. M. Ciro, *Organometallics* **1994**, *13*, 2430.

5. Z. Xie, J. Manning, R. W. Reed, R. Mathur, P. D. W. Boyd, A. Benesi, and C. A. Reed, *J. Am. Chem. Soc.* **1996**, *118*, 2922.

6. J. B. Lambert and W. Schilf, *Organometallics* **1988**, *7*, 1659.

7. J. B. Lambert and B. Kuhlmann, *J. Chem. Soc., Chem. Commun.* **1992**, 931.

8. M. Kira, T. Oyamada, and H. Sakurai, *J. Organomet. Chem.* **1994**, *471*, C4.

9. J. B. Lambert, Y. Zhao, H. Wu, W. C. Tse, and B. Kuhlmann, *J. Am. Chem. Soc.* **1999**, *121*, 5001.

10. M. Ichinohe, H. Fukui, and A. Sekiguchi, *Chem. Lett.* **2000**, 600.

11. I. Zharov, B. T. King, Z. Havlas, A. Pardi, and J. Michl, *J. Am. Chem. Soc.* **2000**, *122*, 10253.

12. P. Jutzi and A. E. Bunte, *Angew. Chem. Int. Ed. Engl.* **1992**, *31*, 1605.

13. M. Ichinohe, Y. Hayata, and A. Sekiguchi, *Chem. Lett.* **2002**, 1054.

14. (a) A. Sekiguchi, T. Fukawa, V. Ya. Lee, M. Nakamoto, and M. Ichinohe, *Angew. Chem. Int. Ed.* **2003**, *42*, 1143. (b) A. Sekiguchi, T. Fukawa, V. Ya. Lee, and M. Nakamoto, *J. Am. Chem. Soc.* **2003**, *125*, 9250.

15. J. B. Lambert, Y. Zhao, and H. Wu, *J. Org. Chem.* **1999**, *64*, 2729.

16. (a) T. Müller, R. Meyer, D. Lennartz, and H.-U. Siehl, *Angew. Chem. Int. Ed.* **2000**, *39*, 3074. (b) T. Müller, M. Juhasz, and C. A. Reed, *Angew. Chem. Int. Ed.* **2004**, *43*, 1543.

17. (a) C. A. Reed, N. L. P. Fackler, K.-C. Kim, D. Stasko, D. R. Evans, P. D. W. Boyd, and C. E. F. Rickard, *J. Am. Chem. Soc.* **1999**, *121*, 6314. (b) C. A. Reed, K.-C. Kim, R. D. Bolskar, and L. J. Mueller, *Science* **2000**, *289*, 101. (c) C. A. Reed, K.-C. Kim, E. S. Stoyanov, D. Stasko, F. S. Tham, L. J. Mueller, and P. D. W. Boyd, *J. Am. Chem. Soc.* **2003**, *125*, 1796. (d) M. Juhasz, S. Hoffmann, E. Stoyanov, K.-C. Kim, C. A. Reed, *Angew. Chem. Int. Ed.* **2004**, *43*, 5352.

18. (a) D. Stasko and C. A. Reed, *J. Am. Chem. Soc.* **2002**, *124*, 1148. (b) T. Kato and C. A. Reed, *Angew. Chem. Int. Ed.* **2004**, *43*, 2908. (c) T. Kato, E. Stoyanov, J. Geier, H. Grützmacher, and C. A. Reed, *J. Am. Chem. Soc.* **2004**, *126*, 12451.

19. W. P. Neumann, H. Hillgärtner, K. M. Baines, R. Dicke, K. Vorspohl, U. Kobs, and U. Nussbeutel, *Tetrahedron Lett.* **1989**, *45*, 951.

20. (a) M. Johannsen, K. A. Jorgensen, and G. Helmchen, *J. Am. Chem. Soc.* **1998**, *120*, 7637. (b) G. A. Olah, G. Rasul, and G. K. Surya Prakash, *J. Am. Chem. Soc.* **1999**, *121*, 9615.

21. M. Schormann, S. Garratt, D. L. Hughes, J. C. Green, and M. Bochmann, *J. Am. Chem. Soc.* **2002**, *124*, 11266.

22. H.-U. Steinberger, C. Bauch, T. Müller, and N. Auner, *Can. J. Chem.* **2003**, *81*, 1223.

23. J. M. Blackwell, W. E. Piers, and R. McDonald, *J. Am. Chem. Soc.* **2002**, *124*, 1295.

24. (a) S. K. Shin and J. L. Beauchamp, *J. Am. Chem. Soc.* **1989**, *111*, 900. (b) H. Tashiro, K. Kikukawa, K. Ikenaga, N. Shimizu, and M. Mishima, *J. Chem. Soc. Perkin Trans. 2* **1998**, 2435.

25. A. G. Brook and K. H. Pannell, *Can. J. Chem.* **1970**, *48*, 3679.

26. J. Y. Corey, D. Gust, and K. Mislow, *J. Organomet. Chem.* **1975**, *101*, C7.

27. J. B. Lambert and W. J. Schulz, *J. Am. Chem. Soc.* **1983**, *105*, 1671.

28. R. Caputo, C. Ferreri, G. Palumbo, and E. Wenkert, *Tetrahedron Lett.* **1984**, 577.

29. C. Eaborn, P. D. Lickiss, S. T. Najim, and M. N. Romanelli, *J. Chem. Soc., Chem. Commun.* **1985**, 1754.

30. J. B. Lambert, J. A. McConnell, and W. J. Schulz, *J. Am. Chem. Soc.* **1986**, *108*, 2482.

31. Y. Apeloig and A. Stanger, *J. Am. Chem. Soc.* **1987**, *109*, 272.

32. J. B. Lambert, J. A. McConnell, W. Schilf, and W. J. Schulz, *J. Chem. Soc., Chem. Commun.* **1988**, 455.

33. J. B. Lambert, W. J. Schulz, J. A. McConnell, and W. Schilf, *J. Am. Chem. Soc.* **1988**, *110*, 2201.

34. J. B. Lambert and W. Schilf, *J. Am. Chem. Soc.* **1988**, *110*, 6364.

35. G. K. Surya Prakash, S. Keyaniyan, R. Aniszfeld, L. Heiliger, G. A. Olah, R. C. Stevens, H.-K. Choi, and R. Bau, *J. Am. Chem. Soc.* **1987**, *109*, 5123.

36. N. Wang, J. R. Hwu, and W. H. White, *J. Org. Chem.* **1991**, *56*, 471.

37. C. A. Reed, Z. Xie, R. Bau, and A. Benesi, *Science* **1993**, *262*, 402.

38. L. Pauling, *Science* **1994**, *263*, 983.

39. (a) G. A. Olah, G. Rasul, X. Li, H. A. Buchholz, G. Sandford, and G. K. Surya Prakash, *Science* **1994**, *263*, 983. (b) G. A. Olah, G. Rasul, and G. K. Surya Prakash, *J. Organomet. Chem.* **1996**, *521*, 271.

40. F. Cacase, M. Attina, and S. Fornarini, *Angew. Chem. Int. Ed. Engl.* **1995**, *34*, 654.

41. (a) D. Cremer, L. Olsson, and H. Ottosson, *J. Mol. Struct. (Theochem)* **1994**, *313*, 91. (b) L. Olsson, C.-H. Ottosson, and D. Cremer, *J. Am. Chem. Soc.* **1995**, *117*, 7460. (c) M. Arshadi, D. Johnels, U. Edlund, C.-H. Ottosson, and D. Cremer, *J. Am. Chem. Soc.* **1996**, *118*, 5120.

42. P. v. R. Schleyer, P. Buzek, T. Müller, Y. Apeloig, and H.-U. Siehl, *Angew. Chem. Int. Ed. Engl.* **1993**, *32*, 1471.

43. C. A. Reed and Z. Xie, *Science* **1994**, *263*, 985.

44. J. B. Lambert, Y. Zhao, and S. M. Zhang, *J. Phys. Org. Chem.* **2001**, *14*, 370.

45. Z. Xie, D. J. Liston, T. Jelinek, V. Mitro, R. Bau, and C. A. Reed, *J. Chem. Soc., Chem. Commun.* **1993**, 384.

46. Z. Xie, R. Bau, A. Benesi, and C. A. Reed, *Organometallics* **1995**, *14*, 3933.

47. J. B. Lambert, S. M. Ciro, and C. L. Stern, *J. Organomet. Chem.* **1995**, *499*, 49.

48. L. D. Henderson, W. E. Piers, G. J. Irvine, and R. McDonald, *Organometallics* **2002**, *21*, 340.

49. M. Arshadi, D. Johnels, and U. Edlund, *Chem. Commun.* **1996**, 1279.

50. A. Sekiguchi, Y. Murakami, N. Fukaya, and Y. Kabe, *Chem. Lett.* **2004**, *33*, 530.

51. P. Jutzi, S. Keitemeyer, B. Neumann, and H. G. Stammler, *Organometallics* **1999**, *18*, 4778.

52. Y. Ishida, A. Sekiguchi, and Y. Kabe, *J. Am. Chem. Soc.* **2003**, *125*, 11468.

53. I. Zharov, T.-C. Weng, A. M. Orendt, D. H. Barich, J. Penner-Hahn, D. M. Grant, Z. Havlas, and J. Michl, *J. Am. Chem. Soc.* **2004**, *126*, 12033.

54. (a) J. Schuppan, B. Herrschaft, and T. Müller, *Organometallics* **2001**, *20*, 4584. (b) H.-U. Steinberger, T. Müller, N. Auner, C. Maerker, and P. v. R. Schleyer, *Angew. Chem. Int. Ed. Engl.* **1997**, *36*, 626. (c) T. Müller, C. Bauch, M. Ostermeier, M. Bolte, and N. Auner, *J. Am. Chem. Soc.* **2003**, *125*, 2158. (d) T. Müller, C. Bauch, M. Bolte, and N. Auner, *Chem. Eur. J.* **2003**, *9*, 1746. (e) T. Müller, *Angew. Chem. Int. Ed.* **2001**, *40*, 3033. (f) T. Müller, in *Organosilicon Chemistry V, From Molecules to Materials* (Eds. N. Auner and J. Weis), Wiley-VCH, Weinheim, **2003**, p. 34. (g) R. Meyer, K. Werner, and T. Müller, *Chem. Eur. J.* **2002**, *8*, 1163. (h) N. Auner, G. Fearon, and J. Weis, in *Organosilicon Chemistry III, From Molecules to Materials* (Eds. N. Auner and J. Weis), Wiley-VCH, Weinheim, **1998**, p. 1. (i) N. Auner, T. Müller, M. Ostermeier, J. Schuppan, and H.-U. Steinberger, in *Organosilicon Chemistry IV, From Molecules to Materials* (Eds. N. Auner and J. Weis), Wiley-VCH, Weinheim, **2000**, p.127.

55. E. D. Jemmis, G. N. Srinivas, J. Leszczynski, J. Kapp, A. A. Korkin, and P. v. R. Schleyer, *J. Am. Chem. Soc.* **1995**, *117*, 11361.

56. (a) A. Sekiguchi, M. Tsukamoto, and A. Ichinohe, *Science* **1997**, *275*, 60. (b) A. Sekiguchi, M. Tsukamoto, M. Ichinohe, and N. Fukaya, *Phosphorus, Sulfur, and Silicon* **1997**, *124–125*, 323. (c) M. Ichinohe, N. Fukaya, and A. Sekiguchi, *Chem. Lett.* **1998**, 1045. (d) A. Sekiguchi, N. Fukaya, and M. Ichinohe, *Phosphorus, Sulfur, Silicon Relat. Elem.*

1999, *150–151*, 59. (e) A. Sekiguchi, N. Fukaya, M. Ichinohe, and Y. Ishida, *Eur. J. Inorg. Chem.* **2000**, 1155.

57. M. Ichinohe, M. Igarashi, K. Sanuki, and A. Sekiguchi, *J. Am. Chem. Soc.* **2005**, *127*, 9978.

58. G. N. Srinivas, E. D. Jemmis, A. A. Korkin, and P. v. R. Schleyer, *J. Phys. Chem. A* **1999**, *103*, 11034.

59. A. Sekiguchi, T. Matsuno, and M. Ichinohe, *J. Am. Chem. Soc.* **2000**, *122*, 11250.

60. (a) T. Nishinaga, Y. Izukawa, and K. Komatsu, *J. Am. Chem. Soc.* **2000**, *122*, 9312. (b) T. Nishinaga, Y. Izukawa, and K. Komatsu, *Tetrahedron* **2001**, *57*, 3645.

61. S. Ishida, T. Nishinaga, R. West, and K. Komatsu, *Chem. Commun.* **2005**, 778.

62. S. Ishida, T. Nishinaga, and K. Komatsu, *Chem. Lett.* **2005**, *34*, 486.

63. P. Jutzi, A. Mix, B. Rummel, W. W. Schoeller, B. Neumann, and H.-G. Stammler, *Science* **2004**, *305*, 84.

64. A. Sekiguchi, Y. Ishida, Y. Kabe, and M. Ichinohe, *J. Am. Chem. Soc.* **2002**, *124*, 8776.

65. J. B. Lambert and Y. Zhao, *Angew. Chem. Int. Ed. Engl.* **1997**, *36*, 400.

66. T. Müller, Y. Zhao, and J. B. Lambert, *Organometallics* **1998**, *17*, 278.

67. K.-C. Kim, C. A. Reed, D. W. Elliott, L. J. Mueller, F. Tham, L. Lin, and J. B. Lambert, *Science* **2002**, *297*, 825.

68. P. P. Gaspar, *Science* **2002**, *297*, 785.

69. J. B. Lambert and L. Lin, *J. Org. Chem.* **2001**, *66*, 8537.

70. M. Nakamoto, T. Fukawa, and A. Sekiguchi, *Chem. Lett.* **2004**, *33*, 38.

71. D. Cremer, L. Olsson, F. Reichel, and E. Kraka, *Isr. J. Chem.* **1994**, *33*, 369.

72. J. B. Lambert, L. Lin, S. Keinan, and T. Müller, *J. Am. Chem. Soc.* **2003**, *125*, 6022.

73. Reviews on the chemistry of the heavier group 14 elements centered free radicals: (a) H. Sakurai, in *Free Radicals, Vol. 2* (Ed. J. K. Kochi), Wiley, New York, **1973**, pp. 741–808. (b) M. F. Lappert and P. W. Lednor, *Adv. Organomet. Chem.* **1976**, *14*, 345. (c) C. Chatgilialoglu, *Acc. Chem. Res.* **1992**, *25*, 188. (d) C. Chatgilialoglu, *Chem. Rev.* **1995**, *95*, 1229. (e) J. Iley, in *The Chemistry of Organic Germanium, Tin and Lead Compounds* (Ed. S. Patai), Wiley, Chichester, **1995**, Chapter 5. (f) C. Chatgilialoglu, C. Ferreri, and T. Gimisis, in *The Chemistry of Organic Silicon Compounds, Vol. 2, Part 2* (Eds. Z. Rappoport and Y. Apeloig), Wiley, Chichester, **1998**, Chapter 25. (g) C. Chatgilialoglu and M. Newcomb, *Adv. Organomet. Chem.* **1999**, *44*, 67. (h) C. Chatgilialoglu and C. H. Schiesser, in *The Chemistry of Organic Silicon Compounds, Vol. 3* (Eds. Z. Rappoport and Y. Apeloig), Wiley, Chichester, **2001**, Chapter 4. (i) M. W. Carland and C. H. Schiesser, in *The Chemistry of Organic Germanium, Tin and Lead Compounds, Vol. 2, Part 2* (Ed. Z. Rappoport), Wiley, Chichester, **2002**, Chapter 19.

74. Reviews on the chemistry of the persistent and stable heavier group 14 elements centered free radicals: (a) P. P. Power, *Chem. Rev.* **2003**, *103*, 789. (b) V. Ya. Lee and A. Sekiguchi, *Eur. J. Inorg. Chem.* **2005**, 1209.

75. M. A. Nay, G. N. C. Woodall, O. P. Strausz, and H. E. Gunning, *J. Am. Chem. Soc.* **1965**, *87*, 179.

76. H. Sakurai, A. Hosomi, and M. Kumada, *Bull. Chem. Soc. Jpn.* **1967**, *40*, 1551.

77. S. W. Bennett, C. Eaborn, A. Hudson, H. A. Hussain, and R. A. Jackson, *J. Organomet. Chem.* **1969**, *16*, P36.

78. H. Sakurai, K. Mochida, and M. Kira, *J. Am. Chem. Soc.* **1975**, *97*, 929.

79. M. Lehnig, *Tetrahedron Lett.* **1977**, 3663.

80. H. U. Buschhaus, M. Lehnig, and W. P. Neumann, *J. Chem. Soc., Chem. Commun.* **1977**, 129.

81. S. K. Tokach and R. D. Koob, *J. Am. Chem. Soc.* **1980**, *102*, 376.

82. M. Lehnig, H. U. Buschhaus, W. P. Neumann, and T. Apoussidis, *Bull. Soc. Chim. Belg.* **1980**, *89*, 907.

83. B. J. Cornett, K. Y. Choo, and P. P. Gaspar, *J. Am. Chem. Soc.* **1980**, *102*, 377.

84. A. F. El-Farargy, M. Lehnig, and W. P. Neumann, *Chem. Ber.* **1982**, *115*, 2783.

85. A. F. El-Farargy and W. P. Neumann, *J. Organomet. Chem.* **1983**, *258*, 15.

86. K. Mochida, *Bull. Chem. Soc. Jpn.* **1984**, *57*, 796.

87. I. Safarik, A. Jodhan, O. P. Strausz, and T. N. Bell, *Chem. Phys. Lett.* **1987**, *142*, 115.

88. G. S. Jackel and W. Gordy, *Phys. Rev.* **1968**, *176*, 443.

89. R. V. Lloyd and M. T. Rogers, *J. Am. Chem. Soc.* **1973**, *95*, 2459.

90. S. A. Fieldhouse, A. R. Lyons, H. C. Starkie, and M. C. R. Symons, *J. Chem. Soc., Dalton Trans.* **1974**, 1966.

91. R. J. Booth, S. A. Fieldhouse, H. C. Starkie, and M. C. R. Symons, *J. Chem. Soc., Dalton Trans.* **1976**, 1506.

92. S. K. Tokach and R. D. Koob, *J. Phys. Chem.* **1979**, *83*, 774.

93. E. Bastian, P. Potzinger, A. Ritter, H.-P. Schuchmann, C. von Sonntag, and G. Weddle, *Ber. Bunsen-Ges. Phys. Chem.* **1980**, *84*, 56.

94. L. Gammie, I. Safarik, O. P. Strausz, R. Roberge, and C. Sandorfy, *J. Am. Chem. Soc.* **1980**, *102*, 378.

95. K. Matsumoto, M. Koshi, K. Okawa, and H. Matsui, *J. Phys. Chem.* **1996**, *100*, 8796.

96. J. A. Connor, R. N. Haszeldine, G. I. Leigh, and R. D. Sedgwick, *J. Chem. Soc. (A)* **1967**, 768.

97. G. B. Watts and K. U. Ingold, *J. Am. Chem. Soc.* **1972**, *94*, 491.

98. H.-U. Buschhaus, W. P. Neumann, and T. Apoussidis, *Liebigs Ann. Chem.* **1981**, 1190.

99. M. Lehnig, T. Apoussidis, and W. P. Neumann, *Chem. Phys. Lett.* **1983**, *100*, 189.

100. K. Mochida, M. Wakasa, Y. Nakadaira, Y. Sakaguchi, and H. Hayashi, *Organometallics* **1988**, *7*, 1869.

101. G. W. Slugget and W. L. Leigh, *Organometallics* **1992**, *11*, 3731.

102. C. Grugel, M. Lehnig, W. P. Neumann, and J. Sauer, *Tetrahedron Lett.* **1980**, *21*, 273.

103. J. E. Bennett and J. A. Howard, *Chem. Phys. Lett.* **1972**, *15*, 322.

104. (a) K. Mochida, M. Wakasa, S. Ishizaka, M. Kotani, Y. Sakaguchi, and H. Hayashi, *Chem. Lett.* **1985**, 1709. (b) K. Mochida, M. Wakasa, Y. Sakaguchi, and H. Hayashi, *Chem. Lett.* **1986**, 773.

105. M. Westerhausen and T. Hildenbrand, *J. Organomet. Chem.* **1991**, *411*, 1.

106. P. J. Davidson, A. Hudson, M. F. Lappert, and P. W. Lednor, *Chem. Commun.* **1973**, 829.

107. M. J. S. Gynane, D. H. Harris, M. F. Lappert, P. P. Power, P. Riviére, and M. Riviére-Baudet, *J. Chem. Soc., Dalton Trans.* **1977**, 2004.

108. R. J. Boczkowski and R. S. Bottei, *J. Organomet. Chem.* **1973**, *49*, 389.

109. V. A. Petrosyan, M. E. Niyazymbetov, S. P. Kolesnikov, and V. Ya. Lee, *Bull. Acad. Sci. USSR, Div. Chem. Sci.* **1988**, *37*, 1690.

110. R. D. Johnson III and J. W. Hudgens, *Chem. Phys. Lett.* **1987**, *141*, 163.

111. R. D. Johnson III, B. P. Tsai, and J. W. Hudgens, *J. Chem. Phys.* **1988**, *89*, 4558.

112. P. D. Lightfoot, R. Becerra, A. A. Jemi-Alade, and R. Lesclaux, *Chem. Phys. Lett.* **1991**, *180*, 441.

113. M. Tanimoto and S. Saito, *J. Chem. Phys.* **1999**, *111*, 9242.

114. (a) R. L. Morehouse, J. J. Christiansen, and W. Gordy, *J. Chem. Phys.* **1966**, *45*, 1751. (b) G. S. Jackel, J. J. Christiansen, and W. Gordy, *J. Chem. Phys.* **1967**, *47*, 4274. (c) G. S. Jackel and W. Gordy, *Phys. Rev.* **1968**, *176*, 443.

115. T. Katsu, Y. Yatsurugi, M. Sato, and Y. Fujita, *Chem. Lett.* **1975**, 343.

116. (a) K. Nakamura, N. Masaki, S. Sato, and K. Shimokoshi, *J. Chem. Phys.* **1985**, *83*, 4504. (b) K. Nakamura, N. Masaki, S. Sato, and K. Shimokoshi, *J. Chem. Phys.* **1986**, *85*, 4204. (c) K. Nakamura, M. Okamoto, T. Takayanagi, T. Kawachi, K. Shimokoshi, and S. Sato, *J. Chem. Phys.* **1989**, *90*, 2992.

117. L. Pauling, *J. Chem. Phys.* **1969**, *51*, 2767.

118. K. Nakamura, T. Takayanagi, M. Okamoto, K. Shimokoshi, and S. Sato, *Chem. Phys. Lett.* **1989**, *164*, 593.

119. M. V. Merritt and R. W. Fessenden, *J. Chem. Phys.* **1972**, *56*, 2353.

120. (a) J. Roncin, *Mol. Cryst.* **1967**, *3*, 117. (b) C. Hesse, N. Leray, and J. Roncin, *J. Chem. Phys.* **1972**, *57*, 749.

121. (a) See ref. 77. (b) S. W. Bennett, C. Eaborn, A. Hudson, R. A. Jackson, and K. D. J. Root, *J. Chem. Soc. (A)* **1970**, 348. (c) P. J. Krusic and J. K. Kochi, *J. Am. Chem. Soc.* **1969**, *91*, 3938.

122. R. A. Jackson, *J. Chem. Soc., Perkin Trans. II* **1983**, 523.

123. R. A. Jackson and H. Weston, *J. Organomet. Chem.* **1984**, *277*, 13.

124. (a) J. Cooper, A. Hudson, and R. A. Jackson, *Mol. Phys.* **1972**, *23*, 209. (b) R. A. Jackson and C. J. Rhodes, *J. Organomet. Chem.* **1987**, *336*, 45.

125. H. Sakurai, K. Ogi, A. Hosomi, and M. Kira, *Chem. Lett.* **1974**, 891.

126. H. Sakurai, K. Mochida, and M. Kira, *J. Organomet. Chem.* **1977**, *124*, 235.

127. K. Mochida, *Bull. Chem. Soc. Jpn* **1984**, *57*, 796.

128. M. Lehnig, W. P. Neumann, and E. Wallis, *J. Organomet. Chem.* **1987**, *333*, 17.

129. T. Berclaz and M. Geoffroy, *Chem. Phys. Lett.* **1977**, *52*, 606.

130. F. M. Bickelhaupt, T. Zigler, and P. v. R. Schleyer, *Organometallics* **1996**, *15*, 1477.

131. F. K. Cartledge and R. V. Piccione, *Organometallics* **1984**, *3*, 299.

132. K. N. Houk and M. N. Paddon-Row, *J. Am. Chem. Soc.* **1981**, *103*, 5046.

133. K. Morokuma, L. Pedersen, and M. Karplus, *J. Chem. Phys.* **1968**, *48*, 4801.

134. J. Moc, Z. Latajka, and H. Ratajczak, *Z. Physik D. Atoms, Molecules and Clusters* **1986**, *4*, 185.

135. J. Moc, Z. Latajka, and H. Ratajczak, *Chem. Phys. Lett.* **1987**, *136*, 122.

136. J. Moc, J. M. Rudzinski, and H. Ratajczak, *Chem. Phys.* **1992**, *159*, 197.

137. M. Guerra, *J. Am. Chem. Soc.* **1993**, *115*, 11926.

138. (a) B. Giese, *Radicals in Organic Synthesis: Formation of Carbon-Carbon Bonds*, Pergamon Press, Oxford, **1986**. (b) D. P. Curran, *Synthesis* **1988**, 417. (c) D. P. Curran,

Synthesis **1988**, 489. (d) M. Malacria, *Chem. Rev.* **1996**, *96*, 289. (e) I. Ryu, N. Sonoda, and D. P. Curran, *Chem. Rev.* **1996**, *96*, 177. (f) W. B. Motherwell and D. Crich, *Free Radical Chain Reactions in Organic Synthesis*, Academic Press, London, **1992**. (g) D. P. Curran, N. A. Porter, and B. Giese, *Stereochemistry of Radical Reactions*, VCH, Weinheim, **1995**. (h) D. P. Curran, in *Comprehensive Organic Synthesis, Vol. 4* (Eds. B. M. Trost and I. Fleming), Pergamon Press, Oxford, **1991**, pp. 715–831. (i) C. P. Jasperse, D. P. Curran, and T. L. Fevig, *Chem. Rev.* **1991**, *91*, 1237.

139. E. Lee, S. B. Ko, K. W. Jung, and M. H. Chang, *Tetrahedron Lett.* **1989**, *30*, 827.

140. G. Stork and R. Mook, Jr., *J. Am. Chem. Soc.* **1987**, *109*, 2829.

141. I. Ryu, A. Kurihara, H. Muraoka, S. Tsunoi, N. Kambe, and N. Sonoda, *J. Org. Chem.* **1994**, *59*, 7570.

142. (a) S. F. Wnuk and M. J. Robins, *J. Am. Chem. Soc.* **1996**, *118*, 2519. (b) S. F. Wnuk, J. M. Rios, J. Khan, and Y.-L. Hsu, *J. Org. Chem.* **2000**, *65*, 4169.

143. M. Nishida, A. Nishida, and N. Kawahara, *J. Org. Chem.* **1996**, *61*, 3574.

144. M. Chareyron, P. Devin, L. Fensterbank, and M. Malacria, *Synlett* **2000**, 83.

145. (a) T. Naito, D. Fukumoto, K. Takebayashi, and T. Kiguchi, *Heterocycles* **1999**, *51*, 489. (b) T. Kiguchi, M. Okazaki, and T. Naito, *Heterocycles* **1999**, *51*, 2711. (c) T. Naito, K. Nakagawa, T. Nakamura, A. Kasei, I. Ninomiya, and T. Kiguchi, *J. Org. Chem.* **1999**, *64*, 2003. (d) T. Naito, J. S. Nair, A. Nishiki, K. Yamashita, and T. Kiguchi, *Heterocycles* **2000**, *53*, 2611. (e) H. Miyabe, M. Ueda, N. Yoshioka, K. Yamakawa, and T. Naito, *Tetrahedron* **2000**, *56*, 2413. (f) T. Kiguchi, K. Tajiri, I. Ninomiya, and T. Naito, *Tetrahedron* **2000**, *56*, 5819. (g) H. Miyabe, H. Tanaka, and T. Naito, *Chem. Pharm. Bull.* **2004**, *52*, 74.

146. A. Alberti and C. Chatgilialoglu, *Tetrahedron* **1990**, *46*, 3963.

147. (a) C. Ferreri, M. Ballestri, and C. Chatgilialoglu, *Tetrahedron Lett.* **1993**, *34*, 5147. (b) C. Chatgilialoglu, M. Ballestri, C. Ferreri, and D. Vecchi, *J. Org. Chem.* **1995**, *60*, 3826.

148. J. Cossy and L. Sallé, *Tetrahedron Lett.* **1995**, *36*, 7235.

149. The first solution ESR spectra of the transient radicals were observed in 1969 and 1972. Silyl radicals: see refs. 121a,c. Germyl radicals: see ref. 121a. Stannyl radicals: ref. 97.

150. (a) See ref. 106. (b) J. D. Cotton, C. S. Cundy, D. H. Harris, A. Hudson, M. F. Lappert, and P. W. Lednor, *J. Chem. Soc., Chem. Commun.* **1974**, 651. (c) A. Hudson, M. F. Lappert, and P. W. Lednor, *J. Chem. Soc., Dalton Trans.* **1976**, 2369.

151. R. A. Jackson, *Chem. Soc. Spec. Publ.* **1970**, 295.

152. J. Roncin and R. Debuyst, *J. Chem. Phys.* **1969**, *51*, 577.

153. S. Kyushin, H. Sakurai, T. Betsuyaku, and H. Matsumoto, *Organometallics* **1997**, *16*, 5386.

154. C. Chatgilialoglu and S. Rossini, *Bull. Soc. Chim. Fr.* **1988**, 298.

155. S. Kyushin, H. Sakurai, and H. Matsumoto, *Chem. Lett.* **1998**, 107.

156. M. Kira, T. Obata, I. Kon, H. Hashimoto, M. Ichinohe, H. Sakurai, S. Kyushin, and H. Matsumoto, *Chem. Lett.* **1998**, 1097.

157. Y. Apeloig, D. Bravo-Zhivotovskii, M. Yuzefovich, M. Bendikov, and A. I. Shames, *Appl. Magn. Reson.* **2000**, *18*, 425.

158. D. Bravo-Zhivotovskii, M. Yuzefovich, N. Sigal, G. Korogodsky, K. Klinkhammer, B. Tumanskii, A. Shames, and Y. Apeloig, *Angew. Chem. Int. Ed.* **2002**, *41*, 649.

159. D. Bravo-Zhivotovskii, I. Ruderfer, S. Melamed, M. Botoshansky, B. Tumanskii, and Y. Apeloig, *Angew. Chem. Int. Ed.* **2005**, *44*, 739.

160. D. Bravo-Zhivotovskii, I. Ruderfer, M. Yuzefovich, M. Kosa, M. Botoshansky, B. Tumanskii, and Y. Apeloig, *Organometallics* **2005**, *24*, 2698.

161. M. M. Olmstead, L. Pu, R. S. Simons, P. P. Power, *Chem. Commun.* **1997**, 1595.

162. A. Sekiguchi, T. Matsuno, M. Ichinohe, *J. Am. Chem. Soc.* **2001**, *123*, 12436.

163. Y. Ishida, A. Sekiguchi, K. Kobayashi, S. Nagase, *Organometallics* **2004**, *23*, 4981.

164. A. Sekiguchi, T. Fukawa, M. Nakamoto, V. Ya. Lee, M. Ichinohe, *J. Am. Chem. Soc.* **2002**, *124*, 9865.

165. B. Gehrhus, P. B. Hitchcock, L. Zhang, *Angew. Chem., Int. Ed.* **2004**, *43*, 1124.

166. M. Weidenbruch, K. Kramer, A. Schäfer, J. K. Blum, *Chem. Ber.* **1985**, *118*, 107.

167. M. Kira, T. Iwamoto, *J. Organomet. Chem.* **2000**, *611*, 236.

168. A. Sekiguchi, S. Inoue, M. Ichinohe, Y. Arai, *J. Am. Chem. Soc.* **2004**, *126*, 9626.

169. (a) T. Fukawa, V. Ya. Lee, M. Nakamoto, A. Sekiguchi, *J. Am. Chem. Soc.* **2004**, *126*, 11758. (b) V. Ya. Lee, T. Fukawa, M. Nakamoto, A. Sekiguchi, B. L. Tumanskii, M. Karni, Y. Apeloig, *J. Am. Chem. Soc.* **2006**, *128*, 11643.

170. M. M. Olmstead, R. S. Simons, P. P. Power, *J. Am. Chem. Soc.* **1997**, *119*, 11705.

171. L. Pu, S. T. Haubrich, P. P. Power, *J. Organomet. Chem.* **1999**, *582*, 100.

172. J. A. Berson, *Acc. Chem. Res.* **1978**, *11*, 446.

173. H. Grützmacher, F. Breher, *Angew. Chem., Int. Ed.* **2002**, *41*, 4006.

174. M. J. S. Dewar and E. F. Healy, *Chem. Phys. Lett.* **1987**, *141*, 521.

175. (a) R. Jain, G. J. Snyder, and D. A. Dougherty, *J. Am. Chem. Soc.* **1984**, *106*, 7294. (b) R. Jain, M. B. Sponsler, F. D. Coms, and D. A. Dougherty, *J. Am. Chem. Soc.* **1988**, *110*, 1356.

176. (a) E. Niecke, A. Fuchs, F. Baumeister, M. Nieger, and W. W. Schoeller, *Angew. Chem., Int. Ed.* **1995**, *34*, 555. (b) T. Baumgartner, D. Gudat, M. Nieger, E. Niecke, and T. J. Schiffer, *J. Am. Chem. Soc.* **1999**, *121*, 5953. (c) E. Niecke, A. Fuchs, and M. Nieger, *Angew. Chem., Int. Ed.* **1999**, *38*, 3028. (d) E. Niecke, A. Fuchs, M. Nieger, O. Schmidt, and W. W. Schoeller, *Angew. Chem., Int. Ed.* **1999**, *38*, 3031. (e) E. Niecke, A. Fuchs, O. Schmidt, and M. Nieger, *Phosphorus, Sulfur, Silicon Relat. Elem.* **1999**, *144–146*, 41. (f) H. Sugiyama, S. Ito, and M. Yoshifuji, *Angew. Chem., Int. Ed.* **2003**, *42*, 3802. (g) D. Scheschkewitz, H. Amii, H. Gornitzka, W. W. Schoeller, D. Bourissou, and G. Bertrand, *Science* **2002**, *295*, 1880. (h) W. W. Schoeller, A. Rozhenko, D. Bourissou, and G. Bertrand, *Chem. Eur. J.* **2003**, *9*, 3611. (i) D. Scheschkewitz, H. Amii, H. Gornitzka, W. W. Schoeller, D. Bourissou, and G. Bertrand, *Angew. Chem., Int. Ed.* **2004**, *43*, 585. (j) H. Amii, L. Vranicar, H. Gornitzka, D. Bourissou, and G. Bertrand, *J. Am. Chem. Soc.* **2004**, *126*, 1344.

177. C. Cui, M. Brynda, M. M. Olmstead, P. P. Power, *J. Am. Chem. Soc.* **2004**, *126*, 6510.

178. H. Cox, P. B. Hitchcock, M. F. Lappert, L. J.-M. Pierssens, *Angew. Chem., Int. Ed.* **2004**, *43*, 4500.

179. Reviews on the chemistry of the heavier group 14 elements centered anions: (a) K. Tamao and A. Kawachi, *Adv. Organomet. Chem.* **1995**, *38*, 1. (b) P. D. Lickiss and C. M. Smith, *Coord. Chem. Rev.* **1995**, *145*, 75. (c) P. Riviere, M. Riviere-Baudet, and J. Satgé, in *Comprehensive Organometallic Chemistry II, Vol. 2* (Ed. G. Wilkinson), Pergamon Press, Oxford, **1995**, Chapter 5. (d) A. G. Davies and P. J. Smith, in *Comprehensive Organometallic Chemistry II, Vol. 2* (Ed. G. Wilkinson), Pergamon Press,

Oxford, **1995**, Chapter 6. (e) P. G. Harrison, in *Comprehensive Organometallic Chemistry II, Vol. 2* (Ed. G. Wilkinson), Pergamon Press, Oxford, **1995**, Chapter 7. (f) J. Belzner and U. Dehnert, in *The Chemistry of Organic Silicon Compounds, Vol. 2* (Eds. Z. Rappoport and Y. Apeloig), Wiley, Chichester, **1998**, Chapter 14. (g) A. Sekiguchi, V. Ya. Lee, and M. Nanjo, *Coord. Chem. Rev.* **2000**, *210*, 11. (h) P. Riviere, A. Castel, and M. Riviere-Baudet, in *The Chemistry of Organic Germanium, Tin and Lead Compounds, Vol. 2, Part 1* (Ed. Z. Rappoport), Wiley, Chichester, **2002**, Chapter 11. (i) H.-W. Lerner, *Coord. Chem. Rev.* **2005**, *249*, 781. (j) M. Saito and M. Yoshioka, *Coord. Chem. Rev.* **2005**, *249*, 765.

180. W. C. Still, *J. Org. Chem.* **1976**, *41*, 3063.

181. H. Sakurai, A. Okada, M. Kira, and K. Yonezawa, *Tetrahedron Lett.* **1971**, 1511.

182. R. J. P. Corriu and C. Guerin, *J. Chem. Soc., Chem. Commun.* **1980**, 168.

183. (a) T. F. Schaaf and J. P. Oliver, *J. Am. Chem. Soc.* **1969**, *91*, 4327. (b) W. H. Ilsley, M. J. Albright, T. J. Anderson, M. D. Glick, and J. P. Oliver, *Inorg. Chem.* **1980**, *19*, 3577. (c) N. S. Vyazankin, G. A. Razuvaev, E. N. Gladyshev, and S. P. Korneva, *J. Organomet. Chem.* **1967**, *7*, 353. (d) E. N. Gladyshev, E. A. Fedorova, L. O. Yuntila, G. A. Razuvaev, and N. S. Vyazankin, *J. Organomet. Chem.* **1975**, *96*, 169. (e) E. Hengge, N. Holtschmidt, *J. Organomet. Chem.* **1968**, *12*, P5.

184. A. R. Claggett, W. H. Ilsley, T. J. Anderson, M. D. Glick, and J. P. Oliver, *J. Am. Chem. Soc.* **1977**, *99*, 1797.

185. (a) N. Wiberg, G. Fisher, and P. Karampatses, *Angew. Chem., Int. Ed. Engl.* **1984**, *23*, 59. (b) N. Wiberg, H. Schuster, A. Simon, and K. Peters, *Angew. Chem., Int. Ed. Engl.* **1986**, *25*, 79.

186. (a) M. V. George, D. J. Peterson, and H. Gilman, *J. Am. Chem. Soc.* **1960**, *82*, 403. (b) H. Wagner and U. Schubert, *Chem. Ber.* **1990**, *123*, 2101. (c) J. Meyer, J. Willnecker, and U. Schubert, *Chem. Ber.* **1989**, *122*, 223.

187. E. W. Colvin, *Silicon in Organic Synthesis*, Butterworth, Sevenoaks, **1981**, pp. 134–140.

188. (a) H. Gilman and T. C. Wu, *J. Am. Chem. Soc.* **1951**, *73*, 4031. (b) A. G. Brook and H. Gilman, *J. Am. Chem. Soc.* **1954**, *76*, 278.

189. A. G. Brook, A. Baumegger, and A. J. Lough, *Organometallics* **1992**, *11*, 310.

190. (a) H. Gilman and C. L. Smith, *J. Organomet. Chem.* **1968**, *14*, 91. (b) G. Gutekunst and A. G. Brook, *J. Organomet. Chem.* **1982**, *225*, 1.

191. A. Sekiguchi, M. Nanjo, C. Kabuto, and H. Sakurai, *J. Am. Chem. Soc.* **1995**, *117*, 4195.

192. H. Gilman and R. L. Harrell, Jr., *J. Organomet. Chem.* **1967**, *9*, 67.

193. H. Gilman, D. J. Peterson, A. W. Jarvie, and H. J. S. Winkler, *J. Am. Chem. Soc.* **1960**, *82*, 2076.

194. J. Belzner, U. Dehnert, and D. Stalke, *Angew. Chem., Int. Ed. Engl.* **1994**, *33*, 2450.

195. O. W. Steward, G. L. Heider, and J. S. Johnson, *J. Organomet. Chem.* **1979**, *168*, 33.

196. D. M. Roddick, R. H. Heyn, and T. D. Tilley, *Organometallics* **1989**, *8*, 324.

197. H. Gilman and W. Steudel, *Chem. & Ind. (London)* **1959**, 1094.

198. (a) P. Boudjouk, U. Samaraweera, R. Sooriyakumaran, J. Chrusciel, and K. R. Anderson, *Angew. Chem., Int. Ed. Engl.* **1988**, *27*, 1355. (b) T. Tsumuraya, S. A. Batcheller, and S. Masamune, *Angew. Chem., Int. Ed. Engl.* **1991**, *30*, 902. (c) R. Corriu, G. Lanneau,

C. Priou, F. Soulairol, N. Auner, R. Probst, R. Conlin, and C. J. Tan, *J. Organomet. Chem.* **1994**, *466*, 55.

199. (a) K. Tamao, A. Kawachi, and Y. Ito, *J. Am. Chem. Soc.* **1992**, *114*, 3989. (b) K. Tamao, A. Kawachi, and Y. Ito, *Organometallics* **1993**, *12*, 580.

200. K. Tamao and A. Kawachi, *Organometallics* **1995**, *14*, 3108.

201. K. Tamao and A. Kawachi, *Angew. Chem., Int. Ed. Engl.* **1995**, *34*, 818.

202. J.-H. Hong and P. Boudjouk, *J. Am. Chem. Soc.* **1993**, *115*, 5883.

203. (a) J.-H. Hong, P. Boudjouk, and S. Castellino, *Organometallics* **1994**, *13*, 3387. (b) R. West, H. Sohn, U. Bankwitz, J. Calabrese, Y. Apeloig, and T. Mueller, *J. Am. Chem. Soc.* **1995**, *117*, 11608.

204. R. West, H. Sohn, D. R. Powell, T. Müller, and Y. Apeloig, *Angew. Chem., Int. Ed. Engl.* **1996**, *35*, 1002.

205. W. P. Freeman, T. D. Tilley, G. P. A. Yap, and A. L. Rheingold, *Angew. Chem., Int. Ed. Engl.* **1996**, *35*, 882.

206. W. P. Freeman, T. D. Tilley, L. M. Liable-Sands, and A. L. Rheingold, *J. Am. Chem. Soc.* **1996**, *118*, 10457.

207. A. Castel, P. Rivière, J. Satgé, Y. H. Ko, and D. Desor, *J. Organomet. Chem.* **1990**, *397*, 7.

208. A. Castel, P. Rivière, J. Satgé, D. Desor, M. Ahbala, and C. Abdenadher, *Inorg. Chim. Acta* **1993**, *212*, 51.

209. (a) W. Reimann, H. G. Kuivila, D. Farah, and T. Apoussidis, *Organometallics* **1987**, *6*, 557. (b) T. S. Kaufman, *Synlett* **1997**, 1377.

210. A. Heine, R. Herbst-Irmer, G. M. Sheldrick, and D. Stalke, *Inorg. Chem.* **1993**, *32*, 2694.

211. K. W. Klinkhammer and W. Schwarz, *Z. Anorg. Allg. Chem.* **1993**, *619*, 1777.

212. (a) U. Edlund, T. Lejon, P. Pyykkö, T. K. Venkatachalam, and E. Buncel, *J. Am. Chem. Soc.* **1987**, *109*, 5982. (b) U. Edlund, T. Lejon, T. K. Venkatachalam, and E. Buncel, *J. Am. Chem. Soc.* **1985**, *107*, 6408.

213. (a) A. Sekiguchi, M. Nanjo, C. Kabuto, and H. Sakurai, *Organometallics* **1995**, *14*, 2630. (b) M. Nanjo, A. Sekiguchi, and H. Sakurai, *Bull. Chem. Soc. Jpn.* **1998**, *71*, 741.

214. R. H. Cox, E. G. Janzen, and W. B. Harrison, *J. Magn. Reson.* **1971**, *4*, 274.

215. H. J. Reich, J. P. Borst, and R. R. Dykstra, *Organometallics* **1994**, *13*, 1.

216. (a) T. F. Schaaf, W. Butler, M. D. Glick, and J. P. Oliver, *J. Am. Chem. Soc.* **1974**, *96*, 7593. (b) W. H. Ilsley, T. F. Schaaf, M. D. Glick, and J. P. Oliver, *J. Am. Chem. Soc.* **1980**, *102*, 3769.

217. B. Teclé, W. H. Ilsley, and J. P. Oliver, *Organometallics* **1982**, *1*, 875.

218. H. V. R. Dias, M. M. Olmstead, K. Ruhlandt-Senge, and P. P. Power, *J. Organomet. Chem.* **1993**, *462*, 1.

219. G. Becker, H.-M. Hartmann, E. Hengge, and F. Schrank, *Z. Anorg. Allg. Chem.* **1989**, *572*, 63.

220. (a) See ref. 213a. (b) M. Nanjo, A. Sekiguchi, and H. Sakurai, *Bull. Chem. Soc. Jpn.* **1999**, *72*, 1387.

221. S. Freitag, R. Herbst-Irmer, L. Lameyer, and D. Stalke, *Organometallics* **1996**, *15*, 2839.

222. D. Reed, D. Stalke, and D. S. Wright, *Angew. Chem. Int. Ed. Engl.* **1991**, *30*, 1459.

223. D. R. Armstrong, M. G. Davidson, D. Moncrieff, D. Stalke, and D. S. Wright, *J. Chem. Soc., Chem. Commun.* **1992**, 1413.

224. A. R. Glaggett, W. H. Ilsley, T. J. Anderson, M. D. Glick, and J. P. Oliver, *J. Am. Chem. Soc.* **1977**, *99*, 1797.

225. R. Goddard, C. Krüger, N. A. Ramadan, and A. Ritter, *Angew. Chem. Int. Ed. Engl.* **1995**, *34*, 1030.

226. L. Rösch, C. Krueger, and A. P. Chiang, *Z. Naturforsch. B* **1984**, *39B*, 855.

227. M. Westerhausen, *Angew. Chem. Int. Ed. Engl.* **1994**, *33*, 1493.

228. A. Sekiguchi, M. Ichinohe, and S. Yamaguchi, *J. Am. Chem. Soc.* **1999**, *121*, 10231.

229. N. Tokitoh, K. Hatano, T. Sadahiro, and R. Okazaki, *Chem. Lett.* **1999**, 931.

230. A. Sekiguchi, R. Izumi, S. Ihara, M. Ichinohe, and V. Ya. Lee, *Angew. Chem. Int. Ed.* **2002**, *41*, 1598.

231. M. Ichinohe, Y. Arai, A. Sekiguchi, N. Takagi, and S. Nagase, *Organometallics* **2001**, *20*, 4141.

232. A. Sekiguchi, R. Izumi, V. Ya. Lee, and M. Ichinohe, *J. Am. Chem. Soc.* **2002**, *124*, 14822.

233. A. Sekiguchi, R. Izumi, V. Ya. Lee, M. Ichinohe, *Organometallics* **2003**, *22*, 1483.

234. N. Nakata, R. Izumi, V. Ya. Lee, M. Ichinohe, and A. Sekiguchi, *J. Am. Chem. Soc.* **2004**, *126*, 5058.

235. N. Nakata, R. Izumi, V. Ya. Lee, M. Ichinohe, and A. Sekiguchi, *Chem. Lett.* **2005**, *34*, 582.

236. M. Ichinohe, M. Igarashi, K. Sanuki, and A. Sekiguchi, *J. Am. Chem. Soc.* **2005**, *127*, 9978.

237. V. Ya. Lee, H. Yasuda, M. Ichinohe, and A. Sekiguchi, *Angew. Chem. Int. Ed.* **2005**, *44*, 6378.

238. (a) M. Nakamoto, T. Fukawa, V. Ya. Lee, and A. Sekiguchi, *J. Am. Chem. Soc.* **2002**, *124*, 15160. (b) T. Fukawa, M. Nakamoto, V. Ya. Lee, and A. Sekiguchi, *Organometallics* **2004**, *23*, 2376.

239. (a) D. M. Jenkins, W. Teng, U. Englich, D. Stone, and K. Ruhland-Senge, *Organometallics* **2001**, *20*, 4600. (b) W. Teng and K. Ruhland-Senge, *Organometallics* **2004**, *23*, 2694. (c) W. Teng and K. Ruhland-Senge, *Organometallics* **2004**, *23*, 952.

240. F. Stabenow, W. Saak, and M. Weidenbruch, *Chem. Commun.* **2003**, 2342.

241. (a) Y. Apeloig, G. Korogodsky, D. Bravo-Zhivotovskii, D. Bläser, and R. Boese, *Eur. J. Inorg. Chem.* **2000**, 1091. (b) Y. Apeloig, M. Yuzefovich, M. Bendikov, D. Bravo-Zhivotovskii, D. Bläser, and R. Boese, *Angew. Chem. Int. Ed.* **2001**, *40*, 3016. (c) D. Bravo-Zhivotovskii, M. Yuzefovich, N. Sigal, G. Korogodsky, K. Klinkhammer, B. Tumanskii, A. Shames, and Y. Apeloig, *Angew. Chem. Int. Ed.* **2002**, *41*, 649. (d) D. Bravo-Zhivotovskii, I. Ruderfer, S. Melamed, M. Botoshansky, B. Tumanskii, and Y. Apeloig, *Angew. Chem. Int. Ed.* **2005**, *44*, 739.

242. (a) C. Marschner, *Eur. J. Inorg. Chem.* **1998**, 221. (b) C. Kayser, R. Fischer, J. Baumgartner, and C. Marschner, *Organometallics* **2002**, *21*, 1023. (c) R. Fischer, T. Konopa, J. Baumgartner, and C. Marschner, *Organometallics* **2004**, *23*, 1899. (d) R. Fischer, T. Konopa, S. Ully, J. Baumgartner, and C. Marschner, *J. Organomet. Chem.* **2003**, *685*, 79. (e) C. Kayser, G. Kickelbick, C. Marschner, *Angew. Chem. Int. Ed.* **2002**, *41*,

989. (f) R. Fischer, D. Frank, W. Gaderbauer, C. Kayser, C. Mechtler, J. Baumgartner, and C. Marschner, *Organometallics* **2003**, *22*, 3723. (g) R. Fischer, J. Baumgartner, G. Kickelbick, and C. Marschner, *J. Am. Chem. Soc.* **2003**, *125*, 3414.

243. M. Weidenbruch, S. Willms, W. Saak, and G. Henkel, *Angew. Chem. Int. Ed. Engl.* **1997**, *36*, 2503.

244. D. Scheschkewitz, *Angew. Chem. Int. Ed.* **2004**, *43*, 2965.

245. M. Ichinohe, K. Sanuki, S. Inoue, and A. Sekiguchi, *Organometallics* **2004**, *23*, 3088.

246. S. Inoue, M. Ichinohe, and A. Sekiguchi, *Chem. Lett.* **2005**, *34*, 1564.

247. L. Pu, M. O. Senge, M. M. Olmstead, and P. P. Power, *J. Am. Chem. Soc.* **1998**, *120*, 12682.

248. H. Schäfer, W. Saak, and M. Weidenbruch, *Angew. Chem. Int. Ed.* **2000**, *39*, 3703.

249. T. Matsuno, M. Ichinohe, and A. Sekiguchi, *Angew. Chem. Int. Ed.* **2002**, *41*, 1575.

250. (a) M. Saito, R. Haga, and M. Yoshioka, *Chem. Commun.* **2002**, 1002. (b) M. Saito, R. Haga, and M. Yoshioka, *Chem. Lett.* **2003**, *32*, 912. (c) M. Saito, R. Haga, M. Yoshioka, K. Ishimura, and S. Nagase, *Angew. Chem. Int. Ed.* **2005**, *44*, 6553.

251. V. Ya. Lee, M. Ichinohe, and A. Sekiguchi, *J. Am. Chem. Soc.* **2000**, *122*, 12604.

252. V. Ya. Lee, R. Kato, M. Ichinohe, and A. Sekiguchi, *J. Am. Chem. Soc.* **2005**, *127*, 13142.

253. V. Ya. Lee, K. Takanashi, M. Ichinohe, and A. Sekiguchi, *J. Am. Chem. Soc.* **2003**, *125*, 6012.

254. V. Ya. Lee, K. Takanashi, T. Matsuno, M. Ichinohe, and A. Sekiguchi, *J. Am. Chem. Soc.* **2004**, *126*, 4758.

255. A. Sekiguchi, T. Matsuo, and H. Watanabe, *J. Am. Chem. Soc.* **2000**, *122*, 5652.

256. K. Takanashi, V. Ya. Lee, T. Matsuno, M. Ichinohe, and A. Sekiguchi, *J. Am. Chem. Soc.* **2005**, *127*, 5768.

257. V. Ya. Lee, K. Takanashi, M. Ichinohe, and A. Sekiguchi, *Angew. Chem. Int. Ed.* **2004**, *43*, 6703.

258. S. Kyushin, H. Kawai, and H. Matsumoto, *Organometallics* **2004**, *23*, 311.

259. (a) G. N. Srinivas, T. P. Hamilton, E. D. Jemmis, M. L. McKee, and K. Lammertsma, *J. Am. Chem. Soc.* **2000**, *122*, 1725. (b) G. N. Srinivas, L. Yu, and M. Schwartz, *Organometallics* **2001**, *20*, 5200.

PART 2

METHODS AND APPLICATIONS

■■■■■ **CHAPTER 3**

An Introduction to Time-Resolved Resonance Raman Spectroscopy and Its Application to Reactive Intermediates

DAVID LEE PHILLIPS, WAI MING KWOK, AND CHENSHENG MA

Department of Chemistry, The University of Hong Kong, Hong Kong

Email: phillips@hkucc.hku.hk, kwokwm@hkucc.hku.hk, macs@hkucc.hku.hk

Reviews of Reactive Intermediate Chemistry. Edited by Matthew S. Platz, Robert A. Moss, Maitland Jones, Jr.
Copyright © 2007 John Wiley & Sons, Inc.

3.1. INTRODUCTION

Since many reactive intermediates in chemical reactions are very short-lived, a number of time-resolved spectroscopic methods have been developed to directly study these reactive intermediates. Time-resolved electronic absorption and emission spectroscopies are among the most popular time-resolved methods and can provide a great deal of information about the kinetics of chemical reactions. We note that Professor J. C. Scaiano presented a very nice chapter on nanosecond laser flash photolysis in the first volume of *Reactive Intermediate Chemistry*.[1] Because the electronic absorption and emission spectra of polyatomic molecules in room temperature liquids are usually broad and structureless, it is hard to obtain detailed information about the structure of chemical reaction intermediates from these spectra. In addition, the changes observed in the time-resolved electronic spectra do not clearly distinguish between changes in the electronic structure and vibrational energy distribution. In order to learn more about the structure of reactive intermediates in chemical reactions, time-resolved vibrational spectroscopic methods like time-resolved infrared (TRIR) and time-resolved resonance Raman (TR^3) have been developed to acquire vibrational spectra of these short-lived intermediates. These time-resolved vibrational spectroscopy techniques are able to provide a great deal of information about the vibrational structure and dynamics of chemical and biological systems. By monitoring the frequencies of the vibrational bands, structural changes can be observed. Similarly, monitoring the vibrational band intensities as a function of time enables kinetic information to be obtained. Microscopic molecular interactions can also be investigated from spectra obtained in different environmental conditions (such as hydrogen bonding interactions or over a range of solvents with varying polarity or viscosity). Time-resolved infrared absorption spectroscopy will be introduced by Professor John P. Toscano in another chapter in this book. In this chapter, we will focus on an introduction to time-resolved resonance Raman spectroscopy (TR^3) and some examples of its application to study reactive intermediates in chemical reactions. This review is intended for upper level undergraduates, graduate students, and others who would like to explore the basics of time-resolved resonance Raman spectroscopy. This review is not intended to be comprehensive but serves to provide a modest starting point to learn about time-resolved resonance Raman spectroscopy and its utility in studying chemical reactions.

Time-resolved resonance Raman (TR^3) spectroscopy experiments were first reported in 1976 and used a 30 ns pulse radiolytic source to generate the intermediates that were then probed on the microsecond time-scale by a laser source to generate the TR^3 spectrum.[2] TR^3 spectroscopy was then extended to study intermediates generated from pulsed laser photolysis.[3–5] The first picosecond TR[3] experiments reported in 1980 employed a single laser pulse to both photolyze and probe the sample.[3,4] Picosecond TR[3] experiments were then extended to a two laser pulse, pump-probe variation by Gustafson and coworkers in 1983.[5] Since these early TR[3] experiments in the mid 1970s to early 1980s, the application of TR[3] spectroscopy to study transient species has exploded and many new variations of TR[3] spectroscopy have been developed and applied to numerous excited states and chemical reaction intermediates. Pulse radiolysis, laser photolysis, electrochemical or chemical reactions can all be used to generate the reactive intermediates to be studied by TR[3] spectroscopy. In this chapter, we will focus only on typical TR[3] methods that employ laser pulses to initiate the reaction of interest and to probe the reactive intermediates.

3.1.1. Raman and Resonance Raman Scattering

Raman scattering occurs when a photon of light is scattered by a molecule and the outgoing photon energy is different from that of the incident photon due to a transfer of energy from the photon to the molecule or from the molecule to the photon. A frequency shift of the incident photon energy to lower energy is called a Stokes shift. A frequency shift of the indicant photon energy to higher energy is called an anti-Stokes shift and can occur when a vibrationally excited molecule transfers some of its vibrational energy to the incident photon.

The probability of Raman scattering from an arbitrary initial state I to a final state F is related to the Raman cross section of a molecule $\sigma_{I \rightarrow F}$ that is described as follows:

$$\sigma_{I \rightarrow F} = \frac{8\pi e^4}{9\hbar^4 c^4} E_s^3 E_L \left| \sum_{\rho\lambda} (\alpha_{\rho\lambda})_{I \rightarrow F} \right|^2 \tag{3.1}$$

where c is the velocity of light, E_S is the energy of the scattered photon of light and E_L is the energy of the incident photon of light, and $\alpha_{\rho\lambda}$ is the transition polarizability tensor with the polarization of the incident and scattered photons indicated by ρ and λ, respectively. Applying second-order perturbation theory leads to the following description of the polarizability:

$$(\alpha_{\rho\lambda})_{I \rightarrow F} = \sum_{V} \left(\frac{\langle F | m_\rho | V \rangle \langle V | m_\lambda | I \rangle}{E_V - E_I - E_L - i\Gamma} + \frac{\langle F | m_\lambda | V \rangle \langle V | m_\rho | I \rangle}{E_V - E_F + E_L - i\Gamma} \right) \tag{3.2}$$

where E_L is the incident photon energy, $|I>$ is the initial vibronic state, $|V>$ is the intermediate vibronic state, and $|F>$ is the final vibronic state. m_ρ and m_λ are the dipole moment operators and Γ is the homogeneous line width of the electronic transition. Employing the Born-Oppenheimer and Condon approximations provides the following description for the resonance Raman scattering cross section:

$$\sigma_{i \to f} = 5.87 \times 10^{-19} M^4 E_s^{\,3} E_L \left| \sum_V \frac{\langle f \mid v \rangle \langle v \mid i \rangle}{\varepsilon_v - \varepsilon_i + E_0 - E_L - i\Gamma} \right|^2 \quad (3.3)$$

where E_0 is the zero–zero energy separation between the lowest vibrational energy levels of the ground and excited states, $|v\rangle$ and $|i\rangle$ are the vibrational states with energies ε_v and ε_i (please see Fig. 3.1). The units for the energies, E_L, E_S, E_0, ε_v and ε_i are cm^{-1}. In this particular case $\varepsilon_i = 0$. The units for M (transition moment) and σ are angstroms (Å) and angstroms squared per molecule (Å^2/molecule), respectively. Equation 3.3 is the Albrecht A-term expression and is the predominant contribution to resonance Raman scattering from allowed electronic transitions.[6]

Resonance Raman scattering occurs when the incident photon of light (E_L) has enough energy to approach or become resonant with an electronic transition of the molecule ($E_0 + \varepsilon_v - \varepsilon_i$) so that the first term in equation (3.2) becomes predominant. This resonance effect or enhancement is due to the coupling between the electronic and vibrational modes of the molecule. The vibrational modes localized on the chromophore (or group of atoms responsible for the electronic transition) of the molecule are the vibrational modes that are resonantly enhanced in the resonance Raman spectra. For fundamental A-term scattering, only totally symmetric vibrational modes are able to contribute. The relative intensities of the vibrational modes

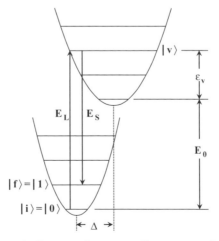

Figure 3.1. Simple schematic diagram of resonance Raman scattering in a harmonic system. The initial state $|i\rangle$ is excited by the incident photon of energy E_L that is resonant with the set of energy levels $|v\rangle$. The scattering of the photon with energy E_S takes the system from $|i\rangle$ to $|f\rangle$.

observed in a resonance Raman spectrum are dependent on both the Franck-Condon overlaps and the oscillator strength of the electronic transition.

The resonance enhancement associated with a chromophore of a molecule gives rise to several useful properties for resonance Raman spectroscopy. Raman cross sections can become several orders of magnitude larger (typically 10^3 to 10^6) owing to the resonance enhancement effect. This enables species of interest to be detected with concentrations as low as 10^{-6} M. Choosing a wavelength that is directly resonant with an allowed electronic transition selects vibrational modes associated with it and helps reject unwanted scattering. In addition, the selectivity of the resonance enhancement is useful for studying complicated molecules composed a number of chromophores. Because there are 3N-6 normal modes of vibration for non-linear molecules, 3N-5 for linear molecules (where N is the number of atoms in the molecule), the interpretation of complicated molecules like proteins can be very difficult. However, using appropriate excitation wavelengths to selectively enhance the Raman scattering associated with different chromophores (or electronic transitions) enables different parts of the molecule to be examined separately and more easily interpret the vibrational spectra associated with resonance Raman scattering. For more detailed information on the theory of Raman and resonance Raman scattering, one can refer to several review articles and textbooks.[7–11]

3.1.2. Time-Resolved Resonance Raman (TR³) Spectroscopy

Time-resolved resonance Raman (TR³) spectroscopy experiments can be labeled as single-pulse or two-pulse experiments. Both single- and two-pulse experiments can investigate short-lived photogenerated species. However, only photogenerated species that appear during the duration of the laser pulse can readily be investigated in single-pulse experiments. In a single-pulse experiment, the same laser pulse excites (or pumps) the sample and probes the sample to make the Raman scattering. In this type of experiment, a Raman spectrum is first made with a low-power laser beam and this Raman spectrum contains little photoproduct or intermediate species. Next, a Raman spectrum is then obtained using a high-power laser beam and this Raman spectrum contains appreciable photoproduct or intermediate species on the time scale of the laser pulse. A "difference" Raman spectrum is then produced by subtracting the low-power Raman spectrum from the high-power Raman spectrum. This "difference" Raman spectrum contains mainly the Raman scattering of the photogenerated species that appears during the time of the laser pulse used in the experiment. Figure 3.2 presents an example for dimethylamino benzonitrile (DMABN) of a low-power Raman spectrum (b), a high-power Raman spectrum (c) and the difference spectrum of the transient Raman spectrum (d) produced by subtracting the low-power and solvent Raman spectra from the high-power Raman spectrum. The single-pulse method has been used to obtain the transient resonance Raman spectra of early time intermediates in a variety of photochemical reactions ranging from triplet organic and inorganic intermediates to intermediates in the photocycle of bacteriorhodopsin.[12,13]

The single-pulse experiments have several disadvantages. The time-delay between the "pump" and "probe" pulses cannot be adjusted since they are part of the

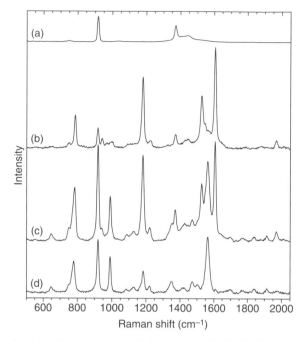

Figure 3.2. Single-pulse Raman spectra of the acetonitrile (a), low laser power DMABN in acetonitrile (b) high laser power DMABN in acetonitrile (c) and (d)=(c)-(b)-(a) obtained using 300 nm 10 ns laser pulse excitation.

same laser pulse and one can only investigate those intermediate species that appear during the duration of the laser pulse. In addition, one needs to be careful to not produce artifacts when acquiring the high-power spectrum due to unwanted interactions between the molecules and the high peak-power laser pulse (some examples include thermal heating, multiphoton absorption and re-pumping the photoproduct or intermediate species). The single-pulse difference Raman spectra are also known as transient Raman spectroscopy because they have limited time-resolution ability.

The two-pulse TR^3 experiment is one of the best techniques to obtain the kinetic and structural insight required in studying the dynamical processes involved in a photoinitiated reaction. In the two-pulse (or pump-probe) experiments, the pump pulse excites the sample to initiate the photochemical reaction and the probe pulse interrogates the sample and the intermediate species produced by the pump pulse. The pump and probe pulses are temporally separate and this enables the time-delay between the pump and probe pulses to be independently varied. The pump and probe pulses are also spectrally separate and typically have different wavelengths although in some cases the same wavelength may be used. The use of separate wavelengths for the pump and probe pulses allows them to be independently varied to optimize the pump wavelength to best excite the sample and the probe wavelength to achieve the best resonance enhancement of the Raman signal from the intermediate or reactive species of interest.

The two-pulse TR^3 experiments allow one to readily follow the dynamics and structural changes occurring during a photo-initiated reaction. The spectra obtained in these experiments contain a great deal of information that can be used to clearly identify reactive intermediates and elucidate their structure, properties and chemical reactivity. We shall next describe the typical instrumentation and methods used to obtain TR^3 spectra from the picosecond to the millisecond time-scales. We then subsequently provide a brief introduction on the interpretation of the TR^3 spectra and describe some applications for using TR^3 spectroscopy to study selected types of chemical reactions.

3.2. TYPICAL INSTRUMENTATION AND METHODS FOR DOING TIME-RESOLVED RESONANCE RAMAN (TR^3) EXPERIMENTS

Since there are a large number of different experimental laser and detection systems that can be used for time-resolved resonance Raman experiments, we shall only focus our attention here on two common types of methods that are typically used to investigate chemical reactions. We shall first describe typical nanosecond TR^3 spectroscopy instrumentation that can obtain spectra of intermediates from several nanoseconds to millisecond time scales by employing electronic control of the pump and probe laser systems to vary the time-delay between the pump and probe pulses. We then describe typical ultrafast TR^3 spectroscopy instrumentation that can be used to examine intermediates from the picosecond to several nanosecond time scales by controlling the optical path length difference between the pump and probe laser pulses. In some reaction systems, it is useful to utilize both types of laser systems to study the chemical reaction and intermediates of interest from the picosecond to the microsecond or millisecond time-scales.

An important consideration when employing high-energy laser systems to do pulsed TR^3 spectroscopy experiments is to make sure the molecule or state is not perturbed. For pulsed excitation, the fraction of molecules photolyzed is described by a photoalternation parameter F that can be expressed as

$$F = (2303 \, E \, \varepsilon \, \varphi)/(\pi \, r^2 \, N_A) \qquad (3.4)$$

where F is the fraction of molecules photolyzed by a single pulse with energy E (photons), ε is the molar extinction coefficient (M^{-1}cm^{-1}), φ is the photochemical quantum yield and r is the radius of the focused laser beam (cm).[14] One must keep the probe pulse intensity low (e.g. $F < 0.2$) in order to have less than 10% of the molecules in the illuminated volume absorb a photon as shown by equation (3.4). For molecular systems that do not relax back to their initial state (such as those that make photoproducts different from the molecule that is photoexcited), it is useful to flow or rotate the sample so that it can be replaced between pump-probe cycles. This is done so that the photoproducts do not interfere with obtaining the desired TR^3 spectrum. For pump pulses, a photoalternation parameter (F) with $F > 1$ is required to do a significant amount of photolysis so that a large concentration of transients can be generated. Improved S/N may be obtained by using higher repetition rates for the laser pulses but if the repetition rate is too high then one cannot generate enough energy per pulse to sufficiently photolyze the sample. The preceding considerations should be kept in mind when preparing to do nanosecond TR[3] and ultrafast TR[3] experiments using the instrumentation that is described in the next two sub-sections.

3.2.1. Nanosecond Time-Resolved Resonance Raman (TR[3])

Figure 3.3 displays a simple schematic diagram of the basic components of a typical two-pulse nanosecond TR[3] experimental apparatus.[15–18] Nanosecond lasers are used to supply the light for the pump and probe laser pulses and are the first major components needed for the nanosecond TR[3] experiments. While a variety of nanosecond lasers (Nd:YAG, Nd:YLF, nitrogen and others) can be used for nanosecond TR[3] experiments, Nd:YAG laser systems are commonly used for helping to generate the pump and probe laser pulses with 5–10 ns pulse widths and 10–100 Hz pulse repetition rates. The 1064 nm fundamental wavelength of the Nd:YAG lasers can be frequency doubled (SHG), tripled (THG), and quadrupled (FHG) to make 532 nm, 355 nm, and 266 nm laser pulses respectively. In order to generate other wavelengths of light for use as pump and probe pulses in the nanosecond TR[3] experiments, these harmonics (532 nm, 355 nm, and 266 nm) of the Nd:YAG laser can pump optical parametric oscillators (OPOs) or dye lasers to produce continuously tunable wavelengths of laser light in the visible region. The output of the OPOs or dye lasers can

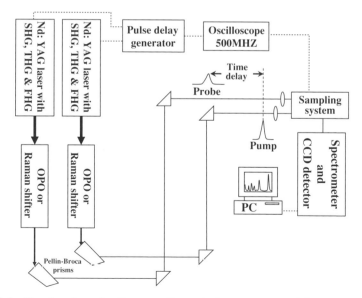

Figure 3.3. Simple schematic diagram of a typical apparatus used for nanosecond TR[3] experiments. See text for more details.

also be frequency doubled or mixed with the output of the Nd:YAG laser to generate laser pulses in the near ultraviolet and ultraviolet wavelength regions. Thus, nanosecond laser pulses can be generated through out the ultraviolet and visible spectral regions. One can also use Raman shifters to generate discrete wavelengths of laser pulses in the ultraviolet and visible spectral regions from the harmonics of the Nd:YAG lasers. Raman shifters are less expensive but are not as versatile in selecting pump and probe wavelengths since they only generate discrete Raman shifted laser lines and are not continuously tunable.

The use of two laser systems enables the pump and probe laser pulses to be independently varied for both their spectral wavelengths and their relative timing. In order to control the relative timing of the pump and probe laser pulses, an electronic pulse generator is used to trigger both the flashlamps and Q-switches of the two laser systems. The relative timing of the pump and probe laser pulses is usually monitored by a fast photodiode connected to a fast oscilloscope to display the timing of the laser pulses. A small part of the pump and probe lasers are directed to a fast photodiode and the rest of the pump and probe laser pulses are directed onto the sample of interest. The pump and probe pulses are usually loosely focused onto the sample so that they spatially overlap in the sample and this overlap region can be imaged into the monochromator by the light collection optics.

To help prevent photoproduct buildup and laser damage to the sample, the sample will usually be moving at a rate fast enough so that new sample will be exposed to each set of pump-probe pulses. A typical moving sample may be a flowing gas stream, a flowing stream of liquid, a liquid in a spinning NMR tube, or a spinning or moving solid. Figure 3.4 shows a simple schematic diagram for the sample handling and light collection system used for a typical nanosecond TR^3 experiment for a flowing liquid stream of sample. The pump and probe laser beams are loosely focused and spatially overlapped onto the flowing liquid stream of sample at the

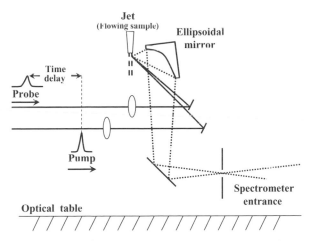

Figure 3.4. Simple diagram for the sample handling and light collection system used for a typical nanosecond TR^3 experiment for a flowing liquid stream of sample. See text for more details.

region where the optics collect the light to image into the monochromator. A back-scattering geometry as shown in Figure 3.4 can be used for both the excitation of the sample (by the pump and probe beams) and collection of light from the excited region of sample. The backscattering geometry helps to minimize sample reabsorption of the resonance Raman scattered signal and allows a larger amount of Raman scattered light to be collected. The collection optics can be lenses or mirrors. Figure 3.4 illustrates a system that employs mirrors (or reflective optics) to collect the Raman scattered light. The use of mirrors helps to minimize chromatic aberrations that may moderately distort the relative intensities of the resonance Raman spectrum. Next, the collection optics image the Raman scattered light onto the entrance slit of the monochromator. The monochromator system may be a single, double, or triple grating system depending on how much stray light rejection one wants to achieve and how much money one is willing to spend. The larger number of gratings, the more stray light rejection and lower Raman shift bands can be obtained but a lower amount of light signal will reach the detector due to losses of light from a greater number of reflective surfaces. For resonance Raman scattering, the resonance enhancement is usually sufficient so that high quality resonance Raman spectra can be obtained with a single grating monochomator system. We have found that a single grating monochromator system has been adequate for most of the nanosecond TR^3 experiments done in our laboratory. The grating(s) of the monochromator disperse the resonance Raman light onto a multichannel detector such as a charge coupled device (CCD) or a photodiode array that detects the light signal and converts it to an electrical signal exported to interfaced electronics and computer system. The detector typically collects the signal for 60–300 s before being readout to an interfaced PC computer and a number of these readouts (5 to 10 for example) are summed to find the resonance Raman spectrum at each time-delay. The number of readouts collected depends on how much signal averaging one wants to do to acquire higher signal to noise spectra. A pump only, a probe only, and pump-probe spectra are obtained for each time-delay. A background scan can also be obtained before and after each experimental trial. The known solvent Raman bands can be used to calibrate the wavenumber shifts of the resonance Raman spectra. The solvent and parent sample Raman bands can be deleted from the pump-probe spectra by subtracting the probe only and pump only Raman spectra so as to find the time-resolved resonance Raman spectra.

3.2.2. Ultrafast Time-Resolved Resonance Raman (TR^3) Spectroscopy

Ultrafast time-resolved resonance Raman (TR^3) spectroscopy experiments need to consider the relationship of the laser pulse bandwidth to its temporal pulse width since the bandwidth of the laser should not be broader than the bandwidth of the Raman bands of interest. The change in energy versus the change in time Heisenberg uncertainty principle relationship can be applied to ultrafast laser pulses and the relationship between the spectral and temporal widths of ultrafast transform-limited Gaussian laser pulse can be expressed as

$$\Delta\omega\,\Delta t \cong 14.7\,\text{cm}^{-1}\,\text{ps} \tag{3.5}$$

where $\Delta\omega$ (in cm^{-1}) is the spectral width and Δt (in ps) is the pulse width. Thus, a 1.5 ps laser pulse would have a spectral width of about 10 cm^{-1} for a near transform-limited Gaussian ultrafast laser pulse. It is now relatively common to use near transform-limited Gaussian laser pulses with time-resolution in the 1–3 ps range to perform picosecond TR^3 spectroscopy experiments where the spectral widths are similar to or somewhat less than the Raman bandwidths of samples in liquid solvents.[14,19,20] There are a wide range of methods that can be used to generate ultrafast laser pulses and that have been used to do picosecond TR^3 spectroscopy experiments (for examples see laser systems described in references 19 and 20). However, with advances in solid-state laser technology, many laser systems being used today for picosecond TR^3 spectroscopy are based on all solid state Ti:Sapphire oscillator/amplifier systems because of their greater stability and ease of use.[21–23] Therefore, we will only describe a commonly used versatile Ti:Sapphire oscillator/amplifier laser system that can be used for picosecond TR^3 experiments.

Figure 3.5 presents a simple schematic diagram of the basic components of a typical two-pulse picosecond TR^3 experimental apparatus based on a Ti:Sapphire oscillator/amplifier laser system. The 532 nm CW output from a diode pumped Nd:YVO$_4$ laser pumps a femtosecond Ti:Sapphire oscillator that is tunable from about 720 nm to 850 nm with a 80 MHz repetition rate and 5–15 nJ per pulse. The output from the femtosecond Ti:Sapphire oscillator can be used to seed a picosecond Ti:Sapphire regenerative amplifier system that is pumped by the 527 nm output from a CW diode pumped Nd:YLF intra-cavity doubled kHz Q-switched pump laser. The output

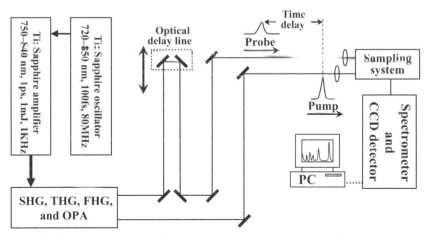

Figure 3.5. A simple schematic diagram of the basic components of a typical two-pulse picosecond TR^3 experimental apparatus based on a Ti:Sapphire oscillator/amplifier laser system. See text for more details

from the picosecond Ti:Sapphire regenerative amplifier system typically operates at 1 kHz with 1–2 mJ per pulse and a 1 ps pulse width and is tunable from 750–840 nm. This output from the picosecond Ti:Sapphire regenerative amplifier system can be frequency doubled and/or tripled to supply pump/probe wavelengths in the 375–420 nm and 250–280 nm regions. The output from the picosecond Ti:Sapphire regenerative amplifier system (around 800 nm) or its frequency doubled component (around 400 nm) can also be used to pump one or two optical parametric amplifier(s) (OPA) to generate a range of laser wavelengths in the near infrared (900–2700 nm) or visible regions (740 nm to 470 nm). The output from the OPAs can also be frequency doubled or mixed with the residual fundamental output from the Ti:Sapphire amplifier to extend the laser wavelengths into the visible and ultraviolet regions. Using appropriate combinations of the harmonics of the picosecond Ti:Sapphire regenerative amplifier system as well as the output from the OPAs and their harmonics or mixing with the fundamental picosecond Ti:Sapphire regenerative amplifier output can be used to generate laser wavelengths over much of the 200–1100 nm region for use as the pump and probe wavelengths for use in picosecond TR^3 experiments.

The characterization of the laser pulse widths can be done with commercial autocorrelators or by a variety of other methods that can be found in the ultrafast laser literature.[24] For example, we have found it convenient to find time zero delay between the pump and probe laser beams in picosecond TR^3 experiments by using fluorescence depletion of trans-stilbene.[25] In this method, the time zero was ascertained by varying the optical delay between the pump and probe beams to a position where the depletion of the stilbene fluorescence was halfway to the maximum fluorescence depletion by the probe laser. The accuracy of the time zero measurement was estimated to be ±0.5 ps for 1.5 ps laser pulses.[25] A typical cross correlation time between the pump and probe pulses can also be measured by the fluorescence depletion method.

A computer-controlled motorized translation stage mounted with a retro-reflector is used to vary the pump laser beam path relative to the probe laser beam path and this controls the relative timing between the pump and probe laser beams. Note that a one-foot difference in path length is about 1 ns time delay difference. The picosecond TR^3 experiments are done essentially the same way as the nanosecond TR^3 experiments except that the time-delay between the pump and probe beams are controlled by varying their relative path lengths by the computer-controlled motorized translation stage. Thus, one can refer to the last part of the description of the nanosecond TR^3 experiments in the preceding section and use the pump and probe picosecond laser beams in place of the nanosecond laser beams to describe the picosecond TR^3 experiments.

3.3. SOME BASICS FOR INTERPRETING TIME-RESOLVED RESONANCE RAMAN (TR^3) SPECTRA

3.3.1. Vibrational Spectroscopy and Functional Groups

To gain structural information from the experimentally recorded Raman spectrum, it is necessary to determine what vibrational mode corresponds to each band in the spectrum. This assignment can be rather difficult due to the large number of closely spaced peaks even in fairly simple molecules. Practically, there are three ways to make the vibrational assignments.

   (i) Correlate Raman bands appearing at characteristic frequency regions to the vibration(s) of a specific functional group.
  (ii) Perform Raman measurements on isotopically labeled compounds and make assignments based on the observed isotopic shifts of the Raman bands.
 (iii) Do computer simulations to calculate the Raman spectrum of the molecule and make assignments based on direct comparison between the calculated and experimental spectra.

In some cases, a combination of the three methods is needed for explicit attribution of the Raman spectrum; this is especially true for analysis of excited state Raman spectra.

For (i): Different motions of a molecule will have different vibrational frequencies. As a general rule, bond stretches are the highest energy vibrations; bond bending vibrational modes are somewhat lower in energy and torsional motions are even lower still in energy. The lowest frequencies are usually torsions between subgroups of large molecules and breathing modes in very large molecules. As the first step to make a preliminary assignment to a Raman spectrum, especially for a spectrum of a ground state species, it is helpful to consider that a certain functional group could be associated with Raman bands appearing at characteristic frequencies. Extensive compilations of Raman vibrational frequencies are available.[26,27] Just as an example, a few frequencies for selected functional groups in organic compounds are listed in Table 3.1.

Besides the vibrational frequency, symmetry is another important feature to describe a vibrational motion. The symmetry of a normal vibration can be determined by measuring the depolarization ratio of the corresponding Raman band, which is defined as

TABLE 3.1. Example of Raman frequency for several functional groups.

	Frequency (cm^{-1})
Alkane C−C stretch	1040–1100
C−Cl strech	650–660
C−Br stretch	565–560
C−I stretch	500–510
C−S stretch	580–704
Cyclopentane ring breath	889
Benzene ring breath	990
Aliphatic CH_3 deformation	1380–1350
Aromatic center C=C	1580–1620
C=O	1730–1740
C≡N	2210–2230

$$\rho = I_{\perp}/I_{//} \tag{3.6}$$

where $I_{\perp}$ and $I_{//}$ are the measured perpendicular and parallel polarization intensity, respectively, relative to the fixed polarization direction of the incident laser beam. The depolarization ratio may vary from near zero for a highly symmetrical vibration to a theoretical maximum of ~0.75 for a totally nonsymmetric vibration. The band depolarization ratio is useful not only to make band assignments but also to identify and detect weak Raman bands that are overlapped by strong Raman bands.

For (ii): The simplest description of a vibration is a harmonic oscillator, which has been found to work reasonably well for most systems. Within this model, the oscillation frequency is given by

$$\omega_{osc.} = \frac{1}{2\pi c}\sqrt{\frac{k}{\mu}}\ \mathrm{cm}^{-1} \tag{3.7}$$

where k is the force constant of the relevant bond, μ is the reduced mass of the related group and c is the velocity of light. From this, one can see that change in the mass of the atom undergoing vibration within a group (i.e. change μ) will cause an alteration in the frequency of the associated vibrational mode. That is, an increase of the atom weight leads to a frequency down shift of the associated vibration while a decrease of the atom weight results in a frequency up-shift of the associated vibration. This is the theoretical basis for making a vibrational assignment based on the isotopic labeling. As an example to show how the functional group and isotopic frequency shifts can be used to help with the vibrational analysis of a Raman spectrum, we display in Figure 3.6 a comparison of the resonance Raman spectra obtained for dimethylaminobenzonitrile (DMABN) and three of its isotopomers.[28–30]

From Figure 3.6, it is obvious that the isotopic insensitive band appearing at ~2219 cm^{-1} is due to the characteristic stretching vibration of the cyano functional group. Isotopic substitution of the amino group leads to frequency downshifts for bands dominated by the amino motions, such as the band with contributions from the N_{amino}-$(CH_3)_2$ stretching and the methyl deformation motions that display ~7 cm^{-1} and ~111 cm^{-1} downshifts in the frequency upon ^{15}N and methyl deuteration, respectively; on the contrary, the ring deutration can induce extensive frequency shifts for many of ring related vibrations: for example, the Raman band with contributions from the ring center carbon-carbon stretch motion (the Wilson 8a mode) and that associated with the ring C−H in-plane bending vibration (the Wilson 9a mode) display, respectively, ~30 cm^{-1}, and ~300 cm^{-1} frequency down shifts upon deuteration. On the basis of the isotopic specified frequency shifts, most of the vibrational bands can be assigned clearly. However, it must be mentioned that for some bands, particularly those with combined contributions from various local vibrations, certain isotopic labeling may, in some cases, lead to an alteration in the nature and composition of the associated normal modes,[31–33] the explanation of such isotopic induced spectral changes are not straightforward and theoretical calculations are required to provide further independent evidence for their identification.

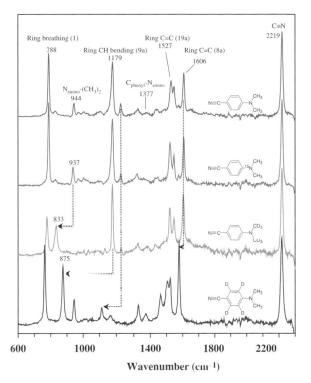

Figure 3.6. Resonance Raman spectra of the ground state of DMABN, DMABN-^{15}N, DMABN-d_6, DMABN-d_4 obtained with 330 nm excitation in methanol.

For (iii): In recent years, ab initio and density functional theory computational methods, such as Hartree-Fock (HF), density functional theory (DFT) and Møller-Plesset (MP2) combined with various basis sets, have been extensively employed to determine the vibrational frequencies for the purpose of not only vibrational analysis but also identification of experimentally observed short-lived transient species.[34] The theoretically predicted frequency and vibrational spectrum may serve as fingerprints for making assignments of the experimental spectrum, determining the structure for the species of interest and helping make an attribution of the reactive intermediate. Since the normal vibrational modes are deduced from the equilibrium geometry, to calculate the frequencies, the program must first compute the geometry of the molecule and then compute the harmonic frequencies at exactly the same level of theory used to optimize the geometry. In these computations, the vibration of a molecule is described by a quantum harmonic oscillator and the calculated harmonic frequencies are typically larger than the fundamentals observed experimentally due mainly to the neglect of anharmonicity effects in the theoretical treatment. However, systematic studies have found that the overestimation of the calculated frequencies are, in most cases, relatively uniform so that, by multiplying the resulting frequency by a scaling factor, a good overall agreement between the scaled theoretical frequencies

TABLE 3.2. Summary of Recommended Frequency Scaling
Factors. [Reprinted in part with permission from reference
[34]. Copyright (1996) American Chemical Society].

Level of theory	Scaling factor
HF/3-21G	0.9085
HF/6-31G(d)	0.8953
HF/6-31+G(d)	0.897
HF/6-31G(d,p)	0.8992
HF/6-311G(d,p)	0.9051
HF/6-311G(df,p)	0.9054
B-LYP/6-31G(d)	0.9945
B-LYP/6-311G(df,p)	0.9986
B-p86/6-31G(d)	0.9914
B3-LYP/6-31G(d)	0.9614
B3-P86/6-31G(d)	0.9558
B3-PW91/6-31G(d)	0.9573
MP2-fu/6-31G(d)	0.9427
MP2-fc/6-31G(d)	0.9434
MP2-fc/6-31G(d,p)	0.937
MP2-fc/6-311G(d,p)	0.9496

and the experimental frequencies can usually be obtained. A list of recommended
frequency scaling factors for typically used methods and basis set combinations are
given in Table 3.2.

Among the various methods, the B3-LYP based DFT procedure appears to
provide a very cost-effective, satisfactory and accurate means of determining the
vibrational frequencies.[34] As an example, Figures 3.7 and 3.8 display direct comparisons between the ground state experimental and DFT B3-LYP/6-31G* calculated
Raman spectra for DMABN and its ring deuterated isotopmer DMABN-d$_4$.[30] The
experimental spectra are normal Raman spectra recorded in solid phase with 532 nm
excitation. For the calculated spectra, a Lorentzian function with a fixed band width
of $\sim$10 cm^{-1} was used to produce the vibrational band and the computed frequencies
were scaled by a factor of 0.9614.

It can be seen from Figures 3.7 and 3.8 that the calculations reproduce very well
not only the experimental spectra but also the experimentally observed isotopic
shifts indicating a high reliability of the computational method. According to this
comparison, definite attribution can be made for even the difficult Raman bands that
cannot be assigned based solely on the experimental results. It is, however, necessary to mention at this point that the calculated Raman spectrum provided directly
by the ab initio computations correspond to the normal Raman spectrum with the
band intensity determined by the polarizability of the correlating vibration. Since
the intensity pattern exhibited by the experimentally recorded resonance Raman
spectrum is due to the resonance enhancement effect of a particular chromophore,
with no consideration of this effect, the calculated intensity pattern may, in many

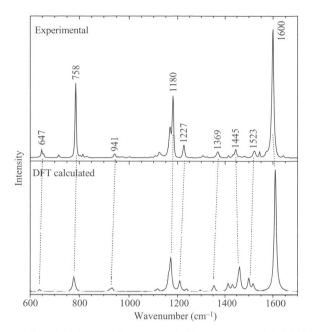

Figure 3.7. Comparison of the ground state normal Raman spectra of DMABN obtained by experimental measurement (with 532 nm excitation in solid phase) and the spectrum obtained from a DFT B3-LYP/6-31G* calculation. (from reference [30] - Reproduced by permission of the PCCP Owner Societies.)

cases, not be directly comparable to the experimental one. For the theoretical calculation to consider the resonance enhancement effect, extra information about the relative changes in the potential surface of the upper state associated with the resonance enhance excitation is needed. Examples for Raman calculations involving the resonance enhancement effect can be found in several papers performed by several groups.[35,36]

Compared with the structure and frequency calculations for the ground state species, the calculations for the excited state are more challenging and relatively less accurate. Commonly used theoretical methods for the excited state calculations include the HF level single-excitation configuration interaction (CIS),[37] the complete active space self-consistent field (CASSCF)[38] and time-dependent density functional theory (TDDFT)[39] for both the singlet, and triplet excited state calculations and open shell DFT calculations for the triplet state calculations.[40–43] Electronic excitation can lead to remarkable structural changes. As a result of this, the frequencies and normal modes revealed by the excited state Raman spectrum can be significantly different from those of the ground state spectrum. Comparative studies based on experimental isotopic labeling TR^3 measurements and theoretical calculations provide a reasonably reliable way to carry out an analysis on the excited state spectrum and determine the excited state structure and electronic property.

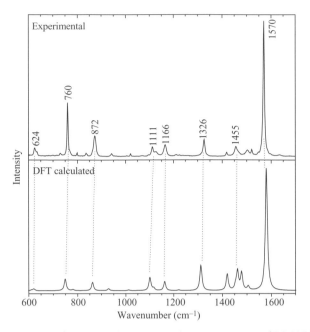

Figure 3.8. Comparison of the ground state normal Raman spectra of DMABN-d_4 obtained by experimental measurement (with 532 nm excitation in the solid phase) and the spectrum obtained from a DFT B3-LYP/6-31G* calculation. (from reference [30] Reproduced by permission of the PCCP Owner Societies.)

3.3.2. Environmental and Solvent Effects on TR³ spectra

The Raman spectrum of a molecule not only depends on the strengths of the molecular bonds, but may also be markedly influenced by environmental factors. Such intermolecular interactions modify the Raman spectra in a number of ways: the wavenumber of the normal vibrational modes of a molecule may be shifted to higher or lower values, the Raman band intensity can be altered, and the half-width of Raman bands may be greatly increased. The wavenumber displacement and alteration in the band-width of a solute vibration is a complex function of both solute and solvent properties and can be explained in terms of weak nonspecific electrostatic interaction (dipole-dipole, dipole-induced dipole, etc.) and of strong specific association of solute with solvent molecules, usually of the hydrogen-bond type.[44] One example showing such an effect is illustrated in Figure 3.9 for the solvent effect on the Raman band corresponding to the carbonyl C=O stretching motion of *p*-methoxyacetophone (MAP).[40]

It is obvious from Figure 3.9 that both the frequency and bandwidth of the C=O stretching mode displays remarkable sensitivity to the polarity and the H-bonding ability of the solvent. The frequency downshift and slight bandwidth broadening observed when changing the solvent from cyclohexane to MeCN is due to the nonspecific electrostatic solvent–solute interaction affected mainly by the solvent

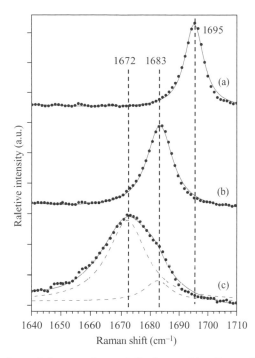

Figure 3.9. Comparison of the experimental C=O stretching Raman band obtained for *p*-methoxyacetophenone in cyclohexane (a), MeCN (b) and 50% H_2O/50% MeCN (v:v) (c) solvents with 532 nm excitation. (Reprinted with permission from reference [42]. Copyright (2005) American Chemical Society.)

property such as polarity and polarizability. The extra frequency downshift and substantial broadening of the C=O band observed in the water containing solvent compared to the non-hydrogen-bonding cyclohexane and acetonitrile is a characteristic reflection of specific solute-solvent hydrogen-bonding interaction. Detailed examination of the Raman C=O band seen in the water-organic mixed solvent reveals two sub-bands that can be associated, respectively, with the free C=O stretching as seen also in the other two solvents and the hydrogen-bonded C=O stretching displayed exclusively in the hydrogen-bonding donating (HBD) solvent. The electron-withdrawing effect induced by the hydrogen-bond formation leads to a weakening of the associated C—O bond and is responsible for the excessive frequency downshift exhibited by the hydrogen-bonded mode relative to that of the free mode. The significant broadening of the hydrogen-bonded mode is caused by a decrease in the vibrational dephasing time.[45–48] The observed co-existence of the free and hydrogen-bonded modes is due to the multiple hydrogen-bonding equilibria associated with the strong aggregation nature of the water solvent.[44,45] The predominant contribution of the hydrogen-bonded mode to the overall Raman band indicates that most of molecules exist with the carbonyl group hydrogen-bonded with the surrounding water molecules.

In most cases, the specific solute-solvent hydrogen-bonding effect is local in nature affecting only the vibrational mode of the bond involved in this interaction and it represents only a small perturbation with little influence on the intrinsic electronic and structural nature of the molecular system. It is however, interesting to note that, under certain circumstances, the influence of the hydrogen-bonding effect can be so profound in modifying both the electronic and structural configuration that the hydrogen-bonded complex can be considered as a distinct species from the corresponding free molecules. One such example is that of the triplet state of MAP.[40] Because of the increase of basicity for the MAP carbonyl oxygen, the solute-solvent hydrogen-bonding strength becomes stronger in the triplet excited state relative to the ground state. By employing a combination of TR[3] spectroscopy and DFT calculations, it has been revealed that the strengthened hydrogen-bonding interaction and the resulting electronic and structural changes are responsible for the differences observed in the triplet resonance Raman spectra (displayed in Figure 3.10) and the

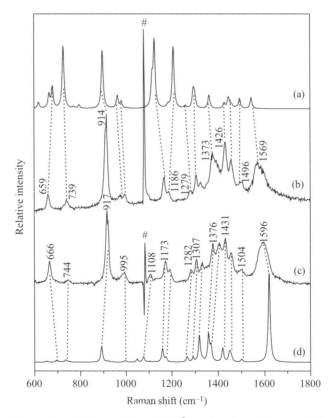

Figure 3.10. Comparison of the experimental TR[3] spectra of p-methoxyacetophenone (MAP) obtained in MeCN (b) and 50% H_2O/50% MeCN (v:v) (c) with the DFT calculated spectra for the free triplet state (a) and triplet of the carbonyl hydrogen-bonded complex (d). (Reprinted with permission from reference [42]. Copyright (2005) American Chemical Society.)

triplet lifetime (displayed in Figure 3.11) in the water containing MeCN and neat MeCN solvents.

The DFT calculations based on the B3-LYP method and the 6-311G** basis set reveals that the free MAP triplet has a slightly twisted geometry with a delocalized $\pi\pi^*$ character and a single-bond-like carbonyl group, whereas the triplet state of the hydrogen-bond complex has a planar structure with a significant ring localized biradical $\pi\pi^*$ state, and a quinoidal ring and a double-bond-like carbonyl group. The calculated structures and their relevant structural parameters are presented in Figure 3.12. As shown in Figure 3.10, the DFT calculated frequencies reproduce reasonably well not only the experimental spectra but also the spectral differences observed in the two solvents and help justify the validity of the calculated results. By facilitating the out-of-plane distortion and increasing the ISC non-radiative rate, the slightly twisted structure observed for the free triplet is one of the main factors accounting for its shorter lifetime than the planar hydrogen-bonded triplet state complex. One characteristic spectral manifestation for the increased quinoidal feature induced by the enhanced carbonyl hydrogen-bonding effect is the TR^3 frequency up-shift (by $\sim27\,cm^{-1}$) of the ring center C=C stretching vibration in the mixed solvent ($1596\,cm^{-1}$) from the neat MeCN ($1569\,cm^{-1}$). This Raman band can thus be taken as a "marker" to signify the carbonyl hydrogen-bonding effect on the triplet structure for a wide range of aromatic carbonyl compounds.[43,49]

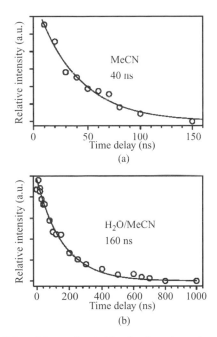

Figure 3.11. Lifetime of the triplet estimated by the time dependence of the $\sim660\,cm^{-1}$ TR^3 band area obtained in MeCN (a) and 50% H$_2$O/50% MeCN (b) solvents. (Reprinted in part with permission from reference [42]. Copyright (2005) American Chemical Society.)

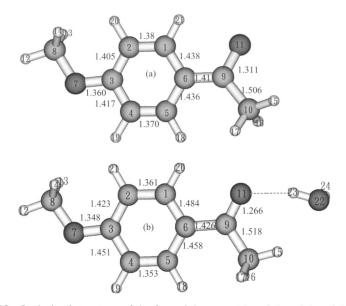

Figure 3.12. Optimized structure of the free triplet state (a) and the triplet of the carbonyl H-bond complex (b) calculated from the DFT calculations using the UB3LYP method with a 6-311G** basis set. (Reprinted with permission from reference [42]. Copyright (2005) American Chemical Society.)

Besides the effect due to the above-mentioned nonspecific and specific inter-solute-solvent interaction, solution phase excess energy relaxation dynamics are another important source of an environmental factor influencing the early time picosecond TR^3 spectra. Following the photoexcition of a molecule in solution, excess excitation energy is redistributed internally and then transferred to surrounding solvent environment. For large polyatomic molecules in solution, intramolecular redistribution of high frequency vibrational energy on the sub-picosecond time scale (IVR) is followed by a much slower excess energy relaxation on the first tens of picosecond time regime attributed to cooling and solvent reorganization.[50–52] The Resonance Raman spectrum is a sensitive probe of the relative long time-scale relaxation dynamics. As indicated in previous studies on *trans*-stilbene[53–56] and several dye molecules,[57–59] the characteristic spectral manifestation of such a process is that some bands in the corresponding excited state TR^3 spectra exhibit time-dependent intensity changes accompanied by simultaneously frequency up-shifts and band width narrowing with ~10–20 ps time constants. Figure 3.13 presents an example for such Raman band changes. These dynamical changes in the Raman spectra have been linked to many factors such as the solution phase heat transport (thermal diffusion) and the relaxation induced time-dependent change of the relevant electronic potential including the electronic resonance shifts, width changes and relaxation of the Franck-Condon displacement.[53–60]

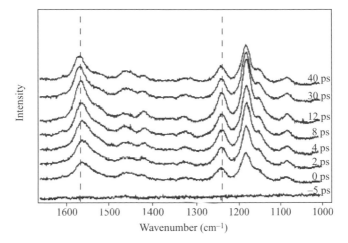

Figure 3.13. Resonance Raman spectra of S_1 excited state trans-stilbene in decane at delay times indicated. The pump wavelength was 292.9 nm and the probe wavelength was 585.8 nm. The vertical dashed lines illustrated the substantial spectral evolution of the 1565 cm⁻¹ compared to the 1239 cm⁻¹ band. (Reprinted with permission from reference [56]. Copyright (1993) American Chemical Society.)

3.4. SOME APPLICATIONS OF TIME-RESOLVED RESONANCE RAMAN (TR³) SPECTROSCOPY TO STUDY SELECTED CHEMICAL REACTIONS AND THEIR REACTIVE INTERMEDIATES

In this section, we will very briefly describe selected examples of the application of time-resolved resonance Raman (TR³) spectroscopy to the study of chemical reactions and the reactive intermediates which participate in those transformations.

3.4.1. Charge Transfer Reactions

Example: Charge Transfer in 4-Dimethylaminobenzonitrile (DMABN)

4-Dimethylaminobenzonitrile (DMABN) is an archetypal electronic donor-acceptor molecule that undergoes an excited state intramolecular charge transfer (ICT) reaction accompanied by structural reorganization in the solution phase.[61,62] Since its discovery in 1951, the solvent dependent dual fluorescence phenomenon of DMABN has been rationalized in terms of a nonradiative ICT conversion between a locally excited LE state and an ICT state. The ICT rate has been found to be highly solvent polarity dependent; such as being $\sim 2 \times 10^{11}$ s⁻¹ in MeCN but $\sim 4 \times 10^{10}$ s⁻¹ in dioxane.[63] There have been, however, fierce debates about the structure of the ICT state over the past 50 years. Monomolecular models interpreting the nature and structure of the ICT state fall into four main classes: TICT (Twisted Intramolecular Charge Transfer),[64] WICT (Wagged ICT),[65] PICT (Planar ICT),[66] and RICT (Rehybridisation ICT).[67] All have an ICT mechanism in common but they differ in the detail nucleic coordinates related to the LE-to-ICT conversion. Many experimental and theoretical studies have been devoted to investigate and evaluate the proposed models. Time-resolved vibrational spectroscopy is a powerful tool to provide direct structural information for excited state molecules and both picosecond time-resolved resonance Raman (TR^3)[68,69] and time-resolved infrared ($TRIR$)[70-72] spectroscopy has been employed to study the DMABN ICT structure. Vibrational spectra obtained from the two methods are generally consistent in terms of the frequencies and assignments of the commonly observed bands. However, due to different selection rules, it turns out that, in relation to the TRIR results, the ps-TR^3 experiment allows noticeably more vibrational bands to be observed and identified. The rich information provided by the ps-TR^3 spectra enables a better comparison with the relevant theoretical results and this leads to an explicit determination of the DNABN ICT structure.

Ps-K-TR^3 spectra of the ICT state of DMABN and two of its isotopomers DMABN-N^{15} and DMABN-d_6 (the isotopic substitution was in the $-N(CH_3)_2$ group) at 50 ps delay time recorded in methanol solvent are displayed in Figure 3.14. The probe wavelength of 330 nm is resonant with the strong transition from the ICT state identified by Okada *et al.*[73] The Kerr gate technique was used to reject the ICT fluorescence and extract the Raman signal. The experiment on the isotopic labeled DMABN compounds are used to help make explicit assignments to the observed Raman bands. Figure 3.15 shows

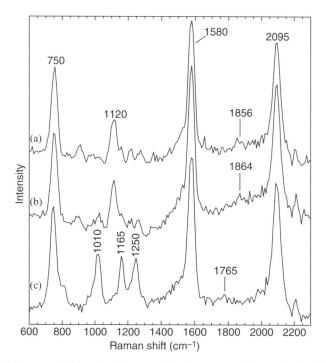

Figure 3.14. Picosecond Kerr gated time-resolved resonance Raman (ps-K-TR^3) spectra of the ICT state of DMABN (a), DMABN-N^{15} (b) and DMABN-d_6 (c) obtained by 267 nm pump, 330 nm probe in methanol at 50 ps delay time. (Reprinted with permission from reference [28]. Copyright (2001) American Chemical Society.)

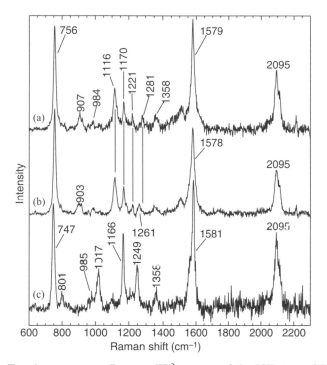

Figure 3.15. Transient resonance Raman (TR²) spectra of the ICT state of DMABN (a), DMABN-N¹⁵ (b) and DMABN-d₆ (c) obtained by 330 nm excitation in methanol solvent. (Reprinted with permission from reference [28]. Copyright (2001) American Chemical Society.)

the corresponding transient Raman spectra for the ICT state of the three compounds obtained using a 330 nm single-color pump probe method. The transient Raman data offer superior spectral resolution allowing better vibrational analysis.

It can be seen from the spectra that transient Raman bands common to all the three compounds are observed at ~756, ~984, ~1170, ~1221, ~1358, ~1580, and ~2095 cm⁻¹. IR measurements of the benzonitrile radical anion yielded bands at 760, 991, 1178, 1268, 1283, 1592, 2093 cm⁻¹.[74] The similarity of these two groups of frequencies is striking and implies that the benzonitrile subgroup in the ICT state of DMABN resembles the benzonitrile radical anion. This is consistent with a full charge transfer mechanism from the dimethylamino to the benzonitrile group and this lends support to the electronic decoupled TICT model. It is certain that, except for the ~2095 cm⁻¹ $C\equiv N$ stretching vibration, the other bands observed here belong to local phenyl ring modes. The bands at 1281, 1116, 907/1261, 1115, 903/1249, 1017, and 801 cm⁻¹ in the spectra of DMABN/DMABN-N¹⁵/DMABN-d₆, are sensitive to isotropic substitution. Comparison in spectra of DMABN and DMABN-N¹⁵ indicates that the ~1281 and 907 cm⁻¹ DMABN band are exclusively sensitive to the N¹⁵ substitution. On the basis of the observed isotopic shifts and its comparison with the ground state counterparts, the two bands can be attributed explicitly to vibrational modes dominated by the C_{phenyl}-N_{aminio} stretching and N_{amino}-$(CH_3)_2$ symmetric stretching

TABLE 3.3. Frequencies and tentative assignments of the Raman bands observed for the DMABN ICT state and a comparison with their ground state counterparts.

Tentative assignment	ICT state	Ground state
C≡N stretch	2095	2219
Ring C=C stretch (8a)	1580	1606
C_{phenyl}-N_{amino} stretch (13)	1281	1377
C_{phenyl}-C_{cyano} stretch (7a)	1221	1227
Ring C−H in-plane-bending (9a)	1170	1170
Deformation of the amino methyl	1116	1166
Ring C−H in plane bending (18a)	984	1003
N_{amino}-$(CH_3)_2$ stretch	907	944
Ring breathing (1)	756	788

motion, respectively. The remaining $1116/1115/1017\,cm^{-1}$ bands in the spectra of the three compounds are consistent with their assignments as modes dominated by vibrations of the amino methyl. Table 3.3 lists the frequencies and assignments of the Raman bands observed for the ICT state of DMABN. The frequencies of the ground state counterparts are also listed for comparison purposes.

Among the observed Raman bands, frequencies of the two modes, C_{phenyl}-N_{amino} and C≡N stretch, are crucial in determining the electronic and geometrical properties of the ICT state.[31–33] The $\sim96\,cm^{-1}$ and $\sim124\,cm^{-1}$ downshift of the C_{phenyl}-N_{amino} and C≡N mode from their respective ground state counterpart agree well with the corresponding theoretical values for the twisted structure obtained by CASSCF calculation on the DMABN ICT state. These results can thus be taken as convincing evidence lending strong support to the TICT model. The extent of the frequency downshifts of the two modes also helps to rule out the PICT and RICT models since the theoretical downshift of the C≡N vibration are estimated to be $\sim36\,cm^{-1}$ and $\sim800\,cm^{-1}$, respectively, for the PICT and RICT models and a frequency up-shift, rather than the observed downshift has been predicted for the PICT C_{phenyl}-N_{amino} mode. As to the WICT model, its validity is doubted since calculations show that a simple wagging motion of the dimethylamino group cannot, by itself, lead to the required highly polarity of the ICT state. Sophisticated theoretical work on DMABN provides detailed structural parameters for the TICT state.[39] It has been found that the associated vibrational frequencies of the various vibrational modes (not only the C_{phenyl}-N_{amino} and C≡N modes but also the modes associated with the aromatic ring) are in general agreement with the experimental Raman frequencies. This justifies further the validity of the TICT description for the DMABN ICT state.

3.4.2. Excited States and Energy Relaxation

Example: Singlet Excited State of *Trans*-Stilbene

Electronic excited states are often reactive intermediates in many photochemical reactions. In a number of cases, the excited state may undergo energy relaxation. The photoisomerization reaction of *trans*-stilbene provides a well-studied example.[5,19,53–56,75–78] Photoexcitation of trans-stilbene to the first singlet excited state (S_1) leads to twisting about the olefinic C=C bond while the molecule stays on the singlet surface.[79] The S_1 potential energy surface has several areas of interest, namely the *trans*, *cis*, and twisted conformations. Isomerization of S_1 *trans*-stilbene has a barrier of about $1000\,cm^{-1}$ and its lifetime is about 70 ps in room temperature hexane solution.[80,81] Two groups (Gustafson and co-workers and Hamaguchi and co-workers) independently reported the first TR³ spectra for *trans*-stilbene in the S_1 excited state following ultraviolet excitation.[5,75] Figure 3.16 shows a picosecond TR³ spectrum of S_1 *trans*-stilbene in hexane solution at 25 ps time delay between the pump and probe pulses.[76] The strong Raman band at about 1567 to $1570\,cm^{-1}$ is attributed to the olefinic C=C stretch vibrational mode of the S_1 state and this value is lower than the $1639\,cm^{-1}$ observed for the ground state of *trans*-stilbene.[77,82] This indicates the olefinic central double bond is noticeably weaker in the S_1 state than the ground state.

Figure 3.17 presents ps-TR³ spectra of the olefinic C=C Raman band region (a) and the low wavenumber anti Stokes and Stokes region (b) of S_1-*trans*-stilbene in chloroform solution obtained at selected time delays upto 100 ps.[78] Inspection of Figure 3.17 (a) shows that the Raman bandwidths narrow and the band positions up-shift for the olefinic C=C stretch Raman band over the first 20–30 ps. Similarly, the ratios of the Raman intensity in the anti-Stokes and Stokes Raman bands in the low frequency region also vary noticeably in the first 20–30 ps. In order to better understand the time dependent changes in the Raman band positions and anti-Stokes/Stokes intensity ratios, a least squares fitting of Lorentzian band shapes to the spectral bands of interest was performed to determine the Raman band positions for the olefinic

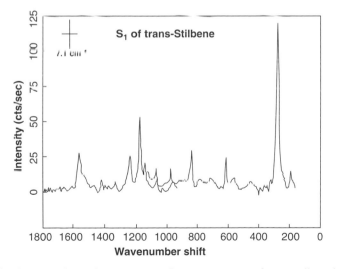

Figure 3.16. Picosecond transient resonance Raman spectrum of *trans*-stibene in hexane at a delay of 25 ps. See text for more details. (Reprinted with permission from reference [76]. Copyright (1984), American Institute of Physics.)

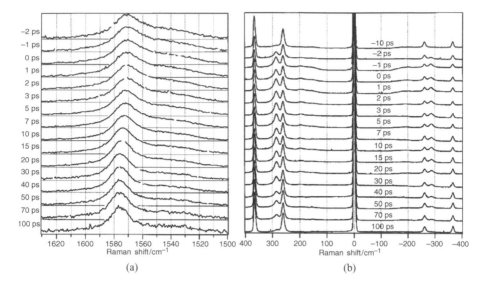

Figure 3.17. (a) C=C stretch region of the time-resolved Raman spectra of S_1 *trans*-stilbene in chloroform. The pump and probe wavelengths were 294 and 588 nm. The intensity of the Raman band at each time delay is normalized. (b) Low wavenumber region of the time-resolved Raman spectra of S_1 *trans*-stilbene in chloroform. The left half is for the Stokes scattering, while the right half is for the anti-Stokes scattering. The Rayleigh light is located at the center. (Reprinted with permission from reference [78]. Copyright (1997) American Chemical Society.)

C=C stretch 1570 cm^{-1} band and intensity ratios of the anti-Stokes/Stokes 285 cm^{-1} Raman band.[78] Results from this analysis are shown in Figure 3.18.[78]

The relative intensities of the Stokes band to its corresponding anti-Stokes Raman band can be used to measure the temperature of the vibrational mode since the anti-Stokes/Stokes intensity ratio is related to the Boltzmann factor $\exp(-E_v/kT)$ where E_v is the vibrational energy level spacing under off-resonance conditions.[78] Inspection of Figure 3.18 reveals the anti-Stokes/Stokes intensity ratio for the 285 cm^{-1} band (open circles) has essentially the same time dependent changes as the Raman band position of the nominal 1570 cm^{-1} olefinic C=C stretch band (filled triangles).[78,83] This agreement between the anti-Stokes/Stokes intensity ratios and the band positions of the 1570 cm^{-1} Raman band indicate the ps-TR^3 spectra can be used as a thermometer to follow the energy relaxation of the initially excited S_1 state of trans-stilbene.[78] The time-dependent changes observed for the anti-Stokes/Stokes intensity ratios and the band positions of the 1570 cm^{-1} Raman band can be fit to single exponential function with a time-constant of about 12 ps for the energy relaxation of the S_1 state.[78] Similar changes for Raman band positions, bandwidths and intensity ratios of anti-Stokes/Stokes bands have been observed for a number of other species in unrelaxed electronic excited states over the first tens of ps after formation.[40,41,84,85]

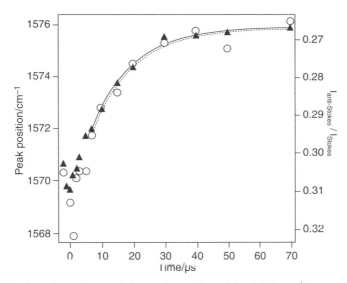

Figure 3.18. Time dependence of the peak position of the 1570 cm⁻¹ Raman band of S_1 *trans*-stilbene in chloroform solution (filled triangle). The time dependence of the anti-Stokes/Stokes intensity ratio is also shown with open circles. The best fit of the peak position change with a single-exponential function is shown with a solid curve, while the best fit of the anti-Stokes/Stokes intensity ratio is shown with a dotted curve. The obtained lifetime for both single-exponential decay functions was 12 ps. (Reprinted with permission from reference [78]. Copyright (1997) American Chemical Society.)

3.4.3. Bimolecular Reactions

3.4.3.1. Hydrogen Abstraction

Photoexcited ketone intermolecular hydrogen atom abstraction reactions are an interesting area of research because of their importance in organic chemistry and due to the complex reaction mechanisms that may be possible for these kinds of reactions. Time resolved absorption spectroscopy has typically been used to follow the kinetics of these reactions but these experiments do not reveal much about the structure of the reactive intermediates.[86–90] Time resolved resonance Raman spectroscopy can be used to examine the structure and properties of the reactive intermediates associated with these reactions. Here, we will briefly describe TR³ experiments reported by Balakrishnan and Umapathy[91] to study hydrogen atom abstraction reactions in the fluoranil/isopropanol system as an example.

Photolysis of fluoranil in 2-propanol leads to very fast formation of a triplet state (within a few tens of ps) via intersystem crossing from the initially excited singlet state.[88] The triplet state has a strong transient absorption at about 485 nm.[88] TR³ experiments were done using a 355 nm pump wavelength and a 485 nm probe wavelength to probe the structure and dynamics of the triplet state of fluoranil in 2-propanol solvent and Figure 3.19 (left-side) show selected spectra obtained from these measurements.[91] In the 1400–1750 cm⁻¹ region, the TR³ spectra of the triplet

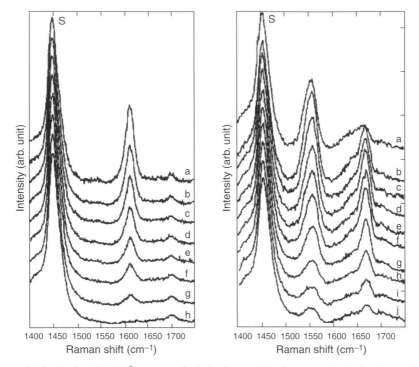

Figure 3.19. (Left side) TR^3 spectra of triplet fluoranil in 2-propanol obtained at various time delays (λ_{pump} 355 nm, λ_{probe} 485 nm): (a) 10 ns, (b) 40 ns, (c) 70 ns, (d) 100 ns, (e) 150 ns, (f) 200 ns, (g) 300 ns and (h) 600 ns (the solvent band is indicated as S). (Right side) TR^3 spectra of fluoranil in 2-propanol obtained at various time delays (λ_{pump} 355 nm, λ_{probe} 416 nm): (a) 10 ns, (b) 150 ns, (c) 250 ns, (d) 500 ns, (e) 800 ns, (f) 1.3 µs, (g) 4.0 µs, (h) 17.0 µs, (i) 50.0 µs and (j) 124.0 µs (the solvent band is indicated as S). See text for more details. (Reprinted from reference [91]. Copyright (1997), with permission from Elsevier.)

state of fluoranil shows a strong resonance Raman band at 1605 cm^{-1} that decays with a time constant of about 6.7×10^6 s^{-1} which is consistent with the value obtained from previous transient absorption experiments.[88] The triplet excited states of p-benzoquinone and its halogenated derivatives like fluoranil are known to undergo hydrogen abstraction reactions to produce ketyl radicals in protonated solvents like water and alcohols such as 2-propanol.[86,90] The ketyl radical has a reasonably strong transient absorption at 420 nm with an extinction coefficient of 5600 M^{-1}cm^{-1}. TR^3 experiments using 355 nm pump and 416 nm probe wavelengths were done to examine the formation of the ketyl radical and Figure 3.19 (right-side) show selected spectra from these experiments.[91] In the 1400–1750 cm^{-1} region, the TR^3 spectra of the ketyl radical show two bands at early times (50 ns) with the 1550 cm^{-1} resonance Raman band being significantly stronger than the 1668 cm^{-1} resonance Raman band. However, the relative intensities of these Raman bands change noticeably as the time delay proceeds to the microsecond time-scale and this suggests the presence of more than one species contributing to the TR³ spectra. The fluoranil radical anion also has an absorption maximum nearby at 435 nm with an extinction coefficient of 7600 M^{-1}cm^{-1} in water[90] and likely also absorbs significantly at the 416 nm used in the TR³ experiments that probe the ketyl radical. Inspection of the TR³ spectra in Figure 3.19 (right-side) reveals that the Raman band shapes also change as the relative intensities of the two main features at about 1550 cm^{-1} and 1668 cm^{-1} and this is consistent with more than one species contributing to the TR³ spectra.

In order to extract the contributions and dynamics of the ketyl radical and fluoranil anion from the TR³ spectra obtained with the 416 nm probe wavelength, a deconvolution of the Raman bands were done using a fitting procedure employing a Lorentzian lineshape for the Raman bands of the two intermediates. Figure 3.20 shows a comparison of the best fit (lines) to the experimental TR³ spectra (dots) in the left-side spectra and the deconvolution extracted from this best fit for the ketyl radical spectra

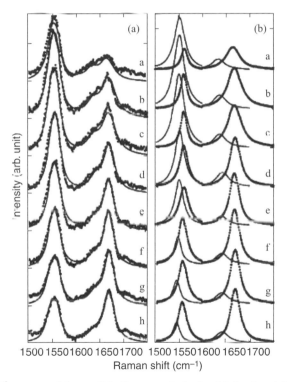

1500 1550 1600 1650 1700 1500 1550 1600 1650 1700
Raman shift (cm^{-1})

Figure 3.20. TR³ spectra of fluoranil in 2-propanol obtained (after band fitting and deconvolution) at various time delays (λ_{pump} 355 nm, λ_{probe} 416 nm): (a) 10 ns, (b) 50 ns, (c) 100 ns, (d) 150 ns, (e) 250 ns, (f) 500 ns, (g) 1.3 μs, (h) 3.0 μs, [a] dots - original spectra and line-fitted spectra, [b] after the deconvolution of each band - the bands indicated as lines belong to the ketyl radical and the bands indicated as dots belong to the radical anion. See text for more details. (Reprinted from reference [91]. Copyright (1997), with permission from Elsevier.)

(lines) and the fluoranil anion spectra (dots) are shown in the right-side spectra.[91] The deconvolution reveals Raman bands at 1550, 1560, 1638, and 1668 cm^{-1} attributed to the ketyl radical and the fluoranil anion intermediates. The Raman bands at 1550 cm^{-1} and 1638 cm^{-1} due to the ketyl radical increase initially with time up to about 100 ns and then decrease. The Raman bands at 1560 cm^{-1} and 1668 cm^{-1} due to the fluoranil anion increase in intensity up to about 500 ns and then decay.[91] The assignments of the TR^3 spectra deconvolution spectra for the ketyl radical and the fluoranil anion were confirmed by obtaining a TR^3 spectrum of the ketyl radical and resonance Raman spectrum of the fluoranil anion.[91,92] For example, the TR^3 spectrum obtained for fluoranil in CHCl$_3$ solvent at 20 ns delay in Figure 3.21 (middle) is primarily due to the ketyl radical due to the highly acidic pKa value. This TR^3 spectrum for the ketyl radical (Figure 3.21 middle) is in good agreement with the deconvoution spectra attributed to the ketyl radical in Figure 3.20 (right side-spectra shown as lines).[91] Similarly,

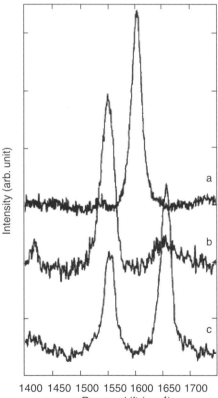

Figure 3.21. Time resolved resonance Raman spectra of fluoranil (a) in acetonitrile (5.0 mM) at 20 ns time delay between pump and probe (λ_{pump} 355 nm, λ_{probe} 485 nm), (b) in chloroform (5.0 mM) at 20 ns delay between pump and probe (λ_{pump} 355 nm, λ_{probe} 416 nm). (c) Resonance Raman spectra of fluoranil radical anion in acetone (λ_{probe} 441 nm). See text for more details. (Reprinted from reference [91]. Copyright (1997), with permission from Elsevier.)

the 441 nm resonance Raman spectrum of the chemically prepared fluoranil radical anion in Figure 3.21 (bottom) is in good agreement with the deconvolution spectra attributed to the fluoranil anion in Figure 3.20 (right-side-spectra shown as dots).

The results from the TR^3 experiments presented in Figures 3.19–3.21 above were used to determine single exponential rate constants of 5.1×10^6 s^{-1} for the decay of the ketyl radical, 1.6×10^6 s^{-1} for the formation of the fluoranil anion and 8.4×10^4 s^{-1} for the decay of the fluoranil radical anion. These rate constants suggest that the fluoranil radical anion is forming from the ketyl radical. A reaction mechanism for the intermolecular abstraction reaction can be described as follows:

Reaction step 5 in Scheme 3.1 can be ruled out because the fluoranil ketyl radical (FAH$^•$) reaches a maximum concentration within 100 ns as the triplet state (^{3}FA) decays by reaction step 2 while the fluoranil radical anion (FA$^{•-}$) takes more than 500 ns to reach a maximum concentration. This difference suggests that the fluoranil radical anion (FA$^{•-}$) is being produced from the fluoranil ketyl radical (FAH$^•$). Reaction steps 1 and 2 are the most likely pathway for producing the fluoranil ketyl radical (FAH$^•$) from the triplet state (^{3}FA) and is consistent with the TR^3 results above and other experiments in the literature.[86,88,92] The kinetic analysis of the TR^3 experiments indicates the fluoranil radical anion (FA$^{•-}$) is being produced with a first order rate constant and not a second order rate constant. This can be used to rule out reaction step 4 and indicates that the fluoranil radical anion (FA$^{•-}$) is being produced by reaction step 3. Therefore, the reaction mechanism for the intermolecular hydrogen abstraction reaction of fluoranil with 2-propanol is likely to predominantly occur through reaction steps 1 to 3.

Comparison of the TR^3 spectra of the fluoranil ketyl radical (FAH$^•$), the triplet state (^{3}FA) and the fluoranil radical anion (FA$^{•-}$) to the results of density functional theory (DFT) calculations for these intermediates provides additional insight into

$$FA + h\nu \rightarrow {}^1FA \rightarrow {}^3FA \qquad \qquad \text{(Step 1)}$$

$$^3FA + (CH_3)_2CHOH \rightarrow FAH^• + (CH_3)_2C^•OH \qquad \text{(Step 2)}$$

$$FAH^• \rightarrow FA^{•-} + H^+ \qquad \qquad \text{(Step 3)}$$

$$^3FA + (CH_3)_2C^•OH \rightarrow FA^{•-} + H^+ + (CH_3)_2CO \qquad \text{(Step 4)}$$

$$^3FA + (CH_3)_2CHOH \rightarrow FA^{•-} + (CH_3)_2C^•OH + H^+ \qquad \text{(Step 5)}$$

Scheme 3.1 Possible reaction steps in the hydrogen abstraction reaction of fluoranil with 2-propanol. Note: FA= fluoranil, $(CH_3)_2CHOH$ = 2-propanaol, FAH$^•$ = fluoranil ketyl radical, FA$^{•-}$ = fluoranil radical anion.

their structures and properties.[93] The ground state of the fluoranil molecule has a substantial quinoidal character with C=C bonds with a bond order of about 1.6 and with C=O bonds with a bond order of about 1.79. In the T_1 triplet state (^{3}FA), the quinoidal character decreases so that the C=C bond order drops to about 1.2 and the C=O bond order decreases to about 1.57. For the fluoranil ketyl radical (FAH$^•$), the C=C bond order is about 1.45 and the C=O bond order is about 1.5. For the fluoranil radical anion (FA$^{•-}$), the C=C bond order is about 1.5 and the C=O bond order is about 1.5. The TR^3 and DFT results indicate the quinoidal character and the C=C and C=O bond orders vary somewhat in the fluoranil ketyl radical (FAH$^•$), the triplet state (^{3}FA) and the fluoranil radical anion (FA$^{•-}$) and appear to be correlated with their reactivity.[93]

3.4.3.2. Protonation Reactions   The photochemistry of aryl azides has been extensively studied using time-resolved transient absorption spectroscopy[94–102] and more recently by time-resolved vibrational spectroscopic methods like TRIR and TR^3.[103–111] Photolysis of aryl azides in room temperature solutions typically leads to formation of a singlet nitrene reactive intermediate that can subsequently undergo different reactions like ring expansion to produce ketenimines, intersystem crossing to form the triplet nitrene, or undergo reactions with nucleophiles.[94–102] Some arylnitrenes such as singlets 2-fluorenylnitrene or 4-biphenylnitrene can react with water to form the corresponding arylnitrenium ion.[99,100] Here, we will show the use of picosecond TR^3 spectroscopy to directly examine the protonation reaction of the singlet 2-fluorenylnitrene species with a water molecule to produce the 2-fluorenyl-nitrenium ion (see Scheme 3.2).

Photolysis of the 2-fluorenyl azide precursor compound by the 267 nm pump laser pulse releases a nitrogen molecule and produces the singlet 2-fluorenylnitrene intermediate as shown in Scheme 3.2. In the presence of appreciable amounts of water, this singlet 2-fluorenylnitrene species can react with the water molecule to form a singlet 2-fluorenylnitrenium ion and an OH$^-$ species as shown in Scheme 3.2.

Figure 3.22 presents an overview of selected ps-TR^3 spectra obtained after 267 nm photolysis of 2-fluorenyl azide in 50% water/50% acetonitrile solution using a 342 nm (left) or a 400 nm (right) probe wavelength.[25] The time-delays between the pump and probe pulses are indicated to the right of each spectrum. Inspection of Figure 3.22 shows that a species attributed to the singlet 2-fluorenylnitrene intermediate is produced within several picoseconds in the spectra for both probe wavelengths. This species reacts further with water to form the singlet 2-fluorenylnitrenium ion

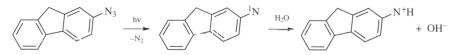

Scheme 3.2 Photolysis of 2-fluorenyl azide to produce singlet 2-fluorenylnitrene that then reacts with water to form a singlet 2-fluorenylnitrenium ion. See text for more details. (Reprinted in part with permission from reference [25]. Copyright (2004) American Chemical Society.)

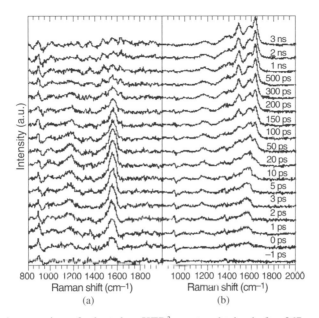

Figure 3.22 An overview of selected ps-KTR³ spectra obtained after 267 nm photolysis of 2-fluorenyl azide in 50% water/50% acetonitrile solvent using (a) 342 nm and (b) 400 nm probe excitation wavelengths. The time delays are indicated to the right of each spectrum. (Reprinted with permission from: reference [25]. Copyright (2004) American Chemical Society.)

species on the hundred(s) of picosecond time-scale as seen in the 400 nm probe spectra of Figure 3.22. The spectra obtained with the 342 nm probe wavelength are more in resonance with the transient absorption of the 2-fluorenylnitrene species while the spectra obtained with the 400 nm probe wavelength are more in resonance with the 2-fluorenylnitrenium ion. The Raman band integration of the strong 1550–1600 cm^{-1} feature in the 342 nm spectra can be used to follow the kinetics of the 2-fluorenyl nitrene species. Similarly, integration of the strong 1630 cm^{-1} Raman band area in the 400 nm spectra can be used to track the kinetics of the 2-fluorenylnitrenium ion intermediate.[25]

Figure 3.23 shows the results of how the intensities of these Raman band features decay and grow as a function of time for spectra obtained in 25% water/75% acetonitrile and 50% water/50% acetonitrile solvent systems.[25] Examination of Figure 3.23 shows that the decay of the first species (singlet 2-fluorenylnitrene) directly corresponds to the growth of the second species (singlet 2-fluorenylnitrenium ion). The decay of the first species and the growth of the second species were simultaneously fit by a common time constant exponential decay and growth functions, respectively where the best-fits are shown by the solid lines in Figure 3.23.[25] The time constants of the decay of the first species and the growth of the second species are 586 ps in the 25% water/75% acetonitrile solvent and 167 ps in the 50% water/50% acetonitrile solvent.[25] The decay

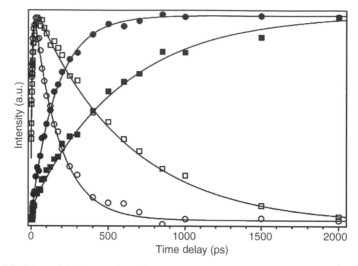

Figure 3.23. Plots of the Raman band integration of the strong 1550–1600 cm^{-1} feature associated with the first species in the 342 nm ps-KTR^3 spectra (open squares and circles) and the strong 1630 cm^{-1} region feature associated with the second species in the 400 nm ps-KTR^3 spectra (solid squares and circles). Data are shown for ps-KTR^3 spectra obtained in 25% water/75% acetonitrile (squares) and 50% water/50% acetonitrile solvent (circles) systems. The lines present best-fit exponential decay and growth curves to the data. (Reprinted with permission from reference [25]. Copyright (2004) American Chemical Society.)

of the first species and the growth of the second species become much faster as the water concentration increases and this indicate that the first species is reacting with water to produce the second species. This and the assignments of the first species to 2-fluorenylnitrene and the second species to the singlet 2-fluorenylnitrenium ion indicates the ps-TR^3 spectra are directly observing the 2-fluorenylnitrene reaction with water to produce the singlet 2-fluorenylnitrenium ion.[25]

The assignment of the TR^3 spectra were based on the known photochemistry of the aryl azides and comparison of the TR^3 spectra vibrational frequencies to those predicted by density functional theory calculations for the likely photochemical intermediates.[25,108–110] The good agreement between the experimental TR^3 vibrational frequencies to those predicted by the density functional theory calculations for the singlet 2-fluorenylnitrene and 2-fluorenylnitrenium ion species helps one assign the TR^3 spectra to these intermediates and also enables structural details and properties to be inferred about these intermediate species.[25,108–110] For example, the vibrational frequency of the aromatic C=C stretch vibrational mode of the 2-fluorenylnitrenium ion mode (1633 cm^{-1} in the TR^3 spectra and 1640 cm^{-1} in the DFT calculations) indicates that this species has substantial iminocyclohexadienyl character and noticeable charge delocalization into the phenyl rings.[109,110] A similar comparison between the vibrational frequencies of the TR^3 spectra and DFT calculations for the singlet 2-fluorenylnitrene species indicates that this intermediate has a moderate amount of cyclohexadienyl character and fairly strong imine character.[108] For this protonation reaction, the TR^3 spectra reveal that the degree of the cyclohexadienyl character and charge delocalization into the phenyl rings becomes substantially stronger when the 2-fluorenylnitrene species is protonated to produce the 2-fluorenylnitrenium ion while the imine character (or C—N bond length) does not change significantly.[108,110]

3.4.4. Ring-opening Reactions (Pericyclic Reactions)

Pericyclic (ring-opening) reactions are an important kind of chemical reaction in chemistry owing to the precise stereo- and regioselectivity of these types of reactions.[112–117] A large number and diverse range of pericyclic photochemical reactions are known and some examples include electrolytic ring-openings and sigmatropic shift reactions.[115,116,118] Here, the application of TR^3 spectroscopy to study the prototypical 1,3-cyclohexadiene (CHD) photochemical ring-opening reaction will be very briefly described. The photochemistry of CHD is interesting because it is the photoreactive part in the conversion of 7-dehydrocholesterol to pre-vitamin D.[116,119] Photoexcitation of CHD in solution first results in excitation to a singlet state ($^1B_2$) that depopulates very efficiently in about 10 fs to a nearby dark state ($^2A_1$) that then undergoes a fast and efficient conrotatory ring-opening to produce a ground state cis-hexatriene product on the picosecond time-scale.[120–122] Reid et al. used ps-TR^3 spectroscopy to directly examine the CHD ring-opening reaction in room temperature solutions.[122] Figure 3.24 displays Stokes and anti-Stokes ps-TR^3 spectra of CHD obtained after ultraviolet excitation in a cyclohexane solvent.[122] The spectra in Figure 3.24 were obtained by subtracting the probe-only resonance Raman spectrum from the pump-probe resonance Raman spectrum. Inspection of the Stokes TR^3 spectra in Figure 3.24 shows negative intensity Raman bands at $1578 \, cm^{-1}$ and $1323 \, cm^{-1}$ in the 0 ps time delay spectrum that can be attributed to depletion of ground state CHD by the pump laser pulse. The $801 \, cm^{-1}$ cyclohexane solvent band also display negative intensity in the 0 ps TR^3 spectrum and this indicates the optical absorbance of the sample has increased. The 1610, 1236, and $390 \, cm^{-1}$ bands have positive intensity in the 4 ps TR^3 spectrum due to formation of the ground state cis-hexatriene product.[122] A kinetic analysis of the formation of the ethylenic $1610 \, cm^{-1}$ Raman feature of cis-hexatriene found that this product is produced with a single-exponential time constant of $6 \pm 1 \, ps$ and that this is a direct measurement of the time required to complete the CHD photochemical ring-opening reaction in a room temperature solution.[122]

Additional information can be obtained for the rate of formation and vibrational relaxation of the cis-hexatriene species.[121,122] The changes in the anti-Stokes ps-TR^3 Raman band intensities contains information on the amount and rate of excess vibrational energy relaxation.[122] In the anti-stokes ps-TR^3 spectra at 0 ps time delay, there is minimal intensity observed in the cis-hexatriene Raman bands consistent with the absence of cis-hexatriene on its ground state surface. Cis-hexatriene Raman bands at 1614 and $1240 \, cm^{-1}$ appear in the 4 ps time delay anti-stokes ps-TR^3 spectra in Figure 3.24 and are in good agreement with the band positions in the corresponding Stokes ps-TR^3 spectrum. In addition, calculations for the Raman cross sections

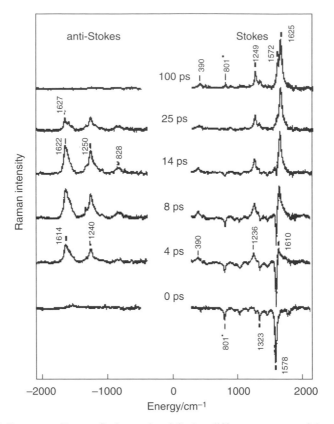

Figure 3.24 Resonance Raman Stokes and anti-Stokes difference spectra of the photochemical ring opening of 1,3-cyclohexadiene. Anti-Stokes spectra were obtained with 284-nm pump and probe wavelengths, while the two-color Stokes spectra were generated with a 284-nm probe and a 275-nm pump. The line at 801 cm^{-1} is due to the cyclohexane solvent. (Reprinted with permission from reference [122]. Copyright (1994) American Chemical Society.)

for CHD and *cis*-hexatriene imply that the *cis*-hexatriene should be the predominant species contributing to the TR^3 spectrum at this probe wavelength.[121,122] In Figure 3.24, the anti-Stokes Raman intensity increases up to about 14 ps and then decays by 100 ps. The formation and decay times of the 1614 cm^{-1} anti-stokes Raman band could be fit reasonably well by a double-exponential kinetics with a best-fit formation time constant of 8 ± 2 ps and a decay time constant of 9 ± 2 ps. This is in good agreement with the Stokes ps-TR^3 data that determined that the ground state *cis*-hexatriene product is formed with about a 6 ps time constant and then undergoes further vibrational relaxation with a time constant of about 9 ps. The changes in frequency with time of the 1610–1625 cm^{-1} and 1236–1249 cm^{-1} regions in the anti-Stokes and Stokes ps-TR^3 spectra are consistent with single-bond isomerization.[121,123,124] Normal-mode calculations and symmetry considerations can be used to indicate the two ethylenic Raman bands (1572 cm^{-1} and 1625 cm^{-1}) in the 100 ps stokes ps-TR³ spectra (Figure 3.24) are only consistent with the lower symmetry of the *s-cis, cis, s-trans*-hexatriene conformer.[121,122] The intensity in the anti-Stokes 828 cm⁻¹ Raman band from 4 to 25 ps gives evidence for an *all-cis* conformer precursor for the *s-cis, cis, s-trans*-hexatriene conformer. Taken together, the increase in the ethylenic Raman band frequency, the appearance of two ethylenic Raman bands, and the presence of the 828 cm⁻¹ Raman band attributed only to the *all-cis* hexatriene conformer indicate that the *cis*-hexatriene photoproduct initially forms on the ground state in its *all-cis* conformer (about 6 ps time constant) and quickly undergoes conformational relaxation to produce the *s-cis, cis, s-trans*-hexatriene conformer (about 7 ps time constant). Figure 3.25 shows an overview of the reaction mechanism of the CHD ring-opening reaction.

Ultraviolet excitation of CHD initially populates the light $^1B_2$ state that then depopulates very fast (about 10 fs) to the dark $^2A_1$ state (as determined from a resonance Raman intensity analysis study).[120-122] The $^2A_1$ state then undergoes the ring-opening reaction (time constant of about 6 ps) to initially populate a vibrationally excited *all-cis* hexatriene conformer that can then vibrationally relax (time constant of about 9 ps) or conformationally relax to the *s-cis, cis, s-trans*-hexatriene conformer (time constant of about 7 ps) as determined from the ps-TR³ study.[121,122] The TR³ experiments for CHD and several related systems have shown that the ≈10 ps formation time of the photoproduct is a general feature of pericyclic rearrangements and have given the first measurements containing structural information of the production and relaxation of pericyclic ring-opening reactions.[121,122]

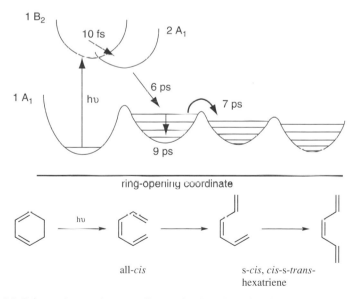

ring-opening coordinate

all-*cis*

s-cis, cis-s-trans-hexatriene

Figure 3.25 Schematic reaction coordinates for the photochemical ring opening reactions of 1,3-cyclohexadiene. (Reprinted with permission from reference [122]. Copyright (1994) American Chemical Society.)

3.4.5. Photodeprotection Reactions

There is great interest in developing efficient phototriggers for real time observation of physiological responses in biological systems.[125–129] The *p*-hydroxyphenacyl (*p*HP) protecting group has drawn great attention because of its practical potential as a very fast and efficient "cage" for the release of a variety of biological stimulants.[130–132] The photodeprotection reaction to release the biological stimulant in *p*HP caged compounds is very solvent dependent and appears to only take place in aqueous or aqueous containing solvents and the photolysis also produces a benign *p*-hydoxyphenylacetic acid (HPAA) final product in addition to release of the biological stimulant.[49,132–134] Scheme 3.3 gives an overview of the photodeprotection reaction for a couple of *p*-hydroxyphenacyl caged phosphate compounds.

A very brief description of using TR^3 spectroscopy to study the photodeprotection and photosolvolytic rearrangement reactions of the *p*-hydroxyphenacyl diethylphosphate (HPDP) compound will be given here as an example.[49,134]

Figure 3.26 present ps-TR^3 spectra obtained with 342 nm (a) and 400 nm (b) probe wavelengths after 267 nm photolysis of HPDP in 50% H_2O/50% MeCN mixed solvent. The 342 nm probe ps-TR^3 spectra in Figure 3.26 show that one species (attributed to the S_1 excited state of HPDP) evolves into another species (attributed to the T_1 state of HPDP) within a few picoseconds.[134] This is consistent with results from femtosecond time-resolved fluorescence and transient absorption spectra.[49] Figure 3.26 (a) provides a direct TR^3 observation of the intersystem crossing (ISC) reaction of S_1 HPDP to T_1 HPDP.

Figure 3.27 presents selected ps-TR^3 (a) and ns-TR^3 (b) obtained with a 400 nm probe wavelength after 267 nm photolysis of HPDP in acetonitrile (MeCN) solvent. Inspection of the early time ps-TR^3 spectra in Figures 3.26 and 3.27 (a) reveals that the triplet state HPDP Raman bands become more intense, shift in frequency and their bandwidths become narrower during the first 20 to 30 ps. These changes in the ps-TR^3 spectra are characteristic of excess energy relaxation processes[19,40,41,53,55,56,78,83–85] and indicate the triplet state is initially formed with substantial excess energy following the very fast ISC from the S_1 singlet state.

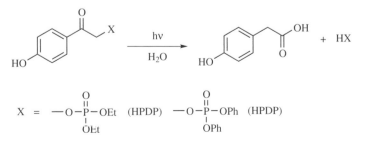

Scheme 3.3 Overall photodeprotection reaction of two *p*-hydroxyphenacyl caged phosphate compounds. See text for more details. (Reprinted in part with permission from reference [49]. Copyright (2006) American Chemical Society.)

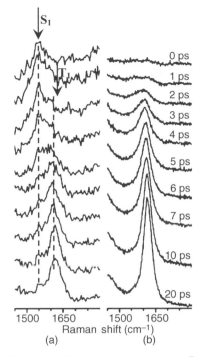

Figure 3.26. Picosecond Kerr gated time-resolved resonance Raman spectra obtained for HPDP in 50% H_2O/50% MeCN mixed solvent with 267 nm pump and 342 nm probe (a) and 400 nm probe (b) wavelengths, respectively. (Reprinted with permission from reference [134]. Copyright (2005) American Chemical Society.)

Employing the intense Raman band in the 1573–1603 cm^{-1} region in the ps-TR³ spectra, the changes in the intensity, band position and bandwidth can be found quantitatively with time and these data are plotted in Figure 3.27 (c) to (e). Examination of Figure 3.27 (e) shows that the changes in the Raman band positions and band widths correlate well with one another and have time constants of about 10 ps. Comparison of the data in Figure 3.27 (d) and (e) shows that the growth of the Raman band intensity has a similar time constant and also correlates well with the changes in the Raman band positions and bandwidths. Figure 3.27 (b) show selected ns-TR³ spectra obtained using a 416 nm probe wavelength following 266 nm photolysis of HPDP in MeCN solvent under open air conditions. A plot of the Raman intensity observed for the triplet state in MeCN under open air conditions reveals it has a decay time constant of about 150 ns (see Figure 3.27 (c)).

Figure 3.28 (a) presents ps-TR³ spectra obtained with a 400 nm probe wavelength following 267 nm photolysis of HPDP in 50% H_2O/50% MeCN mixed solvent (left) and neat MeCN (right). Inspection of Figure 3.28 reveals that the triplet state of HPDP is strongly quenched in the presence of appreciable amounts of water. A plot of the Raman intensity for the ps-TR³ spectra (using the strong Raman band in the

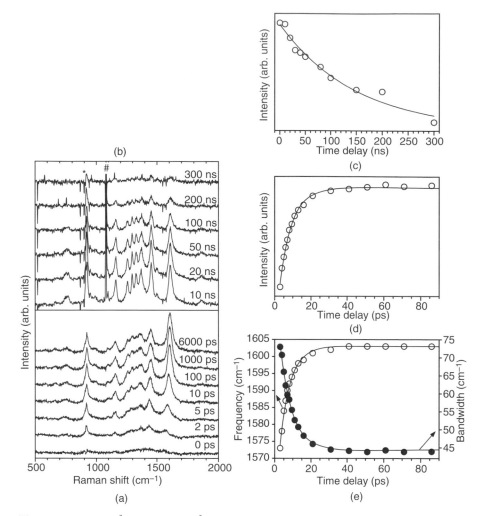

Figure 3.27. Ps-TR[3] (a) and ns-TR[3] (b) obtained with a 400 nm probe wavelength after 267 nm photolysis of HPDP in acetonitrile (MeCN) solvent. (Reprinted with permission from reference [41]. Copyright (2004) American Chemical Society.) See text for more details.

1600 cm^{-1} region) is given in Figure 3.28 (b) and this indicates that the triplet state has a decay time constant of about 400 ps in 50% H$_2$O/50% MeCN mixed solvent. This decay time in the 50% H$_2$O/50% MeCN mixed solvent is much faster than the about 150 ns triplet decay in neat MeCN (see Figure 3.27 (c)) and suggests that water is reacting with the triplet state of HPDP.[134]

Comparison of the TR[3] spectra for HPDP in MeCN solvent to results from density functional theory calculations for the triplet state of HPDP indicates the triplet state has a quinoidal structure with the carbonyl group about 16° out of the quinoidal plane and a delocalized ππ* character.[134] Figure 3.29 compares ps-TR[3] spectra of

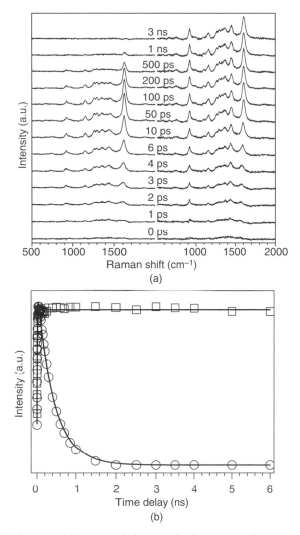

Figure 3.28 (a) Picosecond Kerr gated time-resolved resonance Raman spectra of HPDP obtained with 267 nm pump and 400 nm probe wavelengths in 50% H_2O/50% MeCN mixed solvent (left) and neat MeCN (right). (b) Temporal dependence of the triplet ~1600 cm^{-1} band areas for HPDP in 50% H_2O/50% MeCN mixed solvent (circles) and neat MeCN (squares) obtained in 400 nm probe ps-KTR³ spectra. Solid lines show an exponential fitting of the experimental data. (Reprinted with permission from reference [134]. Copyright (2005) American Chemical Society.)

the triplet state of HPDP obtained in neat MeCN (b) and in 50% H_2O/50% MeCN mixed solvent (a). Inspection of Figure 3.29 shows that there are distinct differences between the spectra obtained in the two solvent systems. In particular, the frequency upshift of the ring center C—C stretch Raman band to about 1620 cm^{-1} in the 50%

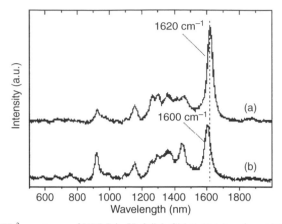

Figure 3.29. Ps-TR^3 spectrum of HPDP in H$_2$O/MeCN (1:1) (a) and neat MeCN (b) obtained with 267 nm excitation and a 400 nm probe wavelength at a 50 ps delay time. (Reprinted with permission from reference [49]. Copyright (2006) American Chemical Society.)

H$_2$O/50% MeCN mixed solvent relative to 1600 cm^{-1} in the neat MeCN solvent indicates there is an increased quinoidal character of the triplet state in the 50% H$_2$O/50% MeCN mixed solvent.[49,134] Comparison of these TR^3 spectra to results from density functional theory calculations that explicitly consider hydrogen bonding to water molecules indicates that hydrogen bonding of the water molecule(s) to the carbonyl group leads to this observed up-shift of the ring center C−C stretch Raman band in the 50% H$_2$O/50% MeCN mixed solvent.[49] The hydrogen bonded triplet state of HPDP has an increased quinoidal structure with the carbonyl group in the quinoidal plane and a more phenyl ring localized $\pi\pi^*$ biradical character in a water containing solvent like the 50% H$_2$O/50% MeCN mixed solvent.[49]

Figure 3.30 (a) presents the ultraviolet absorption spectra of HPDP (dashed line) and HPAA (solid line) in a 50% H$_2$O/50% MeCN mixed solvent. The HPAA rearrangement product is weakly absorbing at most wavelengths relative to the HPDP parent molecule except in the 200 nm region. Therefore, a 200 nm probe wavelength was employed in ps-TR^3 experiments used to determine the formation dynamics of the HPAA rearrangement product. Figure 3.30 (b) presents selected ps-TR^3 spectra (solid line spectra with time delays indicated to the right of each spectrum) obtained using a 200 nm probe wavelength following 267 nm photolysis of HPDP in 50% H$_2$O/50% MeCN mixed solvent. A 200 nm resonance Raman spectrum of an authentic sample of HPAA is shown as a dashed line spectrum at the top of Figure 3.30 (b) for comparison purposes. Examination of Figure 3.30 (b) shows that the HPAA rearrangement product appears with a time constant of about 1100 ps after HPDP photolysis in a 50% H$_2$O/50% MeCN mixed solvent. This was the first direct time-resolved detection of the solvolytic rearrangement reaction product for pHP phototrigger compounds.[49]

Employing all of the TR^3 results in conjunction with results from femtosecond time-resolved transient absorption (fs-TA) and femtosecond time-resolved Kerr gated fluorescence (fs-KTRF) experiments enables a reaction mechanism to be developed

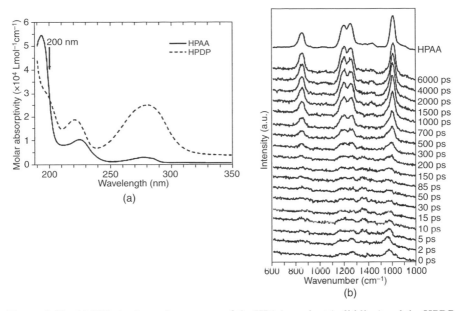

Figure 3.30. (a) UV-vis absorption spectra of the HPAA product (solid line) and the HPDP substrate (dash line) in a H_2O/MeCN (1:1) mixed solvent. (b) Picosecond time-resolved resonance Raman (ps-TR³) spectra of HPDP obtained with a 267 nm pump and 200 nm probe wavelengths in a H_2O/MeCN (1:1) mixed solvent. Resonance Raman spectrum of an authentic sample of HPAA recorded with 200 nm excitation is displayed at the top. (Reprinted with permission from reference [49]. Copyright (2006) American Chemical Society.)

for the deprotection and solvolytic rearrangement reactions of pHP caged phosphates like HPDP.[49,134] Scheme 3.4 presents the reaction mechanism(s) consistent with all of the TR³, fs-TA and fs-KTRF experiments. The ps-TR³ experiments directly probed the structure and character of the S_1 and T_1 states and also observed the ISC reaction from S_1 to T_1 occurs within a couple of picoseconds. The ps-TR³ spectra also indicate the T_1 state initially contains substantial excess energy that relaxes with a time constant of about 10 ps.[134] The TR³ spectra of T_1 in neat MeCN and in 50% H_2O/50% MeCN mixed solvent in conjunction with results from density functional theory calculations indicate that the hydrogen bonded triplet state of HPDP has an increased quinoidal structure with the carbonyl group in the quinoidal plane and a more phenyl ring localized $\pi\pi^*$ biradical character in a water containing solvent like the 50% H_2O/50% MeCN mixed solvent.[49,134] The solvent and water concentration dependence of the ps-TR³ and fs-TA spectra indicate the deprotection reaction occurs with a time constant on the order of 400 ps for HPDP via a solvent assisted heterolytic cleavage pathway.[49,134] The solvolytic rearrangement reaction to produce the HPAA product takes longer than the deprotection and this indicates the deprotection and rearrangement reactions occur sequentially with a short lived intermediate in between them. This intermediate was proposed to be a water solvated contact ion pair species as shown in Scheme 3.4 and is described in reference 49.

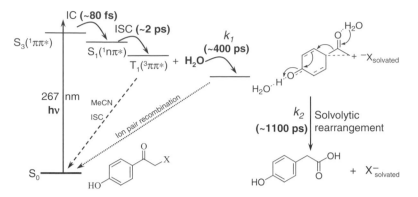

Scheme 3.4 Proposed reaction mechanism for the deprotection and solvolytic rearrangement reactions of *p*HP caged phosphates like HPDP consistent with the TR^3, fs-TA and fs-KTRF experiments described here and in references 49 and 134.

By utilizing different probe wavelengths and time-scales, several different intermediates and their reactions were characterized by TR^3 spectroscopy. These results in combination with fs-TA and fs-KTRF experiments provide important kinetics and structural information that enable an overall mechanistic characterization for the photophysical and photochemical events taking place after photolysis of *p*HP caged phosphates in various solvent environments.

3.4.6. Biologically Related Reactions

TR^3 spectroscopy can also be applied to study complex biological reactive intermediates. In this section, the application of TR^3 spectroscopy to investigate some of the reactive intermediates involved in the bacteriorhodopsin photocycle and the hemoglobin-CO photocycle are briefly presented as examples.

3.4.6.1. Bacteriorhodopsin   Bacteriorhodopsin (BR) is a retinal-protein complex that acts as a proton-translocating system in the cell membrane of *Halobacterium halobium*.[135] The all-*trans* retinal chromophore is the light sensitive moiety of BR and is attached to a lysine group in the inner part of the protein via a protonated Schiff base.[135] Absorption of light causes this all-*trans* retinal chromophore to quickly isomerize about the C13=C14 double bond to produce a 13-cis form.[136,137] This process stores about 30% of the initial energy of the absorbed photon of light and BR subsequently changes through a series of structurally distinct intermediates on its way to return to its initial state about 5 ms after absorbing the photon of light.[137–144] The photocycle of BR is presented in Figure 3.31 and is based on time-resolved transient absorption and TR^3 experiments done to observe the intermediates involved in the BR photocycle and determine the kinetics of the different reaction steps.[137–144]

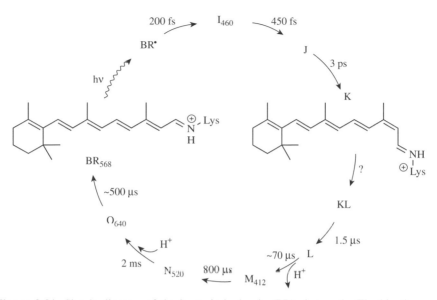

Figure 3.31. Simple diagram of the bacteriorhodopsin (BR) photocycle. The kinetics constants were based on data from time-resolved absorption or TR³ spectra given in references 137, 143 and 144. (Reprinted with permission from reference [145]. Copyright (1991) American Chemical Society.)

The application of picosecond TR³ spectroscopy by Doig et al. to study the J, K, and KL intermediates of the BR photocycle will be briefly described here.[145] The 550 nm pump and 589 nm probe pulses in the TR³ experiments were chosen to be near the absorption maxima of ground state BR (568 nm) and K (590 nm) respectively.[137,145] Stokes TR³ spectra were obtained with time delays varying from 0 ps to 13 ns between the pump and probe pulses in order to examine the structure and kinetics of the J → K → KL sequence of the BR photocycle.[145]

Figure 3.32 shows selected Stokes TR³ spectra obtained at ambient temperature with time-delays of about 0 ps, 3 ps and 3.7 ns that correspond to the J, K, and KL intermediates respectively.[145] The ethylenic stretch Raman band has the same $1518\,cm^{-1}$ vibrational frequency in both the J and K intermediates and this suggests that they have a very similar electronic structure. The strong hydrogen out-of-plane (HOOP) Raman bands at 956 and $1000\,cm^{-1}$ in the J spectrum of Figure 3.32 indicate that the J intermediate has a strongly twisted 13-cis structure since planar protonated Schiff bases have very small or zero HOOP intensity.[146] The HOOP Raman bands decrease in intensity within 3 ps and this suggests that J conformationally relaxes quickly to form the planar K intermediate. The 3 ps TR³ spectrum shown in Figure 3.32 was assigned to the K intermediate owing to its uncrowded nature and the strong single narrow Raman band at $1189\,cm^{-1}$ that is characteristic a 13-cis chromophore.[147] The distinctive $1189\,cm^{-1}$ Raman band of the K intermediate also indicates that isomerization is finished within 3 ps. The lack of appreciable HOOP

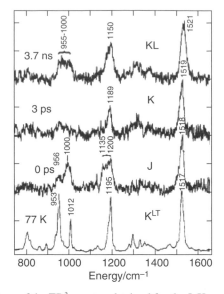

Figure 3.32. Comparison of the TR³ spectra obtained for the J, K, and KL species at ambient temperature (from reference 145) and a Raman spectrum of a photoproduct (K^LT) trapped at 77 K from reference 150. (Reprinted with permission from reference [145]. Copyright (1991) American Chemical Society.)

Raman intensity in the K spectrum suggests the chromophore has relaxed about its single and/or double bonds. The HOOP Raman bands begin to increase in intensity again after 100 ps when twisting in the chromophore is regained owing to formation of the KL intermediate.[145]

Anti-Stokes picosecond TR³ spectra were also obtained with pump-probe time delays over the 0 to 10 ps range and selected spectra are shown in Figure 3.33.[145] The anti-Stokes Raman spectrum at 0 ps indicates that hot, unrelaxed, species are produced. The approximately 1521 cm⁻¹ ethylenic stretch Raman band vibrational frequency also suggests that most of the 0 ps anti-Stokes TR³ spectrum is mostly due to the J intermediate. The 1521 cm⁻¹ Raman band's intensity and its bandwidth decrease with a decay time of about 2.5 ps, and this can be attributed the vibrational cooling and conformational relaxation of the chromophore as the J intermediate relaxes to produce the K intermediate.[145] This very fast relaxation of the initially hot J intermediate is believed to be due to strong coupling between the chromophore the protein bath that can enable better energy transfer compared to typical solute-solvent interactions.[148,149]

The picosecond TR³ experiments described above for BR reveal that a hot un-relaxed J intermediate with a highly twisted structure forms and then vibrationally cools and conformationally relaxes within 3 ps to form the K intermediate. Subsequently, an isomerization induced protein conformational change takes place during 20–100 ps to produce the KL inermediate.[145]

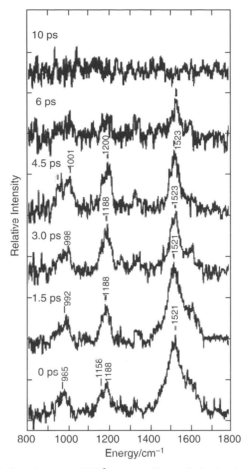

Figure 3.33. Anti-Stokes picosecond TR³ spectra of bacteriorhodopsin and its photoproducts for time delays from 0 to 10 ps from reference 145. (Reprinted with permission from reference [145]. Copyright (1991) American Chemical Society.)

3.4.6.2. Heme Proteins

Hemoglobin (Hb) experiences a series of structural and molecular changes on the nanosecond and microsecond times-scales after photodissociation of CO ligands.[151–155] Heme deligation leads to a substantial change in the absorption spectrum and subsequent smaller changes in the absorption spectra of the deoxy-heme species are due to alterations of the protein surroundings of the heme during the geminate recombination and recombination from solution processes.[151,153] Figure 3.34 presents a kinetic scheme for successive relaxations in the HbCO photocycle derived from an analysis of time-resolved transient absorption spectra.[151,153,154,156]

TR³ spectroscopy can be used to investigate the changes in the molecular structure associated with the transient absorption spectral changes observed during the

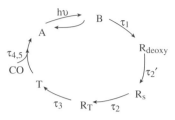

Figure 3.34. Kinetic scheme for successive relaxations in the HbCO photocycle. (Reprinted from reference [156]. Copyright (2004), with permission from Elsevier.)

HbCO photocycle. Spiro and co-workers have employed TR^3 spectroscopy to examine the structural and molecular changes occurring during the HbCO photocycle and a very brief description of one recent study will be described as an example of the application of TR^3 spectroscopy to study heme proteins.[156]

Figure 3.35 presents TR^3 spectra obtained with varying time delays between the pump and probe pulses following photolysis of HbCO.[156] The Raman band features have been attributed to tyrosine (Y) and tryptophan (W) vibrational modes as indicated at the top of Figure 3.35. These Raman bands change with time and provide information about the environment of the typrosine and tryptophan residues in the heme.[157–159] Inspection of Figure 3.35 reveals that the negative difference Raman

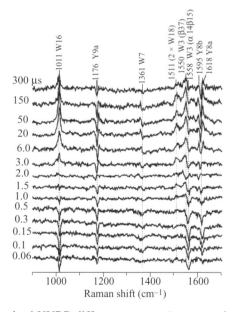

Figure 3.35. Time-resolved UVRR difference spectra (pump+probe minus probe only) at the indicated times (μs) following the photolysis of HbCO. For display purposes, the top four traces were multiplied by a scale factor of 0.5. Band positions and assignments are marked at the top. (Reprinted from reference [156]. Copyright (2004), with permission from Elsevier.)

bands gain intensity and reach a maximum at about 300 ns and then decay. These Raman bands are assigned to the R_{deoxy} difference spectrum since it is also observed for Hb constructs that have the R conformation but where one or more subunit is unligated.[157,160] In order to better analyze the TR³ spectra in Figure 3.35, four Raman bands were chosen to do a kinetic analysis of the data (the strongest tyrosine [Y8a at 1618 cm⁻¹] and tryptophan [W16 at 1011 cm⁻¹] bands and the two components of the W3 band attributed to Trp residues β37 at 1550 cm⁻¹ and α14 + β15 at 1558 cm⁻¹).

Figure 3.36 shows the determined difference intensities (photoproduct minus HbCO) of these Raman bands versus time and these Raman intensities were fitted to a consecutive series of exponential functions. This fitting found time constants that lie in five narrow ranges with average time constants of 0.065, 0.74, 2.9, 20.5, and 497 μs.[156]

These average time constant values were used to decompose the difference time-resolved resonance Raman spectra of Figure 3.35 into component spectra for a set of intermediates employing the kinetic model of Figure 3.36 and assuming the initial

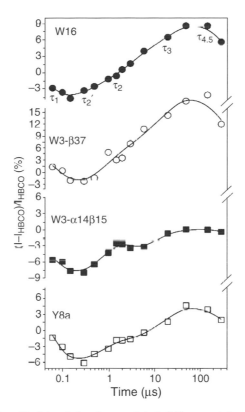

Figure 3.36. Log-time T plots of the deconvoluted difference intensities, expressed as a percentage of the HbCO intensity, for the indicated UVRR bands. On the basis of reproducibility of the perchlorate internal standard, 1%, the uncertainty in the difference intensities is estimated to be 10%. (Reprinted from reference [156]. Copyright (2004), with permission from Elsevier.)

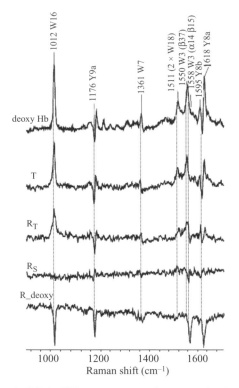

Figure 3.37. Computed UVRR difference spectra for successive intermediates in the kinetic scheme of Figure 3.36. The deoxyHb minus HbCO spectrum is shown at the top for comparison. (Reprinted from reference [156]. Copyright (2004), with permission from Elsevier.)

state B has the same spectrum as HbCO. This analysis for the TR^3 spectra of Figure 3.35 enables the component difference spectra for the R_{deoxy}, R_S, R_T, and T intermediates to be determined and these spectra are shown in Figure 3.37.[156] These difference TR^3 spectra, in conjunction with other data for these intermediates, enables a working model for the allosteric reaction coordinate in the HbCO photocycle to be developed as shown in Figure 3.38 and the reader is referred to reference 156 for details. Photolysis of HbCO (A) results in the geminate state, B, where forces are stored in the heme and these forces subsequently lead to rotation of the E and helices that then breaks the interhelical H-bonds in the R_{deoxy} intermediate. These H-bonds become restored in the R_S intermediate and subsequently produce a hinge contact formation in the R_T intermediate that is then followed by formation of the switch contact to result in the T intermediate.[156]

While the working model for the allosteric reaction coordinate for the HbCO photocycle deduced from the TR^3 data is only a partial description of a very complex process, it does provide new insight into this process and demonstrates that there are distinct intermediates along the R to T reaction pathway.[156]

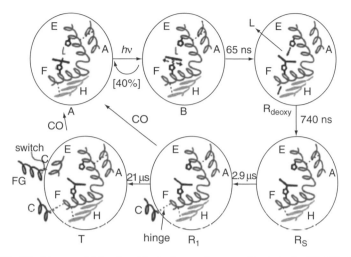

Figure 3.38. Working model for the allosteric reaction coordinate in the HbCO photocycle. Photolysis of HbCO (A) produces the geminate state, B, in which forces are stored at the heme. These forces compel rotation of the E and F helices, which break interhelical H-bonds in the Rdeoxy intermediate. These H-bonds are restored in RS, followed by hinge contact formation, in RT. Formation of the switch contact complete the transition to the T state. (Reprinted from reference [156]. Copyright (2004), with permission from Elsevier.)

3.5. CONCLUSION AND OUTLOOK

Time-resolved resonance Raman spectroscopy (TR^3) can be employed to investigate a very large range and number of chemical reactions and their reactive intermediates. TR^3 spectroscopy is able to provide detailed information about the structure, identity, properties, and chemical reactivity of reactive intermediates. In some of the applications described, TR^3 spectroscopy was able to provide information about the reaction system that would be difficult or impossible to extract from time-resolved emission or absorption spectroscopies. The selective nature of TR^3 spectroscopy allows one to excite different chromophores of the molecule and investigate different intermediates and/or structural changes occurring in different parts of the molecule. This selectivity can also help in simplifying the interpretation of reactions that have different competitive reaction pathways as well as in very complex systems like proteins. The application of TR^3 spectroscopy can suffer from its inherently weak Raman signal and may be difficult or impossible to use in highly luminescent reactions or for studying intermediates with very weak concentrations.

Advances in technology are enabling the development of new techniques that can greatly extend the utility of TR^3 spectroscopy to a greater number of reactions systems. Two particularly exciting recent developments include using Kerr-gating to reject fluorescence in picosecond TR^3 spectra[68,161,162] and femtosecond-stimulated Raman spectroscopy (FSRS)[163–165] that can provide an order of magnitude improvement in time-resolution (<100 fs) while maintaining very good spectral resolution

($<15\,cm^{-1}$). Matousek et al. at the Rutherford Appleton Laboratory in the U.K. have developed an efficient picosecond Kerr-gated fluorescence version of TR^3 spectroscopy that can be used to obtain Raman and TR^3 spectra of very highly luminescent samples.[68,161,162] This technique can be extended to longer time scales by using a combination of a nanosecond photolysis laser that employs an electronic time-delay relative to the ultrafast laser system that supplies the Kerr gating and probe pulses. Thus, this optical Kerr-gating fluorescence rejection technique can potentially be used to investigate reactive intermediates from the picosecond to the microsecond time-scales for highly luminescent systems. This development can greatly extend the range of reactive intermediates that can be investigated using TR^3 spectroscopy. Mathies and co-workers have recently developed femtosecond-stimulated Raman spectroscopy (FSRS)[163–165] and applied it to obtain time-resolved Raman spectra of rhodopsin from 200 fs to 1 ps.[165] This work directly probed the temporal sequence of geometric changes that take place in the retinal backbone that activates this visual receptor.[165] This new FSRS technique promises to be especially useful for time-resolved vibrational spectroscopy on the femtosecond time-scale.[163–165]

SUGGESTED READING

G. H. Atkinson in *Time-Resolved Vibrational Spectroscopy*, Ed. G. H. Atkinson, Academic Press, New York, **1983**, p. 179.

D. E. Morris and W. H. Woodruff, in *Advances in Infrared and Raman Spectroscopy*, R. J. H. Clark and R. E. Hester (Eds.), Wiley, New York, **1987**, vol. 14, p. 285.

H. Hamaguchi, in *Vibrational Spectra and Structure*, J. Durig (Ed.), Elsevier, Amsterdam, **1987**, vol. 16, p. 227.

G. N. R. Tripathi in *Advances in Infrared and Raman Spectroscopy*, R. J. H. Clark and R. E. Hester (Eds.), Wiley, New York, **1989**, vol. 18, p. 157.

H. Hamaguchi and T. L. Gustafson, *Annu. Rev. Phys. Chem.* **1994**, *45*, 593.

P. A. Thompson and R. A. Mathies in Laser Techniques in Chemistry, A. B. Myers and T. R. Rizzo (Eds.), Wiley, New York, **1995**, p. 185.

S. E. J. Bell, *Analyst* **1996**, *121*, 107R.

REFERENCES

1. J. C. Scaiano, in *Reactive Intermediate Chemistry*, (Eds. R. A. Moss, M. S. Platz, M. Jones, Jr.) **2004**, WileyHoboken, New Jersey, U.S.A., p. 847.

2. P. Pagsberg, R. Wilbrandt, K. V. Hansen, and C. V. Weisberg, *Chem. Phys. Lett.* **1976**, *39*, 538.

3. J. Terner, T. G. Spiro, M. Nagumo, M. F. Nicol, and M. A. El-Sayed, *J. Am. Chem. Soc.* **1980**, *102*, 3238.

4. M. Coopey, H. P. Valat, and B. Alpert, *Nature* **1980**, 284, 568.

5. T. L. Gustafson, D. M. Roberts, and D. A. Chernoff, *J. Chem. Phys.* **1983**, *79*, 1559.

6. A. C. Albrecht, *J. Chem. Phys.* **1961**, *34*, 1476.

7. A. Warshel, *Annu. Rev. Biophys. Bioeng.* **1977**, *6*, 273.

8. T. G. Spiro and P. Stein, *Annu. Rev. Phys. Chem.* **1977**, *28*, 501.

9. E. J. Heller, R. L. Sunberg, and D. Tannor, *J. Phys. Chem.* **1982**, *86*, 1822.

10. P. M. Champion and A. C. Albrecht, *Annu. Rev. Phys. Chem.* **1982**, *33*, 353.

11. A. B. Myers and R. A. Mathies, in *Biological Applications of Raman Spectroscopy*: Vol. 2-Resonance Raman Spectra of Polyenes and Aromatics, T. G. Spiro, (ed.). Wiley, New York, **1987**, p. 1.

12. R. van den Berg, D. J. Jang, H. C. Bitting, and M. A. El-Sayed, *Biophys. J.* **1990**, *58*, 135.

13. R. van den Berg and M. A. El-Sayed, *Biophys. J.* **1990**, *58*, 931.

14. R. A. Mathies, in *Chemical and Biochemical Applications of Lasers*, B. B. Moore (Ed.) Academic, New York, **1979**, p. 55.

15. G. H. Atkinson, in *Time-Resolved Vibrational Spectroscopy*, ed. G. H. Atkinson, Academic Press, New York, **1983**, p. 179.

16. H. Hamaguchi, in *Vibrational Spectra and Structure*, J. Durig (Ed.), Elsevier, Amsterdam, **1987**, vol. 16, p. 227.

17. S. E. J. Bell, *Analyst* **1996**, *121*, 107R.

18. Y.-L. Li, K. H. Leung, and D. L. Phillips, *J. Phys. Chem. A* **2001**, *105*, 10621.

19. H. Hamaguchi and T. L. Gustafson, *Annu. Rev. Phys. Chem.* **1994**, *45*, 593.

20. P. A. Thompson and R. A. Mathies, in Laser Techniques in Chemistry, A. B. Myers and T. R. Rizzo (Eds.), Wiley, New York, **1995**, p. 185.

21. J. E. Kim, D. W. McCamat, L. Zhu, and R. A. Mathies, *J. Phys. Chem. B* **2001**, *105*, 1240.

22. D. J. Liard, M. Busby, I. R. Farrell, P. Matousek, M. Towrie, and A. Vlćek Jr., *J. Phys. Chem. A* **2004**, *108*, 556.

23. W. M. Kwok, C. Zhao, Y.-L. Li, X. Guan, D. Wang, and D. L. Phillips, *J. Am. Chem. Soc.* **2004**, *126*, 3119.

24. L. Sarger and J. Oberle, in *Femtosecond Laser Pulses: Principles and Experiments*, C. Rullicre, Eds., Springer, New York, **2005**, p. 195

25. W. M. Kwok, P. Y. Chan, and D. L. Phillips, *J. Phys. Chem. B* **2004**, *108*, 19068.

26. G. Socrates, *Infrared and Raman characteristic group frequencies*, John Wiley & Sons LTD, Baffins Lane, Chichester, West Susses PO19 1UD, England, **2001**, p. 1–347.

27. G. Varsanyi, *Assignments for vibrational spectra of seven hundred benzene derivatives* L. Lang, Ed., Adam Hilger: London, **1974**, Vol. I. p. 1.

28. W. M. Kwok, C. Ma, P. Matousek, A. W. Parker, D. Phillips, W. T. Toner, and M. Towrie, *J. Phys. Chem. A.* **2002**, *105*, 984.

29. C. Ma, W. M. Kwok, P. Matousek, A. W. Parker, D. Phillips, W. T. Toner, and M. Towrie, *J. Photochem. Photobio. A.* **2001**, *142*, 177.

30. W. M. Kwok, I. Gould, C. Ma, M. Puranik, S. Umapathy, P. Matousek, A. W. Parker, D. Phillips, W. T. Toner, and M. Towrie, *Phys. Chem. Chem. Phys.* **2001**, *3*, 2424.

31. A. M. Brouwer and R. Wilbrandt, *J. Phys. Chem.* **1996**, *100*, 9678.

32. O. Poizat and V. Guichard, *J. Chem. Phys.* **1989**, *90*, 4697.

33. V. Guichard, A. Bourkba, M.-F. Lautie, and O. Poizat, *Spectrochimica Acta* **1989**, *45A*, 187.

34. A. P. Scott and L. Radom, *J. Phys. Chem.* **1996**, *100*, 16502 and references therein.

35. M. Puranik, S. Umapathy, and J. G. Snijders, *J. Chem. Phys.* **2001**, *115*, 6106.

36. M. S. C. Foley, D. A. Braden, B. S. Hudson, and M. Z. Zgierski, *J. Phys. Chem. A* **1997**, *101*, 1455.

37. C. Ma, W. M. Kwok, P. Matousek, A. W. Parker, D. Phillips, W. T. Toner, and M. Towrie, *J. Phys. Chem. A* **2001**, *105*, 4648.

38. J. Dreyer and A. Kummrow, *J. Am. Chem. Soc.* **2000**, *122*, 2577.

39. D. Rappoport and F. Furche, *J. Am. Chem. Soc.* **2004**, *126*, 1277.

40. C. Ma, W. S. Chan, W. M. Kwok, P. Zuo, and D. L. Phillips, *J. Phys. Chem. B* **2004**, *108*, 9264.

41. C. Ma, P. Zuo, W. M. Kwok, W. S. Chan, J. T. W. Kan, P. H. Toy, and D. L. Phillips, *J. Org. Chem.* **2004**, *69*, 6641.

42. W. S. Chan, C. Ma, W. M. Kwok, and D. L. Phillips, *J. Phys. Chem. A.* **2005**, *109*, 3454.

43. P. Zuo, C. Ma, W. M. Kwok, W. S. Chan, and D. L. Phillips, *J. Org. Chem.* **2005**, *70*, 8661.

44. C. Reichardt, *Solvent and Solvent Effects in Organic Chemistry*, **2003**, Wiley-VCH Verlag GmbH&Co. KgaA, Weinheim, p. 1.

45. M. Besnard and M. I. Cabaco, *Chem. Phys.* **1992**, *163*, 103.

46. S. Woutersen, Y. Mu, G. Stock, and P. Hamm, *Chem. Phys.* **2001**, *266*, 137.

47. D. W. Oxtoby, D. Levesque, and J. J. Weis, *J. Chem. Phys.* **1978**, *68*, 5528.

48. J. Stenger, D. Madsen, P. Hamm, E. T. J. Nibbering, and T. Elsaesser, *Phys. Rev. Lett.* **2001**, *87*, 027401-1-4.

49. C. Ma, W. M. Kwok, W. S. Chan, Y. Du, J. T. W. Kan, P. H. Toy, and D. L. Phillips, *J. Am. Chem. Soc.* **2006**, *128*, 2558.

50. A. Laubereau and W. Kaiser, *Rev. Mod. Phys.* **1978**, *50*, 607.

51. T. Elsaesser and W. Kaiser *Annu. Rev. Phys. Chem.* **1991**, *42*, 83.

52. R. M. Stratt and M. Maroncelli, *J. Phys. Chem.* **1996**, *100*, 12981.

53. W. L. Weaver, L. A. Houston, K. Iwata, and T. L. Gustafson, *J. Phys. Chem.* **1992**, *96*, 8956.

54. K. Iwata and H. Hamaguchi, *Chem. Phys. Lett.* **1992**, *196*, 462.

55. R. E. Hester, P. Matousek, J. N. Moore, A. W. Parker, W. T. Toner, and M. Towrie, *Chem. Phys. Lett.* **1993**, *208*, 471.

56. J. Quian, S. Schultz, G. R. Bradburn, and J. M. Jean, *J. Phys. Chem.* **1993**, *97*, 10638.

57. R. M. Butler, M. A. Lynn, and T. L. Gustafson, *J. Phys. Chem.* **1993**, *97*, 2609.

58. M. Towrie, P. Matousek, A. W. Parker, W. T. Toner, and R. E. Hester, *Spectrochim. Acta A* **1995**, *51*, 2491.

59. K. Iwata and H. Hamaguchi, *J. Raman Spectrosc.* **1994**, *25*, 615.

60. P. Matousek, A. W. Parker, M. Towrie, and W. T. Toner, *J. Chem. Phys.* **1997**, *107*, 9807.

61. E. Lippert, W. Rettig, V. Bonacic-Koutecky, F. Heisel, and J. A. Miehe, *Adv. Chem. Phys.* **1987**, *68*, 1.

62. Z. R. Grabowski, K. Rotkiewicz, and W. Rettig, *Chem. Rev.* **2003**, *103*, 3899.

63. P. Changenet, P. Plaza, M. M. Martin, and Y. H. Meyer, *J. Phys. Chem. A* **1997**, *101*, 8186.

64. K. Rotkiewicz, K. H. Grellmann, and Z. R. Grabowski *Chem. Phys. Lett.* **1973**, *19*, 315.

65. W. Schuddeboom, S. A. Jonker, J. M. Warman, U. Leinhos, W. Kuhnle, and K. A. Zachariasse, *J. Phys. Chem.* **1992**, *96*, 10809.

66. K. A. Zachariasse, M. Grobys, T. Von der Haar, A. Hebecker, Y. V. Ilchev, Y.-B. Jiang, O. Morawski, and W. Kuhnle, *J. Photochem. Photobiol. A: Chem.* **1996**, *102*, 59.

67. A. L. Sobolewski, W. Sudholt, and W. Domcke, *Chem. Phys. Lett.* **1996**, *259*, 119.

68. W. M. Kwok, C. Ma, D. Phillips, P. Matousek, A. W. Parker, and M. Towrie, *J. Phys. Chem. A* **2000,** *104*, 4189.

69. W. M. Kwok, C. Ma, P. Matousek, A. W. Parker, D. Phillips, W. T. Toner, and M. Towrie, *Chem. Phys. Lett.* **2000**, *322*, 395.

70. M. Hashimoto and H. Hamaguchi, *J. Phys. Chem.* **1995**, *99*, 7875.

71. C. Chudoba, A. Kummrow, J. Dreyer, J. Stenger, E. T. J. Nibbering, T. Elsaesser, and K. A. Zachariasse, *Chem. Phys. Lett.* **1999**, *309*, 357.

72. H. Okamoto, *J. Phys. Chem. A* **2000**, *104*, 4182.

73. T. Okada, M. Uesugi, G. Kohler, K. Rechthaler, K. Rotkiewicz, W. Rettig, and G. Grabner, *Chem. Phys.* **1999**, *241*, 327.

74. I. Juchnovski, C. Tsvetanov, and I. Panayotov, *Monatsh. Chem.* **1969**, *100*, 1980.

75. H. Hamaguchi, C. Kato, and M. Tasumi, *Chem. Phys. Lett.* **1983**, *100*, 3.

76. T. L. Gustafson, D. M. Roberts, and D. A. Chernoff, *J. Chem. Phys.* **1984**, *81*, 3438.

77. T. Urano, H. Hamaguchi, M. Yamanouchi, S. Tsuchiya, and T. L. Gustafson, *J. Chem. Phys.* **1989**, *91*, 3884.

78. K. Iwata and H. Hamaguchi, *J. Phys. Chem. A* **1997**, *101*, 632.

79. J. Saltiel, J. D'Agostino, E. D. Megarity, L. Metts, K. R. Neuberger, M. Wrighton, and O. C. Zafirou, *Org. Photochem.* (O. L. Chapman, ed., Marcel Dekker, New York) Vol. 3 **1971**, 1.

80. R. M. Hochstrasser, *Pure Appl. Chem.* **1980**, *52*, 2683.

81. J. A. Syage, W. R. Lambert, P. M. Felker, A. H. Zewail, and R. M. Hochstrasser, *Chem. Phys. Lett.* **1982**, *88*, 266.

82. A. Bree and M. Edelson, *Chem. Phys.* **1980**, *51*, 77.

83. P. Matousek, A. W. Parker, W. T. Toner, M. Towric, D. L. A. de Faria, R. E. Hester, and J. N. Moore, *Chem. Phys. Lett.* **1995**, *237*, 373.

84. W. M. Kwok, C. Ma, A. W. Parker, D. Phillips, M. Towrie, P. Matousck, and D. L. Phillips, *J. Chem. Phys.* **2000**, *113*, 7471.

85. C. Ma, W. M. Kwok, P. Matousek, A. W. Parker, D. Phillips, W. T. Toner, and M. Towrie, *J. Raman Spectrosc.* **2001**, *32*, 115.

86. S. M. Beck and L. E. Brus, *J. Am. Chem. Soc.* **1982**, *104*, 4789.

87. P. J. Wagner, R. J. Truman, A. E. Puchalski, and R. Wake, *J. Am. Chem. Soc.* **1986**, *108*, 7727.

88. A. P. Darmanyan and S. C. Foote, *J. Phys. Chem.* **1992**, *96*, 6317.

89. L. C. T. Shoute and J. P. Mittal, *J. Phys. Chem.* **1993**, *97*, 8630.

90. L. C. T. Shoute and J. P. Mittal, *J. Phys. Chem.* **1994**, *98*, 1094.

91. G. Balakrishnan and S. Umapathy, *Chem. Phys. Lett.* **1997**, *270*, 557.

92. G. N. R. Tripathi and R. H. Schuler, *J. Phys. Chem.* **1983**, *87*, 3101.

93. G. Balakrishnan, P. Mohandas, and S. Umapathy, *J. Phys. Chem. A* **2001**, *105*, 7778.

94. G. B. Schuster and M. S. Platz, *Adv. Photochem.* **1992**, *17*, 69.

95. M. S. Platz, *Acc. Chem. Res.* **1995**, *28*, 487.

96. W. T. Borden, N. P. Gritsan, C. M. Hadad, W. L. Karney, C. R. Kemnitz, and M. S. Platz, *Acc. Chem. Res.* **2000**, *33*, 765.

97. M. S. Platz, in *Reactive Intermediates*, R. A. Moss, M. S. Platz, M. Jones Jr. **2004**, Wiley-Interscience, pp. 501.

98. E. Leyva, M. S. Platz, G. Persy, and J. Wirz, *J. Am. Chem. Soc.* **1986**, *108*, 3783.

99. G. B. Anderson and D. E. Falvey, *J. Am. Chem. Soc.* **1993**, *115*, 9870.

100. R. A. McClelland, M. J. Kahley, P. A. Davidse, and G. Hadzialic, *J. Am. Chem. Soc.* **1996**, *118*, 4794.

101. N. P. Gritsan, T. Yuzawa, and M. S. Platz, *J. Am. Chem. Soc.* **1997**, *119*, 5059.

102. R. Born, C. Burda, P. Senn, and J. Wirz, *J. Am. Chem. Soc.* **1997**, *115*, 5061.

103. C. J. Shields, D. R. Chrisope, G. B. Schuster, A. J. Dixon, M. Poliakoff, and J. J. Turner, *J. Am. Chem. Soc.* **1987**, *109*, 4723.

104. X.-Z. Sun, I. G. Virrels, M. W. George, and H. Tomioka, *Chem. Lett.* **1996**, 1089.

105. S. Srivastava, J. P. Toscano, R. J. Moran, and D. E. Falvey, *J. Am. Chem. Soc.* **1997**, *119*, 11552.

106. S. Srivastava, P. H. Ruane, J. P. Toscano, M. B. Sullivan, C. J. Cramer, D. Chiapperino, E. C. Reed, and D. E. Falvey, *J. Am. Chem. Soc.* **2000**, *122*, 8271.

107. M.-L. Tsao, N. Gritsan, T. R. James, M. S. Platz, D. Hrovat, and W. T. Borden, *J. Am. Chem. Soc.* **2003**, *125*, 9343.

108. S. Y. Ong, P. Zhu, Y. F. Poon, K.-H. Leung, W. H. Fang, and D. L. Phillips, *Chem. Eur. J.* **2002**, *8*, 2163.

109. P. Zhu, S. Y. Ong, P. Y. Chan, K.-H. Leung, and D. L. Phillips, *J. Am. Chem. Soc.* **2001**, *123*, 2645.

110. P. Zhu, S. Y. Ong, P. Y. Chan, Y. F. Poon, K.-H. Leung, and D. L. Phillips, *Chem. Eur. J.* **2001**, *7*, 4928.

111. P. Y. Chan, W. M. Kwok, and D. L. Phillips, *J. Am. Chem. Soc.* **2005**, *127*, 8246.

112. R. B. Woodward and R. Hoffmann, *The Conservation of Orbital Symmetry*, 3rd ed., Verlag Chemie International: Weinheim, **1981**, pp. 1–177.

113. K. Fukui, *Acc. Chem. Res.* **1971**, *4*, 57.

114. H. E. Zimmerman, *Acc. Chem. Res.* **1971**, *4*, 272.

115. H. J. C. Jacobs and E. Havinga *Adv. Photochem.* **1979**, *11*, 305.

116. W. G. Dauben, E. L. McInnis, and D. M. Michno, in *Rearrangements in Ground and Excited States*, Academic Press: New York, **1980**, Vol. 3, p. 91.

117. M. J. S. Dewar and C. Jie, *Acc. Chem. Res.* **1992**, *25*, 537.

118. K. J. Crowley, *J. Org. Chem.* **1968**, *33*, 3679.

119. N. Gottfried, W. Kaiser, M. Braun, W. Fuss, and K. L. Kompa, *Chem. Phys. Lett.* **1984**, *110*, 335.

120. M. O. Trulson, G. D. Dollinger, and R. A. Mathies, *J. Chem. Phys.* **1989**, *90*, 4274.

121. P. J. Reid, S. J. Doig, S. D. Wickham, and R. A. Mathies, *J. Am. Chem. Soc.* **1993**, *115*, 4754.

122. P. J. Reid, M. K. Lawless, S. D. Wickham, and R. A. Mathies, *J. Phys. Chem.* **1994**, *98*, 5597.

123. R. J. Hemley, B. R. Brooks, and M. Karplus, *J. Chem. Phys.* **1986**, *85*, 6550.

124. H. Yoshida, Y. Furukawa, and M. Tasumi, *J. Mol. Struct.* **1989**, *194*, 279.

125. R. S. Givens and L. W. Kueper, *Chem. Rev.* **1993**, *93*, 55 and references therein.

126. R. S. Rock and S. I. Chan, *J. Am. Chem. Soc.* **1998**, *120*, 10766.

127. K. Lee and D. E. Falvey, *J. Am. Chem. Soc.* **2000**, *122*, 9361.

128. C. S. Rajesh, R. S. Givens, and J. Wirz, *J. Am. Chem. Soc.* **2000**, *122*, 611.

129. Y. V. Il'chev, M. A. Schworer, and J. Wirz, *J. Am. Chem. Soc.* **2004**, *126*, 4581.

130. R. S. Givens, J. F. W. Weber, P. G. Conrad II, G. Orosz, S. L. Donahue, and S. A. Thayer, *J. Am. Chem. Soc.* **2000**, *122*, 2687–2697 and references therein.

131. P. G. Conrad II, R. S. Givens, J. F. W. Weber, and K. Kandler, *Org. Lett.* **2000**, *2*, 1545.

132. P. G. Conrad II, R. S. Givens, B. Hellrung, C. S. Rajesh, M. Ramseier, and J. Wirz, *J. Am. Chem. Soc.* **2000**, *122*, 9346.

133. K. Zhang, J. E. T. Corrie, V. R. N. Munasunghe, and P. Wan, *J. Am. Chem. Soc.* **1999**, *121*, 5625.

134. C. Ma, W. M. Kwok, W. S. Chan, P. Zuo, J. T. W. Kan, P. H. Toy, and D. L. Phillips, *J. Am. Chem. Soc.* **2005**, *127*, 1463.

135. D. Oesterhelt and W. Stoechenius, *Proc. Natl. Acad. Sci. USA* **1973**, *70*, 2853.

136. R. R. Birge, *Biochim. Biophys. Acta* **1990**, *1016*, 293.

137. R. A. Mathies, S. W. Lin, J. B. Ames, and W. T. Pollard, *Annu. Rev. Biophys. Biophys. Chem.* **1991**, *20*, 491.

138. A. V. Sarkov, A. V. Pakulev, S. V. Chekalin, and Y. A. Matveetz, *Biochim. Biophys. Acta* **1985**, *808*, 94.

139. H.-J. Pollard, M. A. Franz, W. Zinth, W. Kaiser, E. Koelling, and D. Oesterhelt, *Biochim. Biophys. Acta* **1986**, *851*, 407.

140. J. Dobler, W. Zinth, W. Kaiser, and D. Oesterhelt, *Chem. Phys. Lett.* **1988**, *144*, 215.

141. R. A. Mathies, C. H. Brito Cruz, W. T. Pollard, and C. V. Shank, *Science* **1988**, *240*, 777.

142. S. Milder and D. Kliger, *Biophys. J.* **1988**, *53*, 465.

143. G. Varo and J. K. Lanyi, *Biochemistry* **1990**, *29*, 6858.

144. J. B. Ames and R. A. Mathies, *Biochemistry* **1990**, *29*, 7181.

145. S. J. Doig, P. J. Reid, and R. A. Mathies, *J. Phys. Chem.* **1991**, *95*, 6372.

146. Curry, I. Palings, A. D. Broek, J. A. Pardoen, J. Lugtenburg, and R. Mathies, *Adv. Infrared Raman Spectrosc.* **1985**, *12*, 115.

147. S. O. Smith, J. A. Pardoen, J. Lugtenburg, and R. A. Mathies, *J. Phys. Chem.* **1987**, *91*, 804.

148. W. Zinth, C. Kolmeder, B. Benna, A. Ingrens-Defregger, S. F. Fischer, and W. Kaiser, *J. Chem. Phys.* **1983**, *78*, 3916.

149. W. Wild, A. Seilmeier, N. H. Gottfried, and W. Kaiser, *Chem. Phys. Lett* **1985**, *119*, 259.

150. M. Braiman and R. Mathies, *Proc. Natl. Acad. Sci. U.S.A.* **1982**, *79*, 403.

151. J. Hofrichter, J. H. Sommer, E. R. Henry, and W. A. Eaton, *Proc. Natl. Acad. Sci. U.S.A.* **1983**, *80*, 2235.

152. V. Jayaraman, K. R. Rodgers, I. Mukerji, and T. G. Spiro, *Science* **1995**, *269*, 1843.

153. R. A. Goldbeck, S. J. Paquette, S. C. Bjorling, and D. S. Kliger, *Biochemistry* **1996**, *35*, 8628.

154. R. A. Goldbeck, S. J. Paquette, and D. S. Kliger, *Biophys. J.* **2001**, *81*, 2919.

155. R. P. Chen and T. G. Spiro, *J. Phys. Chem. A* **2002**, *106*, 3413.

156. G. Balakrishnan, M. A. Chase, A. Pevsner, X. Zhao, C. Tengroth, G. L. McLendon, and T. G. Spiro, *J. Mol. Biol.* **2004**, *340*, 843.

157. K. R. Rodgers, C. Su, S. Subramaniam, and T. G. Spiro, *J. Am. Chem. Soc.* **1992**, *114*, 3697.

158. X. H. Hu and T. G. Spiro, *Biochemistry* **1997**, *36*, 15701.

159. D. J. Wang, X. J. Zhao, and T. G. Spiro, *J. Phys. Chem. A* **2000**, *104*, 4149.

160. K. R. Rodgers and T. G. Spiro, *Science* **1994**, *265*, 1697.

161. P. Matousek, M. Towrie, A. Stanley, and A. W. Parker, *Appl. Spectrosc.* **1999**, *53*, 1485.

162. P. Matousek, M. Towrie, C. Ma, W. M. Kwok, D. Phillips, W. T. Toner, and A. W. Parker, *J. Raman Spectrosc.* **2001**, *32*, 983.

163. S. Y. Lee, D. H. Zhang, D. W. McCamant, P. Kukura, and R. A. Mathies, *J. Chem. Phys.* **2004**, *121*, 3632.

164. D. W. McCamant, P. Kukura, and R. A. Mathies, *J. Phys. Chem. A* **2003**, *107*, 8208.

165. P. Kukura, D. W. McCamant, S. Yoon, D. B. Wandschneider, and R. A. Mathies, *Science* **2005**, *310*, 1006.

CHAPTER 4

Time-Resolved Infrared (TRIR) Studies of Organic Reactive Intermediates

JOHN P. TOSCANO

Department of Chemistry, Johns Hopkins University, Baltimore, MD

4.1. INTRODUCTION

Time-resolved spectroscopic techniques are important and effective tools for mechanistic photochemical studies. The most widely used of these tools, time-resolved UV-VIS absorption spectroscopy, has been applied to a variety of problems since its introduction by Norrish and Porter almost 60 years ago.[1] Although a great deal of information about the reactivity of organic photochemical intermediates (e.g., excited states, radicals, carbenes, and nitrenes) in solution at ambient temperatures has been amassed with this technique, only limited structural information can be extracted from

Reviews of Reactive Intermediate Chemistry. Edited by Matthew S. Platz, Robert A. Moss, Maitland Jones, Jr.

such investigations because absorption bands are usually quite broad and featureless. Questions of bonding, charge distribution, and solvation (in addition to those of dynamics) are more readily addressed with time-resolved vibrational spectroscopy.

Indeed, time-resolved resonance Raman (TR^3) spectroscopy has been successfully employed to study the structure and dynamics of many short-lived molecular species and is the topic of a separate chapter by D. L. Phillips in this book. Like TR^3 spectroscopy, TRIR spectroscopy gives one the ability to monitor directly both the structure and dynamics of the reactants, intermediates, and products of photochemical reactions. The time-resolved Raman and IR experiments, along with their transient UV-VIS absorption predecessor, are of course all complementary, and a combination of these techniques can give a very detailed picture of a photochemical reaction.

Experimental limitations initially limited the types of molecular systems that could be studied by TRIR spectroscopy. The main obstacles were the lack of readily tunable intense IR sources and sensitive fast IR detectors. Early TRIR work focused on gas phase studies because long pathlengths and/or multipass cells[2] could be used without interference from solvent IR bands. Pimentel and co-workers first developed a rapid scan dispersive IR spectrometer[3] (using a carbon arc broadband IR source) with time and spectral resolution on the order of $10\,\mu s$ and $1\,cm^{-1}$, respectively, and reported the gas phase IR spectra of a number of fundamental organic intermediates (e.g., CH_3, CD_3, and CF_2).[4] Subsequent gas phase approaches with improved time and spectral resolution took advantage of pulsed IR sources.

Weitz and co-workers extended gas phase TRIR investigations to the study of coordinatively unsaturated metal carbonyl species.[5] Metal carbonyls are ideally suited for TRIR studies owing to their very strong IR chromophores. Indeed, initial TRIR work in solution, beginning in the early 1980s, focused on the photochemistry of metal carbonyls for just this reason.[6] Since that time, instrumental advances have significantly broadened the scope of TRIR methods and as a result the excited state structure and photoreactivity of organometallic complexes in solution have been well studied from the microsecond to picosecond time scale.[7]

TRIR methods have also found utility in the elucidation of reaction mechanisms involved in biological systems, most notably photosynthetic and respiratory proteins.[8] In addition, TRIR spectroscopy has also been used to enhance our understanding of the dynamics of protein folding processes.[9]

In contrast to gas phase, organometallic, and biological studies, until recently, relatively few organic systems had been examined by TRIR methods.[10] This chapter will begin with a brief survey of experimental approaches to TRIR spectroscopy and will follow with a discussion of several representative studies of organic reactive intermediates that demonstrate the significant utility of this technique.

4.2. INSTRUMENTAL APPROACHES TO TRIR SPECTROSCOPY

Recent technical advances have greatly expanded the applicability of TRIR spectroscopy, making measurements over wide temporal and spectral ranges now feasible. The relative merits of different experimental approaches have been discussed previously.[7a–c,g]

4.2.1. Ultrafast Methods

Ultrafast (sub-nanosecond) experiments in general rely on a "pump-probe" scheme in which the time resolution is obtained by spatial delay of the probe pulse relative to the pump pulse (1 ps = 0.30 mm). Thus, ultrafast TRIR methods[11] require the generation of short IR pulses, which is commonly accomplished by nonlinear mixing schemes in crystals such as $LiNbO_3$, $LiIO_3$, $AgGaS_2$, or GaSe. Photons of one frequency (typically in the visible) are converted, under the proper phase-matching conditions and with conservation of energy and momentum, into two photons, one at a lower visible frequency and the other at an IR frequency equal to the energy difference between the two visible photons. Since a 100 fs IR pulse has a bandwidth of approximately $150\,cm^{-1}$, uncertainty broadening is a potential concern with ultrashort IR pulses. However, spectral resolution lost to uncertainty broadening can be recovered by dispersing a spectrally broad pulse through a monochromator after it has passed through the sample.[12] In addition, broadband approaches that take advantage of multichannel IR detectors have been developed[13] and are usually employed in modern ultrafast TRIR spectrometers.[11f,g]

4.2.2. Nanosecond/Microsecond Methods

Although very detailed, fundamental information is available from ultrafast TRIR methods, significant expertise in femtosecond/picosecond spectroscopy is required to conduct such experiments. TRIR spectroscopy on the nanosecond or slower timescale is a more straightforward experiment. Here, mainly two alternatives exist: step-scan FTIR spectroscopy and conventional pump-probe dispersive TRIR spectroscopy, each with their own strengths and weaknesses. Commercial instruments for each of these approaches are currently available.

The step-scan FTIR experiment, initially used to investigate time-dependent phenomena in the mid-1970s,[14] has previously been described in detail.[15] Briefly, the moving mirror of an FTIR spectrometer is translated in discrete steps rather than continuously as is normally done in a static FTIR experiment. At each mirror position, the photochemical process of interest is triggered (e.g., by a nanosecond laser pulse) and the temporal changes in detector response are recorded. Data from each mirror position are transposed into a series of interferograms that correspond to different time delays following the trigger. These interferograms are then Fourier transformed to provide a series of time-resolved IR spectra. Kinetic data at a particular frequency can then be derived from these spectra.

In conventional nanosecond pump-probe dispersive TRIR experiments, also described previously,[16] kinetic data are collected at one frequency at a time. These data can then be used to construct a series of time-resolved IR spectra. Thus, in the dispersive experiment *kinetic* data are used to construct spectra, and in the step-scan experiment *spectral* data are used to derive kinetics.

The pump source in the dispersive experiment is typically a nanosecond laser; the probe source can be broadband IR light from a globar or tunable IR light from a CO laser or a semiconductor diode laser. Although CO and diode lasers can produce high-intensity IR light, these probe sources are somewhat limited in their spectral coverage. The CO laser is restricted to the range 2000–1500 cm^{-1} and a typical IR diode laser has a maximum scanning range of 100–150 cm^{-1}.[7b,c] Of course, a series of diode lasers will cover a broader range of frequencies, but such a setup can quickly become cumbersome and expensive. Normal globar sources are not limited in their spectral range, but unfortunately produce relatively low-intensity IR light that typically leads to signal-to-noise problems. Thus, in dispersive nanosecond TRIR pump-probe experiments one is usually faced with a choice between probe sources with high intensity or broad spectral coverage.

One solution to this dilemma has been advanced by Hamaguchi and co-workers[17] who have made use of a MoSi$_2$ IR source newly developed by JASCO that provides approximately twice the emissive intensity of conventional globar sources. This probe source was incorporated into a dispersive TRIR spectrometer that allows access to the entire mid-IR spectrum with high sensitivity ($\Delta A < 10^{-5}$) and sufficient time (50 ns) and frequency (4–16 cm^{-1}) resolution to probe a wide range of transient intermediates in solution.

TRIR spectral data are usually obtained in the form of a difference spectrum. Thus, depletion of reactant gives rise to negative signals, and formation of transient intermediates or products leads to positive bands. In ideal cases, the depletion of reactants, the growth and decay of intermediates, and the growth of products of a photoinitiated reaction can be monitored. Since the detection of transient species is more problematic in regions with strong solvent bands as a result of low transmission of IR light, available spectral windows are limited by solvent absorbance and path length. Major spectral windows available for some solvents typically employed in TRIR studies (for a 0.5 mm pathlength) are as follows: Freon-113: 2200–1420 cm^{-1}, hexane: 2550–1500 cm^{-1} and 1300–900 cm^{-1}, acetonitrile: 2200–1500 cm^{-1} and 1300–1100 cm^{-1}, acetonitrile-d_3: 2200–1200 cm^{-1}.

In order to obtain TRIR spectra with sufficient sensitivity, substantial signal averaging is typically required. A flow cell, therefore, is necessary to prevent excessive sample decomposition, especially when monitoring photo-irreversible processes. To maintain sample integrity for noncyclic systems, one is usually forced in the dispersive TRIR experiment to acquire data in a series of short (e.g., 100–200 cm^{-1}) scans rather than in one complete scan. Thus, a substantial amount of sample (and patience!) may be required. Sample integrity is also of significant concern in the step-scan FTIR experiment since data must be collected at each mirror position. In order to address this concern very large reservoirs of solution are required; alternatively, a sample changing wheel[18] or very focused pump/probe beams in combination with sample translation[19] have been used with thin film samples.

Although the dispersive TRIR experiment does not have the spectral multiplexing advantage of the step-scan FTIR experiment, both experiments require that the transient process being probed is reliably repeatable. (The dispersive experiment demands substantial signal averaging; the step-scan experiment requires data collection at a series of mirror positions.) The step-scan FTIR method, however, can be more sensitive to any variation in measuring conditions compared with the dispersive TRIR experiment. For example, a noise spike in an interferogram will affect the entire spectrum in the step-scan experiment, whereas it will distort only a single point in the dispersive method. Indeed, early step-scan FTIR experiments were plagued by spectral artifacts caused by fluctuating measuring conditions (especially mirror position stability)[20] that have since been addressed in modern instruments. An additional concern with both experiments is the possible contribution of thermal artifacts (resulting from a transient temperature rise in the sample caused by the excitation pulse) to the observed time-resolved signal.[17b,20,21] These thermal artifacts are usually of most concern in thin-film samples or in regions of high-background absorbance by solvent.

4.2.3. Millisecond Methods

Since modern FTIR spectrometers can operate in a rapid scan mode with approximately 50 ms time resolution, TRIR experiments in the millisecond time regime are readily available. Recent advances in ultra-rapid scanning FTIR spectroscopy[22] have improved the obtainable time resolution to 5 ms. Alternatively, experiments can be performed at time resolutions on the order of 1–10 ms with the planar array IR technique, which utilizes a spectrograph for wavelength dispersion and an IR focal plane detector for simultaneous detection of multiple wavelengths.[23]

4.3. APPLICATIONS OF TRIR SPECTROSCOPY TO THE STUDY OF ORGANIC REACTIVE INTERMEDIATES

The following representative examples of TRIR studies are not meant to be an exhaustive treatment of the various organic reactive intermediates that have been investigated by TRIR methods, but rather to demonstrate the unique insight that such studies can provide. The direct observation of organic intermediates in solution at room temperature by IR spectroscopy can reveal fundamental information related both to bonding and structure of reactive intermediates as well to mechanisms of product formation.

4.3.1. The Structure of α-Lactones

α-Lactones (oxiranones) **1** are three-membered heterocyclic rings that have been invoked as intermediates in a variety of organic transformations[24] as well as in enzymatic glycosyltransferase reactions.[25] Most α-lactones are unstable at room temperature;[26] those that are stable require bulky[27] or strongly electron-withdrawing substitution[28] at the α-carbon. Much of the chemistry of α-lactones has been explained by invoking a higher energy ring-opened zwitterionic form **2**. The relative stability of ring-closed form **1** versus ring-opened form **2** is dependent on the substituents R. Calculations have predicted that for the parent α-lactone (R = H), ring-closed form **1** is significantly lower in energy than ring-opened form **2**.[29–31] Calculations have also shown that, as expected, substitution with electron-donating R groups, as well as polar solvation, preferentially stabilizes **2**.[32]

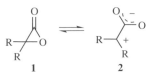

Nucleophilic attack of alcohols (e.g., ethanol) on α-lactones gives different regiochemistry depending on the substituent R. In cases where R is alkyl or phenyl, ethanol reacts with α-lactones to produce α-ethoxyacids **3**, presumably via zwitterionic form **2**.[33-35] When R is the strongly electron-withdrawing trifluoromethyl group (such as in **4**), however, ethanol reacts to produce α-hydroxyester **5**, which presumably is formed by attack on ring-closed form **1**.[28] The trifluoromethyl groups obviously destabilize the dipolar ring-opened structure.

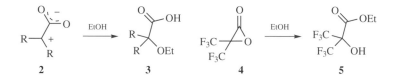

α-Lactones readily undergo decarbonylation and also polymerization reactions. Loss of carbon monoxide most likely occurs through ring-closed form **1**, while polymerization occurs through ring-opened form **2**. Thus, structures **1** and **2** both play important roles in α-lactone chemistry; however, in essentially all cases where ring-opened form **2** dominates the observed reactivity, it has not been detected spectroscopically. For example, the reactivity of α-lactones **6** and **7**, even at temperatures as low as −100°C, is consistent with ring-opened form **2**.[33-35] However, low-temperature (77 or 10 K) matrix IR spectroscopy of **6** ($v_{C=O}$ = 1900 cm^{-1})[34] and **7** ($v_{C=O}$ = 1890 cm^{-1})[36,37] has revealed carbonyl stretching modes at considerably high frequencies, indicative of ring-closed form **1**.

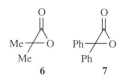

These observations have been rationalized in terms of a small energy gap between structure **1** and its more reactive, but higher energy counterpart **2**, and are consistent with previous computational work. Since the IR signatures of structures **1** and **2** are expected to be quite different, TRIR spectroscopy was used to examine the influence of substituents (R) and solvent on the relative stability of **1** and **2**;[38] the results of these studies are summarized in this section. The related iminooxirane and α-lactam intermediates have also been recently examined by TRIR spectroscopy.[39]

α-Lactones were generated from the reaction of carbenes with carbon dioxide. This method of α-lactone generation had been used by Kistiakowsky in the gas phase,[40]

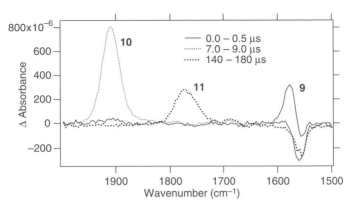

Figure 4.1. TRIR difference spectra averaged over the timescales indicated following 355 nm laser photolysis of diazirine **8** (15.7 mM) in CO_2-saturated dichloromethane. Reprinted with permission from B. M. Showalter and J. P. Toscano, *J. Phys. Org. Chem.* **2004**, *14*, 743. Copyright 2004, John Wiley & Sons Limited.

by Bartlett in solution,[35] and by both Jacox[41] and Sander[36,37] in low temperature matrices. In addition, Kovacs and Jackson recently reported a comprehensive computational investigation of the reaction of methylene with CO_2.[42] Representative TRIR data observed for the reaction of phenylchlorocarbene (**9**), produced by laser photolysis of phenylchlorodiazirine (**8**), with CO_2 in dichloromethane are shown in Figure 4.1.[38] Depletion of reactants gives rise to negative signals and the formation of transient intermediates or products leads to positive bands. The negative signal at 1566 cm⁻¹ was assigned to depletion of diazirine **8**, while the positive bands at 1586, 1910, and 1776 cm⁻¹ were assigned to carbene **9**,[37] α-lactone **10**,[37] and acid chloride **11**,[43] respectively, based on previous studies (Scheme 4.1).[37,44]

The observed TRIR data are consistent with Scheme 4.1. Depletion of the diazirine and formation of the carbene occurs within the time resolution (50 ns) of the experiment. Subsequent decay of the carbene ($k_{osbd} = 3.0 \times 10^5$ s⁻¹) is observed at the same rate within experimental error (±10%) that the α-lactone is produced ($k_{osbd} = 3.2 \times 10^5$ s⁻¹) and the final decay of the α-lactone ($k_{osbd} = 2.0 \times 10^4$ s⁻¹) occurs at the same rate as the acid chloride product is formed ($k_{osbd} = 1.8 \times 10^4$ s⁻¹). The position of the α-lactone band at 1910 cm⁻¹ is clearly indicative of ring-closed form **1** and in very good agreement with the signal observed at 10 K (1920 cm⁻¹) by Sander and co-workers.[37]

Sander and co-workers have also previously examined the effects of carbene spin state and philicity on the carboxylation reaction in low-temperature matrices.[37]

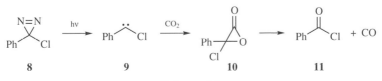

Scheme 4.1.

They concluded that the reactivity of carbenes toward CO_2 is determined by their philicity (more nucleophilic carbenes are more reactive) and that carbene spin state interestingly has little effect. Kovacs and Jackson have suggested that this reactivity pattern may be explained by a nonequilibrium surface crossing mechanism.[42]

Using the pseudo-first-order equation $k_{obsd} = k_0 + k_{CO2}$ [CO_2](where k_{CO2} is the second-order rate constant for the reaction of carbene with CO_2 and k_0 is the rate of carbene decay in the absence of CO_2), solution-phase values of k_{CO2} for phenylchlorocarbenes **9** and **12**, and diphenylcarbenes **14** and **15** in dichloromethane were estimated (Table 4.1).[38] (The concentration of CO_2 in saturated dichloromethane solution at 25°C and 1 atm is 196 mmol/L.[45]) The trend of these estimated second-order order rate constants agrees with that observed in low-temperature matrices by Sander and co-workers.[36,37]

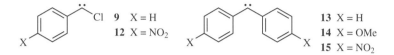

$$\begin{array}{ll} \textbf{9} & X = H \\ \textbf{12} & X = NO_2 \end{array} \qquad \begin{array}{ll} \textbf{13} & X = H \\ \textbf{14} & X = OMe \\ \textbf{15} & X = NO_2 \end{array}$$

Although carbene **15** is unreactive with CO_2, TRIR data were obtained for the reaction of carbenes **9** and **12–14** with CO_2; the experimentally observed C–O stretching frequencies are given in Table 4.1. Spectral data for the α-lactone derived from carbene **12**, although significantly weaker in intensity, is analogous to that derived from carbene **9**. The TRIR spectra for the α-lactones derived from carbenes **13** and **14**, however, are dramatically different from each other (Figure 4.2).[38] Whereas the product of CO_2 reaction with **13** in dichloromethane is clearly the ring-closed α-lactone **16** ($\nu_{C=O} = 1880$ cm^{-1}, again in good agreement with that observed by low-temperature matrix IR spectroscopy[36] at 1890 cm^{-1}), the carboxylation of **14** in the same solvent leads to a structure that appears to be best described by zwitterion **17**. In this latter case (Figure 4.2b), the spectral region between 1800 and 2000 cm^{-1} is devoid of any signal that may be attributed to the ring-closed form. Instead, intense IR bands are detected at 1620 and 1576 cm^{-1}, consistent with B3LYP/6-31G(d) optimized geometries and calculated frequencies (scaled by 0.96) (Table 4.2) indicative of ring-opened form **2**.[38] Indeed, even geometry minimizations starting from the ring-closed form of **17** resulted in a ring-opened minimum. In addition,

TABLE 4.1. Estimated second-order rate constants for carbene reactions with CO_2 in dichloromethane and IR frequencies of the corresponding α-lactones.[38]

Carbene	k_{CO_2} (M^{-1}s^{-1})	α-Lactone (cm^{-1})
9	1×10^6	1910
12	2×10^5	1920
13	a	1880
14	2×10^7	1620, 1576
15	$< 10^4$	b

[a]A carbene band was not observed in dichloromethane due to overlap with a diazo precursor depletion band. [b]Carbene **15** was unreactive with CO_2; no α-lactone signals were observed.

B3LYP/6-31G(d) calculations indicate that the corresponding triplet-biradical **18** is 16.1 kcal/mol higher in energy than zwitterion **17**.

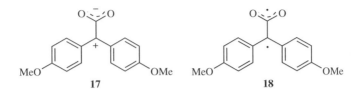

In a related study, Lee-Ruff, Johnston, and co-workers[46] photolyzed 9-hydroxy-9-fluorenecarboxylic acid in hexafluoro-2-propanol (HFIP) and detected a transient (τ = ca. 20 μs) with λ_{max} = 495 nm and strong IR bands at 1575, 1600, and 1620 cm^{-1}, which was assigned to zwitterion **20**, in excellent agreement with the measured IR

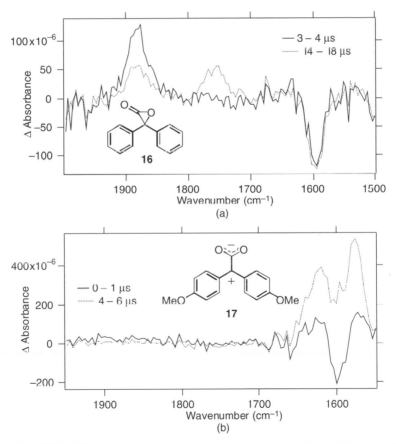

Figure 4.2. TRIR difference spectra averaged over the timescales indicated following 266 nm photolysis of (a) diphenyl diazomethane (3.6 mM) and (b) bis-(4-methoxyphenyl)-diazomethane (1.1 mM) in CO$_2$-saturated dichloromethane. Reprinted with permission from B. M. Showalter and J. P. Toscano, *J. Phys. Org. Chem.* **2004**, *14*, 743. Copyright 2004, John Wiley & Sons Limited.

TABLE 4.2. Selected experimental and B3LYP/6-31G(d) calculated IR frequencies (scaled by 0.96) for diphenyloxiranone (16) and *bis*-(4-methoxyphenyl)oxiranone (17).[38]

α-Lactone	Experimental freq. (cm^{-1})	Calculated freq. (cm^{-1})	Assignment
16	1880	1907	Carbonyl str.
17	1576	1587	Phenyl C–C str.
		1597	Phenyl C–C str.
	1620	1658	Asym. C–O str.

bands for ring-opened form **17** (Scheme 4.2).[38] They examined the region 1988 to 1855 cm^{-1} and reported that no signals were observed for ring-closed form **21**.

The difference in structure between **16** and **17** is readily understood in terms of the addition of strongly electron-donating substituents, but the contrast between **16** and **20** is less easily rationalized. Photolysis of **19** was carried out in HFIP (dielectric constant (ε) = 16.75), while TRIR experiments with diphenyl diazomethane (**22**) were carried out in dichloromethane (ε = 9.08), suggesting that α-lactone structure may be dependent on solvent polarity.

Consistent with this suggestion, TRIR data observed following photolysis of diazo compound **22** in acetonitrile-d_3 (Figure 4.3) contain signals that can be attributed to both ring-closed form **16** and ring-opened form **23**.[38] SCRF B3LYP/6-31G(d) calculations using the Onsager model for acetonitrile (ε = 35.9) predict that ring-closed form **16** is 2.6 kcal/mol more stable than ring-opened form **23**. Strong IR bands are predicted at 1886 cm^{-1} for **16** (in good agreement with the experimentally observed 1880 cm^{-1} band) and at 1314 cm^{-1} for **23** (in reasonable agreement with the experimentally observed 1356 cm^{-1} band). Less intense IR bands are also predicted at 1638 and 1577 cm^{-1} for **23**. These calculated bands are likely obscured by an overlapping diazo depletion band at 1592 cm^{-1} in the experimental data.

The kinetics observed for the 1880 cm^{-1} band of **16** and the 1356 cm^{-1} band of **23** are in good agreement with each other, indicating that these two species are in

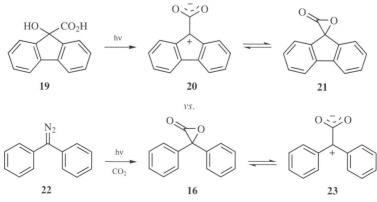

Scheme 4.2.

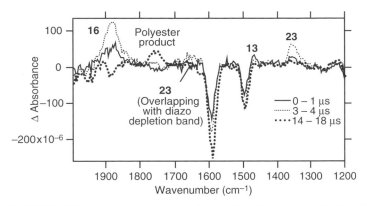

Figure 4.3. TRIR difference spectra averaged over the timescales indicated following 266 nm laser photolysis of diphenyl diazomethane (9.8 mM) in CO_2-saturated acetonitrile-d_3. Reprinted with permission from B. M. Showalter and J. P. Toscano, *J. Phys. Org. Chem.* **2004**, *14*, 743. Copyright 2004, John Wiley & Sons Limited.

equilibrium (**16**: $k_{growth} = 1.5 \times 10^6$ s^{-1}, $k_{decay} = 1.1 \times 10^5$ s^{-1}; **23**: $k_{growth} = 1.2 \times 10^6$ s^{-1}, $k_{decay} = 1.2 \times 10^5$ s^{-1}).[38] Given that an equilibrium mixture of **16** and **23** is observed, the calculated energy difference of 2.6 kcal/mol between the two is almost certainly overestimated.

TRIR data observed following photolysis of **22** in cyclohexane (Figure 4.4) contains a band easily attributed to ring-closed form **16**, but in contrast to data obtained in acetonitrile, no signals indicating the presence of ring-opened form **23**.[38] SCRF B3LYP/6-31G(d) calculations using the Onsager model for cyclohexane ($\varepsilon = 2.02$)

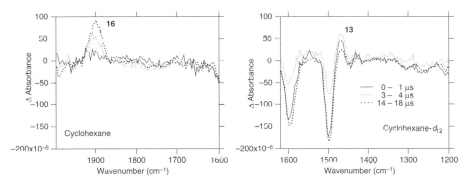

Figure 4.4. TRIR difference spectra averaged over the timescales indicated following 266 nm laser photolysis of diphenyl diazomethane (6.3 mM) in CO_2-saturated cyclohexane and cyclohexane-d_{12}. Since the detection of transient species is more problematic in regions with strong solvent bands due to the low transmission of IR light, cyclohexane-d_{12} was required for the spectral region below 1600 cm^{-1}. Reprinted with permission from B. M. Showalter and J. P. Toscano, *J. Phys. Org. Chem.* **2004**, *14*, 743. Copyright 2004, John Wiley & Sons Limited.

predict a strong IR band at $1900\,cm^{-1}$ in excellent agreement with the experimentally observed band. These calculations, however, indicate that ring-opened form **23** is not a minimum on the potential energy surface, but rather a transition state structure (one imaginary frequency $= 72\,cm^{-1}$). The calculated frequencies for this structure are, however, similar to those obtained in the acetonitrile calculations.

Thus, the structure of α-lactones, as reflected in their observed TRIR spectra, is dependent both on substituents at the α-carbon and on solvent polarity, with electron-donating substituents and polar solvents favoring a zwitterionic ring-opened structure. B3LYP calculations using SCRF methods to account for solvent polarity are consistent with these experimental conclusions.

4.3.2. The Kinetics of Carbene Rearrangements

Chemists have long appreciated that photolysis of diazirines or diazo compounds often leads to product mixtures different from those observed on thermolysis of the same carbene precursors.[47] More recent photochemical studies of these nitrogenous carbene precursors have confirmed that rearrangement products can originate from a short-lived precursor singlet excited state as well as from the carbene.[48] (Alternatives to the excited state, such as a carbene-olefin complex in the case of olefin trapping agent,[49] unstable diazo isomers in the case of diazirines,[50] and excited carbenes[51] have also been proposed.) With the exception of recent diazirine fluorescence studies, which indicate that diazirines less prone to rearrangement fluoresce more strongly,[48] evidence for the involvement of a precursor excited state in rearrangement chemistry has been based mainly on product studies. For example, even in the presence of a large excess of a carbene trap such as an olefin or alcohol, rearrangement products are still observed.[48,49] Analysis of the ratios of trapping product to rearrangement product can provide the extent to which rearrangement occurs from a precursor excited state.

One example of photoinduced rearrangement chemistry that has received significant attention[52] is the Wolff rearrangement of diazocarbonyl compounds to ketenes, a reaction discovered at the beginning of the 20th century.[53] In the mid-1960s Kaplan and Meloy demonstrated that acyclic diazocarbonyl compounds exist as equilibrium mixtures of *syn* and *anti* forms (Scheme 4.3), and suggested that the relative contribution of these conformations could have a significant

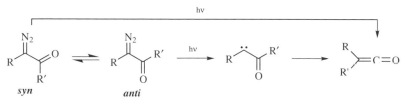

Scheme 4.3.

impact on the mechanism of photoinduced ketene formation.[54] Subsequent work has confirmed the important role that conformation plays in this rearrangement process.[55] The mechanistic picture that has emerged from these investigations is that concerted rearrangement to ketene from a singlet excited state is facile from the *syn* conformation, but carbene production occurs from the *anti* form.[56] (As will be shown below, carbene formation can also occur from the *syn* form.) The evidence for involvement of precursor excited states in such rearrangement chemistry has been based mainly on product studies. For example, even in the presence of a large excess of a carbene trap such as alcohol, ketene rearrangement products are still observed for diazocarbonyls with significant population of the *syn* conformer.[57]

Nanosecond TRIR spectroscopy is very well suited to examining this issue by monitoring directly the kinetics of carbene rearrangement process. Because the lifetimes of diazocarbonyl singlet excited states are likely much less than 1 ns and certainly well below the time resolution of nanosecond TRIR spectroscopy, ketene production from only the excited state would be indicated by a fast, unresolvable increase in ketene IR absorbance following laser photolysis. Production only from the carbene would be revealed by a rate of ketene growth equal to that of carbene decay. In those cases where both the excited state and the carbene contribute, the rearrangement product ketene growth rate should be biexponential and consist of two components, viz., a fast, unresolvable component (production from a short-lived excited state) and a slower, resolvable component (production from carbene). In this section, we consider the photoinduced Wolff rearrangement of methyl 2-diazo-(2-naphthyl) acetate (**24**) and the related cyclic analog (phenyl version), 4-diazo-3-isochromanone (**25**), both of which have recently been examined by TRIR spectroscopy.[58,59] Analogous TRIR studies of the photo-Curtius rearrangement of nitrenes to isocyanates have also been recently reported.[60]

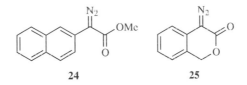

24   **25**

TRIR kinetic data of the decay of carbene **26** and the growth of ketene **27** following laser photolysis of ^{15}N-labeled **24** are shown in Figure 4.5.[58] The use of ^{15}N-labeled **24** was required to shift a strong negative depletion signal in the unlabeled diazoester away from the positive ketene IR band of interest. These data indicate that the rate of ketene growth is equivalent to the rate of carbene decay, clearly demonstrating that ketene **27** arises entirely from carbene **26** (Scheme 4.4).

These results are in excellent agreement with the alcohol trapping experiments of Platz and co-workers, who isolated essentially only carbene derived adducts,[61]

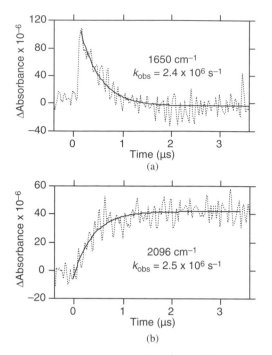

Figure 4.5. Kinetic traces observed at (a) $1650\,cm^{-1}$ and (b) $2096\,cm^{-1}$ following 266 nm photolysis (10 ns, 0.4 mJ) of ^{15}N-labeled **24** (3.1 mM) in argon-saturated Freon-113. The dotted curves are experimental data; the solid curves are the calculated best fit to a single exponential function. Reprinted with permission from Y. Wang, T. Yuzawa, H. Hamaguchi, and J. P. Toscano, *J. Am. Chem. Soc.*, **1999**, *121*, 2875. Copyright 1999, American Chemical Society.

and the calculations of Bally, McMahon, and co-workers.[62] Those calculations indicate that naphthyldiazoester (**24**) is planar and exists almost entirely (99%) in conformations in which the diazo and carbonyl groups are *anti*. Thus, the preferred conformation of **24** leads to efficient carbene production upon photolysis with very little if any direct rearrangement to ketene **27**.

The lack of excited state involvement for **24** is in contrast to behavior in systems that have substantial equilibrium concentrations of the *syn* conformer as discussed

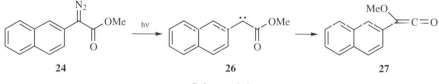

Scheme 4.4.

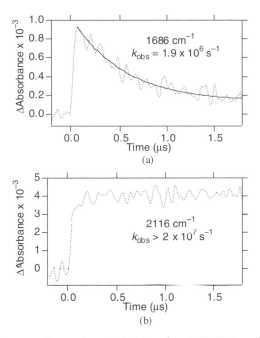

Figure 4.6. Kinetic traces observed at (a) 1686 cm^{-1} and (b) 2116 cm^{-1} following 266 nm photolysis (5 ns, 4 mJ) of ^{15}N-labeled **25** (2 mM) in argon-saturated Freon-113. The dotted curves are experimental data; the solid curve is the calculated best fit to a single exponential function. Reprinted with permission from Y. Wang and J. P. Toscano, *J. Am. Chem. Soc.*, **2000**, *122*, 4512. Copyright 2000, American Chemical Society.

above. To examine the effect that conformation has on ketene growth kinetics, the cyclic diazoester **25**, which is locked in the *syn* conformation, was also examined by TRIR methods.[59] In this case (Figure 4.6), laser photolysis of ^{15}N-labeled **25** (again used to shift a strong negative depletion signal in the unlabeled diazoester away from the positive ketene IR band of interest) results in formation of carbene **28**, which decays with a lifetime of 526 ns in Freon-113 (comparable with the lifetime observed for **26**). The ketene, however, in dramatic contrast to the data observed with acyclic diazocarbonyl **24**, is produced faster than the time resolution (50 ns) of the TRIR spectrometer. Thus, ketene **29** is clearly not formed from carbene **28**; it was suggested that ketene **29** is formed entirely from the excited state of **25** (Scheme 4.5). In agreement with this hypothesis, oxygen and methanol quench the carbene derived from **25**, but leave the initial intensity of the ketene IR band unaffected.

Additional evidence for a photochemically produced noncarbene precursor to ketene **29** is provided by product analysis. Photolysis of **25** in neat methanol leads to both carbene derived (i.e., **30** in 75% absolute yield) and ketene-derived adducts (i.e., **31** in 18% absolute yield), but thermolysis of **25** in neat methanol (sealed tube at 170°C) provides only carbene-derived adduct **30** (91% absolute yield).[59] The ratio

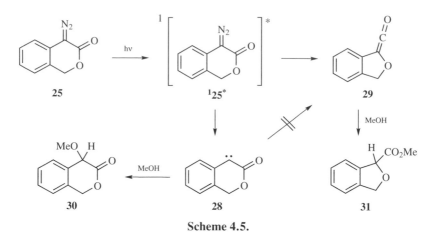

Scheme 4.5.

of carbene- to ketene-derived adducts in the photolysis experiment suggests that the diazoester excited state ^{1}**25*** partitions between carbene production and ketene formation in an approximately 4 to 1 ratio. This observation confirms that *syn* diazoesters can form carbene intermediates.[63]

4.3.3. Specific Solvation of Carbenes

As demonstrated in the two previous sections, TRIR spectroscopy can be used to provide direct structural information concerning organic reactive intermediates in solution as well as kinetic insight into mechanisms of product formation. TRIR spectroscopy can also be used to examine solvent effects by revealing the influence of solvent on IR band positions and intensities. For example, TRIR spectroscopy has been used to examine the solvent dependence of some carbonylcarbene singlet–triplet energy gaps.[58,64] Here, we will focus on TRIR studies of specific solvation of carbenes.

Given the zwitterionic nature of single carbenes, the possibility exists for coordinating solvents such as ethers or aromatic compounds to associate weakly with the empty p-orbital of the carbene.[65] Several experimental studies have revealed dramatic effects of dioxane[66] or aromatic solvents[67,68] on product distributions of carbene reactions. Computational evidence has also been reported for carbene-benzene complexes.[68,69] Indeed, picosecond optical grating calorimetry studies have indicated that singlet methylene and benzene form a weak complex with a dissociation energy of 8.7 kcal/mol.[69]

Based on this work, it has been proposed that a specifically solvated carbene (Scheme 4.6, Reaction 2) undergoes bimolecular reactions at slower rates than a free carbene (Scheme 4.6, Reaction 1). Other alternatives that must be considered are participation of rapid and reversible ylide formation with the ylide acting as a

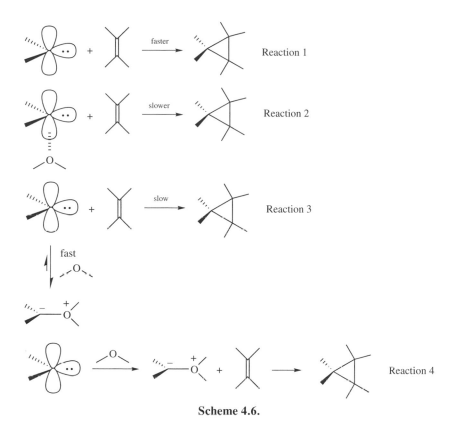

Scheme 4.6.

reservoir that slowly releases free carbene (Scheme 4.6, Reaction 3) and reaction of the ylide itself (Scheme 4.6, Reaction 4).

Since the most direct evidence for specific solvation of a carbene would be a spectroscopic signature distinct from that of the free carbene and also from that of a fully formed ylide, TRIR spectroscopy has been used to search for such carbene-solvent interactions. Chlorophenylcarbene (**32**) and fluorophenylcarbene (**33**) were recently examined by TRIR spectroscopy in the absence and presence of tetrahydrofuran (THF) or benzene.[70] These carbenes possess IR bands near 1225 cm^{-1} that largely involve stretching of the partial double bond between the carbene carbon and the aromatic ring. It was anticipated that electron pair donation from a coordinating solvent such as THF or benzene into the empty carbene p-orbital might reduce the partial double bond character to the carbene center, shifting this vibrational frequency to a lower value. However, such shifts were not observed, perhaps because these halophenylcarbenes are so well stabilized that interactions with solvent are too weak to be observed. The bimolecular rate constant for the reaction of carbenes **32** and **33** with tetramethylethylene (TME) was also unaffected by THF or benzene, consistent with the lack of solvent coordination in these cases.[70]

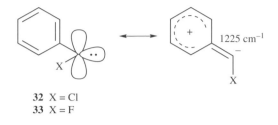

32 X = Cl
33 X = F

Further studies were carried out with halocarbene amides **34** and **35**.[71] Although again no direct spectroscopic signatures for specifically solvated carbenes were found, compelling evidence for such solvation was obtained with a combination of laser flash photolysis (LFP) with UV-VIS detection via pyridine ylides, TRIR spectroscopy, density functional theory (DFT) calculations, and kinetic simulations. Carbenes **34** and **35** were generated by photolysis of indan-based precursors (Scheme 4.7) and were directly observed by TRIR spectroscopy in Freon-113 at 1635 and 1650 cm^{-1}, respectively. The addition of small amounts of dioxane or THF significantly *retarded* the rate of biomolecular reaction with both pyridine and TME in Freon-113. Also, the addition of dioxane *increased* the observed lifetime of carbene **34** in Freon-113. These are both unprecedented observations.

In either neat dioxane or THF, carbene-ether ylides are observed as a broad IR absorption band between 1560 and 1610 cm^{-1}, distinct from the IR bands of the free carbenes. With discrete spectroscopic signatures for the free carbene and its corresponding ether ylides, TRIR spectroscopy was used to confirm that the effects described above with dilute ether in Freon-113 were due to specific solvation of the carbene (Scheme 4.6, Reaction 2) rather than a pre-equilibration with the coordinating solvent (Scheme 4.6, Reaction 3) or reactivity of the ylide itself (Scheme 6, Reaction 4). In Freon-113 containing 0.095 M THF simultaneous TRIR observation of both the free carbene (τ = ca. 500 ns) and the carbene-THF ylide (τ = ca. 5 μs) was possible.[71] The observation that lifetimes of these species were observed to be so different conclusively demonstrates that the free carbene and the carbene-THF ylide are not in rapid equilibrium and that Reaction 3 of Scheme 4.6 is not operative. By examining the kinetics of the carbene **34** at 1635 cm^{-1} directly in Freon-113 with small amounts of added dioxane, it was observed that the rate of reaction with TME was reduced, consistent with Reaction 2 (and not Reaction 4) of Scheme 4.6.

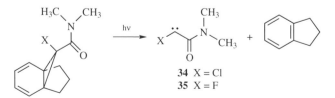

34 X = Cl
35 X = F

Scheme 4.7.

4.4. CONCLUSIONS AND OUTLOOK

The wealth of structural and kinetic information available from TRIR spectroscopy makes it a powerful tool for establishing important structure/reactivity relationships for short-lived reactive intermediates. As demonstrated by the representative investigations discussed in this chapter, recent technical advances have greatly expanded the applicability of this technique so that organic systems can now be routinely examined in the nanosecond timescale. As the necessary instrumentation becomes even more refined and easier to implement, application of TRIR spectroscopy to a range of photochemical and photobiological problems with enhanced time resolution promises to become more common. Given its obvious potential, future TRIR work will continue to advance our fundamental understanding of reactive intermediates just as the application of time-resolved UV-VIS absorption spectroscopy began to do almost 60 years ago.[1]

ACKNOWLEDGMENT

The generous support of research in the Toscano laboratory by the National Science Foundation and the donors of the Petroleum Research Fund, administered by the American Chemical Society, is gratefully acknowledged.

SUGGESTED READING

J. P. Toscano in *Advances in Photochemistry, Vol. 26* (Eds. D. C. Neckers, G. von Bünau, and W. S. Jenks), John Wiley & Sons, New York, **2001**, p. 41.

B. M. Showalter and J. P. Toscano, *J. Phys. Org. Chem.* **2004**, *17*, 743.

A. D. Cohen, B. M. Showalter, and J. P. Toscano, *Org. Lett.* **2004**, *6*, 401.

Y. Wang, T. Yuzawa, H. Hamaguchi, and J. P. Toscano, *J. Am. Chem. Soc.* **1999**, *121*, 2875.

Y. Wang and J. P. Toscano, *J. Am. Chem. Soc.* **2000**, *122*, 4512.

E. A. Pritchina, N. P. Gritsan, A. Maltsev, T. Bally, T. Autrey, Y. Liu, Y. Wang, and J. P. Toscano, *Phys. Chem. Chem. Phys.* **2003**, *5*, 1010.

Y. Wang, C. M. Hadad, and J. P. Toscano *J. Am. Chem. Soc.* **2002**, *124*, 1761.

E. M. Tippmann, M. S. Platz, I. B. Svir, and O. V. Klymenko *J. Am. Chem. Soc.* **2004**, *126*, 5750.

REFERENCES

1. R. G. W. Norrish and G. Porter, *Nature* **1949**, *164*, 658.

2. (a) J. U. White, *J. Opt. Soc. Am.* **1942**, *32*, 285. (b) D. Herriott, H. Kogelnik, and R. Kompfner, *Appl. Opt.* **1964**, *3*, 523. (c) J. S. Pilgrim, R. T. Jennings, and C. A. Taatjes, *Rev. Sci. Instrum.* **1997**, *68*, 1875.

3. K. C. Herr and G. C. Pimentel, *Appl. Opt.* **1965**, *4*, 25.

4. (a) A. S. Lefohn and G. C. Pimentel, *J. Chem. Phys.* **1971**, *55*, 1213. (b) L. Y. Tan, A. M. Winer, and G. C. Pimentel, *J. Chem. Phys.* **1972**, *57*, 4028.

5. For a review see: E. Weitz, *J. Phys. Chem.* **1987**, *91*, 3945.

6. (a) H. Hermann, F.-W. Grevels, A. Henne, and K. Schaffner, *J. Phys. Chem.* **1982**, *86*, 5151. (b) S. P. Church, H. Hermann, and F.-W. Grevels, *J. Chem. Soc., Chem. Commun.* **1984**, 785. (c) B. D. Moore, M. B. Simpson, M. Poliakoff, and J. J. Turner, *J. Chem. Soc., Chem. Commun.* **1984**, 972. (d) B. D. Moore, M. Poliakoff, M. B. Simpson, and J. J. Turner, *J. Phys. Chem.* **1985**, *89*, 850.

7. For reviews see: (a) M. W. George, M. Poliakoff, and J. J. Turner, *Analyst* **1994**, *119*, 551. (b) P. C. Ford, J. S. Bridgewater, and B. Lee, *Photochem. Photobiol.* **1997**, *65*, 57. (c) K. McFarlane, B. Lee, J. S. Bridgewater, and P. C. Ford, *J. Organomet. Chem.* **1998**, *554*, 49. (d) M. W. George and J. J. Turner, *Coord. Chem. Rev.* **1998**, *177*, 201. (e) J. R. Schoonover and G. F. Strouse, *Chem. Rev.* **1998**, *98*, 1335. (f) D. C. Grills, J. J. Turner, and M. W. George, *Comp. Coord. Chem.* **2003**, *2*, 91. (g) M. K. Kuimova, W. Z. Alsindi, J. Dyer, D. C. Grills, O. S. Jina, P. Matousek, W. W. Parker, P. Portius, X. Z. Sun, M. Towrie, C. Wilson, J. Yang, and M. W. George, *J. Chem. Soc., Dalton Trans.* **2003**, 3996.

8. For reviews see: (a) F. Siebert, *Methods Enzymol.* **1995**, *246*, 501. (b) R. M. Slayton and P. A. Anfinrud, *Curr. Opin. Struct. Biol.* **1997**, *7*, 717. (c) R. Vogel and F. Siebert, *Curr. Opin. Chem. Biol.* **2000**, *4*, 518. (d) A. Barth and C. Zscherp, *Quart. Rev. Biophys.* **2002**, *35*, 369. (e) C. Kötting and K. Gerwert, *Chem. Phys. Chem.* **2005**, *6*, 881.

9. For reviews see: (a) R. H. Callender, R. B. Dyer, R. Gilmanshin, and W. H. Woodruff, *Annu. Rev. Phys. Chem.* **1998**, *49*, 173. (b) R. B. Dyer, F. Gai, W. H. Woodruff, R. Gilmanshin, and R. H. Callender, *Acc. Chem. Res.* **1998**, *31*, 709.

10. For a previous review see: J. P. Toscano in *Advances in Photochemistry*, *Vol. 26* (Eds. D. C. Neckers, G. von Bünau, and W. S. Jenks), John Wiley & Sons, New York, 2001, p. 41.

11. For reviews see: (a) P. O. Stoutland, R. B. Dyer, and W. H. Woodruff, *Science* **1992**, *257*, 1913. (b) J. C. Owrutsky, D. Raftery, and R. M. Hochstrasser, *Ann. Rev. Phys. Chem.* **1994**, *45*, 519. (c) K. Wynne and R. M. Hochstrasser, *Chem. Phys.* **1995**, *193*, 211. (d) G. C. Walker and Hochstrasser, R. M. in *Laser Techniques in Chemistry* (Eds. A. B. Myers and T. R. Rizzo), John Wiley & Sons, New York, **1995**, p. 385. (e) B. Akhremitchev, C. Wang, and G. C. Walker, *Rev. Sci. Instrum.* **1996**, *67*, 3799. (f) D. C. Grills and George, M. W. in *Handbook of Vibrational Spectroscopy*, *Vol. 1* (Eds. J. M. Chalmers and P. R. Griffiths), John Wiley & Sons, New York, **2002**, p. 677 (g) M. Lim, and P. A. Anfinrud in *Methods Molecular Biology*, *Vol. 305* (Ed. G. U. Nienhaus), Humana Press, Totowa, NJ, **2005**, p. 243. (h) E. T. J. Nibbering, H. Fidder, and E. Pines, *Ann. Rev. Phys. Chem.* **2005**, *56*, 337.

12. J. D. Berkerle, R. R. Cavanagh, M. P. Casassa, E. J. Heilweil, and J. C. Stephenson, *J. Chem. Phys.* **1991**, *95*, 5403.

13. (a) E. J. Heilweil, *Opt. Lett.* **1989**, *14*, 551. (b) T. P. Dougherty and E. J. Heilweil, *Opt. Lett.* **1994**, *19*, 129. (c) P. Hamm, S. Wiemann, M. Zurek, and W. Zinth, *Opt. Lett.* **1994**, *19*, 1642. (d) S. M. Arrivo, V. D. Kleiman, T. P. Dougherty, and E. J. Heilweil, *Opt. Lett.* **1997**, *22*, 1488.

14. (a) R. E. Murphey, F. H. Cook, and H. Sakai, *J. Opt. Soc. Am.* **1975**, *65*, 600. (b) H. Sakai and R. E. Murphy, *Appl. Opt.* **1978**, *17*, 1342.

15. For a recent review see: G. D. Smith and R. A. Palmer, in *Handbook of Vibrational Spectroscopy*, *Vol. 1* (Eds. J. M. Chalmers and P. R. Griffiths), John Wiley & Sons, New York, **2002**, p. 625.

16. For a recent review see: M. Hashimoto, T. Yuzawa, C. Kato, K. Iwata, and H. Hamaguchi in *Handbook of Vibrational Spectroscopy*, *Vol. 1* (Eds. J. M. Chalmers and P. R. Griffiths), John Wiley & Sons, New York, **2002**, p. 666.

17. (a) K. Iwata and H. Hamaguchi, *Appl. Spectrosc.* **1990**, *44*, 1431. (b) T. Yuzawa, C. Kato, M. W. George, and H. Hamaguchi, *Appl. Spectrosc.* **1994**, *48*, 684.

18. C. Rödig and F. Siebert, *Vib. Spectrosc.* **1999**, *19*, 271.

19. R. Rammelsberg, S. Boulas, H. Chronongiewski, and K. Gerwert, *Vib. Spectrosc.* **1999**, *19*, 143.

20. (a) C. J. Manning and P. R. Griffiths, *Appl. Spectrosc.* **1997**, *51*, 1092. (b) C. Rödig and F. Siebert, *Appl. Spectrosc.* **1999**, *53*, 893. (c) C. Rödig, C. and F. Siebert, *Handbook of Vibrational Spectroscopy*, *Vol. 1* (Eds. J. M. Chalmers and P. R. Griffiths), John Wiley & Sons, New York, **2002**, p. 651.

21. C. Rödig, H. Georg, F. Siebert, I. Rousso, and M. Sheves, *Laser Chem.* **1999**, *19*, 169.

22. (a) P. R. Griffiths, B. L. Hirsche, and C. J. Manning, *Vib. Spectrosc.* **1999**, *19*, 165. (b) H. Yang, P. R. Griffiths, C. J. Manning, *Appl. Spectrosc.* **2002**, *56*, 1281.

23. (a) C. Pellerin, C. M. Snively, D. B. Chase, and J. F. Rabolt, *Appl. Spectrosc.* **2004**, *58*, 639. (b) C. M. Snively, C. Pellerin, J. F. Rabolt, and D. B. Chase, *Anal. Chem.* **2004**, *76*, 1811.

24. (a) S. Niwayama, H. Noguchi, M. Ohno, and S. Kobayashi, *Tetrahedron Lett.* **1993**, *34*, 665. (b) M. Schmittel and H. von Seggern, *Liebigs Ann.* **1995**, 1815. (c) B. Strijtveen and R. M. Kellogg, *Trav. Chim. Pays-Bas* **1987**, *106*, 539. (d) W. Adam and L. Blancafort, *J. Am. Chem. Soc.* **1996**, *118*, 4778.

25. (a) M. Ashwell, X. M. Guo, and M. L. Sinnott, *J. Am. Chem. Soc.* **1992**, *114*, 10158. (b) A. K. J. Chong, M. S. Pegg, N. R. Taylor, and M. Vonitzstein, *Eur. J. Biochem.* **1992**, *207*, 335. (c) X. M. Guo and M. L. Sinnott, *Biochem. J.* **1993**, *296*, 291. (d) X. M. Guo, W. G. Laver, E. Vimr, and M. L. Sinnott, *J. Am. Chem. Soc.* **1994**, *116*, 5572. (e) B. A. Horenstein and M. Bruner, *J. Am. Chem. Soc.* **1996**, *118*, 10371. (f) B. A. Horenstein, *J. Am. Chem. Soc.* **1997**, *119*, 1101.

26. G. L'abbé, *Angew. Chem., Int. Ed. Engl.* **1980**, *19*, 276.

27. P. L. Coe, A. Sellers, J. C. Tatlow, G. Whittaker, and H. C. Fielding, *J. Chem. Soc., Chem. Commun.* **1982**, 362.

28. W. Adam, J. Liu, and O. Rodriguez, *J. Org. Chem.* **1973**, *38*, 2269.

29. J. F. Liebman and A. Greenberg, *J. Org. Chem* **1974**, *39*, 123.

30. C. S. C. Chung, *J. Mol. Struct.* **1976**, *30*, 189.

31. D. Antolovic, V. J. Shiner, and E. R. Davidson, *J. Am. Chem. Soc.* **1988**, *110*, 1357.

32. (a) S. Firth-Clark, C. F. Rodriquez, and I. H. Williams, *J. Chem. Soc., Perkin Trans. 2* **1997**, 1943. (b) G. D. Ruggiero and I. H. Williams, *J. Chem. Soc., Perkin Trans. 2* **2001**, 733. (c) J. G. Buchanan, M. H. Charlton, M. F. Mahon, J. J. Robinson, G. D. Ruggiero, and I. H. Williams, *J. Phys. Org. Chem.* **2002**, *15*, 642.

33. W. Adam and R. Rucktäschel, *J. Am. Chem. Soc.* **1971**, *93*, 557.

34. O. L. Chapman, P. W. Wojtkowski, W. Adam, and R. Rucktäschel, *J. Am. Chem. Soc.* **1972**, *94*, 1365.

35. R. Wheland and P. D. Bartlett, *J. Am. Chem. Soc.* **1970**, *92*, 6057.

36. W. W. Sander, *J. Org. Chem.* **1989**, *54*, 4265.

37. S. Wierlacher, W. Sander, and M. T. H. Liu, *J. Org. Chem.* **1992**, *57*, 1051.

38. B. M. Showalter and J. P. Toscano, *J. Phys. Org. Chem.* **2004**, *17*, 743.

39. A. D. Cohen, B. M. Showalter, and J. P. Toscano, *Org. Lett.* **2004**, *6*, 401.

40. G. B. Kistiakowsky and K. Sauer, *J. Am. Chem. Soc.* **1958**, *80*, 1066.

41. D. E. Milligan and M. E. Jacox, *J. Chem. Phys.* **1962**, *36*, 2911.

42. D. Kovacs and J. E. Jackson, *J. Phys. Chem. A* **2001**, *105*, 7579.

43. (Ed. C. J. Pouchert), *The Aldrich Library of Infrared Spectra, Edition III*, Aldrich Chemical: Milwaukee, **1981**, p. 1057.

44. G. A. Ganzer, R. S. Sheridan, and M. T. H. Liu, *J. Am. Chem. Soc.* **1986**, *108*, 1517.

45. (Ed. P. G. T. Fogg), *Carbon Dioxide in Non-Aqueous Solvents at Pressures Less than 200 KPa*, Pergamon Press, Oxford, **1992**, p. 259.

46. C. S. Q. Lew, B. D. Wagner, M. P. Angelini, E. Lee-Ruff, J. Lusztyk, and J. L. Johnston, *J. Am. Chem. Soc.* **1996**, *118*, 12066.

47. For a review see: M. S. Platz in *Advances in Carbene Chemistry, Vol 2* (Ed. U. H. Brinker), JAI Press, Greenwich, CT, **1998**, p. 133 and references therein.

48. For example, see (a) D. A. Modarelli, S. Morgan, and M. S. Platz, *J. Am. Chem. Soc.* **1992**, *114*, 7034. (b) W. R. White, III and M. S. Platz, *J. Org. Chem.* **1992**, *57*, 2841. (c) J. S. Buterbaugh, J. P. Toscano, W. L. Weaver, J. R. Gord, T. L. Gustafson, C. M. Hadad, and M. S. Platz, *J. Am. Chem. Soc.* **1997**, *119*, 3580.

49. (a) H. Tomioka, N. Hayashi, Y. Izawa, and M. T. H. Liu, *J. Am. Chem. Soc.* **1984**, *106*, 454. (b) R. Bonneau, M. T. H. Liu, K. C. Kim, and J. L. Goodman, *J. Am. Chem. Soc.* **1996**, *118*, 3829 and references therein.

50. R. Bonneau and M. T. H. Liu, *J. Am. Chem. Soc.* **1996**, *118*, 7229 and references therein.

51. P. M. Warner, *Tetrahedron Lett.* **1984**, *25*, 4211.

52. W. Kirmse, *Eur. J. Org. Chem.* **2002**, 2193.

53. (a) L. Wolff, *Justus Liebigs Ann. Chem.* **1902**, *325*, 129. (b) L. Wolff, *Justus Liebigs Ann. Chem.* **1912**, *394*, 23.

54. (a) F. Kaplan and G. K. Meloy, *Tetrahedron Lett.* **1964**, 2427. (b) F. Kaplan and G. K. Meloy, *J. Am. Chem. Soc.* **1966**, *88*, 950.

55. For examples see: (a) F. Kaplan and M. L. Mitchell, *Tetrahedron Lett.* **1979**, 759. (b) H. Tomioka, H. Okuno, and Y. Izawa, *J. Org. Chem.* **1980**, *45*, 5278. (c) J. P. Toscano, M. S. Platz, and V. Nikolaev, *J. Am. Chem. Soc.* **1995**, *117*, 4712.

56. Alternative explanations have also been proposed: (a) M. Torres, J. Ribo, A. Clement, and O. P. Strausz, *Can. J. Chem.* **1982**, *61*, 996. (b) C. Marfisi, P. Verlaque, G. Davidovics, J. Pourcin, L. Pizzala, J.-P. Aycard, and H. Bodot, *J. Org. Chem.* **1983**, *48*, 533. (c) A. Bogdanova and V. V. Popik, *J. Am. Chem. Soc.* **2004**, *126*, 11293. (d) V. V. Popik, *Can. J. Chem.* **2005**, *83*, 1382.

57. (a) R. R. Rando, *J. Am. Chem. Soc.* **1970**, *92*, 6706. (b) R. R. Rando, *J. Am. Chem. Soc.* **1972**, *94*, 1629. (c) H. Tomioka, H. Kitagawa, and Y. Izawa, *J. Org. Chem.* **1979**, *44*, 3072.

58. Y. Wang, T. Yuzawa, H. Hamaguchi, and J. P. Toscano, *J. Am. Chem. Soc.* **1999**, *121*, 2875.

59. Y. Wang and J. P. Toscano, *J. Am. Chem. Soc.* **2000**, *122*, 4512.

60. E. A. Pritchina, N. P. Gritsan, A. Maltsev, T. Bally, T. Autrey, Y. Liu, Y. Wang, and J. P. Toscano, *Phys. Chem. Chem. Phys.* **2003**, *5*, 1010.

61. J.-L. Wang, I. Likhotvorik, and M. S. Platz, *J. Am. Chem. Soc.* **1999**, *121*, 2883.

62. Z. Zhu, T. Bally, L. L. Stracener, R. J. McMahon, *J. Am. Chem. Soc.* **1999**, *121*, 2863.

63. (a) R. J. McMahon, O. L. Chapman, R. A. Hayes, T. C. Hess, and H.-P. Krimmer, *J. Am. Chem. Soc.* **1985**, *107*, 7597. (b) J. J. M. Vleggaar, A. H. Huizer, P. A. Kraakman, W. P. M. Nijssen, R. J. Visser, and C. A. G. O. Varma, *J. Am. Chem. Soc.* **1994**, *116*, 11754.

64. (a) Y. Wang, C. M. Hadad, and J. P. Toscano, *J. Am. Chem. Soc.* **2002**, *124*, 1761. (b) C. M. Geise, Y. Wang, O. Mykhaylova, B. T. Frink, J. P. Toscano, and C. M. Hadad, *J. Org. Chem.* **2002**, *67*, 3079.

65. For a recent review see: W. Kirmse, *Eur. J. Org. Chem.* **2005**, 237.

66. H. Tomioka, Y. Ozaki, and Y. Izawa, *Tetrahedron* **1985**, *41*, 4987.

67. R. T. Ruck and M. Jones, Jr., *Tetrahedron Lett.* **1998**, *39*, 2277.

68. (a) R. A. Moss, S. Yan, and K. Krogh-Jesperson, *J. Am. Chem. Soc.* **1998**, *120*, 1088. (b) K. Krogh-Jesperson, S. Yan, and R. A. Moss, *J. Am. Chem. Soc.* **1999**, *121*, 6269.

69. M. I. Khan and J. L. Goodman, *J. Am. Chem. Soc.* **1995**, *117*, 6635.

70. Y. Sun, E. M. Tippmann, and M. S. Platz, *Org. Lett.* **2003**, *5*, 1305.

71. E. M. Tippmann, M. S. Platz, I. B. Svir, and O. V. Klymenko, *J. Am. Chem. Soc.* **2004**, *126*, 5750.

███████ **CHAPTER 5**

Studies of the Thermochemical Properties of Reactive Intermediates by Mass Spectrometric Methods

PAUL G. WENTHOLD

Department of Chemistry, Purdue University, West Lafayette, IN

Reviews of Reactive Intermediate Chemistry. Edited by Matthew S. Platz, Robert A. Moss, Maitland Jones, Jr.
Copyright © 2007 John Wiley & Sons, Inc.

5.1. INTRODUCTION AND HISTORICAL PERSPECTIVE

Thermochemical properties of molecules provide valuable insight into their structure and reactivity. The relative stabilities of molecules and ions are reflected by such parameters as bond dissociation energies, ionization potentials, and electron affinities. However, determination of thermochemical quantities for reactive neutral species such as radicals, carbenes, and diradicals is a challenge because the lifetimes of these species are usually too short to investigate using traditional condensed-phase calorimetric methods. Mass spectrometry provides a means for the investigation of reactive molecules under dilute, gas-phase conditions where complications due to the enhanced reactivity are minimized. By its nature, mass spectrometry is especially well-suited to the studies of ions, including cations and anions. Thus, it is not surprising that studies of ion reactivity and properties are commonly carried out. However, despite the natural connection between mass spectrometry and studies of ions, mass spectrometric techniques are also amenable for the study of *neutral* reactive molecules. Although neutral species cannot be detected directly using mass spectrometry, their formation can be inferred in many gas-phase reactions, such as proton or electron transfer, or dissociation. For example, Pollack and Hehre observed proton transfer in the reaction of phenyl cation with ethyl-*N,N*-dimethylcarbamate.[1] Although the resulting neutral product was not observed, they concluded that it was likely *ortho*-benzyne (Eq. 5.1) on the basis of deuterium labeling studies and typical characteristics of gas-phase reactions.

$$\text{(5.1)}$$

Gas-phase reagents generally do not have enough energy to overcome significant energy barriers, and many reactions that are commonly observed in solution, such as rearrangements, are too slow to occur in the gas-phase. It does not matter if the energy barrier originates from enthalpic or entropic constraints, as either can slow the rate of reaction. As a consequence, reactions that occur under thermal conditions in the gas phase generally have small (<12 kJ/mol) enthalpic barriers and loose transition states. Thus, the reaction of phenyl cation with base is likely to occur by simple proton transfer to form the more reactive dehydrobenzene product, as opposed to proton transfer accompanied by rearrangement to give a more energetically stable, ring-opened product. This aspect of gas-phase reactions is often utilized for thermochemical investigations, as the occurrence of the reaction indicates that it is thermodynamically accessible. Therefore, the reactivity can be used to assess the relative thermochemical properties of reagents, and the determination of thermochemical properties has been among mass spectrometry's most important contribution to the field of reactive neutral molecules.

Of the many thermochemical properties that can be defined for any substrate, perhaps the most important is the bond dissociation energy (BDE). Although different types of bond dissociation energies can be considered, the most relevant for neutral substrates is the homolytic BDE. For convenience, it is generally the homolytic C—H BDE that is of interest. In principle, hypovalent organic systems can be considered to arise from a bond-dissociation process.[2] Radicals, for example, are formed by homolytic bond dissociation of closed-shell precursors. Breaking a second bond in the precursor (or breaking a bond in the radical) leads to the formation of didehydro systems such as carbenes and diradicals. Triradicals result from breaking bonds in diradicals, and can dissociate to give tetraradicals, and the process continues. However, a discussion of radicals in the context of bond dissociation does more than provide an idealized genesis for their formation, it also provides a basis for understanding many of their unusual characteristics. Because bond dissociation is an energetically costly process, radicals are generally high energy species, and, consequently, will undergo transformations normally considered to be disfavored, as the energy benefit is greater than the cost

In molecules with more than one unpaired electron, electron–electron interactions can have a significant role in the stabilization of the system.[3] Bond formation that results from direct overlap is highly favorable and, thus is an overriding consideration in all low-spin polyradicals, even to the extent that the system sometimes adopts a strained, closed-shell state as opposed to a polyradical. In cases in which bonding cannot occur, indirect interactions that are usually insignificant, such as electron exchange and spin-polarization, can have significant impact. The presence of these interactions is often reflected in the thermochemical properties.

The utility of mass spectrometry for the investigation of thermochemical properties of reactive intermediates, and, in particular, homolytic BDEs comes from the fact that the homolytic BDEs need not be measured directly, but can be obtained indirectly, either by using simple thermochemical cycles or from enthalpies of formation measured with ion-based approaches. Indeed, mass spectrometry has been used to investigate reactive organic species for over 50 years. Among the earliest studies of reactive species are those by Lossing et al. who used mass spectrometry to measure the concentrations of methyl radicals formed by decomposition of dimethyl mercury.[4] In 1954, Lossing et al. reported the measurement of the ionization energies of methyl, allyl, and benzyl radicals.[5] In these studies, the radicals were generated by pyrolysis, and the ionization energies were obtained from electron impact appearance energies. It is a testament to the quality of these early studies that the ionization energies obtained, 9.95 and 8.16 eV for methyl and allyl radicals, respectively, are in reasonable agreement with the most recent high-resolution measurements of 9.8384^6 and $8.1535^{7,8}$ eV. More significant disagreement between the early measured ionization energy of benzyl (7.73 eV)[5] and the modern value (7.2491 eV),[9] however, highlights the challenges that are encountered when studying large systems, prone to rearrangement upon ionization.

Significant progress in experimental techniques has been made in the time since Lossing's initial reports. In addition to technical improvements in the experimental methods that have been used, there has also been creative development of new strategies for measuring thermochemical properties of a wider range of targets.

Section 5.3 of this chapter highlights the most commonly used mass spectrometric methods for the investigation of the thermochemical properties of reactive molecules, focusing on the experimental strategies that have been employed, including the direct measurement of thermochemical properties by using energetics measurements, and the indirect measurements carried out by using chemical reactivity studies.

A second role for mass spectrometry in the investigation of reactive intermediates involves the use of spectroscopy. Although an important use of ion spectroscopy is the determination of thermochemical properties, including ionization energies (addition or removal of an electron), as in photoelectron or photodetachment spectroscopy, and bond dissociation energies in ions, as in photodissociation methods, additional spectroscopic data can also often be obtained, including structural parameters such as frequencies and geometries.

The utility of the experimental methods are illustrated in this chapter by considering their applications to the study of reactive molecules, including radicals, carbenes and diradicals, carbynes and triradicals, and even transition states. These are provided in Section 5.4, which includes results for representative bond dissociation energies and an extensive list of thermochemical results for carbenes, diradicals, carbynes, and triradicals. Section 5.5 provides a comparison and assessment of the results obtained for selected carbenes and diradicals, whereas spectroscopic considerations are addressed in Section 5.6.

5.2. DEFINITIONS OF THERMOCHEMICAL PROPERTIES

Before delving into experimental methodology, it is useful to recall the definitions of common thermochemical parameters. A detailed description of thermochemical properties is available.[10]

> *Proton Affinity* (PA): The PA is the 298 K enthalpy for the dissociation reaction of a cation
>
> $$BH^+ \rightarrow B + H^+ \quad PA(B) = \Delta H_{298}$$

The free energy change for this reaction, ΔG, is the *gas-phase basicity*.

> *Ionization Energy* (IE) (formerly IP): The 0 K energy for the removal of an electron from a neutral molecule
>
> $$A \rightarrow A^+ + e^- \quad IE(A) = \Delta E_0$$

> *Appearance Energy* (AE): The cation appearance energy is the 298 K enthalpy required to form a cationic fragment from a neutral precursor.
>
> $$AB \rightarrow A + B^+ + e^- \quad AE(B^+, AB) = \Delta H_{298}$$

The appearance energy for any fragment critically depends on the precursor from which it is formed.

Gas-phase Acidity (ΔH_{acid}): The enthalpy for heterolytic cleavage to form an anion and a proton. Unless otherwise specified, the temperature is 298 K.

$$AH \rightarrow A^- + H^+ \quad \Delta H_{acid}(AH) = \Delta H_{298}$$

The free energy for this reaction, ΔG_{acid}, is also called the gas-phase acidity.

Electron Affinity (EA): The electron affinity of a neutral molecule is the 0 K energy for removing an electron from the corresponding anion

$$A^- \rightarrow A + e^- \quad EA(A) = \Delta E_0$$

Hydride Affinity (HA): The 298 K enthalpy required to remove a hydride to form the neutral molecule

$$AH^- \rightarrow A + H^- \quad HA(A) = \Delta H_{298}$$

Other Ion Affinities: Binding affinities for many different types of ions to neutrals are defined analogously to hydride affinity, as the 298 K enthalpy required to dissociate the complexed species. The ion can be a cation or an anion. Conversely, ion affinities can be described in terms of the dissociation.

cation

$AB^+ \rightarrow A + B^+ \quad \Delta H_{298} = $ "B^+ affinity of A" or "the dissociation energy (enthalpy) of AB^+"

anion

$AB^- \rightarrow A + B^- \quad \Delta H_{298} = $ "B^- affinity of A" or "the dissociation energy (enthalpy) of AB^-"

Although this discussion refers to properties of neutral molecules, the substrates A and AH can in principle be ionic. For example, the hydride affinity can also be described for a cation

$$AH \rightarrow A^+ + H^- \quad \Delta H_{298} = HA(A^+)$$

Because of this, the thermochemistry of many physical processes can be described in different ways. Thus, the ionization energy of neutral A is the same as the electron affinity of A^+, the proton affinity of B is also the gas-phase acidity of BH^+, and the gas-phase acidity of AH is the same as the proton affinity of A^-.

5.3. EXPERIMENTAL APPROACHES FOR MEASURING THERMOCHEMICAL PROPERTIES

In general, two types of approaches are used for thermochemical measurements. These include thermal reactivity based methods, in which thermochemical properties are deduced from reactivity, as mentioned above, or are measured directly for translationally driven processes, and spectroscopic methods. This section provides a brief overview of the experimental methodology, and its applications for determining the thermochemical properties of positive and negative ions. A more detailed review has been provided by Ervin.[11]

It is always important in thermochemical studies to be aware of the temperature at which the thermochemical properties are determined, and to combine only those properties at the same temperature. Temperature corrections can be made by using integrated heat capacities over the temperature ranges in question.[12] However, it is often assumed that the temperature corrections for ionization energies and electron affinities are small ($<1\,kJ/mol$) and therefore can be neglected.

5.3.1. Reaction Equilibrium Measurements

When reactions are reversible, it is possible to measure the equilibrium constant. As noted above, reactions are only reversible when the energy change for the reaction is small, such that that endothermic process can be observed. It is generally the case that the energy change needs to be less than *ca.* $+10\,kJ/mol$ for the reaction to be observed under thermal conditions. A larger energy range is accessible with high pressure instruments that are capable of achieving higher temperatures.[13,14]

The equilibrium constant for a reaction such as

$$AH + B^- \rightleftharpoons A^- + BH$$

can be determined either by direct measurements of the concentrations of AH, B^-, A^-, and BH under equilibrium conditions, or by measurement of the forward and reverse rates of the reaction. The choice of which approach to use depends predominately on the available experimental apparatus. The forward and reverse reaction rate approach is most often used with flowing afterglow or selected-ion flow tube (SIFT) instruments,[15] whereas direct equilibria measurements are generally carried out using high-pressure mass spectrometry (HPMS)[13,14] or ion cyclotron resonance (ICR).

Equilibrium measurements measure the relative ΔG, and thermochemical studies generally are interested in enthalpy values, ΔH. The enthalpy can be obtained from ΔG by using the relation $\Delta G = \Delta H - T\Delta S$. The entropy of proton transfer can either be estimated, reliably calculated using electronic structure calculations, or can be measured directly by using a Van't Hoff approach. Measuring the quantity ΔS requires a variable temperature study.

In principle, the equilibrium approach can be used to measure any of the thermochemical properties listed above. However, in practice, it is most commonly used for the determination of gas-phase acidities, proton affinities, and electron affinities. In addition, equilibrium measurements are used for measuring ion affinities, including halide (F^-, Cl^-) and metal ion (alkali and transition metal) affinities.

5.3.2. Bracketing

Because gas-phase reactions do not occur if there is a barrier, the occurrence of a reaction can be interpreted to indicate that the reaction is energetically favorable. Thus, the occurrence or non-occurrence of proton transfer reactions, with appropriate references, can be used to bracket thermochemical parameters. For example, if proton transfer is observed between B$^-$ and HA but not between B$^-$ and HC, then the acidity of HB is assigned to be between those for HA and HC. Preferably, the assignment can be confirmed by examining the reverse reactions, those between A$^-$ and C$^-$ with HB. Carrying out the bracketing in both directions is more accurate because slightly endothermic reactions can occur, which can be revealed by carrying out the reverse reaction. The quality of a bracketed value depends highly on the quality of the bracketing references. More precise values can be obtained by using more references, with closely spaced acidities, and preferably with small uncertainties.

Although there has been debate over what energy is being measured in bracketing experiments, there are cases that are clearly free energy driven. For example, Squires showed that protonation of HCO$_3^-$ or HSO$_3^-$ occurs with acids where proton transfer is endothermic, but there is a favorable free energy change.[16] Therefore, if bracketing gives the value of ΔG_{acid}, then ΔH_{acid} must be obtained by calculating or estimating ΔS_{acid}.

The bracketing approach is probably the most commonly used approach for measuring thermochemical properties of reactive molecules because of its versatility. It can be used to determine almost any type of thermochemical property, and because it can carried out by examining a reaction in only a single direction, it is amenable to the study of highly reactive molecules, albeit with a loss in accuracy.

Modifications have been made to the single-direction bracketing method, wherein the rates of reaction are measured in order to further clarify and refine the thermochemical assignment.[17]

5.3.3. Kinetic Methods

The "kinetic method," developed by Cooks et al.,[18] is commonly used to measure acidities for systems that readily form proton-bound cluster ions. For example, acidities are determined from the branching ratios for dissociation of proton-bound cluster ions of anions and reference. In the simplest form, there is assumed to be a semi-logarithmic relationship between the CID branching ratio, r, and the difference in the free energies for the two processes (Eq. 5.2).

$$-RT\ln r = \delta\Delta G \qquad (5.2)$$

Although the kinetic method as a thermochemical tool has been debated,[19–21] it is ultimately a refined bracketing approach, using branching ratios to interpolate a more precise location of the thermochemical property (as opposed to just between those of HA and HC). Therefore, the kinetic method should be at least as reliable as bracketing for the measurement of gas-phase thermochemistry. The kinetic method has a wide variety of applications, and whereas it has been typically used to determine gas-phase acidities and proton affinities, it has also been applied to the determination of electron affinities,[22–24] metal ion,[25–29] halogen ion,[30] and even halide affinities.[31]

The silane-cleavage method developed by DePuy et al.[32] is a variation of the kinetic method for the measurement of acidities. In this experiment, alkyl- and arylsilanes undergo reaction with hydroxide, leading to the competitive formation of siloxide ions. As shown in Eq. 5.3, the reaction is presumed to proceed via initial formation of a hydroxysiliconate

$$
RSiMe_3 \ + \ OH^- \ \longrightarrow \ \left[\begin{array}{c} Me \diagdown \ \overset{OH}{\underset{|}{Si}} - R \\ Me \diagup \ | \\ Me \end{array} \right]^- \quad \begin{array}{l} \nearrow \ RMe_2SiO^- \ + \ CH_4 \\ \\ \searrow \ Me_3SiO^- \ + \ RH \end{array} \tag{5.3}
$$

intermediate,[33] which undergoes competitive dissociation by loss of hydrocarbons to form the siloxides. As with the kinetic method, a semi-logarithmic relationship is assumed between the branching ratios for product formation and the difference in the acidities of the groups lost. Instead of using metastable ions or CID, as in the kinetic method, the activation energy required to dissociate the reactant ion is provided by the exothermic formation of the silicon-hydroxide bond.[34]

Recent work has shown that there are many assumptions required in order to use the silane-cleavage method.[33] However, an advantage of this approach is that it can be used to determine acidities at regiospecific positions in very weak acids, including those for which the conjugate base anion is unstable with respect to electron detachment.

Other variations of the kinetic method have been proposed.[35]

5.3.4. Appearance Energy Measurements

The experimental approaches described above are examples of *relative* methods, wherein a thermochemical property is measured with respect to that of a standard, or an "anchor." The quality of these measurements ultimately depends on the quality of the anchor. Alternatively, there are methods of determining thermochemical properties, in which the energy for a chemical process is measured on an *absolute* basis. Among the more common of these are the appearance energy measurements, in which the threshold energy for formation of an ionic fragment from an activated precursor is measured. There are many different methods of activation that can be used. Some of these are discussed here.

5.3.4.1. Electron Impact[36]   The yield of ionic product is monitored as a function the energy of ionizing electrons. This approach can be used to measure either ionization energies, generally referring to formation of the positive ion,

$$ R + e^- \ \rightarrow \ R^+ + 2e^- $$

or the appearance energy for a fragment,

$$AB + e^- \;\rightarrow\; A^+ + B + 2e^-$$

$$AB + e^- \;\rightarrow\; A^+ + B^- + e^-$$

$$AB + e^- \;\rightarrow\; A + B^-$$

A large number of the measured ionization energies for stable neutral molecules come from electron impact appearance energy studies, but it has also been adapted for the direct study of reactive neutral molecules.[37,38] If the reactant molecule is RH, then the appearance energy for R^+ from RH, designated $AE(R^+, RH)$,

$$RH + e^- \;\rightarrow\; R^+ + H + 2e^-$$

is related to the hydride affinity, HA, by the electron affinity of hydrogen atom,

$$AE(R^+, RH) = HA(R^+) + EA(H)$$

5.3.4.2. Collision-induced Dissociation/Translationally-driven

Reactions[39]   Tandem-mass spectrometry can be used to determine the threshold energy for collision-induced dissociation (CID),

$$AB^+ \;\rightarrow\; A^+ + B$$

$$AB^- \;\rightarrow\; A^- + B$$

or for bimolecular reactions,

$$A^+ + B \;\rightarrow\; C^+ + D$$

$$A^- + B \;\rightarrow\; C^- + D$$

Energy-resolved CID can be used to measure bond dissociation energies directly, and therefore is readily applicable for the determination of ion affinities. However, Graul and Squires have also described a method for measuring gas-phase acidities using CID of carboxylates.[40] Upon CID, carboxylate ions, RCO_2^-, undergo decarboxylation to form the alkyl anions, R^-,

$$RCO_2^- \;\rightarrow\; R^- + CO_2$$

Threshold CID can be used to measure the energy required for decarboxylation, in order to determine the enthalpy of formation of R^-, which can be used to calculate the gas-phase acidity. While nominally straight-forward, the decarboxylation approach is limited to systems that have a bound anion, R^-, and requires an instrument with the capability of carrying out energy-resolved CID. However, it does have an advantage of being a regiospecific approach.

Similarly, Ervin and co-workers have measured acidities of organic molecules by measuring the energy for endothermic proton transfer reactions between acids and anionic bases.[41,42] Alternatively, it is possible to use competitive CID of proton-bound dimer ions.[43] Nominally, these are relative approaches for measuring acidities, as the measured acidities depend on the properties of the reference acids or bases. However, it is usually possible to select references with very accurately known acidities (such as HF, HCN, or HCl), such that the accuracy of the final measurement depends predominantly on the accuracy of the threshold energy determination.

5.3.4.3. Photodissociation

It is also possible to use light as the activation method. Laser sources are especially useful because of their high intensity and narrow wavelength bandwidth. Photodissociation can be used to determine bond dissociation energies in ions directly, similar to what is done with threshold CID, or, alternatively, can be used in conjunction with direct ionization.

$$AB + h\nu \rightarrow AB^+ + e^- \rightarrow A^+ + B + e^-$$

In these experiments, the initial ionization occurs rapidly, and the dissociation energy of AB^+ is determined from the kinetics for its decomposition, which can be measured in a time-resolved experiment, typically within an ion trap.

5.3.4.3.1 Photodetachment/Photoionization

Photodetachment and photoionization refer to the processes of removing an electron from either a negative ion or a neutral molecule.

$$A^- + h\nu \rightarrow A + e^-$$

$$A + h\nu \rightarrow A^+ + e^-$$

The threshold energies for these processes correspond directly to the electron affinity (EA) or ionization energy (IE).

The ionization energy can be determined by monitoring the appearance of the cation, A^+, as a function of the photon wavelength. Threshold photoionization is challenging because the ionization energies of neutral molecules are so high that ionization requires tunable vacuum UV irradiation. By contrast, molecular electron affinities are approximately 1 eV (100 kJ/mol), which corresponds to photon wavelengths near the visible region of the spectrum. Photodetachment threshold energies are typically measured by monitoring the depletion of negative ion signal.

5.3.4.3.2 Photoelectron Spectroscopy

Photoelectron spectroscopy (PES) is a variation of photodetachment/photoionization. Instead of using a tunable light source, photoelectron spectroscopy uses a fixed-frequency source to eject the electron. Spectroscopic information is obtained by measuring the energies of the electrons that are ejected. An energy diagram illustrating the photoelectron experiment is shown in Figure 5.1. The origin of the photoelectron band is defined as the transition to the lowest energy level of the upper electronic state ($v = 0, J = 0$). For photoelectron spectroscopy of neutral species, the upper state is the cation, and the IE can be obtained from the electron kinetic energy (eKE) of the origin, where hv is the photon energy.

$$IE = hv - eKE$$

Photoelectron spectroscopy has routinely been used to determine the ionization energies of stable molecules. It has also been adapted for the investigation of reactive species, including radicals, biradicals, and carbenes, usually generated chemically or by using pyrolysis.[44]

In negative ion photoelectron spectroscopy (NIPES), the reactant is a negative ion, and the upper state is neutral. In this case, the origin can be used to determine the electron affinity.

$$EA = hv - eKE$$

The selection rule for a transition in photoelectron spectroscopy is that $\Delta S = \pm 1$. Therefore, when starting with a singlet reactant state, the resulting state is doublet. Similarly, if the reactant state is doublet, then the upper state can be either singlet or triplet. For this reason, photoelectron spectroscopy is well-suited for determining state term energies in reactive molecules, including, for example, singlet-triplet energy splittings in carbenes and biradicals.

The upper state can also be formed in energetically excited ro-vibrational states. Most photoelectron experiments do not have enough resolution to observe rotational levels, except in rare cases, but vibrational resolution is commonly achieved. Therefore, it is possible to carry out limited vibrational spectroscopy of cations and reactive neutral molecules using this approach.

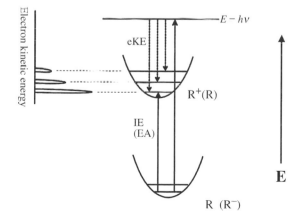

Figure 5.1. Schematic potential energy surfaces for the photoelectron spectroscopy experiment. Labels in parentheses refer to negative ion photoelectron spectroscopy.

TABLE 5.1. Ionization energies of organic radicals measured
by using ZEKE spectroscopy.

Organic radical	Formula	Ionization energy (eV)	Reference
Methylene	CH_2	10.3864 ± 0.0004	56
Methyl	CH_3	9.8380 ± 0.0004	6
Propargyl	C_3H_3	8.6731 ± 0.0012	117
Allyl	C_3H_5	8.1535 ± 0.0006	7,8
Cyclopentadienyl	C_5H_5	8.4271 ± 0.0005	115
Benzyl	$C_6H_5CH_2$	7.2491 ± 0.0006	9
	$C_6H_5CD_2$	7.2429 ± 0.0006	9
	$C_6D_5CD_2$	7.2389 ± 0.0006	9
Methylthio	CH_3S	9.2649 ± 0.0010	118
Ethylthio	CH_3CH_2S	9.107 ± 0.004	119

5.3.4.3.3 ZEKE Photoelectron Spectroscopy In the last 15 years, there has
been extensive development of Zero Electron Kinetic Energy (ZEKE) photoelectron
spectroscopy.[45,46] In this experiment, the reactant is spectroscopically excited to a
level very close to but not above the ionization threshold, the Rydberg levels of the
molecule. The properties of these Rydberg states are very similar to those of the detachment product (i.e., the cation for spectroscopy of neutral molecules). Ionization
is completed by using a weak, pulsed electric field ("pulsed field ionization," PFI).
The advantage of ZEKE spectroscopy is that the resolution of the measurement is
determined by the linewidth of the laser, leading to much higher resolution than
can be achieved in normal photoelectron spectroscopy measurements. Ionization
energies measured by using ZEKE spectroscopy can be accurate to within 0.001 eV
(0.1 kJ/mol) or less. While nominally utilized for the study of simple, stable systems,
ZEKE spectroscopy has also been applied to selected reactive molecules through the
use of photodissociation and pyrolysis sources. A list of the radical ionization energies that have been measured by using ZEKE- or PFI-PES is provided in Table 5.1.

5.4. APPLICATIONS TO REACTIVE MOLECULES

Thermochemical properties of reactive molecules can be either measured directly,
or obtained indirectly through thermochemical cycles. The thermochemical relationships between standard thermochemical properties are illustrated in Figure 5.2.
From these, many different relationships can be found.

Perhaps the most useful thermochemical property for reactive molecules is the
homolytic bond dissociation energy. Although bond dissociation energies are often
misinterpreted in the assessment of radical stability,[47] they do provide insight into
this aspect and into the nature of the bond being broken. Bond dissociation energies
are especially valuable in the investigation of more unsaturated species. Benson has
defined the π bond energy as the difference between the energy required to break
C−H bonds in the formation of an olefin.[48] More recently, this concept has been

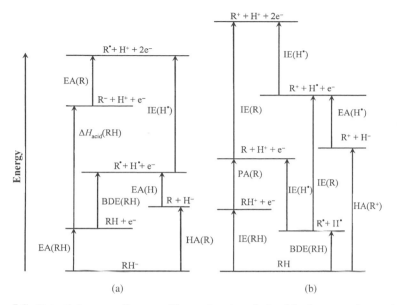

Figure 5.2. Potential energy diagrams illustrating the relationships between thermochemical properties. (a) Relationships for properties involving negative ions, and (b) relationships among properties involving positive ions. See Section 2 for definitions.

generalized as a measure of the *electronic interaction* in other types of open-shell systems as well, including carbenes, diradicals, and triradicals.[44]

The thermochemical schemes in Figure 5.2 reveal many different approaches for obtaining the BDE, by using either positive ion or negative ion based methods. Among these are Eqs 5.4 and 5.5. More in-depth discussion of these relationships can be found elsewhere.[12,49]

positive ion

$$BDE(R-H) = HA(R^+) + EA(H) - IE(R) = AE(R^+, RH) - IE(R) \qquad (5.4a)$$

$$BDE(R-H) = IE(RH) + PA(R) - IE(H) \qquad (5.4b)$$

negative ion

$$BDE(R \; H) = \Delta H_{acid}(RH) + EA(R) - IE(H) \qquad (5.5)$$

The following sections describe applications of these methods to the investigation of the thermochemical properties of radicals, carbenes, biradicals, and triradicals.

5.4.1. Radicals

Ellison and co-workers have provided detailed descriptions of BDE measurements for stable molecules, in which bond breaking leads to the formation of radicals.[12,50]

Very accurate BDE measurements (to within ±4 kJ/mol) can be made by using both positive ion (Eq. 5.4a) and negative ion (Eq. 5.5) approaches. These measurements complement radical abstraction kinetics and equilibrium investigations, and comparisons of the three methods find good agreement.[12] The keys to the mass spectrometric measurements are the ability to measure the cation hydride affinities and radical ionization energies for the positive ion approach, and the ability to determine radical electron affinities for the negative ion approach. As noted in Section 5.1, Lossing was among the earliest to measure radical ionization energies by using electron impact ionization threshold measurements on pyrolytically generated radicals.[5] However, more accurate measurements are made by using photoionization and photoelectron spectroscopy methods, in which the uncertainties can be smaller than ±2 kJ/mol. Similarly, cation hydride affinities can be accurately measured from the appearance energies for the formation of R^+ upon photodissociation of precursors, RH. Electron affinities of radicals required for the negative ion cycle are readily measured by using spectroscopic methods, such as photodetachment or negative ion photoelectron spectroscopy, because the precursor anions required for these measurements correspond to the readily available $[M–H]^-$ ions.

An alternate positive ion approach, similar to that in Eq. 5.4a is to obtain a carbon–halogen BDE, R−X, from which it is possible to obtain the enthalpy of formation of the radical from which the hydrocarbon BDE can be derived. The advantage of this approach is that it is easier to measure the R^+ appearance energy from RX than it is from RH because of the weaker RX bond. However, a limitation of the approach is that the enthalpies of formation of organic halides, required to determine the enthalpies of formation of the cations, are generally not known as accurately as those for hydrocarbons.

Modern reviews of BDEs in closed-shell molecules measured by using mass spectrometric methods have been provided by Ellison and co-workers.[12,50] Additional evaluations have been provided by Ervin and DeTuri.[51] A list of important fundamental hydrocarbon BDEs is given in Table 5.2.

5.4.2. Carbenes and Diradicals

Both negative and positive ion approaches have been used for the investigation of the thermochemical properties of didehydro-substrates, including carbenes, diradicals, and strained rings. Among the biggest challenges is the generation of the appropriate neutral and ionic precursors that can be used for the primary thermochemical measurements. Descriptions of the chemical methods utilized to this end, and the measurements involved, are provided below.

5.4.2.1. Positive Ion Approaches
The positive ion approaches commonly used to determine BDEs in saturated hydrocarbons are less amenable to the investigation of BDEs in radicals mainly because of the challenge in measuring the cation hydride affinities ($HA(R^+)$). However, photodissociation has been used in select cases. Alternatively, cation enthalpies of formation can in principle be obtained by dissociation of disubstituted precursors

TABLE 5.2. Representative BDEs for closed-shell organic molecules measured by using mass spectrometric techniques.[a]

Name	Formula (RH)	BDE	$\Delta H_f(R)$[b]	Reference
Methane	CH_3-H	439.28 ± 0.13	146.42 ± 0.34	51
Ethane	C_2H_5-H	423.0 ± 1.7	121.2 ± 1.7	50
Propane	$i\text{-}C_3H_7-H$	412.5 ± 1.7	90.7 ± 1.8	50
iso-Butane	$t\text{-}C_4H_9-H$	403.8 ± 1.7	51.6 ± 1.8	50
Ethene	C_2H_3-H	463.0 ± 2.7	297.5 ± 2.7	51
Ethyne	$HCC-H$	557.8 ± 0.3	566.5 ± 0.3	51
Propene	C_3H_5-H	369.8 ± 1.7	172.2 ± 1.7	120(Eq. 5.14)
		359.6 ± 1.9	162.0 ± 1.9	121,122
				(Eq. 5.4a)
Benzene	C_6H_5-H	472.2 ± 2.2	336.6 ± 2.2	51
Toluene	$C_6H_5CH_2-H$	377.5 ± 3.3	209.5 ± 3.4	120(Eq. 5.14)
		379 ± 10	211 ± 10	9,123(Eq. 5.4a)
1,3,5-Cyclooctatriene	C_8H_9-H	346 ± 16	311 ± 16	114
1,3,5,7-Cyclooctatetraene	C_8H_7-H	389 ± 10	469 ± 10	114
Cubane	C_8H_7-H	427 ± 17	831 ± 18	124
Cyclopropane	C_3H_5-H	450 ± 11	285 ± 11	24,32
3,3-Dimethyl-cyclopropene	$(CH_3)_2C_3H-H$	446 ± 15		125
Methanol	CH_3O-H	437.7 ± 2.8	18.7 ± 2.8	51
Ethanol	C_2H_5O-H	438.1 ± 3.3	-15.2 ± 3.3	51
2-Propanol	$i\text{-}C_3H_7O-H$	442.3 ± 2.8	-48.5 ± 2.8	51
t-Butanol	$t\text{-}C_4H_9O-H$	444.9 ± 2.8	-85.7 ± 2.9	51
Phenol	C_6H_5O-H	359 ± 8	45 ± 8	126

[a]Values in kJ/mol; calculated from primary measurements using updated reference thermochemical data provided in references 127,128, and 10.
[b]Derived by using enthalpies of formation of R—H obtained from reference 127.

$$XRY + h\nu \rightarrow R^+ + XY + e^-$$

Unfortunately, because the barriers to cation rearrangement are generally very low, the dissociation can also be accompanied by extensive rearrangement of the cation, R^+, during its formation. Therefore, this approach is only reliable for systems in which the structure of the cationic product can be established, either chemically, spectroscopically, or on the basis of thermochemical arguments (e.g., it is the lowest energy isomer). Ionization energies of carbenes and diradicals can be measured by using the same approaches used for radicals, although precursor generation is generally limited to pyrolysis.

An example of a system that can be evaluated using this approach is methylene (CH_2). The ionization energy of CH_2 has been measured directly from electron ionization thresholds to give values of 10.35 ± 0.15[52] and 10.2 ± 0.2 eV.[53] However, very accurate measurements of the ionization energy have been made more recently by using photoionization and ZEKE photoelectron spectroscopy. Ruscic and co-workers have obtained an ionization energy of $10.393 \pm 0.011\,eV$,[54] using photoionization mass spectrometry of CH_2 generated by the photolysis of ketene. They recommend a value of $10.400 \pm 0.005\,eV$.[55] More recently, Willitsch et al. have used ZEKE photoelectron spectroscopy to obtain an ionization energy of $83\,772 \pm 3\,cm^{-1}$ $(10.3864 \pm 0.0004\,eV)$.[56] It is worth noting that the 1.3 kJ/mol disagreement between the ZEKE-measured IE and the value recommended by Ruscic and co-workers is nominally larger than the assigned error limits, but is not chemically significant.

Appearance energies for CH_2^+ have been measured from a large number of precursors, including methane, ketene, and even methyl radical.

$$CH_4 \;\rightarrow\; CH_2^+ + H_2 + e^-$$

The measured appearance energies for methylene cation formation from methane range from 15.06 ± 0.02 to 15.3 eV. One of the most accurate appearance energy measurements for CH_2^+ comes from the photoionization of ketene. The value, $AE(CH_2^+, CH_2CO) = 13.743 \pm 0.005\,eV$,[55] implies an 298 K enthalpy of formation of CH_2^+ of $1392.8 \pm 1.0\,kJ/mol$, which, when combined with the ionization energy, gives $\Delta H_f(CH_2) = 390.6 \pm 1.0\,kJ/mol$. Combining this value with the enthalpy of formation of methyl radical (Table 5.2) gives $BDE(CH_2-H) = 462.4 \pm 1.1\,kJ/mol$.

Methylene is a system in which the BDE can also be determined directly from positive ion measurements. Ruscic and co-workers have measured the CH_2^+ appearance upon ionization of methyl radical.[55] Combining the appearance energy, $15.120 \pm 0.006\,eV$, with the ionization energy of 10.3864 eV according to Eq. 5.4a gives $BDE(CH_2-H) = 462.6 \pm 0.6\,kJ/mol$, after applying the appropriate thermal corrections.[54]

Unfortunately, appearance energy measurements become more complicated with larger substrates, where the cations are more prone to rearrangement during ionization. For example, numerous attempts have been made to measure the energy for formation of o-benzyne cation, $C_6H_4^+$ by using benzonitrile as a precursor (Eq. 5.6).

$$(5.6)$$

The appearance energy for $C_6H_4^+$ from benzonitrile has been measured by using both photoionization[57] and photoionization-photoelectron coincidence studies.[58] However, it has subsequently been shown that the $C_6H_4^+$ products formed upon ionization are predominantly ring-opened isomers.[59] Because the appearance energy measurement detects the lowest energy processes, the measured appearance energy is lower than that required for formation of the benzyne cation, and the enthalpy of formation of o-benzyne that would be obtained by using this approach would be too low.

Alternatively, enthalpies of formation of carbenes and biradicals can be measured by using the approach shown in Eq. 5.4b. The key to the measurement is the determination of the proton affinity of the substrate, PA(R), which can be obtained

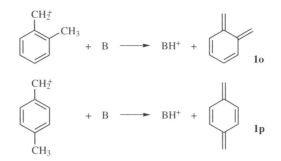

Figure 5.3. Formation of *o*- and *p*-xylylene by deprotonation of benzyl cations.

by bracketing using the RH$^+$ cation. Although the ionization energy of RH is also required, this is nominally the ionization energy for a radical, and can be obtained either directly or derived from supplemental data.

Hehre and co-workers have used this approach for the investigation of biradicals and other reactive neutral molecules. For example, by using the bracketing approach, they were able to determine the proton affinities of *o*- and *p*-xylylene (*o*- and *p*-quinodimethane (**1o** and **1p**) Figure 5.3), from which they were able to determine the enthalpies of formation of the reactive, Kekulé molecules.[60] They found the proton affinity of the *meta* isomer to be too high to be measured directly by bracketing, but were able to assign a lower limit, and subsequently a lower limit to the enthalpy of formation of the *m*-xylylene diradicals.

An unfortunate limitation of the proton affinity approach is that it lacks regioselectivity. The cations under investigation have low barriers to rearrangement, such that the technique is best used for ions that are global energy minima, or metastable species with very high barriers to isomerization. Also, the deprotonation reaction is kinetically controlled, and therefore can be unselective. Hehre and co-workers have used deuterium labeling to distinguish between different sites in structurally intact cations.[61] However, these experiments are subject to label scrambling either before or during bracketing experiments. Examples of carbenes, biradicals, and other reactive species investigated by using these positive ion approaches are provided in Table 5.3.

5.4.2.2. Negative Ion Approaches
The negative ion approach illustrated in Eq. 5.5 is amenable to the investigation of carbenes and diradicals as well. The biggest challenge in the experiments is the generation of the negative ion of the substrate. Once formed, the proton affinity of the ion is typically determined by bracketing and the electron affinity can be found by either bracketing or by using photodetachment techniques.

Different methods have been developed for the generation of carbene and diradical negative ions. One of the most commonly used approaches involves the reaction of an organic substrate with atomic oxygen ion, O$^-$, to form water by H$_2^+$ abstraction (Eq. 5.7).[62]

TABLE 5.3. Enthalpies of formation and BDEs for formation of carbenes, diradicals, carbynes, and triradicals.[a]

R		ΔH_f^b	BDE(R$-$H)[c]	Method[d]	Reference
Carbenes					
Methylene	CH_2	391.0 ± 0.7	462.6 ± 0.6	I	54,56
Fluoro-	CHF	143 ± 13	394 ± 16	IV	87
Difluoro-	CF_2	-182 ± 13	274 ± 9	Evaluated	Table 5.5
Chloro-	CHCl	326 ± 8	426 ± 12	Evaluated	87
Dichloro-	CCl_2	229 ± 11	358 ± 14	Evaluated	Table 5.6
Chlorofluoro	CFCl	31 ± 13	315 ± 17	IV	87
Bromo-	CHBr	373 ± 18	422 ± 18	IV	129
Iodo-	CHI	428 ± 21	418 ± 21	IV	129
Vinyl-	C_2H_3CH	390 ± 14	435 ± 14	IV	88
Phenyl-	C_6H_5CH	430 ± 15	440 ± 15	IV	88
3-Methylphenyl-	$CH_3C_6H_4CH$	411 ± 13	453 ± 14	IV	74
Cyclopropenylidene	C_3H_2	500 ± 9	238	II	130
		504 ± 17	240	I	3
Diazo-	CNN	569 ± 21	331 ± 17	III	131
Vinylidene	$CH_2=C$	428 ± 17	346 ± 15	III	67
Propadienylidene	$CH_2=C=C$	541 ± 17	418 ± 21	III	94
Cyano-	HCCN	484 ± 21	449 ± 23	IV	89
			435 ± 8^e	III	49
Isocyano-	HCNC		444 ± 17^e	III	49
Diradicals					
o-Benzyne	C_6H_4	448 ± 14	329 ± 14	Evaluated	Table 5.7
m-Benzyne	C_6H_4	510 ± 13	391 ± 13	IV	85,95
p-Benzyne	C_6H_4	577 ± 12	458 ± 12	IV	85,95
5-Chloro-*m*-benzyne	C_6H_3Cl	486 ± 15		IV	103
α,2-Dehydrotoluene	C_7H_6	431 ± 13	440 ± 13	IV	90
α,3-Dehydrotoluene	C_7H_6	431 ± 13	440 ± 13	IV	90
α,4-Dehydrotoluene	C_7H_6	431 ± 13	440 ± 13	IV	90
Trimethylenemethane (TMM)	C_4H_6	293 ± 13	377 ± 8	III	99,132
Tetramethylene-ethylene (TME)	C_6H_8	377 ± 21	393 ± 17	III	100
o-Xylylene	$CH_2C_6H_4CH_2$	229 ± 19	271 ± 23	II	60
m-Xylylene	$CH_2C_6H_4CH_2$	338 ± 10	379 ± 12	III,IV	74
p-Xylylene	$CH_2C_6H_4CH_2$	225 ± 19	270 ± 23	II	60
5-Methyl-*m*-xylylene	C_9H_{10}	313 ± 14		IV	102
5-Chloromethyl-*m*-xylylene	C_9H_9Cl	265 ± 13		IV	102
1,2-Dehydrocyclo-octatetraene	*c*-C_8H_6	535 ± 36	284 ± 35	III	114
o-Benzoquinone	OC_6H_4O	-97 ± 17	274 ± 15	III	133
m-Benzoquinone	OC_6H_4O	28 ± 17	377 ± 15	III	133
p-Benzoquinone	OC_6H_4O	-116 ± 13	253 ± 7	III	133

TABLE 5.3. (*Continued*)

α,2-Dehydrophenol	C_6H_4O	290 ± 16	451 ± 16	III	134
α,3-Dehydrophenol	C_6H_4O	288 ± 16	449 ± 16	III	134
α,4-Dehydrophenol	C_6H_4O	290 ± 16	451 ± 16	III	134
2,3-Dehydronaphthalene	$C_{10}H_6$	506 ± 21	319 ± 25	III	134b
2,6-Dehydronaphthalene	$C_{10}H_6$	632 ± 21	445 ± 25	III	134b
benzocyclo-butadiene	C_8H_6	407 ± 18	259 ± 10	III	79
Acenaphthyne	$C_{12}H_6$	668 ± 19	353 ± 11	III	80
Bicyclo[1.1.0]but-1(3)-ene	C_4H_4	545 ± 42	336 ± 27	III	135
Tetrafluoro-*o*-benzyne	C_6F_4	−280 ± 18	322 ± 14	III	136
Dimethylsilene	$(CH_3)_2Si{=}CH_2$	21.0 ± 8.7	222 ± 11[f]	II	137–139
		21.8 ± 8.2	223 ± 11[f]	I	138–141
Cubene	C_8H_6	989 + 24	376 ± 16	III	70
1,5-Dehydro-quadricyclane	C_7H_6	720 ± 36	361 ± 31	III	142
Phenylcyclo-butadiene	$C_6H_5\text{-}C_4H_3$	487 ± 24	318 ± 17	II	143
Carbynes					
Methyne	CH	595 ± 13	426 ± 3	IV	101
Fluoro-	CF	254 ± 14	328 ± 19	IV	101
Chloro-	CCl	443 ± 13	335 ± 15	IV	101
Triradicals					
1,3,5-Trimethylene-benzene	C_9H_9	464 ± 17	369 ± 21	IV	102
1,3,5-Tridehydro-benzene	C_6H_3	749 ± 19	457 ± 23	IV	103
1,2,4-Tridehydro-benzene	C_6H_3	684 ± 21	456 ± 24[g]	IV	144,145
1,2,3-Tridehydro-benzene	C_6H_3	649 ± 16	421 ± 22[g]	IV	144,145
5-dehydro-*m*-xylylene	C_8H_7	590 ± 21	469 ± 17	III	105

[a]Values in kJ/mol; calculated from primary data, using updated reference thermochemical values from references 127,128, and 10.

[b]298 K value.

[c]298 K bond dissociation enthalpy, unless otherwise noted.

[d]I: positive ion approach using Eq. 5.4a; II: positive ion approach using Eq. 5.4b; III: negative ion approach using Eq. 5.5; IV: negative ion approach using Eq. 5.14.

[e]0 K bond dissociation energy.

[f]Corresponds to the C−H bond dissociation energy in the trimethylsilyl radical, implying a π bond strength of 192–193 kJ/mol.

[g]Refers to the BDE in *ortho* benzyne.

$$HRH + O^- \rightarrow R^- + H_2O \qquad (5.7)$$

This reaction can proceed by 1,1-proton abstraction to form a carbene radical anion, but can also occur by 1,n-abstraction to form the negative ion of a diradical. Thus, reaction of O^- with methylene chloride results in the formation of CCl_2^- (Eq. 5.8a),[63] reaction with ethylene gives vinylidene radical anion, H_2CC^- (Eq. 5.8b),[64] and the reaction with acetonitrile gives the radical anion of cyanomethylene, $HCCN^-$ (Eq. 5.8c).[63,65] Investigations of these ions have been used to determine the thermochemical properties of dichlorocarbene, CCl_2,[66] vinylidene,[67] and cyanomethylene.[49]

$$CH_2Cl_2 + O^- \rightarrow CCl_2^- + H_2O \qquad (5.8a)$$

$$H_2C=CH_2 + O^- \rightarrow H_2C=C^- + H_2O \qquad (5.8b)$$

$$CH_3CN + O^- \rightarrow HCCN^- + H_2O \qquad (5.8c)$$

Reaction with atomic oxygen anion can also be used to generate ions of diradicals, including the ions of o-benzyne (Eq. 5.9a),[68] m-xylylene (Eq. 5.9b),[69] cubene (Eq. 5.9c)[70] and tetramethyleneethylene (Eq. 5.9d).[71] In these systems, the BDEs for formation of the diradicals are readily obtained by using Eq. 5.5. However, although the reaction with O^- is successful for many systems, there are also many reactions that do not give the desired diradical product. For example, the reaction of O^- with isobutene (Eq. 5.10) proceeds predominately by deprotonation (Eq. 5.10a), and does not give the desired trimethyl-enemethane (TMM) radical anion product (Eq. 5.10b),[65] despite the fact that reaction of 1,1-dimethylfulvene with O^- does give the corresponding TMM analogue.[72] In addition, the reaction with O^- can be unselective and produce a mixture of anionic products, as occurs in the reactions with benzene, giving both *ortho* and *meta* benzyne anion products,[73] and with *m*-xylene, leading to both methyl- and ring hydrogen abstraction.[74]

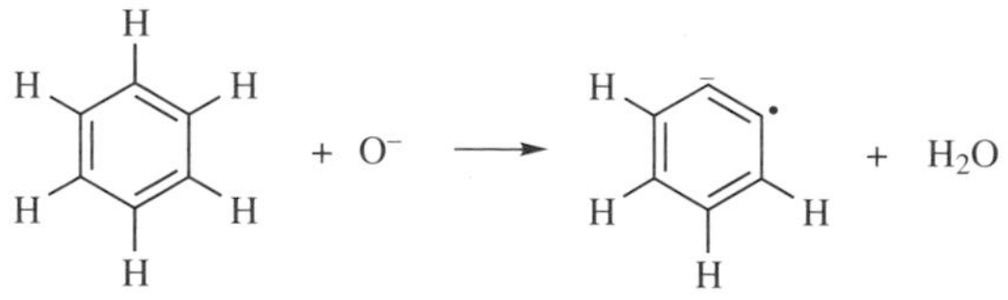

(5.9a)

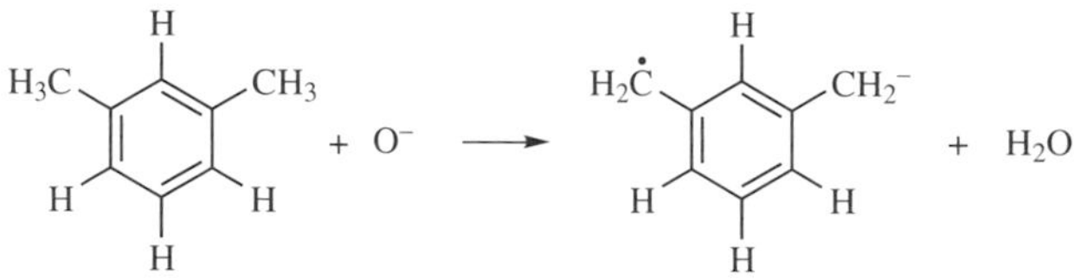

(5.9b)

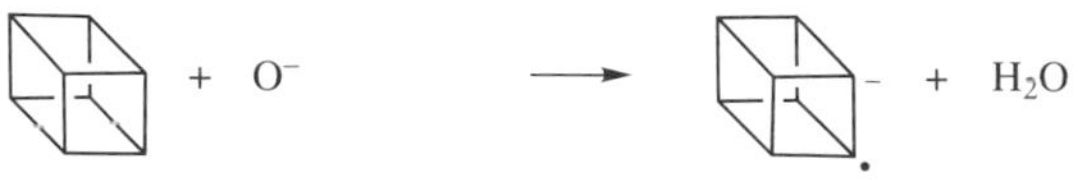

(5.9c)

(5.9d)

$$H_3C \overset{O}{\underset{}{\|}} CH_3 \; + \; O^- \; \Big\langle$$

$$H_3C \overset{O}{\underset{}{\|}} CH_2^- \; + \; OH \qquad (5.10a)$$

$$H_2\overset{\bullet}{C} \overset{O}{\underset{}{\|}} CH_2^- \; + \; H_2O \qquad (5.10b)$$

Because the reaction with atomic oxygen anion can be unpredictable and/or unselective, alternative methods for generating diradical negative ions are useful. Squires and co-workers discovered that reaction between a silyl-substituted carbanion and molecular fluorine, F_2, can be used to generate diradical anions regioselectively.[75,76] The mechanism proposed for the reaction, shown in Scheme 5.1 for p-trimethylsilylphenyl anion, involves an initial dissociative electron transfer from the carbanion to F_2, leading to formation of F and F^- anion, which is capable of undergoing nucleophilic substitution at the silane. The "Squires reaction" has been used to generate a wide variety of regiospecific diradical anions, including m- and p-benzyne radical anions,[75] the acetate radical,[77] and even the TMM anion,[75] which was inaccessible by the O^- approach.

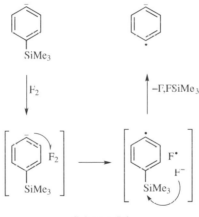

Scheme 5.1

Kass and Broadus have generated diradical anions by decarboxylation of dicarboxylate dianions using ion-cyclotron resonance with electrospray ionization (ESI).[78] Highly stabilized dianions, including dicarboxylates, can be formed by ESI. Decarboxylation by CID, however, results in not only loss of CO_2, but also loss of an electron to form a carboxylate radical. Additional CID can be carried out to achieve a second decarboxylation, if desired, to give the organic radical anion. The process has been used to generate negative ions of substrates such as 2,3-dehydronaphthalene (Eq. 5.11),[78] benzocyclobutadiene (**2**),[79] and acenaphthyne (**3**).[80]

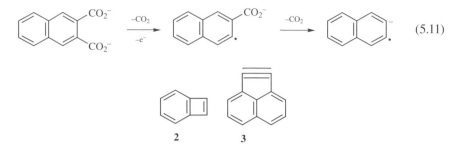

2   3

Finally, in some cases diradical negative ions can even be generated directly upon ionization of appropriate precursors. For example, nitrene and carbene anions can be formed by EI of organic azides, diazo-compounds, and diazirines,[81] whereas Brauman and co-workers have reported the formation of oxyallyl anions by EI of fluorinated acetyl compounds (Eq. 5.12).[82]

$$\underset{H_3C}{\overset{O}{\underset{}{\|}}}CF_3 \; + \; e^- \; \longrightarrow \; \underset{H_2\overset{\bullet}{C}}{\overset{O}{\underset{}{\|}}}CF_2^- \; + \; HF \qquad (5.12)$$

Similarly, Hammad and Wenthold have shown that 1,4-dicyanocyclohexane-1,4-diyl radical anion forms spontaneously upon ionization of 2,5-dicyano-1,5-hexadiene (Eq. 5.13).[83]

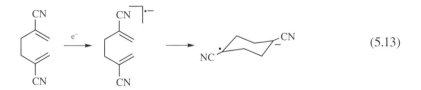

$$(5.13)$$

These experimental methods for preparing diradical negative ions were generally developed for use in reactivity studies and therefore are usually the basis for reactivity-based thermochemical studies, such as bracketing or kinetic method experiments. However, some have been used to generate ions for spectroscopic experiments as well. The atomic oxygen anion, in particular, is conveniently generated in NIPES flowing afterglow sources, and therefore the reaction of substrates with O$^-$ has often been used to prepare ions for NIPES experiments. Similarly, the Squires reaction has also been used for photoelectron measurements, whereas dissociative EI methods have been applied to generate ions for photoelectron and photodetachment experiments. Compatibility with spectroscopic methods is advantageous because it allows for more accurate determinations of electron affinities than can be obtained by using reaction based methods such as bracketing or the kinetic method.

Alternative negative ion-based methods for measuring carbene and diradical enthalpies of formation have been developed, which can give BDEs indirectly. A common approach for this involves the use of *halide affinity* measurements. The relationship between enthalpy of formation and halide affinity is illustrated by Eq. 5.14.

$$\text{HRX} \xrightarrow{\Delta H_{acid}(HRX)} \text{RX}^- \xrightarrow{D(R-X^-)} \text{R} \; + \; \text{X}^-$$

$$\Delta H_f(R) = \Delta(R-X^-) \; + \; \Delta H_{acid}(HRX) \; + \; \Delta H_f(HRX) \; - \; \Delta H_{acid}(HX) \; - \; \Delta H_f(HX)$$

(5.14)

Squires and co-workers have used energy-resolved CID measurements as a means for measuring the halide affinities of carbenes[84] and diradicals,[85] where the regiospecific anions are generated either by deprotonation or by fluoride-induced desilylation.[86] Carbenes that have been investigated using this approach include halocarbenes, such as chloro-,[87] fluoro-,[87] dichloro-,[84] difluoro-,[84] and chlorofluorocarbenes,[87] as well as triplet carbenes, such as methylene[88] and vinyl,[88] phenyl,[88] and cyanocarbenes.[89] Diradicals that have been investigated include the benzynes,[85] α,2-, α,3-, and α,4-dehydrotoluene,[90] and *m*-xylylene.[74] It is critical in these experiments to ensure that the halo-substituted anion precursor is structurally pure, as low-energy isomer impurities can lead to erroneous results. Advantages of the approach include the fact that different halide ions can be used for the dissociation, and each constitutes an independent measure of the thermochemistry. Triplet carbenes are readily investigated by using this approach, but triplet diradical measurements can be systematically affected by slow intersystem crossing (ISC) during the spin-forbidden dissociation.[90] In such cases, larger halides can be used to increase the rate of ISC by increased spin-orbit coupling.

A variation of the halide affinity approach was used by Riveros et al. in the investigation of the enthalpy of formation of *o*-benzyne.[91] Reaction of bromo- or iodobenzene with base in an ICR leads predominantly to the formation the expected M-1 anion, but also leads to the formation of solvated halide ions (Eq. 5.15). By using substrates with known halide affinities, it was possible to assign limits to the enthalpy of formation of the benzyne product. Ultimately, the experiment is comparable to that outlined in Eq. 5.14, although the acidity and halide affinity measurements are made in a single step.

(5.15)

Enthalpies of formation and BDEs for the formation of carbenes and diradicals measured by using negative ion approaches are included in Table 5.3.

5.4.2.3. Singlet-triplet Energy Splittings in Carbenes and Diradicals

An important feature of photodetachment methods is that the selection rule for electronic transition is $\Delta S = \pm 1$. Therefore, photodetachment of (doublet) negative ions derived from carbenes or diradicals can lead to formation of either the singlet or triplet state of the neutral, and these approaches can be used to measure the singlet-triplet energy differences in these important open-shell molecules. Negative ion photoelectron spectroscopy (Figure 5.1)[92] is particularly useful in this regard, because the detachment energy is much higher than the detachment threshold, and often higher than the threshold for the excited state, although state energy splittings have also been estimated from the positions of second onsets in threshold photodetachment experiments.

An important consideration in spectroscopic studies of systems with multiple electronic states is the difference in the geometries between the two states. Because photodetachment is governed by the Franck–Condon principle, very large differences in geometry can result in significant differences in the shapes of the spectral features for the two electronic states. Therefore, it is common to have a case in which detachment to one state is nearly vertical, whereas formation of the other occurs with a very large geometry change. In favorable cases, the singlet and triplet geometries are similar, with similarly shaped spectral bands. This is true for vinylidene, H_2CC,[93] and propadienylidene, H_2CCC,[94] and also to an extent for some rigid systems such as o- and p-benzyne.[95] Carbenes, however, are notorious for having extreme geometry differences between the singlet states, which have bond angles of about 100–105°, similar to those in the ions, and the triplet states, which have very large bond angles of 150° or greater. Therefore, the photoelectron bands for formation of the singlet are usually very simple, whereas those for formation of the triplet have extended vibrations with weak origin peaks, which can lead to difficulties in determining the singlet–triplet energy splittings in these systems.[96–98]

The Franck–Condon question is also an issue in diradical studies. It is particularly relevant in nonrigid systems, such as trimethylenemethane (TMM)[99] and tetramethyleneethylene (TME),[100] in which neutral and/or anion states can be planar or nonplanar. Thus, transitions can occur from planar anions to nonplanar neutral states, or from nonplanar anions to planar neutrals. In these cases, the energy differences between the planar and nonplanar states have generally been estimated by using quality electronic structure calculations.

A list of singlet–triplet energy splittings measured using photoelectron spectroscopy is give in Table 5.4.

5.4.3. Triradicals and Beyond

Some of the negative ion methods described above for the determination of thermochemical properties of diradicals are also amenable to the study of triradicals, including molecular triradicals and carbynes. Again, the challenge is in the formation of the requisite precursor anions. Triradical thermochemical measurements require either generation of the triradical negative ions directly, for use of Eq. 5.5, or generation of halo-substituted diradical/carbene anions, for use in energy-resolved CID measurements, as with Eq. 5.14.

The methods for generating these types of ions are the same as those described above for diradicals and carbenes. For example, the reaction of dichloromethane with O^- leads to the formation of CCl_2^-. Thus, Jesinger and Squires have used CID of halocarbene anions to determine the thermochemical properties of carbynes (Eq. 5.16).[101]

TABLE 5.4. Singlet-triplet energy splittings measured using photoelectron spectroscopy.[a]

Species	Formula	ΔE_{ST}	Reference
Carbenes			
Methylene	CH_2	-62.9 ± 0.6	98
	CD_2	-62.2 ± 0.6	98
Fluoro-	CHF	62.3 ± 1.7	146
Chloro-	CHCl	17.4 ± 10.6	146
Bromo-	CHBr	10.7 ± 9.2	146
Iodo-	CHI	-26 ± 17	146
Difluoro-	CF_2	226 ± 12	147
Dichloro-	CCl_2	12 ± 13^{b}	147
Dibromo-	CBr_2	10 ± 13^{b}	147
Diiodo-	CI_2	-6 ± 13^{b}	147
Cyano-	HCCN	-49.7 ± 1.5	49
Isocyano-	HCNC	4.8 ± 2.7	49
Vinylidene	$CH_2{=}C$	199.2 ± 0.6	93
Fluorovinylidene	$CHF{=}C$	127.4 ± 0.9	148
Difluorovinylidene	$CF_2{=}C$	89.2 ± 0.9	148
Vinylvinylidene	$C_2H_3CH{=}C$	185.5 ± 1.4	149
t-Butylvinylidene	$t\text{-}C_4H_9CH{=}C$	190.6 ± 1.4	150
Propadienylidene	$CH_2{-}C{-}C$	124.3 ± 0.8	94
Diazo-	CNN	-81.6 ± 1.4	131
Diradicals			
Dicarbon	C_2	6.9 ± 1.0	151
o-Benzyne	C_6H_4	54.4 ± 0.7	73,95
m-Benzyne	C_6H_4	88 ± 13	95
p-Benzyne	C_6H_4	15.9 ± 2.1	95
Trimethylenemethane (TMM)	C_4H_6	$-54--67$	99,132
Tetramethyleneethylene (TME)	C_6H_8	8	100
m-Xylylene	C_8H_8	-40.2 ± 0.8	152
Cyclooctatetraene	C_8H_8	50.6 ± 0.6^{c}	110
Dehydrocyclooctatetraene	C_8H_6	68.3 ± 0.6	153
Dehydrobenzoquinone	$C_6H_2O_2$	<100	154
Cyclopentadienyl cation	$C_5H_5^{+}$	-19	115
Other Systems			
Amidogen	NH	-152.4 ± 1.6	155
Methylnitrene	CH_3N	-130.5 ± 1.1	156
Phenylnitrene	C_6H_5N	62.0 ± 1.0^{d}	157
Cyanonitrene	NCN	-97.5 ± 1.0	158
Silylene	SiH_2	<60	159
Silicon carbide cluster	CCCSi	-26.4 ± 1.4	160
Silicon carbide cluster	SiCCCCSi	-10.6 ± 0.4	160

[a]0K values in kJ/mol correspond to $E_{singlet} - E_{triplet}$, such that values less than 0 indicate triplet ground states.
[b]These results are based on extensive modeling of the photoelectron spectra and have been called into question. For example, see reference 161.
[c]Energy difference between the singlet and triplet states in the planar geometry. The energy difference between the ground state (tub) singlet and the triplet is 100 kJ/mol.
[d]Recent results from the author's laboratory suggest that the originally reported value of 75 kJ/mol is too high, and resulted from a misassignment of a phenoxide impurity as the singlet state. The value listed here is obtained by re-analysis of the spectrum in reference 157.

$$CH_2Cl_2 + O^- \xrightarrow{-H_2O} CCl_2^- \xrightarrow{CID} CCl + Cl^- \qquad (5.16)$$

Larger halogenated systems also react with atomic oxygen anion, although the reactions again can be unselective. However, the Squires reaction can be used to prepare regiospecific halogenated precursors for CID measurements. Triradicals that have been investigated using these methods include 1,3,5-trimethylenebenzene, Eq. 5.17a,[102] and 1,3,5-tridehydrobenzene, Eq. 5.17b.[103]

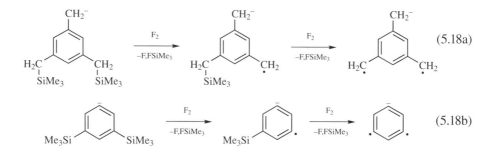

The Squires reaction can also be used to generate radical anions of triradicals by using polysilylated precursors. Therefore, sequential reaction of 3,5-*bis*-trimethylsilylmethylbenzyl or 3,5-*bis*-trimethylsilylphenyl anions with 2 molecules of F_2 leads to the formation of 1,3,5-trimethylenebenzene (Eq. 5.18a) or 1,3,5-tridehydrobenzene (Eq. 5.18b) radical anion.[104]

Wenthold and co-workers have used this approach to generate the negative ion of 5-dehydro-*m*-xylylene (Eq. 5.19).[105,106] Using bracketing experiments to determine the gas-phase acidity at the 5-position of *m*-xylylene, and the kinetic method to determine the triradical electron affinity, the BDE at the 5-position of *m*-xylylene was measured (Eq. 5.20).

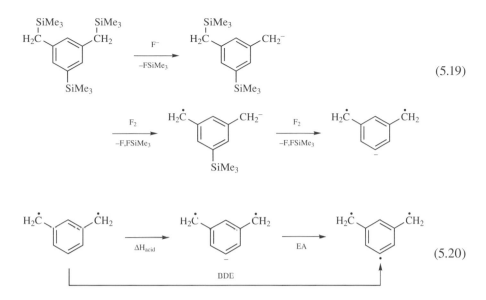

$$(5.19)$$

$$(5.20)$$

Just as the negative ion procedures used for carbenes and diradicals can be applied to carbynes and triradicals, they can in principle be extended to the study of more complex open-shell systems, including tetraradicals and beyond. However, highly open-shell systems are more challenging due to the instability of the precursors, which also must be highly reactive species and therefore can potentially undergo rearrangement. For example, whereas Gronert and DePuy were able to generate the didehydrophenyl anion, $C_6H_3^-$, by CID of deprotonated fluorobenzene (Eq. 5.21),[107] attempts to produce tetradehydrophenyl anion, C_6H^-, from difluorophenyl anion (Eq. 5.22a)

$$(5.21)$$

were unsuccessful, because the product ring-opened to give deprotonated hexatriyne (Eq. 5.22b).

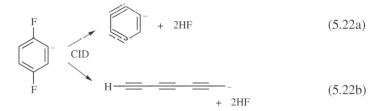

$$(5.22a)$$

$$(5.22b)$$

Few positive ion methods are generally applicable to the investigation of triradicals because of the problem of measuring ionization energies for di- and triradicals and because there are few good methods for preparing structurally intact cations of di- and triradicals for thermochemical studies. Reported thermochemical properties of triradicals are included in Table 5.3.

5.4.4. Transition States

Mass spectrometric studies are not limited to the investigation of stable intermediates; they have also been carried out on reaction transition states. The ultrafast studies by Zewail, for example, are nominally mass spectrometric based, where photoionization is used to detect reactive species on exceedingly short (femtosecond) time scales.[108] Time resolved studies provide insight into the rates of unimolecular reactions, but do not provide direct thermochemical insight.

In select cases, negative ion photoelectron spectroscopy can also be used to investigate reaction transition states.[109] The ideal situation is illustrated in Figure 5.4. If the geometry of the lower (ionic) state is very similar to that of the transition state and is very different from that of the stable neutral states, then the transition state can be generated by vertical photodetachment. For example, as shown in Figure 5.4, the cyclooctatetraene negative ion has a planar geometry, very similar to that for the transition state for ring inversion of the tub-like, neutral molecule. Therefore, because the geometry difference between the ion and the neutral ground state is large, vertical photodetachment leads to formation of the neutral in the transition state geometry.

The photoelectron spectrum of cyclooctatetraene negative ion is shown in Figure 5.5.[110] The peaks at low electron binding energy correspond to formation of the planar singlet state of cyclooctatetraene. From the position of the origin peak in the spectrum, it is possible to determine the *electron binding energy* of the transition state. As indicated in Figure 5.4, the energy difference between the electron binding

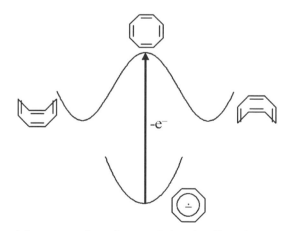

Figure 5.4. Potential energy surfaces for out-of-plane bending of cyclooctatetraene and its negative ion. The arrow indicates a vertical transition upon photodetachment to the transition state of the neutral.

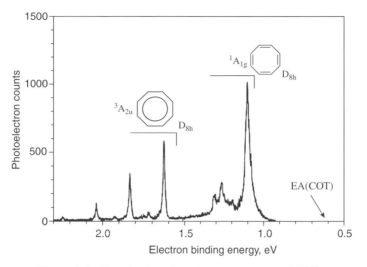

Figure 5.5. Negative ion photoelectron spectrum of COT⁻.

energy of the transition state and the electron affinity of COT is the energy barrier for the ring inversion reaction of COT.

Remarkably, the photoelectron spectrum provides more than just the energy of the transition state. As can be seen in Figure 5.5, the spectrum also contains peaks corresponding to the transition state in excited vibrational levels, where the activated vibrations are orthogonal to the reaction coordinate. Therefore, NIPES can even be used to carry out *vibrational spectroscopy* of reaction transition states.

Recently, Zewail and co-workers have combined the approaches of photodetachment and ultrafast spectroscopy to investigate the reaction dynamics of planar COT.[111] They used a femtosecond photon pulse to carry out ionization of the COT ring-inversion transition state, generated by photodetachment as shown in Figure 5.4. From the photoionization efficiency, they were able to investigate the time-resolved dynamics of the transition state reaction, and observe the ring-inversion reaction of the planar COT to the tub-like D_{2d} geometry on the femtosecond time scale. Thus, with the advent of new mass spectrometric techniques, it is now possible to examine detailed reaction dynamics in addition to traditional state properties.[112]

5.5 COMPARISON OF EXPERIMENTAL METHODS

Comparison of the BDEs and enthalpies of formation obtained from different experimental approaches can be used to assess the quality of the measurements. Berkowitz, et al. have compared the results of positive ion and negative ion methods for determining BDEs of stable molecules with those obtained from kinetics studies.[12] They found that for most systems, the agreement in the results from the three approaches was excellent. In this section, the results that have been obtained for the enthalpies of formation of carbenes and diradicals using the gas-phase techniques described above will be discussed and evaluated. The test systems that will provide a means for comparison are dichloro- and difluorocarbene, and *o*-benzyne.

5.5.1. Difluorocarbene

The enthalpies of formation that have been obtained for difluorocarbene are summarized in Table 5.5. The enthalpy of formation of CF_2 has been calculated by combining an appearance energy with the ionization energy (Eq. 5.4a), calculated from the proton affinity (Eq. 5.4b), and derived from the CID threshold energy for CF_3^- (Eq. 5.14). The weighted average of the values in Table 5.5 is -182 ± 13 kJ/mol, where the uncertainty includes the standard deviation of the values and a statistical uncertainty component. The recommended value agrees best with that obtained by using the appearance energy from C_2F_4, whereas the values obtained by using Eqs. 5.4b and 5.14 are *ca.* 20 kJ/mol lower and higher, respectively. It is not clear as to the origin of the error in the PA bracketing. The discrepancy between the recommended value and that obtained from CID threshold measurements is possibly due to a competitive shift in the threshold energy, although this is not sufficiently established to warrant omitting the data point.

5.5.2. Dichlorocarbene

The enthalpies of formation that have been obtained for dichlorocarbene are listed in Table 5.6. Equations 5.4a, 5.4b, 5.5, and 5.14 have all been used. Generally good agreement among the values listed in Table 5.6 suggest that the true value is probably 215–240 kcal/mol. The weighted average of all the data in Table 5.5 is 238 ± 28 kJ/mol. However, that value is strongly influenced by the single value at 268 kJ/mol, which is significantly out of line with the other reported values. If the highest (268 kJ/mol) and lowest (163 kJ/mol) values are excluded, the average becomes 229 ± 11 kJ/mol, which is the recommended value.

TABLE 5.5. Measured enthalpy of formation of difluorocarbene.[a]

$\Delta H_{f,298}(CF_2)$	Experimental approach	Reference
-175.6 ± 3.9	Eq. 5.4a (C_2F_4)	162[b]
-206 ± 13	Eq. 5.4a (CF_3Br)	163[b]
-196 ± 30	Eq. 5.4a (CF_3Cl)	164[b]
-212 ± 11	Eq. 5.4b	165
-212 ± 22	Eq. 5.4b	166
-165 ± 14	Eq. 5.14	84
-182 ± 13	*recommended*	

[a]298 K values in kJ/mol.
[b]Includes an ionization energy of 11.445 ± 0.012 eV from reference 167.

TABLE 5.6. Measured enthalpy of formation of dichlorocarbene.[a]

$\Delta H_{f,298}(CCl_2)$	Experimental approach	Reference
236	Eq. 5.4a	168
229 ± 16[b]	Eq. 5.4a (CF_2Cl_2)	169
213 ± 14[b]	Eq. 5.4a (CF_2Cl_2)	170
268 ± 6[c]	Eq. 5.4a ($CFCl_3$)	171
234 ± 8	Eq. 5.4b	172
163 ± 13	Eq. 5.4b	165
218 ± 13	Eq. 5.14	84
219 ± 22[d]	Eq. 5.5	173
239 ± 13	Eq. 5.5	66
229 ± 11	*recommended*	

[a]298 K enthalpy of formation in kJ/mol.
[b]Calculated using the enthalpy of formation of CF_2 from Table 5.3.
[c]Calculated using the ionization energy of 9.27 ± 0.04 eV from reference 170.
[d]Calculated using the electron affinity of 1.59 ± 0.07 eV from reference 147.

5.5.3. *ortho*-Benzyne

The enthalpies of formation that have been reported for *o*-benzyne are listed in Table 5.7. The approaches that have been used include Eqs. 5.4a, 5.4b, 5.5, and 5.14. In addition, Riveros and co-workers have estimated the enthalpy of formation by using the approached illustrated in Eq. 5.15. Their value is in reasonable agreement with other recent determinations. However, as noted previously, the values based on the appearance energy of $C_6H_4^+$ are suspect because the $C_6H_4^+$ product is likely a mixture of cyclic and acyclic products. Therefore, the weighted average value listed in Table 5.7 does not include those measurements. The remaining values are in reasonable agreement, ranging from 440–474 kJ/mol. The average value obtained by

TABLE 5.7. Measured enthalpy of formation of *o* Benzyne.[a]

$\Delta H_{f,298}(C_6H_4)$	Experimental approach	Reference
474 ± 21	Eq. 5.4b	1
440 ± 13	Eq. 5.15	91
448 ± 13	Eq. 5.5	68
440 ± 13	Eq. 5.14	85
454 ± 13	Eq. 5.5	174
446 ± 11	Eq. 5.4a	58
433 ± 13	Eq. 5.4a	57
412 ± 16	Eq. 5.4a	175
448 ± 14	*recommended*[b]	

[a]298 K enthalpy of formation in kJ/mol; calculated using updated auxiliary thermochemical values; see reference 95.
[b]Does not include values obtained using Eq. 5.4a because of questions regarding the ion structure in the appearance energy measurement; see text.

using Eq. 5.4a is 430 kJ/mol, lower than the range of the other values and consistent with the assessment that the $C_6H_4^+$ ion ring-opens upon dissociative ionization of neutral precursors.[59]

The following conclusions can be drawn from the data presented in this section: (1) In most instances, negative ion approaches can be successfully used to determine enthalpies of formation for reactive neutral species. However, care should be taken when using appearance energy measurements to account for possible errors. (2) The use of an appearance energy in conjunction with an ionization energy can be used to determine accurate values for enthalpies of formation provided that the structure of the cationic product can be established. (3) Enthalpies of formation can be determined from proton affinity measurements if the proton affinity measurements are sufficiently accurate. The results listed here indicate that the enthalpies of formation of carbenes and diradicals derived from one-way proton affinity bracketing can differ significantly from those obtained with other techniques.

5.6. STRUCTURAL ASSESSMENTS BY MASS SPECTROMETRY

Although mass spectrometry is a powerful tool for the investigation of thermochemical properties of reactive intermediates, it is also possible to obtain structural information, particularly when using photoelectron spectroscopy, as Franck–Condon profiles provide insight into the relative geometries of the substrates. For example, Kohn and Chen used Franck–Condon analysis to establish that the cyclobutadiene radical cation, **4**, has a rectangular geometry.[113] Similarly, Kato et al.[114] used a combination of Franck–Condon analysis and angular distribution measurements to discern that deprotonated cyclooctatetraene anion has an allenic/π-delocalized structure, **5**. Moreover,

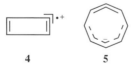

      4   5

even in studies where the focus has been on thermochemistry, general structural insight has been gained, as in the photoelectron study of benzyne where it was concluded that *m*-benzyne was significantly distorted beyond the level predicted by simple MCSCF calculations.[95] Indeed, it is the combination of the empirical results with computational predictions where structural investigations are most insightful.

5.7. CONCLUSION AND OUTLOOK

Advances in experimental methods over the past 50 years have impacted all of chemistry, including the field of reactive intermediates, opening possibilities that until recently could hardly be imagined. Thermochemical studies, fundamental to our understanding of reactive intermediates, are among those aspects that have benefited from the technical and creative developments. The number of known BDEs in stable molecules continues to grow, and the accuracy of measurements continues to improve, with key BDEs known to within 4 kJ/mol. The most significant advances have come in the investigations of more highly reactive species over the last 25 years. Now, thermochemical studies of species such as carbenes and diradicals are common, as exemplified by the data in Table 5.3, and thermochemical studies are even being carried out on triradicals. Thus, thermochemical studies have reached a level beyond what can be readily achieved in reactivity studies, and it has now become a misnomer to refer to these substrates as reactive intermediates. Is there a reaction in which 5-dehydro-*m*-xylylene is formed as an intermediate?

Although it is now possible to investigate highly unusual molecules, the data in Tables 5.2 and 5.3 do reflect the fact that the studies become more difficult as the complexity increases. The average uncertainty in the enthalpies of formation of simple radicals in Table 5.2 is only 5 kJ/mol. While this is undoubtedly biased as a selected sample, it is still the case that simple BDEs are routinely measured to within 8 kJ/mol, such that enthalpies of formation of radicals can generally be known to within 10 kJ/mol. Meanwhile, the average uncertainty in enthalpies of formation of carbenes and diradicals is 16 kJ/mol, whereas those for carbynes and triradicals is 17 kJ/mol. These increases are natural, given the accumulation of uncertainty in each bond dissociation step. On the other hand, BDE measurements themselves do not necessarily have to suffer, especially those that are determined directly via Eqs 5.4 or 5.5. BDEs derived indirectly from enthalpy of formation measurements, however, will likely have higher uncertainties. This will likely improve as more high resolution studies focus on organic systems. The recent photoelectron study of the anti-aromatic cyclopentadienyl cation by Wörner and Merkt[115] is an excellent example of how high-resolution spectroscopy can be used to provide insight into long-standing physical organic questions.

In addition to the natural improvements expected in the accuracy of the measurements, and the increased scope in the types of systems examined, new techniques go beyond the issue of thermochemistry to allow for very detailed studies of reaction dynamics. The investigation by Zewail and co-workers of the reactivity of planar COT[111] on the femtosecond time scale is likely only the beginning. *Time-resolved photoelectron spectroscopy*,[112] for example, has recently been used to map the potential energy surfaces for the dissociation of simple ions IBr^- and I_2^-.[116] Although applications in the field of organic reactive molecules are likely far off, they are now possible.

In the end, mass spectrometry and ion techniques will continue to be powerful tools for the investigation of the structure, bonding, energetics, and reactivity of unusual organic molecules. New sophisticated techniques will continue to be developed and applied to interesting problems in physical organic chemistry. These studies, along with the continued improvements in computational methods (Chapter 9), provide means to obtain very detailed and accurate descriptions of chemical reactions.

SUGGESTED READING

For further information regarding the investigation of reactive intermediate thermochemistry using mass spectrometry, the reader should consider:

P. G. Wenthold, "Toward the Systematic Decomposition of Benzene," *Angew. Chem., Int. Ed.* **2005**, *44*, 7170.

K. M. Ervin, "Experimental Techniques in Gas-Phase Ion Thermochemistry," *Chem. Rev.* **2001**, *101*, 391.

J. Berkowitz, G. B. Ellison, and D. Gutman, "Three Methods to Measure RH Bond Energies," *J. Phys. Chem.* **1994**, *98*, 2744.

P. G. Wenthold and W. C. Lineberger, "Negative Ion Photoelectron Spectroscopy Studies of Organic Reactive Intermediates," *Acc. Chem. Res.* **1999**, *32*, 597.

S. Willitsch, J. M. Dyke, and F. Merkt, "Generation and High-Resolution Photoelectron Spectroscopy of Small Organic Radicals in Cold Supersonic Expansions," *Helv. Chim. Acta* **2003**, *86*, 1152.

REFERENCES

1. S. K. Pollack and W. J. Hehre, *Tetrahedron Lett.* **1980**, *21*, 2483.

2. P. G. Wenthold, *Angew. Chem., Int. Ed.* **2005**, *44*, 7170.

3. H. Clauberg, D. W. Minsek, and P. Chen, *J. Am. Chem. Soc.* **1992**, *114*, 99.

4. F. P. Lossing and A. W. Tickner, *J. Chem. Phys.* **1952**, *20*, 907.

5. F. P. Lossing, K. U. Ingold, and I. H. S. Henderson, *J. Chem. Phys.* **1954**, *22*, 621.

6. J. A. Blush, P. Chen, R. T. Wiedmann, and M. G. White, *J. Chem. Phys.* **1993**, *98*, 3557.

7. T. Gilbert, I. Fischer, and P. Chen, *J. Chem. Phys.* **2000**, *113*, 561.

8. T. Schultz, J. S. Clarke, T. Gilbert, H.-J. Deyerl, and I. Fischer, *Faraday Discuss.* **2000**, *115*, 17.

9. G. C. Eiden, K.-T. Lu, J. Badenhoop, F. Weinhold, and J. C. Weisshaar, *J. Chem. Phys.* **1996**, *104*, 8886.

10. S. G. Lias and J. E. Bartmess, "Gas-Phase Ion Thermochemistry," In *NIST Chemistry WebBook, NIST Standard Reference Database Number 69*; February 2000 ed.; W. G. Mallard and P. J. Linstrom, Eds.; National Institute of Standards and Technology: Gaithersburg, MD, 2005.

11. K. M. Ervin, *Chem. Rev.* **2001**, *101*, 391.

12. J. Berkowitz, G. B. Ellison, and D. Gutman, *J. Phys. Chem.* **1994**, *98*, 2744.

13. T. B. McMahon, *NATO Science Series, Series C: Mathematical and Physical Sciences* **1999**, *535 (Energetics of Stable Molecules and Reactive Intermediates)*, 259.

14. P. Kebarle, *Int. J. Mass Spectrom.* **2000**, *200*, 313.

15. S. T. Graul and R. R. Squires, *Mass Spectrom. Rev.* **1988**, *7*, 1.

16. R. R. Squires, *Int. J. Mass Spectrom. Ion Processes* **1992**, *117*, 565.

17. G. Bouchoux, J. Y. Salpin, and D. Leblanc, *Int. J. Mass Spectrom. Ion Processes* **1996**, *153*, 37.

18. R. G. Cooks, J. S. Patrick, T. Kotiaho, and S. A. McCluckey, *Mass Spectrom. Rev.* **1994**, *13*, 287.

19. P. B. Armentrout, *J. Mass Spectrom.* **1999**, *34*, 74.

20. L. Draho, C. Peltz, and K. Vèkey, *J. Mass Spectrom.* **2004**, *39*, 1016.

21. K. M. Ervin and P. B. Armentrout, *J. Mass Spectrom.* **2004**, *39*, 1004.

22. P. G. Wenthold, J. Hu, and R. R. Squires, *J. Am. Chem. Soc.* **1996**, *118*, 11865.

23. J. W. Denault, G. Chen, and R. G. Cooks, *J. Am. Soc. Mass Spectrom.* **1998**, *9*, 1141.

24. R. A. Seburg and R. R. Squires, *Int. J. Mass Spectrom. Ion Processes* **1997**, *167/168*, 541.

25. V. Ryzhov, R. C. Dunbar, B. Cerda, and C. Wesdemiotis, *J. Am. Soc. Mass Spectrom.* **2000**, *11*, 1037.

26. L. Z. Chen and J. M. Miller, *Org. Mass Spectrom.* **1992**, *27*, 883.

27. C. C. Liou and J. S. Brodbelt, *J. Am. Soc. Mass Spectrom.* **1992**, *3*, 543.

28. J. M. Talley, B. Cerda, G. Ohanessian, and C. Wesdemiotis, *Chem. Eur. J.* **2002**, *8*, 1377.

29. W. Y. Feng, S. Gronert, and C. B. Lebrilla, *J. Am. Chem. Soc.* **1999**, *121*, 1365.

30. M. N. Eberlin, T. Kotiaho, B. J. Shay, S. S. Yang, and R. G. Cooks, *J. Am. Chem. Soc.* **1994**, *116*, 2457.

31. R. Augusti, D. V. Augusti, H. Chen, and R. G. Crooks, *Eur. J. Mass Spectrom.* **2004**, *10*, 847.

32. C. H. DePuy, S. Gronert, S. E. Barlow, V. M. Bierbaum, and R. Damrauer, *J. Am. Chem. Soc.* **1989**, *111*, 1968.

33. I. H. Krouse and P. G. Wenthold, *Organometallics* **2004**, *23*, 2573.

34. P. G. Wenthold and R. R. Squires, *J. Mass Spectrom.* **1995**, *30*, 17.

35. J. J. Hache, J. Laskin, and J. H. Futrell, *J. Phys. Chem. A* **2002**, *106*, 12051.

36. S. Matt, T. Fiegele, G. Hanel, D. Muigg, G. Denifl, K. Becker, H. Deutsch, O. Echt, N. Mason, A. Stamatovic, P. Scheier, and T. D. Mark, *AIP Conference Proceedings* **2000**, *543 (Atomic and Molecular Data and Their Applications)*, 191.

37. C. Aubry, J. L. Holmes, and J. C. Walton, *J. Phys. Chem. A* **1998**, *119*, 9039.

38. H. F. Gruetzmacher and J. Lohmann, *Justus Liebigs Ann. Chem.* **1967**, *705*, 81

39. P. B. Armentrout, *J. Am. Soc. Mass Spectrom.* **2002**, *13*, 419.

40. S. T. Graul and R. R. Squires, *J. Am. Chem. Soc.* **1990**, *112*, 2517.

41. V. F. DeTuri, M. A. Su, and K. M. Ervin, *J. Phys. Chem. A* **1999**, *103*, 1468.

42. V. F. DeTuri and K. M. Ervin, *Int. J. Mass Spectrom. Ion Processes* **1998**, *175*, 123.

43. V. F. DeTuri and K. M. Ervin, *J. Phys. Chem. A* **1999**, *103*, 6911.

44. J. A. Blush, H. Clauberg, D. W. Kohn, D. W. Minsek, X. Zhang, and P. Chen, *Acc. Chem. Res.* **1992**, *25*, 385.

45. I. Fischer, *Int. J. Mass Spectrom.* **2002**, *216*, 131.

46. S. Willitsch, J. M. Dyke, and F. Merkt, *Helv. Chim. Acta* **2003**, *86*, 1152.

47. A. A. Zavitsas, *J. Chem. Ed.* **2001**, *78*, 417.

48. S. W. Benson, *Thermochemical Kinetics*, 2nd ed.; Wiley: New York, 1976.

49. M. R. Nimlos, G. E. Davico, C. M. Geise, P. G. Wenthold, W. C. Lineberger, S. J. Blanksby, C. M. Hadad, G. A. Petersson, and G. B. Ellison, *J. Chem. Phys.* **2002**, *117*, 4323.

50. S. J. Blanksby and G. B. Ellison, *Acc. Chem. Res.* **2003**, *36*, 255.

51. K. M. Ervin and V. F. DeTuri, *J. Phys. Chem. A* **2002**, *106*, 9947.

52. W. Reineke and K. Strein, *Ber. Bunsen-Ges. Phys. Chem.* **1976**, *80*, 343.

53. A. Niehaus, *Z. Naturforsch* **1967**, *22a*, 690.

54. M. Litorja and B. Ruscic, *J. Chem. Phys.* **1998**, *108*, 6748.

55. B. Ruscic, M. Litorja, and R. L. Asher, *J. Phys. Chem. A* **1999**, *103*, 8625.

56. S. Willitsch, L. L. Imbach, and F. Merkt, *J. Chem. Phys.* **2002**, *117*, 1939.

57. A. Maccoll and D. Mathur, *Org. Mass Spectrom.* **1981**, *16*, 261.

58. H. M. Rosenstock, R. Stockbauer, and A. C. Parr, *J. Chim. Phys., Phys.-Chim. Biol.* **1980**, *77*, 745.

59. W. J. van der Hart, E. Oosterveld, T. A. Molenaar-Langeveld, and N. M. M. Nibbering, *Org. Mass Spectrom.* **1989**, *24*, 59.

60. S. K. Pollack, B. C. Raine, and W. J. Hehre, *J. Am. Chem. Soc.* **1981**, *103*, 6308.

61. S. K. Pollack and W. J. Hehre, *J. Am. Chem. Soc.* **1977**, *99*, 4845.

62. J. Lee and J. J. Grabowski, *Chem. Rev.* **1992**, *92*, 1611.

63. J. J. Grabowski and S. J. Melly, *Int. J. Mass Spectrom. Ion Processes* **1987**, *81*, 147.

64. Y. Guo and J. J. Grabowski, *Int. J. Mass Spectrom. Ion Processes* **1990**, *97*, 253.

65. J. H. J. Dawson and K. R. Jennings, *J. Chem. Soc., Faraday Trans. 2* **1976**, *72*, 700.

66. J. J. Grabowski, in *Advances in Gas Phase Ion Chemistry*; N. G. Adams and L. M. Babcock, Eds.; JAI Press: Greenwich, 1992; Vol. 1.

67. K. M. Ervin, S. Gronert, S. E. Barlow, M. K. Gilles, A. G. Harrison, V. M. Bierbaum, C. H. DePuy, W. C. Lineberger, and G. B. Ellison, *J. Am. Chem. Soc.* **1990**, *112*, 5750.

68. Y. Guo and J. J. Grabowski, *J. Am. Chem. Soc.* **1991**, *113*, 5923.

69. A. P. Bruins, A. J. Ferrer-Correia, A. G. Harrison, K. R. Jennings, and R. K. Mitchum, *Adv. Mass Spectrom.* **1978**, *7*, 355.

70. P. O. Staneke, S. Ingemann, P. E. Eaton, N. M. M. Nibbering, and S. R. Kass, *J. Am. Chem. Soc.* **1994**, *116*, 6445.

71. J. Lee, P. K. Chou, P. Dowd, and J. J. Grabowski, *J. Am. Chem. Soc.* **1993**, *115*, 7902.

72. J. Zhao, P. Dowd, and J. J. Grabowski, *J. Am. Chem. Soc.* **1996**, *118*, 8871.

73. D. G. Leopold, A. E. S. Miller, and W. C. Lineberger, *J. Am. Chem. Soc.* **1986**, *108*, 1379.

74. L. A. Hammad and P. G. Wenthold, *J. Am. Chem. Soc.* **2000**, *122*, 11203.

75. P. G. Wenthold, J. Hu, and R. R. Squires, *J. Am. Chem. Soc.* **1994**, *116*, 6961.

76. P. G. Wenthold, J. Hu, B. T. Hill, and R. R. Squires, *Int. J. Mass Spectrom. Ion Processes* **1998**, *117*, 633.

77. P. G. Wenthold and R. R. Squires, *J. Am. Chem. Soc.* **1994**, *116*, 11890.

78. S. R. Kass and K. M. Broadus, *J. Phys. Org. Chem.* **2002**, *15*, 461.

79. K. M. Broadus and S. R. Kass, *J. Am. Chem. Soc.* **2000**, *122*, 10697.

80. K. M. Broadus and S. R. Kass, *J. Am. Chem. Soc.* **2001**, *123*, 4189.

81. R. N. McDonald, *Tetrahedron* **1989**, *45*, 3993.

82. M. Zhong, M. L. Chabinyc, and J. I. Brauman, *J. Am. Chem. Soc.* **1996**, *118*, 12432.

83. L. A. Hammad and P. G. Wenthold, *J. Am. Chem. Soc.* **2003**, *125*, 10796.

84. J. A. Paulino and R. R. Squires, *J. Am. Chem. Soc.* **1991**, *113*, 5573.

85. P. G. Wenthold and R. R. Squires, *J. Am. Chem. Soc.* **1994**, *116*, 6401.

86. C. H. DePuy, V. M. Bierbaum, L. A. Flippin, J. J. Grabowski, G. K. King, R. J. Schmitt, and S. A. Sullivan, *J. Am. Chem. Soc.* **1980**, *102*, 5012.

87. J. C. Poutsma, J. A. Paulino, and R. R. Squires, *J. Phys. Chem. A* **1997**, *101*, 5327.

88. J. C. Poutsma, J. J. Nash, J. A. Paulino, and R. R. Squires, *J. Am. Chem. Soc.* **1997**, *119*, 4686.

89. J. C. Poutsma, S. D. Upshaw, R. R. Squires, and P. G. Wenthold, *J. Phys. Chem. A* **2002**, *106*, 1067.

90. P. G. Wenthold, S. G. Wierschke, J. J. Nash, and R. R. Squires, *J. Am. Chem. Soc.* **1994**, *116*, 7378.

91. J. M. Riveros, S. Ingemann, and N. M. M. Nibbering, *J. Am. Chem. Soc.* **1991**, *113*, 1053.

92. P. G. Wenthold and W. C. Lineberger, *Acc. Chem. Res.* **1999**, *32*, 597.

93. K. M. Ervin, J. Ho, and W. C. Lineberger, *J. Chem. Phys.* **1989**, *91*, 5974.

94. M. S. Robinson, M. L. Polak, V. M. Bierbaum, C. H. DePuy, and W. C. Lineberger, *J. Am. Chem. Soc.* **1995**, *117*, 6766.

95. P. G. Wenthold, R. R. Squires, and W. C. Lineberger, *J. Am. Chem. Soc.* **1998**, *120*, 5279.

96. This aspect was partially responsible for the infamous case of methylene, in which the singlet - triplet splitting initially obtained by using photoelectron spectroscopy, $-82\,kJ/mol$, was substantially higher than the accepted value, $-38\,kJ/mol$ (reference 97). Because the triplet feature was an extended vibrational progression, it was difficult to distinguish the origin from a hot band, expecially given the high vibrational temperature of the ion. The advent of thermal and sub-thermal ion sources (reference 98) has to a large extent eliminated the problems of hot bands, but does not solve the problem of poor Franck–Condon overlap.

97. P. C. Engelking, R. R. Cordeman, J. J. Wendoloski, G. B. Ellison, S. V. ONeill, and W. C. Lineberger, *J. Chem. Phys.* **1981**, *74*, 5460.

98. D. G. Loopold, K. K. Murray, A. E. S. Miller, and W. C. Lineberger, *J. Chem. Phys.* **1985**, *83*, 4849.

99. P. G. Wenthold, J. Hu, R. R. Squires, and W. C. Lineberger, *J. Am. Chem. Soc.* **1996**, *118*, 475.

100. F. P. Clifford, P. G. Wenthold, W. C. Lineberger, G. B. Ellison, C. X. Wang, J. J. Grabowski, F. Vila, and K. D. Jordan, *J. Chem. Soc., Perkin Trans. 2* **1998**, 1015.

101. R. A. Jesinger and R. R. Squires, *Int. J Mass Spectrom.* **1999**, *185/186/187*, 745.

102. L. A. Hammad and P. G. Wenthold, *J. Am. Chem. Soc.* **2001**, *123*, 12311.

103. H. A. Lardin, J. J. Nash, and P. G. Wenthold, *J. Am. Chem. Soc.* **2002**, *124*, 12612.

104. J. Hu and R. R. Squires, *J. Am. Chem. Soc.* **1996**, *118*, 5816.

105. L. V. Slipchenko, T. E. Munsch, P. G. Wenthold, and A. I. Krylov, *Angew. Chem., Int. Ed.* **2004**, *43*, 742.

106. T. E. Munsch, L. V. Slipchenko, A. I. Krylov, and P. G. Wenthold, *J. Org. Chem.* **2004**, *69*, 5735.

107. S. Gronert and C. H. DePuy, *J. Am. Chem. Soc.* **1989**, *111*, 9253.

108. A. H. Zewail, *Faraday Discuss.* **1991**, *91*, 207.

109. D. M. Neumark, *Adv. Ser. Phys. Chem.* **2004**, *14*, 453.

110. P. G. Wenthold, D. A. Hrovat, W. T. Borden, and W. C. Lineberger, *Science* **1996**, *272*, 1456.

111. D. H. Paik, D.-S. Yang, I.-R. Lee, and A. H. Zewail, *Angew. Chem., Int. Ed.* **2004**, *43*, 2830.

112. A. Stolow, A. E. Bragg, and D. M. Neumark, *Chem. Rev.* **2004**, *104*, 1719.

113. D. W. Kohn and P. Chen, *J. Am. Chem. Soc.* **1993**, *115*, 2844.

114. S. Kato, R. Gareyev, C. H. DePuy, and V. M. Bierbaum, *J. Am. Chem. Soc.* **1998**, *120*, 5033.

115. H. J. Wörner and F. Merkt, *Angew. Chem., Int. Ed.* **2006**, *45*, 293.

116. R. Mabbs, K. Pichugin, and A. Sanov, *J. Chem. Phys.* **2005**, *122*, 174305/1.

117. T. Gilbert, R. Pfab, I. Fischer, and P. Chen, *J. Chem. Phys.* **2000**, *112*, 2575.

118. C.-W. Hsu and C. Y. Ng, *J. Chem. Phys.* **1994**, *101*, 5596.

119. Y.-S. Cheung, C.-W. Hsu, and C. Y. Ng, *J. Elect. Spect. Related Phenomena* **1998**, *97*, 115.

120. G. B. Ellison, G. E. Davico, V. M. Bierbaum, and C. H. DePuy, *Int. J. Mass Spectrom. Ion Proc.* **1996**, *156*, 109.

121. C.-W. Liang, C.-C. Chen, C.-Y. Wei, and Y.-T. Chen, *J. Chem. Phys.* **2002**, *116*, 4162.

122. J. C. Traeger, *Int. J. Mass Spectrom. Ion Processes* **1984**, *58*, 259.

123. C. Lifshitz, Y. Gotkis, J. Laskin, A. Iofffe, and S. Shaik, *J. Phys. Chem.* **1993**, *97*, 12291.

124. M. Hare, T. Emrick, P. E. Eaton, and S. R. Kass, *J. Am. Chem. Soc.* **1997**, *119*, 237.

125. A. Fattahi, R. E. McCarthy, M. R. Ahmad, and S. R. Kass, *J. Am. Chem. Soc.* **2003**, *125*, 11746.

126. L. A. Angel and K. M. Ervin, *J. Phys. Chem. A* **2004**, *108*, 8346.

127. H. Y. Afeefy, J. F. Liebman, and S. E. Stein, Neutral Thermochemical Data. In *NIST Chemistry WebBook, NIST Standard Reference Database Number 69*; June 2005 ed.; W. G. Mallard and P. J. Linstrom, Eds.; National Institute of Standards and Technology: Gaithersburg, MD, 20899, June 2005.

128. J. E. Bartmess, Negative Ion Energetics Data. In *NIST Chemistry WebBook, NIST Standard Reference Database Number 69*; February 2000 ed.; W. G. Mallard and P. J. Linstrom, Eds.; National Institute of Standards and Technology: Gaithersburg, MD, 20899, June, 2005.

129. M. Born, S. Ingemann, and N. M. M. Nibbering, *J. Am. Chem. Soc.* **1994**, *116*, 7210.

130. L. J. Chyall and R. R. Squires, *Int. J. Mass Spectrom. Ion Processes* **1995**, *149/150*, 257.

131. E. P. Clifford, P. G. Wenthold, W. C. Lineberger, G. A. Petersson, K. M. Broadus, S. R. Kass, S. Kato, C. H. DePuy, V. M. Bierbaum, and G. B. Ellison, *J. Phys. Chem. A* **1998**, *102*, 7100.

132. P. G. Wenthold, J. Hu, R. R. Squires, and W. C. Lineberger, *J. Am. Soc. Mass Spectrom.* **1999**, *10*, 800.

133. A. Fattahi, S. R. Kass, J. F. Liebman, M. A. R. Matos, M. S. Miranda, and V. M. F. Morais, *J. Am. Chem. Soc.* **2005**, *127*, 6116.

134. (a) D. R. Reed, M. C. Hare, A. Fattahi, G. Chung, M. S. Gordon, and S. R. Kass, *J. Am. Chem. Soc.* **2003**, *125*, 4643. (b) D. R. Reed, M. Hare, and S. R. Kass, *J. Am. Chem. Soc.* **2000**, *122*, 10689.

135. P. K. Chou and S. R. Kass, *J. Am. Chem. Soc.* **1991**, *113*, 697.

136. L. M. Pratt, A. Fattahi, and S. R. Kass, *European J. Mass Spectrom.* **2004**, *10*, 813.

137. W. J. Pietro, S. K. Pollack, and W. J. Hehre, *J. Am. Chem. Soc.* **1979**, *101*, 7126.

138. M. S. Gordon, J. A. Boatz, and R. Walsh, *J. Phys. Chem.* **1989**, *93*, 1584.

139. L. Ding and P. Marshall, *J. Am. Chem. Soc.* **1992**, *114*, 5754.

140. L. E. Gusel'nikov and N. S. Nametkin, *J. Organomet. Chem.* **1979**, *169*, 155.

141. J. M. Dyke, G. D. Josland, R. A. Lewis, and A. Morris, *J. Phys. Chem.* **1982**, *86*, 2913.

142. R. L. Hoenigman, S. Kato, V. M. Bierbaum, and W. T. Borden, *J. Am. Chem. Soc.* **2005**, *127*, 17772.

143. A. Fattahi, L. Lis, and S. R. Kass, *J. Am. Chem. Soc.* **2005**, *127*, 13065.

144. A.-M. C. Cristian, Y. Shao, and A. I. Krylov, *J. Phys. Chem. A* **2004**, *108*, 6581.

145. H. A. Lardin, "The Thermochemistry of Aromatic Radicals, Biradicals, and Triradicals," PhD Thesis, Purdue University, 2003.

146. M. K. Gilles, K. M. Ervin, J. Ho, and W. C. Lineberger, *J. Phys. Chem.* **1992**, *96*, 1130.

147. R. L. Schwartz, G. E. Davico, T. M. Ramond, and W. C. Lineberger, *J. Phys. Chem. A* **1999**, *103*, 8213.

148. M. K. Gilles, W. C. Lineberger, and K. M. Ervin, *J. Am. Chem. Soc.* **1993**, *115*, 1031.

149. R. F. Gunion, H. Köppel, G. W. Leach, and W. C. Lineberger, *J. Chem. Phys.* **1995**, *103*, 1250.

150. R. F. Gunion and W. C. Lineberger, *J. Phys. Chem.* **1996**, *100*, 4295.

151. K. M. Ervin and W. C. Lineberger, *J. Phys. Chem.* **1991**, *95*, 1167.

152. P. G. Wenthold, J. B. Kim, and W. C. Lineberger, *J. Am. Chem. Soc.* **1997**, *119*, 1354.

153. P. G. Wenthold and W. C. Lineberger, *J. Am. Chem. Soc.* **1997**, *119*, 7772.

154. G. E. Davico, R. L. Schwartz, T. M. Ramond, and W. C. Lineberger, *J. Am. Chem. Soc.* **1999**, *121*, 6047.

155. P. C. Engelking and W. C. Lineberger, *J. Chem. Phys.* **1976**, *65*, 4323.

156. M. J. Travers, D. C. Cowles, E. P. Clifford, G. B. Ellison, and P. C. Engelking, *J. Chem. Phys.* **1999**, *111*, 5349.

157. M. J. Travers, D. C. Cowles, E. P. Clifford, and G. B. Ellison, *J. Am. Chem. Soc.* **1992**, *114*, 8699.

158. T. R. Taylor, R. T. Bise, K. R. Asmis, and D. M. Neumark, *Chem. Phys. Lett.* **1999**, *301*, 413.

159. A. Kasdan, E. Herbst, and W. C. Lineberger, *J. Chem. Phys.* **1975**, *62*, 541.

160. G. E. Davico, R. L. Schwartz, and W. C. Lineberger, *J. Chem. Phys.* **2001**, *115*, 1789.

161. M. L. McKee and J. Michl, *J. Phys. Chem. A* **2002**, *106*, 8495.

162. T. A. Walter, C. Lifshitz, W. A. Chupka, and J. Berkowitz, *J. Chem. Phys.* **1969**, *51*, 3531.

163. J. T. Clay, E. A. Walters, J. R. Grover, and M. V. Wilcox, *J. Chem. Phys.* **1994**, *101*, 2069.

164. H. Schenk, H. Oertel, and H. Baumgärtel, *Ber. Bunsen-Ges. Phys. Chem.* **1979**, *83*, 683.

165. S. G. Lias, Z. Karpas, and J. F. Liebman, *J. Am. Chem. Soc* **1985**, *107*, 6089.

166. J. Vogt and J. L. Beauchamp, *J. Am. Chem. Soc.* **1975**, *97*, 6682.

167. T. J. Buckley, R. D. Johnson, III, R. E. Huie, Z. Zhang, S. C. Kuo, and R. B. Klemm, *J. Phys. Chem.* **1995**, *99*, 4879.

168. J. S. Shapiro and F. P. Lossing, *J. Phys. Chem.* **1968**, *72*, 1552.

169. K. Rademann, H.-W. Jochims, and H. Baumgärtel, *J. Phys. Chem.* **1985**, *89*, 3459.

170. D. W. Kohn, E. S. J. Robles, C. F. Logan, and P. Chen, *J. Phys. Chem.* **1993**, *97*, 4936.

171. J. M. Ajello, W. T. Huntress, Jr., and P. Rayermann, *J. Chem. Phys.* **1976**, *64*, 4746.

172. B. A. Levi, R. W. Taft, and W. J. Hehre, *J. Am. Chem. Soc.* **1977**, *99*, 8454.

173. M. Born, S. Ingemann, and N. M. M. Nibbering, *Int. J. Mass Spectrom.* **2000**, *194*, 103.

174. H. E. K. Matimba, A. M. Crabbendam, S. Ingemann, and N. M. M. Nibbering, *J. Chem. Soc., Chem. Comm.* **1991**, 644.

175. M. Moini and G. Leroi, *J. Phys. Chem.* **1986**, *90*, 4002.

CHAPTER 6

Reactive Intermediates
in Combustion

JOHN K. MERLE

National Institute of Standards and Technology, Gaithersburg, MD

CHRISTOPHER M. HADAD

Department of Chemistry, The Ohio State University, Columbus, Ohio

6.1. INTRODUCTION

The fields of combustion and atmospheric chemistry are intimately connected. Both of these fields are dominated by the reactivity of radical intermediates. The oxidation (combustion) of fossil fuels and their derivatives converts chemical energy into heat

Reviews of Reactive Intermediate Chemistry. Edited by Matthew S. Platz, Robert A. Moss, Maitland Jones, Jr.

energy that can be used, for example, to power electrical generators and automobiles. Under ideal conditions, the complete combustion of a hydrocarbon-based fuel will consume molecular oxygen (the oxidant) and produce carbon dioxide and water. For example, the complete oxidation of one methane molecule consumes two units of molecular oxygen and yields one CO_2 and two water molecules as products.

$$CH_4 + 2\ O_2 \rightarrow CO_2 + 2\ H_2O$$

Ideal combustion, therefore, requires a well-mixed "reactor" with fuel and oxidant as well as a nonlimiting amount of oxidant for complete combustion to occur. However, the conditions for combustion are rarely ideal, either by design or by engineering difficulties. Moreover, typical fuels derived from crude oil or coal may contain a variety of chemical functionalities and, in some cases, noncombustible impurities/components, thereby resulting in the incomplete consumption of the fuel, as well as the emission of oxidation intermediates and particles into the atmosphere.[1-3] Typical emissions result from burning solid fossil fuels and solid waste in industrial boilers or incinerators as well as in gasoline- or diesel-powered engines and include volatile organic compounds (VOCs), carbon particles, NO_x, and SO_x; these emissions are commonly considered to be anthropogenic (man-made) sources of pollution. However, some of these emissions may also be derived from natural (biogenic) sources, such as forest fires and volcanic eruptions, VOC emissions from the leaf surfaces of plants, and the degradation of organic material in natural soils, sediments, and waters (e.g., swamps).

Anthropogenic pollution can reach significant levels in highly developed regions, resulting in regionally localized smog (soot or photochemical smog) and correspondingly high levels of ozone (a harmful oxidant). VOCs with low atmospheric reactivity can potentially diffuse over significant distances within the troposphere (lower atmosphere) or be incorporated into atmospheric aerosols, where they can undergo wet or dry deposition and thereby affect terrestrial life cycles.

As previously stated, the chemistry in both combustion and atmospheric environments is dominated by the reactivity of radical intermediates (often called free radicals). Radicals are highly reactive species that contain an unpaired electron centered on an atom, thereby violating the octet rule. Figure 6.1 shows examples of prototypical organic radicals on sp^3, sp^2, and sp hybridized carbons as the radical centers of methyl ($H_3C^\bullet$), ethenyl (vinyl, $H_2C{=}CH^\bullet$), and ethynyl (acetylenyl, $HC{\equiv}C^\bullet$) radicals, respectively. Radicals with an unpaired electron on an atom adjacent to a π bonding network can be stabilized by delocalization of the unpaired electron; as a result, allylic ($H_2C{=}CH{-}CH_2^\bullet$), benzylic ($C_6H_5CH_2^\bullet$), and cyclopentadienyl ($c\text{-}C_5H_5^\bullet$) radicals possess enhanced stability over carbon-centered radicals on saturated molecules. The unpaired electron can adopt one of two degenerate electronic states resulting from the two possible spin angular momentum alignments for an electron (s = 1/2 or −1/2). Electron paramagnetic resonance (EPR) is an important tool for the detection and characterization of stable radicals in the condensed phase.[4] In EPR experiments, the production of a radical species inside a magnetic field removes the degeneracy between the two degenerate spin states, and absorption of electromagnetic radiation, typically in the microwave region, occurs at an energy equivalent to the energy separation between the two spin states.

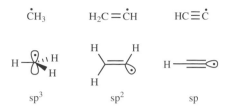

Figure 6.1. Examples of radicals for sp³, sp², and sp hybridized carbon centers as the methyl, ethenyl (vinyl), and ethynyl (acetylenyl) radicals.

Radicals can be formed by either thermal or photochemical processes. In combustion, radical chemistry is primarily initiated by providing sufficient thermal energy for unimolecular, homolytic bond scission to occur; in the atmosphere, radical chemistry is initiated via photolytic processes where high energy (ultraviolet, UV) sunlight breaks chemical bonds. In this chapter, we will discuss the reactive radical intermediates that are responsible for initiating and driving the chemistry of combustion and reactions in the atmosphere. Experimental and theoretical methods commonly utilized to study and to refine our understanding of these processes will also be introduced.

6.2. COMBUSTION CHEMISTRY

Chemical combustion is initiated by the oxidation or thermal decomposition of a fuel molecule, thereby producing reactive radical species by a chain-initiating mechanism. Radical initiation for a particular fuel/oxygen mixture can result from high-energy collisions with other molecules (M) in the system or from hydrogen-atom abstraction by O_2 or other radicals, as expressed in reactions 6.1–6.3:

$$R-H + M \rightarrow R^\bullet + H^\bullet + M^* \tag{6.1}$$

$$R-R' + M \rightarrow R^\bullet + R'^\bullet + M^* \tag{6.2}$$

$$R-H + O_2 \rightarrow R^\bullet + HO_2^\bullet \tag{6.3}$$

The particular mechanism responsible for the initiation reaction is strongly dependent on the fuel's chemical functionality, as well as system temperature and pressure. Typical C–C and C–H bond dissociation energies (enthalpy) are between 320 and 440 kJ/mol. Therefore, reactions 6.1 and 6.2 require considerable thermal energy and pressure to acquire the necessary force and frequency for homolytic bond-breaking collisions and thereby obtain a reasonable probability of occurrence. Such reaction mechanisms are prominent at only higher temperatures (typically T > 1000 K). The specific bond that is broken is strongly dependent on the relative strength of the disparate chemical bonds contained within the fuel molecule. Table 6.1 provides a list of some homolytic bond dissociation energies (BDEs, ΔH_{298}) that are important when considering radical-initiation mechanisms or estimating the temperature dependence of a radical-initiation reaction. Hydrogen-atom abstraction initiations involving O_2 (reaction 6.3) are possibilities only in low and moderate temperature regimes where insufficient thermal energy is available to effect unimolecular, homolytic bond cleavage; however, more efficient hydrogen-atom abstractions from the fuel can occur with $^•OH$, and even with very low concentrations of this reactive radical.

Individually, the radical-initiation reactions 6.1–6.3 are too slow and thermodynamically disfavored, due to their significant endothermicities, to effect combustion on their own. For example, consider the initiation reactions of 6.1–6.3 with ethane (CH_3–CH_3) as the fuel:

$$CH_3CH_2-H + M \rightarrow CH_3CH_2^• + H^• + M^* \quad \Delta H_{298} = +423\,kJ/mol \quad (6.4)$$

$$CH_3-CH_3 + M \rightarrow 2\ ^•CH_3 + M^* \quad\quad\quad \Delta H_{298} = +377\,kJ/mol \quad (6.5)$$

$$CH_3CH_2-H + O_2 \rightarrow CH_3CH_2^• + HO_2^• \quad \Delta H_{298} = +215\,kJ/mol \quad (6.6)$$

The C–C and C–H BDEs for ethane are 377.0 and 423.0 kJ/mol, respectively, and the H–O BDE of hydroperoxyl ($HO_2^•$) radical is only 207.5 kJ/mol. The large reaction endothermicities for reactions 6.4 and 6.5 highlight the unlikeliness of their occurrence at lower temperatures. In pyrolytic (heat, but no oxidant) or high fuel/oxidant ratio conditions, the endothermicities indicate that ethane and other alkanes will

TABLE 6.1. Some important bond dissociation energies (ΔH_{298} kJ/mol) for radical initiation in combustion.

Bond	BDE kJ/mol	Ref.	Bond	BDE kJ/mol	Ref.
H–H	436.0	a	**C–C bonds**		
HO–H	497.1	a	H_3C–CH_3	377.0	a
HO–OH	214.1	b	H_3CCH_2–CH_3	372.4	a
HOO–H	367.4	a	$(H_3C)_2CH$–CH_3	370.7	a
OO–H	207.5	c	$(H_3C)_3C$–CH_3	366.1	a
C–H bonds			H_2C=CH_2	728.4	a
H_3C–H	439.3	a	HC≡CH	965.2	a
H_3CCH_2–H	423.0	a	C_6H_5–CH_3	433.0	a
$(H_3C)_2CH$–H	412.5	a	H_2CCH–CH_3	424.3	a
$(H_3C)_3C$–H	403.8	a	H_2CCHCH_2–CH_3	320.1	a
H_2C=CH–H	463.2	a	$C_6H_5CH_2$–CH_3	324.7	a
HC≡C–H	557.8	a	**C(=O)–H bonds**		
C_6H_5–H	472.4	a	$HC(=O)$–H	368.8	a
H_2C=$CHCH_2$–H	371.5	a	$H_3CC(=O)$–H	374.0	a
$C_6H_5CH_2$–H	375.7	a	**CO–H**		
$HOCH_2$–H	402.1	a	H_3CO–H	437.6	a

[a]Blanksby, S. J.; Ellison, G. B. *Acc. Chem. Res.* **2003**, *36*, 255.
[b]NIST Webbook webpage, http://webbook.nist.gov, visited 01/20/06, June 2005 release.
[c] Howard, C. J. *J. Am. Chem. Soc.* **1980**, *102*, 6937.

undergo C–C homolytic bond scission in favor of C–H bond scission because C–C bond cleavage is ≈36 kJ/mol less endothermic. While the lower reaction endothermicity of reaction 6.6 seems to favor radical initiation via H-atom abstraction over homolytic bond scissions, $HO_2^•$ becomes unstable at elevated temperatures, such that other mechanisms will dominate, thereby minimizing its significance. We will see, however, that this reaction plays an important role at lower temperatures.

Subsequent to initiation, a pool of reactive intermediates is generated via chain-propagation reactions where a radical, formed from an initiation event, reacts with either fuel or oxidant molecules to yield a new radical for further chain propagation. When chain-propagation rates are greater than chain-terminating rates, a radical pool develops. A large radical pool increases the reactivity of the system and increases the probability that a radical chain-branching reaction can occur. In a chain-branching reaction, the number of radical species produced by that specific step is more than that of the reactants for the step. It is the rapid enhancement of radicals brought about by chain-branching events in the system that leads to the uncontrolled reactivity of oxidation or of an explosion.

In a spark-ignition engine, as for a typical gasoline-powered automobile, there are four stages: (1) filling of the combustion chamber with fuel and air (oxidant); (2) compression of the fuel-air mixture; (3) spark-initiated ignition of the fuel; and (4) removal of the burnt gases. The timing of the fuel's combustion, relative to the compression of the fuel/air mixture and then the ignition from the spark plug in a combustion chamber, is critical for (a) the efficient usage of the fuel; (b) maximizing power output for each cycle of stages 1 through 4; and also (c) minimizing engine damage from premature ignition (engine knock). Gasoline fuels are rated with an octane number that is related to the fuel's ability to resist spontaneous ignition (knock) in the compression stage prior to consumption by the flame from the spark ignition. If the fuel ignites too early in the compression cycle, often called auto-ignition, then engine knock is generated. Each fuel component has an octane rating: *iso*-octane (2,2,5-trimethylpentane) defines an octane rating of 100 and is not susceptible to auto-ignition (knock), while *n*-heptane has an octane rating of 0 and is susceptible to auto-ignition. Each fuel component (or mixture) is then related to its ability to auto-ignite under a specific compression ratio and compared with *iso*-octane (assigned as 100) and *n*-heptane (assigned as 0) as reference fuels.[5]

In the following sections, the types of radical chain-propagating and chain-branching reactions responsible for combustion under both low- and high-temperature conditions will be discussed.

6.2.1. Low-Temperature Mechanisms

Semenov proposed the first general mechanism for low-temperature (≈600 K or less) combustion of a hydrocarbon.[6,7] In the Semenov mechanism, following initiation by hydrogen-atom abstraction from an alkane (R—H) to yield an organic radical (R•), as depicted in reaction 6.3, radical chain-propagation and chain-branching reactions are proposed to proceed by the following mechanistic sequence:

$$R^\bullet + O_2 \to \text{alkene} + HO_2{}^\bullet \tag{6.7}$$

$$R^\bullet + O_2 + M \to RO_2{}^\bullet + M \tag{6.8}$$

$$RO_2{}^\bullet + R-H \to RO_2H + R^\bullet \tag{6.9}$$

$$R'CH_2O_2{}^\bullet \to R'CH(=O) + HO^\bullet \tag{6.10a}$$

$$R'R''CHO_2{}^\bullet \to R'CH(=O) + R''O^\bullet \tag{6.10b}$$

$$HO_2{}^\bullet + R-H \to H_2O_2 + R^\bullet \tag{6.11}$$

$$RO_2H \to RO^\bullet + {}^\bullet OH \tag{6.12}$$

$$R'CH(=O) + O_2 \to R'C^\bullet(=O) + HO_2{}^\bullet \tag{6.13}$$

Reactions 6.7–6.11 are the chain-propagating steps, responsible for creating and maintaining the pool of radicals. Of these initially proposed reactions by Semenov, not all are of equal importance in combustion environments. In particular, the formation of alkylperoxy $RO_2{}^\bullet$ radicals (reaction 6.8) is very important, but the subsequent H-atom abstraction by the $RO_2{}^\bullet$ radicals (reaction 6.9) is slow. For primary alkylperoxy ($R'CH_2O_2{}^\bullet$) radicals, reaction 6.10a can proceed by an intramolecular H-atom abstraction from the α-CH_2 group, followed by O–O bond scission to generate an aldehyde and hydroxyl radical as products. For secondary alkylperoxy ($R'R''CHO_2{}^\bullet$) radicals, an O–C bond (for the $R''O^\bullet$ product) can be formed prior to O–O bond scission, as shown in reaction 6.10b. Also, H-atom abstraction by $HO_2{}^\bullet$ from the fuel (reaction 6.11) is less important than the competitive reaction of 2 equivalents of $HO_2{}^\bullet$ to react and then generate H_2O_2 and O_2 as products — this self-destructive reaction of $HO_2{}^\bullet$ radicals is very important in combustion systems. Reactions 6.12 and 6.13 are degenerate chain-branching reactions, in which the nonradical alkylhydroperoxide (RO_2H) and aldehyde ($R'CH(=O)$) components, respectively, decompose in a unimolecular manner or react to yield two distinct radical intermediates. The degenerate chain-branching reactions provide an exponential increase in the number of radicals introduced into the system and can lead to a fuel's ignition.

Reactions 6.9 and 6.10, in the Semenov mechanism, are responsible for producing species capable of undergoing chain branching. However, it is the competition between alkene and alkylperoxy ($RO_2{}^\bullet$) radical generation, via reactions 6.7 and 6.8, that determines whether a chain-branching event will occur at low temperatures. The creation of an alkene and $HO_2{}^\bullet$, as in reaction 6.7, serves essentially as a radical chain-terminating step since the reactivity of $HO_2{}^\bullet$ is insignificant at low temperatures. Indeed, one would expect that as the temperature rises, the time to ignition of the fuel would decrease, and such a decrease in the ignition delay[7] as a function of temperature is shown in Figure 6.2a. However, at moderate temperatures, the ignition delay can actually increase for specific fuels (Fig. 6.2c), and this effect is due to the involvement of the concerted elimination of $RO_2{}^\bullet$ to form an alkene and $HO_2{}^\bullet$, a radical chain-terminating process that inhibits the propagation of combustion.

The yield of alkylperoxy ($RO_2{}^\bullet$) radicals in reaction 6.8, on the contrary, provides a route for chain branching to take place and is both pressure and temperature dependent.[7,9] The pressure dependence exists because the energy-rich alkylperoxy ($RO_2^{\bullet}$) radical formed in the O_2 addition step must be stabilized by collisions with other species to have a significant lifetime. For example, the formation of *tert*-butylperoxy radical (($CH_3)_3C-O_2^{\bullet}$) from *tert*-butyl radical and O_2 is exothermic by 159 kJ/mol.[8] When pressures are sufficiently high, collisional stabilization of the newly formed alkylperoxy radical will minimize both unimolecular decomposition to bimolecular products and the return to reactants, as in the following:

$$R^{\bullet} \ + \ O_2 \ \rightleftharpoons \ RO_2^{\bullet*} \ + \ M \ \rightleftharpoons \ RO_2^{\bullet} \ + \ M^*$$

$$\Updownarrow$$

Products

Thermally, alkylperoxy ($RO_2^{\bullet}$) radicals can become unstable as temperatures approach ≈ 600 K, since equilibrium would favor the alkyl radical and O_2 as reactants.[9] As temperatures increase further, high temperature oxidation mechanisms will begin to prevail.

For longer-chain alkylperoxy ($RO_2^{\bullet}$) radicals, starting with *n*-propylperoxy ($CH_3CH_2CH_2O_2^{\bullet}$) radical, a facile intramolecular 1,5-H-transfer through a six-membered ring transition state (theoretical $E_a = \approx 96$ kJ/mol[10,11]) will yield a new hydroperoxyalkyl radical ($^{\bullet}CH_2CH_2CH_2O_2H$), and the "rearranged" hydroperoxyalkyl radical is often denoted $Q^{\bullet}OOH$. As noted below, further reaction of a hydroperoxyalkyl radical with O_2 provides a low-energy pathway for the generation of alkylhydroperoxides, as a complement to reaction 6.9, allowing for an alternate

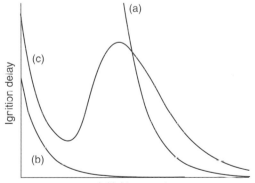

Figure 6.2. Typical ignition delay of an alkane fuel as a function of the initial mixture's temperature. Three different kinetic models are shown: (a) High temperature chemistry only; that is, no peroxy radical chemistry. (b) Same as (a), but the "$Q^{\bullet}OOH$" chain-branching channel of the peroxy radicals has been considered. (c) Same as (b), but the concerted elimination of $RO_2^{\bullet}$ to alkene $+ HO_2^{\bullet}$ has been considered. (Figure courtesy of Timothy Barckholtz, ExxonMobil Research and Engineering.)

pathway to degenerate chain-branching.[12,13] The following scheme shows this sequence of conversions for n-propylperoxy ($CH_3CH_2CH_2O_2{}^\bullet$) radical:

$$CH_3CH_2CH_2O\dot{O} \longrightarrow \dot{C}H_2CH_2CH_2OOH \xrightarrow{\;O_2\;} \dot{O}_2CH_2CH_2CH_2OOH$$

$$\longrightarrow HO_2CH_2CH_2\dot{C}HOOH \longrightarrow HO_2CH_2CH_2CH(=O) + \dot{O}H$$

Recently, however, an alternative mechanism for the formation of unreactive alkenes and $HO_2{}^\bullet$ has been elucidated by theoretical methods in which the reaction proceeds via a concerted 1,4-H-atom abstraction and $HO_2{}^\bullet$ elimination mechanism from alkylperoxy radicals with two or more carbons. This process can compete with the mechanisms above to yield chain-terminating species.[14] This concerted mechanism is estimated theoretically to have a reaction barrier $\approx$36 kJ/mol lower than the stepwise 1,4-H-atom transfer and $HO_2{}^\bullet$ elimination for the generation of propene and $HO_2{}^\bullet$ from n-propylperoxy radical.[10] Furthermore, despite a lower activation barrier for the 1,5-isomerization ($\Delta H^{\ddagger}$(0 K) = +100 kJ/mol, relative to the n-propylperoxy radical), the subsequent steps for $^\bullet CH_2CH_2CH_2O_2H$ decomposition proceed through significantly higher energetic barriers, thereby rendering its unimolecular decomposition products unlikely. Experiments have shown a bimodal production of $HO_2{}^\bullet$ with increasing temperature in which a prompt source is attributed to reaction via the activated complex and a later thermal source at higher temperatures.[15] Since combustion at low temperatures is highly dependent on the character of the fuel molecule and its propensity to yield chain-branching events, low-temperature combustion is often called auto-ignition. One of the most important auto-ignition events occurs in automobiles and can lead to engine knock and can result in engine damage as well as poor fuel efficiency.[13]

6.2.2. High-Temperature Mechanisms

At higher temperatures (T >1000 K), enough thermal energy becomes available to allow species to traverse the high-activation barriers of a number of radical-initiation reactions which were insurmountable at lower temperatures, such as unimolecular decompositions by reactions 6.1 and 6.2. In general, $HO_2{}^\bullet$ and peroxy radicals become increasingly unstable at higher temperatures, causing a shift of equilibrium toward reactants in O_2 abstraction and addition steps, and thereby favoring alternate mechanisms.

The high-temperature creation of hydrogen atom or alkyl radicals is dependent on the strength of the chemical bonds contained in the fuel molecule. The aldehydic C–H bond, for example, has a relatively low bond dissociation energy ($\approx$370 kJ/mol, Table 6.1) and can readily provide the initial H atom. On the contrary, aromatic sp^2 C–H bonds have bond dissociation energies of $\approx$470 kJ/mol and require higher temperatures to yield hydrogen atoms via a unimolecular cleavage. Typical C–H bond strengths for the alkanes range from 404 to 440 kJ/mol (Table 6.1), with alkane BDEs increasing in the order of tertiary, secondary, primary, and methane, respectively.

Similarly, C–C alkane bond strengths range from 366 to 377 kJ/mol, with increasing BDEs in the order of tertiary, secondary, and ethane. In fact, at elevated temperatures, unimolecular C–C bond scission (via thermolysis) becomes more important than C–H bond scission for the initiation events due to their generally lower bond strengths. Some exceptions include species such as propene ($H_2C=CH-CH_3$) and toluene ($C_6H_5CH_3$) for which loss of an H atom results in a resonance-stabilized intermediate, such as allylic ($H_2C=CH-CH_2^{\bullet}$) and benzylic ($C_6H_5CH_2^{\bullet}$) radicals, respectively, due to delocalization of the unpaired electron, thereby making C–H bond scission competitive with C–C bond scission. Also for the alkyl radicals, cleavage reactions can also occur, and many alkyl radicals will undergo a β-scission fragmentation to form an alkene and smaller alkyl radical.

After radical initiation of an alkane fuel, chain branching is immediately accessible via new high-temperature pathways with O_2:

$$H^{\bullet} + O_2 \rightarrow O(^3P) + HO^{\bullet} \tag{6.14}$$

$$R^{\bullet} + O_2 \rightarrow O(^3P) + RO^{\bullet} \tag{6.15}$$

Reactions 6.14 and 6.15 are not accessible at lower temperatures; instead, the formation of $HO_2^{\bullet}$ or $RO_2^{\bullet}$ radicals is preferred. In fact, it is the competition between reactions 6.14 or 6.15 with the thermal stabilities of $HO_2^{\bullet}$ or $RO_2^{\bullet}$ radicals, respectively, that determines the temperature at which the high- and low-temperature combustion regimes exist for a particular fuel. As noted above, 2 equivalents of $HO_2^{\bullet}$ will react to generate H_2O_2 and O_2 as products. Also, H_2O_2 can thermally dissociate to form two $^{\bullet}OH$ radicals, and this reaction is temperature dependent—H_2O_2 is stable up to 800–900 K, but decomposes effectively around 1100 K. Hydroxyl radical is an efficient hydrogen-atom abstracting agent and will propagate subsequent H-atom abstraction and radical-addition reactions.

Further oxidation of an alkoxy radical ($RO^{\bullet}$), via H-atom abstraction at the carbon adjacent to the oxygen's radical center, leads to the formation of an aldehyde.

Given the low bond strength (~370 kJ/mol) of the aldehydic C–H bond relative to the very strong (497 kJ/mol) O–H bond in water, facile H-atom abstraction could occur, followed by expulsion of CO and formation of a smaller alkyl radical. However, such bimolecular pathways between $RO^{\bullet}$ and $^{\bullet}OH$ radicals are unlikely in a propagating flame due to the relative probability of encountering both transient species; instead, $RO^{\bullet}$ radicals can undergo a unimolecular fragmentation to form the corresponding $R'CH(=O)$ aldehyde and hydrogen atom as products. The $R'CH(=O)$ aldehyde can subsequently react with $^{\bullet}OH$, and then lose CO, to eventually yield a completely oxidized alkane by the following sequence of reactions.

$$RO^\bullet \rightarrow R'CH(=O) + H^\bullet$$

$$HO^\bullet + R'CH(=O) \rightarrow R'C(=O)^\bullet + H_2O$$

$$R'C(=O)^\bullet \rightarrow R'^\bullet + CO$$

Unsaturated hydrocarbons and aromatic molecules have much stronger sp² C–H bonds (typically >460 kJ/mol) and even stronger carbon–carbon bonds due to their greater bond order (Table 6.1). Aromatic compounds are used in gasoline formulations to increase resistance to auto-ignition, thereby increasing the fuel's octane rating. In the high-temperature oxidation of benzene (C_6H_6), phenyl ($C_6H_5^\bullet$) radical will be generated by the homolytic cleavage of the sp² C–H bond. At high temperatures, phenyl radical will react with O_2 (reaction 6.15) to yield phenoxy ($C_6H_5O^\bullet$) radical and oxygen (³P) atom.[16] Phenoxy radical is capable of delocalizing the unpaired electron, rendering it a relatively stable radical. Phenoxy radical decomposes unimolecularly by losing a CO molecule via expulsion and a concomitant ring contraction to yield cyclopentadienyl (c-$C_5H_5^\bullet$) radical, which can undergo a repeat of the O_2-addition/CO expulsion sequence.

Under rich fuel conditions, that is, a high fuel-to-air relative stoichiometry, the cyclopentadienyl (c-$C_5H_5^\bullet$) radical is proposed to be an integral intermediate in the formation of polycyclic aromatic hydrocarbons (PAHs) and also soot formation as there is insufficient O_2 for complete oxidation, and cyclopentadienyl radical can react further with other components from the fuel. CO expulsion from cyclopentadienonyl ($C_5H_5O^\bullet$), however, does not involve a ring contraction, due to the high ring strain of the cyclobutenyl radical, so instead generates the 1,3-butadien-1-yl (CH_2=CH−CH=CH$^\bullet$) radical.

Under low-temperature combustion and atmospheric conditions, the phenylperoxy ($C_6H_5O_2^\bullet$) radical is likely to play an important role in the decomposition of phenyl ($C_6H_5^\bullet$) radical. Yu and Lin[17] successfully performed kinetic studies to determine the rate of reaction of phenyl radical with O_2, using cavity-ring down (CRD) spectroscopy, and detected phenylperoxy radical at temperatures as high as 473 K. Semiempirical molecular orbital theory calculations were utilized by Carpenter to elucidate possible decomposition pathways for phenylperoxy radical leading to cyclopentadienyl radical and CO_2. This work identified a novel pathway, highlighting a spirodioxiranyl radical intermediate that could subsequently form the thermodynamically stable, seven-membered ring, 2-oxepinoxy radical (**1**), with an enthalpic barrier of $\approx$109 kJ/mol.[18] Similar dioxarinyl radicals have been computed by Mebel et al. to be involved in the decomposition of the ethenylperoxy (vinylperoxy, $H_2C=CHO_2{}^\bullet$) radical, which was seen in the oxidative decomposition of ethene.[19]

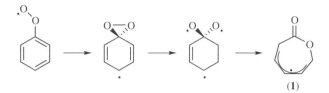

(**1**)

Barckholtz et al.[20] and Fadden et al.[21] have verified the energetics of this pathway using density functional theory (DFT) and high-level ab initio calculations; the spirodioxiranyl radical and a high-energy triradical intermediate were shown to offer the most viable reaction path leading to 2-oxepinoxy radical (**1**). Initial experimental methods to provide a thermodynamic estimate of the heat of formation of 2-oxepinoxy radical (**1**) have been reported by Kroner et al.[22] Furthermore, a low-temperature pathway leading from **1** to form cyclopentadienyl radical was shown to be viable; this route would reproduce some of the experimental products[23] observed in low-temperature combustion of benzene via a ring-opening and subsequent acyclic decomposition.[24] By this decomposition route, the unimolecular decomposition of 2-oxepinoxy radical (**1**) leads to the formation of ethenyl ($H_2C=CH^\bullet$) radical, and ethyne ($HC\equiv CH$), and two CO units.

For the aza (nitrogen-containing) analogs of benzene—that is, pyridine (c-C_5H_5N), pyridazine (1,2-diazabenzene c-$C_4H_4N=N$), and pyrimidine (1,3-diazabenzene, c-C_3H_3NCHN)—the homolytic C–H bond dissociation energies to form the corresponding sp^2 aryl radicals are smaller for the C–H position adjacent to the N atom than for a C–H bond in benzene.[25] Therefore, it is more facile to form the carbon-centered aryl radicals in the six-membered ring azabenzenes. However, in contrast to benzene's oxidation pathway, theoretical calculations indicate that dissociation of the arylperoxy intermediate to form the corresponding aryl radical and O_2 is favored relative to unimolecular O-atom expulsion to form the aryloxy radical.[26] On the contrary, arylperoxy radicals derived from the five-membered ring heterocycles, such as furan (c-C_4H_4O), pyrrole (c-C_4H_4NH), thiophene (c-C_4H_4S), and oxazole (c-C_2H_2NCHO), are predicted to favor unimolecular loss of O atom relative to loss of O_2.[27] However, the C–H bond dissociation energies in the five-membered ring heterocycles are, in general, larger than that for benzene, so it is more difficult to break the required C–H bond in the heterocycle.[25]

6.3. GENERATION OF POLLUTANTS FROM COMBUSTION

6.3.1 PAH and Soot Formation

One of the more significant classes of compounds resulting from and emitted by combustion sources include polycyclic aromatic hydrocarbons (PAHs); these species serve as nuclei for the formation of soot particles. Past studies have concluded that 85% of airborne PAHs are in the form of particles less than 5 μm in diameter.[28] Particles of this size can readily be inhaled into the respiratory airways and enter the lungs. PAHs are known inducers of the CYP1 family of cytochrome P450 enzymes;[29,30] CYP1A2 levels have been identified as biomarkers for PAH exposure and have also provided a link to increased risk of prostate, breast, and bladder cancers.[29–31]

A topic of unsettled debate involves determining the dominant reaction mechanism for the initial seeding of a benzene unit under pyrolytic or sooting flame conditions, a step that is important to allow further PAH formation.[32–36] Sooting occurs primarily when a system lacks sufficient oxygen to completely oxidize all of the fuel present. Nonstoichiometric fuel/oxygen mixtures result either by intentional design or insufficient mixing of fuel and oxidant. Much of the pioneering work into the benzene-forming mechanism has been based on experimental and modeling studies for acetylene (ethyne, $HC\equiv CH$) oxidation, as acetylene is thought to be an essential species in the formation and growth of PAHs. The most accepted mechanisms for the generation of benzene in acetylene or hydrocarbon flames include the reactions of unsaturated C_4 and C_3 molecules with acetylene. When acetylene adds to either 1,3-butadien-1-yl ($^{\bullet}CH=CH-CH=CH_2$) or buta-1-en-3-yn-1-yl ($^{\bullet}CH=CH-C\equiv CH$) radicals, unsaturated C_6 acyclic radials are formed and these can undergo cyclization to yield a benzene and phenyl radical, respectively. These reactions are reminiscent of the processes involved in the Bergman cyclization.[37]

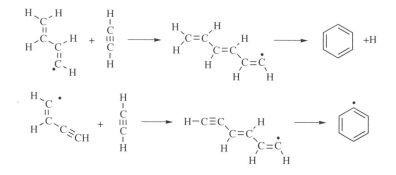

Alternatively, two propargyl radicals ($^{\bullet}CH_2-C\equiv CH$) can combine and undergo H-atom transfers, followed by cyclization, to yield benzene.

In aromatic combustion flames, cyclopentadienyl radicals ($c\text{-}C_5H_5^{\bullet}$) can be precursors for PAH formation.[36,40] At high temperatures, benzene is oxidized by reaction with an oxygen molecule to yield phenylperoxy ($C_6H_5O_2^{\bullet}$) radical, via the initial formation of the phenyl radical (by C–H bond cleavage) and then the rapid addition of O_2 (reaction 6.16). After expulsion of CO from phenylperoxy radical, a resonance-stabilized cyclopentadienyl radical ($c\text{-}C_5H_5^{\bullet}$) is formed (reaction 6.16).

Cyclopentadienyl radical can combine with methyl radicals to form fulvene, along with a subsequent loss of H atom. Fulvene may also be formed by reactions of the *iso* isomer of C_4H_5 with acetylene.[35] Once fulvene adds an H atom, subsequent rearrangement to benzene has been shown to be thermodynamically favorable using quantum chemical methods.[36] Cyclopentadienyl radicals may also combine with each other and then rearrange to form naphthalene.[36,38,40]

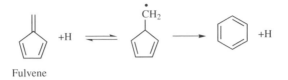

Fulvene

Once the initial benzene molecule is formed, iterations of a sequence of hydrogen-atom abstractions followed by acetylene (C_2H_2) additions result in PAH formation and growth.[7,32] In fact, several intermediates consistent with this mechanism have been isolated in benzene oxidation studies, including phenylacetylene, vinylbenzene, and naphthalene. Frenklach and Wang have delineated a mechanism for this growth, and provided a name as the H-abstraction-C_2H_2-addition (HACA) mechanism.[32,39] Figure 6.3 shows the successive H-atom abstraction/acetylene addition steps that occur

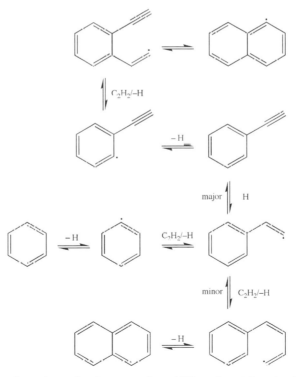

Figure 6.3. Reaction scheme for the successive addition of acetylene to phenyl radical to yield a PAH (naphthalene).

to form naphthalene from benzene. Further growth to larger PAHs can be attained via the HACA mechanism, as well as through PAH–PAH radical-addition reactions.[40]

Many PAHs also include cyclopenta (c-C_5) units as well as benzene units. The integration of cyclopenta units into PAH yields nonplanar PAHs, such as corannulene, which, as the simplest curved PAH, allows for formation of the ball-shaped fullerenes.

Corannulene

Vinylidene ($R_2C{=}C{:}$) carbene intermediates have been proposed to be involved in forming cyclopenta-fused PAH from an ethynyl-substituted PAH via a rearrangement mechanism.[41,42] Following acetylene addition to naphthalene at the 1 position and H-atom loss, 1-ethynylnaphthalene can be formed, in a pathway analogous to that shown in Figure 6.3 for benzene. Migration of the ethynyl H-atom via a 1,2-H-migration yields the 2-(1-naphthyl)ethylidene carbene which can insert the carbenic center into the *peri* C–H of the naphthyl ring to form acenaphthalene, with formation of a cyclic (6,6,5) ring system.

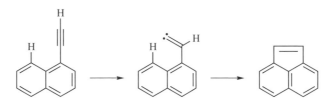

This mechanism was proposed when flash vacuum pyrolysis experiments of more complex ethynyl-substituted PAHs yielded PAH products that were only possible via migration of the ethynyl group and reversible carbenic C–H insertion, to form the more thermodynamically stable PAH resulting from cyclopenta or cyclohexa augmentation.[41,42] Theoretical studies, however, cast some doubt on the intermediacy of the vinylidene structure, suggesting instead that, immediately following the initial H-atom migration, the cyclopenta ring is formed via a concerted single-step process.[43]

Mechanistic understanding of the transition from PAH to soot is less well characterized than the formation of the initial C_6 benzene ring. When PAHs grow to molecular weights of 500–1000 amu, they are thought to take the form of agglomerated particles. These particles serve as nuclei to facilitate further particle growth by the coalescence of PAHs. The particle size also increases via the addition (physical and chemical adsorption) of gas-phase molecules to the particle, possibly by radical mechanisms.[40] Eventually, these large particles undergo collisions and stick together in a coagulation phase to form soot.

6.3.2. NO_x and SO_x

A second important class of pollutant compounds resulting from combustion processes is the general class of the oxides of nitrogen (NO, NO_2, NO_3, N_2O_4, and so forth), typically denoted as NO_x. These NO_x molecules are key intermediates in the atmospheric conversion of VOCs into photochemical smog and ozone. There are three identified sources of NO molecules in combustion systems.

The first and most prominent source is known as *thermal* NO or Zeldovich-NO.[44] The label *thermal* refers to the high temperatures required to break the N_2 triple bond in its reaction with O atom and its location of appearance in a flame.

$$O(^3P) + N_2 \rightarrow NO^\bullet + N(^4S)$$

Subsequent to formation, the N atoms will react with O_2 and $^\bullet OH$ radical as additional sources for thermal NO.

$$O_2 + N(^4S) \rightarrow NO^\bullet + O(^3P)$$

$$^\bullet OH + N(^4S) \rightarrow NO^\bullet + H^\bullet$$

Thermal NO was observed to be generated outside of a flame's combustion region where the highest temperatures exist.

Fenimore had noted an earlier and smaller source of NO produced within the flame combustion region.[45] This *prompt* NO (Fenimore-NO) was proposed to be generated by the reaction of N_2 with methyne (HC), the parent of the monovalent carbon (carbyne) family, with the formation of HCN and $N(^4S)$ atom as important intermediates.

$$HC(^2\Pi) + N_2 \rightarrow HCN + N(^4S) \rightarrow \ldots \rightarrow NO$$

Recently, Moskaleva et al. have proposed a new mechanism based on electronic structure calculations.[46] Earlier experimental studies by Kasdan et al. determined that methyne (HC) has a doublet ground state and with a doublet-quartet energy splitting (ΔE_{DQ}) of 71.5 ± 0.8 kJ/mol.[47] Moskaleva et al. noted that the initially proposed mechanism (for HCN and $N(^4S)$ atom formation) is therefore spin forbidden, and they also proposed a more favorable and spin-allowed reaction on the doublet surface. This new route on the doublet energy surface proceeds through the formation of an NCN intermediate, with concomitant formation of (doublet) hydrogen atom.

$$HC(^2\Pi) + N_2 \rightarrow NCN(^3\Sigma_g^-) + H^\bullet \rightarrow \ldots \rightarrow NO$$

NCN and its related isomers have also been experimentally detected in the gas phase.[48] It is truly remarkable that methyne (HC) is capable of breaking the extremely strong $N\equiv N$ bond, and demonstrates, yet again, the power of reactive intermediates to dictate chemical transformations.

Nitrogen that is integrated within fuel molecules is a final source of NO. Crude oil contains a small amount of nitrogen which may be concentrated within particular distillate fractions from the refining process. Coal will also contain nitrogen in varying amounts, and the content is dependent on the location of its extraction and its rank. The relatively small amount of NO resulting from nitrogen in the fuel source is termed *fuel*-NO.

Similarly, SO_2 and SO_3 (SO_x) compounds are produced in combustion by the oxidation of sulfur compounds within the fuel source. SO_x emitted into the atmosphere can be incorporated into aerosol particles and wet-deposited as corrosive sulfuric acid. Both NO_x and SO_x emissions contribute to acid rain content from wet deposition, due to their participation in the formation of nitric and sulfuric acid, respectively.

The formation of NO_x and SO_x is very dependent on the concentration of molecular oxygen in the combustion environment. Similarly, incomplete combustion, and hence nonoptimal energy output per gram of fuel, will result if the oxidant is provided in a substoichiometric manner relative to the fuel. Furthermore, the formation of soot, including PAHs, may result when the fuel is in excess relative to the oxidant. As a result, the use of fuel rich (high fuel/air ratio), stoichiometric, or fuel lean (low fuel/air ratio) conditions is often an engineering and optimization process for a specific application/engine/boiler so as to maximize power and fuel efficiency, while minimizing deleterious emissions to the environment.

6.4. ATMOSPHERIC CHEMISTRY

Primary volatile organic compounds (VOCs) are emitted into the atmosphere via anthropogenic and biogenic (natural) sources. The fate and persistence of a VOC is largely determined by its reactivity with a group of reactive radicals present in the troposphere: $^{•}OH$, $O(^3P)$, O_3, and $NO_3^{•}$. Hydroxyl radical ($^{•}OH$) is the most significant oxidizer of VOCs in the troposphere and is active during daylight hours. On the contrary, $NO_3^{•}$ radicals, which can be photochemically decomposed by sunlight and are less reactive, predominate at night.[49] Hydroxyl radicals result from the photolysis of ozone (O_3) at wavelengths of 350 nm or less.[50,51]

$$O_3 + hv \ (\lambda \leq 350 \ nm) \rightarrow O(^1D) + O_2 \tag{6.17}$$

$$O(^1D) + M \rightarrow O(^3P) + M^* \tag{6.18}$$

$$O(^1D) + H_2O \rightarrow 2 \ ^{•}OH \tag{6.19}$$

Photolysis of O_3 yields O_2 and electronically excited $O(^1D)$, which can either be collisionally stabilized (reaction 6.18) or react with a water molecule to yield two hydroxyl radicals (reaction 6.19). Atmospheric concentrations of hydroxyl radical on a 24-h seasonal average basis are estimated at 1×10^6 molecules cm^{-3}, while peak daytime concentrations of 46×10^6 molecules cm^{-3} have been observed.[51,52]

Hydroxyl radical reacts with VOCs via either an H-atom abstraction or radical-addition mechanism (the latter is possible if the VOC is an alkene). The nascent radical is transformed, via ensuing reactions with other abundant atmospheric radicals or O$_2$, to yield secondary VOCs in most cases. Figure 6.4 shows a general reaction scheme for the atmospheric oxidation of a primary alkane VOC to secondary VOC. The scheme is similar for unsaturated molecules, wherein hydroxyl radical-addition reactions dominate over H-atom abstraction so as to yield a hydroxy-substituted alkyl radical. Due to high atmospheric concentrations, molecular oxygen is the principal species to react with the new VOC (R$^\bullet$) radical derivative, typically via an addition mechanism to form a peroxy (RO$_2$$^\bullet$) radical. The persistence of peroxy radicals formed via the radical-addition process is pressure-dependent, relying on collisional stabilization by other gas molecules; otherwise, regeneration of reactants is possible.

A peroxy radical can react with other peroxy species; however, the reactions of peroxy radicals with nitric oxide (NO) are most noteworthy when considering atmospheric air quality. Peroxy radical reactions with NO are of central importance in the formation of photochemical smog from anthropogenic VOCs. Peroxy radicals can react with NO, which is produced from high-temperature combustion sources and emitted in an engine's exhaust, and these reactions produce excess NO$_2$ in the lower troposphere. Photolysis of NO$_2$ produces O(^{3}P) and regenerates NO. The oxygen atoms then combine with O$_2$, resulting in increased ozone (O$_3$) concentration according to reactions 6.20 and 6.21:[50,53]

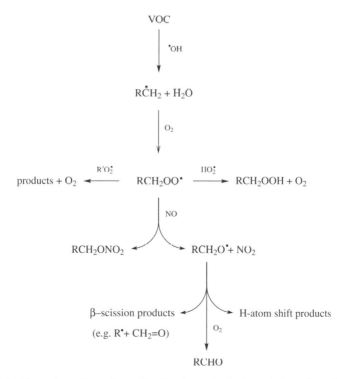

Figure 6.4. General reaction scheme for the atmospheric degradation of an alkane VOC.

$$NO_2 + hv \; (\lambda < 430 \text{ nm}) \rightarrow NO + O(^3P) \tag{6.20}$$

$$O(^3P) + O_2 + M \rightarrow O_3 + M^* \tag{6.21}$$

In a clean troposphere, ozone would react with NO molecules, resulting in no net generation of ozone. However, in a "dirty" atmosphere, excess ozone is generated, thereby resulting in the formation of an oxidant which can lead to health effects, an especially hazardous condition for children and the elders who suffer from asthma and other respiratory challenges.

The other products of the reaction of peroxy radicals with NO are oxygen-centered radicals and alkyl-, aryl-, and acyl-nitrates. Oxygen-centered radicals react with O_2 via abstraction of an H atom β to the oxygen radical center to yield an aldehyde as a secondary VOC (Fig. 6.4). Furthermore, an oxygen-centered radical can react in a unimolecular fashion via β-scission or H-atom migration reactions to form decomposition or isomerization products, respectively. Also, alkyl-, aryl- and acyl-nitrates ($RONO_2$) can act as reservoirs for nitrate ($NO_3^\bullet$) radicals, allowing for their transport away from their origin source for later release. Nitrates formed from aldehydic ($RCH(=O)$) VOCs, which are common secondary VOCs (Fig. 6.4), are peroxyacylnitrates (PANs, $RC(=O)OONO_2$) which are known to be lachrymators.[54] These PANs form via cleavage of the weak aldehydic C–H bond to form the acyl radical ($RC(=O)^\bullet$), followed by subsequent reaction with O_2, and then NO_2.

6.5. DETECTION OF GAS-PHASE TRANSIENT INTERMEDIATES

In the design of an experimental setup to study reactive transient species in gas-phase combustion systems, one needs to generate the reactive species of relevance to the combustion environment and also needs to qualitatively and/or quantitatively analyze their respective concentrations.

Initially, the substrate precursors or reactive intermediates must be prepared. In situ flame analysis has been an important and fruitful method of studying combustion intermediates for many years. Flames are typically categorized as having either laminar or turbulent flows of the fuel and oxidizing gases.[7,55] A Bunsen burner is an example of a laminar flow system, while internal combustion engines (an automobile's cylinder, for example) experience turbulent flows of the gases injected into the cylinder. Moreover, the fuel and oxidizer can be premixed or nonpremixed prior to ignition.[7,55] In situ flame methods have the advantage that they more closely reflect environments in real combustion systems. Flames can also be directly interrogated by a variety of analytical techniques which will be discussed below. In the combustion process itself, the color of a flame is due to the emission of photons by the myriad of reactive intermediates, including the radicals discussed above. The emission of these photons occurs as these energetic species release part of their vibrational and/or electronic energy.

For studying elementary reactions of importance in combustion, a more controlled environment may be required. Radicals or atoms may be generated by photolysis with UV radiation or with a microwave discharge on a suitable molecular precursor.

For example, a microwave discharge on H_2 in an inert diluent, such as argon gas, is an efficient method for producing H atoms as reactants. Subsequent reaction of these H atoms with NO_2 will yield $^\bullet OH$ and NO, and can serve as a useful source of hydroxyl radicals. These methods of reactant formation are well suited for experiments involving either static or flow reactor[56] systems.

A popular method for simulating a high-temperature and high-pressure environment is by the use of a shock tube. In a shock tube, a chamber containing the reactants of interest are isolated, and a second adjacent chamber, separated from the first by a diaphragm, contains an inert gas at considerably greater pressure. Upon rupture of the diaphragm, the high-pressure gases of the second chamber expand and create a pressure wave that is driven through the lower temperature chamber. The pressure wave causes a large increase in temperature to occur, such that the initiation of high activation-barrier reactions can occur. Shock tubes are unique in that they are capable of producing temperatures up to 2000 K and pressures up to 1000 atm and can be combined with laser diagnostics to analyze intermediates for reactions occurring on a submicrosecond timescale. For a complete discussion of shock tubes, see references 57 and 58.

Another popular method for studying combustion species is via crossed molecular beams. In this technique, the reactant molecules of interest are propelled as beams toward an intersection where their molecular collisions bring about reactions. For a more complete discussion of crossed molecular beam experiments, see reference 59.

Once precursors have been generated and incorporated into a controlled environment, one last factor must be considered: choice of analytical method for the detection of the subsequent reactive intermediates. Each of the above experimental systems can be coupled with a variety of analytical techniques for species identification. Several of the more prominent techniques will be discussed here.

Mass spectrometry combined with gas chromatography (GC–MS) can be used for analyzing nearly all species present in a reaction, but this method does require that the neutral species are converted to some ionic signature (for a m/z ratio) for detection. Typically, electron impact (EI) or chemical ionization (CI) methods are used to convert the neutral species into an ion with a corresponding mass-to-charge ratio for detection. Inlets placed at desired positions in a flame or laminar flow can provide information regarding the species present at various flame positions for different reaction times, thereby providing valuable kinetic information. Structural information for species cannot be provided; however, since only mass-to-charge ratios are obtained, thereby resulting in some ambiguity for isobaric (i.e., same mass) isomers. Additionally, the probe may interfere with the flame that is being analyzed, rendering analysis of the results to be complicated.

Laser-based methods of identification are extremely powerful; they are able to provide species and structural information, as well as accurate system temperature values. Spontaneous Raman scattering experiments are useful for detection of the major species present in the system. Raman scattering is the result of an inelastic collision process between the photons and the molecule, allowing light to excite the molecule into a virtual state. The scattered light is either weaker (Stokes shifted) or stronger (anti-Stokes shifted) than the absorbed photon. This phenomenon results from the molecule's polarizability allowing an induced dipole via interaction with the electric field of the laser photons. This method produces a signal that is linear in both laser intensity and molecular concentration. Improved signal can be obtained from coherent anti-Stokes Raman scattering (CARS). This method utilizes multiple photons and multiple pumping events to obtain the resultant scatter. These Raman techniques are beneficial in that they are capable of analyzing a large number of molecular species simultaneously. For a more detailed explanation of Raman spectroscopic methods, see references 60 and 61.

Laser-induced fluorescence (LIF) is a very sensitive technique, allowing for the detection of species at sub-ppb (part-per-billion) concentrations. In LIF, laser light pumps molecules into well-defined and bound excited electronic states (as opposed to Raman techniques that pump into virtual states, which fluoresce and relax back into lower energy states). The need to excite into real excited states requires that the states of the molecule be well characterized. Typically, only small molecules are sufficiently characterized (i.e., $^{\bullet}OH$, NO, and HCN) and have sufficiently narrow Boltzmann distributions at high temperature. A disadvantage of the LIF technique is that the signal is affected by quenching via collisional relaxation, rendering it pressure-dependent. For a more complete discussion regarding these and other analytical laser techniques and their applications, see references 60, 61, and 62.

However, even the best experimental technique typically does not provide a detailed mechanistic picture of a chemical reaction. Computational quantum chemical methods such as the ab initio[63] molecular orbital and density functional theory (DFT)[64] methods allow chemists to obtain a detailed picture of reaction potential energy surfaces and to elucidate important reaction-driving forces. Moreover, these methods can provide valuable kinetic and thermodynamic information (i.e., heats of formation, enthalpies, and free energies) for reactions and species for which reactivity and conditions make experiments difficult, thereby providing a powerful means to complement experimental data.

Computational modeling is a very powerful approach for studying combustion, providing highly detailed analyses of relevant systems. Mechanisms are constructed to account for the kinetic and thermodynamic properties of all important species and reactions for the combustion of a particular fuel. Furthermore, these data are used to construct a set of differential equations constrained to a set of conditions, such as detailed mass balance under isothermal, isobaric, isochoric, or adiabatic conditions and energy conservation. Other factors unique to the characteristics of the system being studied may also be included as constraints. However, reliable solutions require a wealth of accurate thermodynamic and kinetic information for the suspected species and chemical reactions involved in the combustion system. The current state-of-the-art mechanism for modeling natural gas (methane) combustion includes 325 chemical reactions and 53 atomic and molecular species.[65] Fuels containing larger molecular species can require an exponentially increased amount of thermochemical and kinetic data. In order to make a problem more tractable, researchers will often determine the sensitivity of the model toward a species or specific steps of a reaction mechanism and refine these data. Alternatively, a group of similar species can be assigned the same thermodynamic and kinetic data (lumping) to greatly reduce a model's complexity.

6.6. CONCLUSIONS AND OUTLOOK

The reactive intermediates of interest in the chemical processes of combustion and atmospheric reactions are primarily radicals, typically the result of homolytic bond scissions, by either thermal or photolytic means, or by abstraction of a hydrogen atom by another radical, most typically ${}^{\bullet}OH$. The oxidation of organic molecules by molecular oxygen has a strong dependence on the thermal stability of the generated peroxy radical. Therefore, two general mechanisms exist for low-temperature combustion (thermally stable peroxy radicals) and high-temperature combustion (thermally unstable peroxy radicals). Unfortunately, most combustion sources fail to convert the fuel(s) cleanly to CO_2 and H_2O. As a result, under-oxidized and partially oxidized combustion by-products are generated and emitted into the atmosphere. Significant components of under-oxidized fuel include PAH and soot, formed via the aggregation of radicals as combustion intermediates and later coagulation of larger particles. Other, more volatile, combustion intermediates can diffuse into the troposphere where radical decomposition mechanisms initiated via photolytic reactions dominate.

A good deal of the general chemistry of combustion and atmospheric oxidation is fairly well understood. Much of the chemistry that eludes us results from the complex and sometimes unknown character of the fuel itself (e.g., gasoline and coal). Furthermore, a great deal of thermodynamic and kinetic data is needed for the individual reaction mechanisms that cover the ranges of temperature and pressure required to develop accurate combustion models. Significant efforts are being geared toward increasing fuel efficiency and ensuring clean emissions from our power plants and internal combustion engines. Homogeneous charge compression ignition (HCCI) engines,[66] which rely on auto-ignition mechanisms within a homogeneous gas mixture, are being developed with these goals in mind. Chemistry at the gas–solid interface may also provide routes of catalysis for both combustion and emission reduction. Furthermore, new experimental techniques continue to be developed which offer a more resolved description of combustion processes.[62] Certainly, progress in experimental and theoretical capacities will play a large part in our future knowledge of reactive intermediates in combustion and atmospheric processes.

ACKNOWLEDGMENTS

We thank Dr. Timothy Barckholtz (ExxonMobil Research and Engineering) for his many thoughtful and helpful comments on this manuscript, and also for providing Figure 6.2. We extend sincere thanks to Carrigan Hayes for her help with the editing of this manuscript.

SUGGESTED READING

I. Glassman, *Combustion, 3rd Ed.*; Academic Press, San Diego, California, **1996**.

J. Warnatz, U. Maas, and R. W. Dibble, *Combustion: Physical and Chemical Fundamentals, Modeling and Simulation, Experiments, Pollutant Formation, 3rd Ed.*, Springer-Verlag, Berlin, **2001**.

T. J. Wallington, E. W. Kaiser, and J. T. Farrell, "Automotive fuels and internal combustion engines: a chemical perspective," *Chem. Soc. Rev.* **2006**, *35*, 335.

M. Frenklach, "Reaction mechanism of soot formation in flames," *Phys. Chem. Chem. Phys.* **2002**, *4*, 2028.

R. Atkinson and J. Arey "Atmospheric degradation of volatile organic compounds," *Chem. Rev.* **2003**, *103*, 4605.

REFERENCES

1. R. Gorches, M. A. Olivella, and F. X. M. de las Heras, *Org. Geochem.* **2003**, *34*, 1627.

2. J. F. Pankow, W. Luo, D. A. Bender, L. I. Isabelle, J. S. Hollingsworth, C. Chen, W. E. Asher, and J. S. Zogorski, *Atmos. Environ.* **2003**, *37*, 5023.

3. N. V. Heeb, A.-M. Forss, C. J. Saxer, and P. Wilhelm, *Atmos. Environ.* **2003**, *37*, 5185.

4. J. E. Wertz and J. R. Bolton, *Electron Spin Resonance: Elementary Theory and Practical Applications*, Chapman and Hall, New York, **1986**.

5. T. J. Wallington, E. W. Kaiser, and J. T. Farrell, *Chem. Soc. Rev.* **2006**, *35*, 335.

6. N. N. Semenov, *Some Problems in Chemical Kinetics and Reactivity*, Chapter 7. Princeton University Press, Princeton, New Jersey, **1958**.

7. I. Glassman, *Combustion, 3rd Ed.*; Academic Press, San Diego, California, **1996**.

8. S. J. Blanksby and G. B. Ellison, *Acc. Chem. Res.* **2003**, *36*, 255.

9. S. W. Benson, *J. Am. Chem. Soc.* **1965**, *87*, 972.

10. J. K. Merle, C. J. Hayes, S. J. Zalyubovsky, B. G. Glover, T. A. Miller, and C. M. Hadad, *J. Phys. Chem. A* **2005**, *109*, 3637.

11. J. D. DeSain, C. A. Taatjes, J. A. Miller, S. J. Klippenstein, and D. K. Hahn, *Faraday Discuss.* **2001**, *119*, 101–120.

12. J. W. Bozzelli and W. J. Pitz, *Twenty-Fifth Symposium (International) on Combustion*, The Combustion Institute: Pittsburgh, PA., **1994**, 783–791.

13. R. G. Compton and G. Hancock, *Comprehensive Chemical Kinetics, Low-Temperature Combustion and Autoignition, Vol. 35* (Ed. M. J. Pilling), Elsevier, Amsterdam, **1997**.

14. J. C. Rienstra-Kiracofe, W. D. Allen, and H. F. Schaefer III., *J. Phys. Chem. A* **2000**, *104*, 9823.

15. J. D. DeSain, E. P. Clifford, and C. A. Taatjes, *J. Phys. Chem. A* **2001**, *105*, 3205.

16. C. Vencat, K. Brezinsky, and I. Glassman, *Nineteenth Symposium (International) on Combustion*, The Combustion Institute: Pittsburgh, PA., **1982**, 143–152.

17. T. Yu and M. C. Lin, *J. Am. Chem. Soc.* **1994**, *116*, 9571–9576.

18. B. K. Carpenter, *J. Am. Chem. Soc.* **1993**, *113*, 9806–9807.

19. A. M. Mebel, E. W. G. Diau, M. C. Lin, and K. Morokuma, *J. Am. Chem. Soc.* **1996**, *118*, 9759–9771.

20. C. Barckholtz, M. J. Fadden, and C. M. Hadad, *J. Phys. Chem. A* **1999**, *103*, 8108–8117.

21. M. J. Fadden, C. Barckholtz, and C. M. Hadad, *J. Phys. Chem. A* **2000**, *104*, 3004–3011.

22. S. M. Kroner, M. P. DeMatteo, C. M. Hadad, and B. K. Carpenter, *J. Am. Chem. Soc.* **2005**, *127*, 7466–7473.

23. Y. Chai and L. D. Pfefferle, *Fuel* **1998**, *77*, 313–320.

24. M. J. Fadden and C. M. Hadad, *J. Phys. Chem. A* **2000**, *104*, 8121.

25. C. Barckholtz, T. A. Barckholtz, and C. M. Hadad, *J. Am. Chem. Soc.* **1999**, *121*, 491–500.

26. M. J. Fadden and C. M. Hadad, *J. Phys. Chem. A* **2000**, *104*, 6088.

27. M. J. Fadden and C. M. Hadad, *J. Phys. Chem. A* **2000**, *104*, 6324.

28. A. Albagli, H. Oja, and L. Dubois, *Environ. Lett.* **1974**, *6*, 241.

29. S. Pavanello, P. Simioli, S. Lupi, P. Gregorio, and E. Clinfero, *Cancer Epidemiology, Biomarkers & Prevention* **2002**, *11*, 998.

30. J. A. Williams, F. L. Martin, G. H. Muir, A. Hewer, P. L. Grover, and D. H. Phillips, *Carcinogenesis* **2000**, *21*, 1683–1689.

31. WHO (1997) The World Health Report. World Health Organization, Geneva, Switzerland.

32. M. Frenklach, *Phys. Chem. Chem. Phys.* **2002**, *4*, 2028.

33. J. D. Bittner and J. B. Howard, *Proc. Combust. Inst.* **1981**, *18*, 1105.

34. P. R. Westmoreland, A. M. Dean, J. B. Howard, and J. P. Longwell, *J. Phys. Chem.* **1989**, *93*, 8171.

35. J. A. Miller and C. F. Melius, *Combust. Flame* **1992**, *91*, 21.

36. C. F. Melius, M. E. Colvin, N. M. Marinov, W. J. Pitz, and S. M. Senkin, *Proc. Combust. Inst.* **1996**, *26*, 685.

37. (a) R. G. Jones and R. G. Bergman, *J. Am. Chem. Soc.* **1972**, *94*, 660. (b) R. G. Bergman, *Acc. Chem. Res.* **1973**, *6*, 25.

38. A. V. Friderichsen, E.–J. Shin, R. J. Evans, M. R. Nimlos, D. C. Dayton, and G. B. Ellison, *Fuel* **2001**, *80*, 1747–1755.

39. M. Frenklach and H. Wang. *Proc. Combust. Inst.* **1991**, *23*, 1559.

40. H. Richter and J. B. Howard, *Prog. Energy Combust. Sci.* **2000**, *26*, 565.

41. L. T. Scott and A. Necula, *Tetrahedron Lett.* **1997**, *38*, 1877–1880.

42. M. Sarobe, L. W. Jenneskens, A. Kleij, and M. Petroutsa, *Tetrahedron Lett.* **1997**, *38*, 7255.

43. J. Cioslowski, M. Schimeczek, P. Piskorz, and D. Moncrieff, *J. Am. Chem. Soc.* **1999**, *120*, 1695.

44. Y. B. Zeldovich, *Acta Physicochim. USSR* **1946**, *21*, 577.

45. C. P. Fenimore, *Proc. Combust. Inst.* **1979**, *17*, 661.

46. L. V. Moskaleva, W. S. Xia, and M. C. Lin, *Chem. Phys. Lett.* **2000**, *331*, 269–277.

47. A. Kasdan, E. Herbst, and W. C. Lineberger, *Chem. Phys. Lett.* **1975**, *31*, 78.

48. (a) K. R. Jennings and J. W. Linnett, *Trans. Faraday Soc.* **1960**, *56*, 1737. (b) E. P. Clifford, P. G. Wenthold, W. C. Lineberger, G. Petersson and G. B. Ellison, *J. Phys. Chem. A* **1997**, *101*, 4338.

49. J. G. Calvert, R. Atkinson, J. A. Kerr, S. Madronich, G. K. Moortgat, T. J. Wallington, and G. Yarwood, *The Mechanisms of Atmospheric Oxidation of the Alkenes*, Oxford University Press, New York, **2000**.

50. B. J. Finlayson-Pitts and J. N. Pitts Jr., *Chemistry of the Upper and Lower Atmosphere: Theory, Experiments, and Applications*; Academic Press, San Diego, California, **2000**.

51. R. Atkinson and J. Arey, *Atmos. Environ.* **2003**, *37* (Supp. 2), S197.

52. L. A. George, T. M. Hard, and R. J. O'Brien, *J. Geophys. Res.* **1999**, *104*, 11643.

53. T. J. Wallington, P. Dagaut, and M. Kurylo, *Chem. Rev.* **1992**, *92*, 667–710.

54. L. S. Andrew and R. Snyder, *Casarett and Doull's Toxicology: The Basic Science of Poisons* (Eds. O. A. Amdur, J. Doull, and D. K. Klaassen), Pergamon Press, New York, **1991**.

55. J. Warnatz, U. Maas, and R. W. Dibble, *Combustion: Physical and Chemical Fundamentals, Modeling and Simulation, Experiments, Pollutant Formation, 3rd Ed.*, Springer-Verlag, Berlin, **2001**.

56. C. Venkat, K. Brezinsky, and I. Glassman, *Proc. Combust. Inst.* **1982**, *19*, 143.

57. W. Tsang and A. Lifshitz, *Annu. Rev. Phys. Chem.* **1990**, *41*, 559.

58. A. G. Gaydon and I. R. Hurle, *The Shock Tube in High-Temperature Chemical Physics*, Reinhold, New York, **1963**.

59. P. Casavecchia, *Rep. Prog. Phys.* **2000**, *63*, 355.

60. D. R. Crosley, *Laser Probes for Combustion Chemistry*, American Chemical Society, ACS Symposium Series, 134, Washington, DC, **1980**.

61. A. C. Eckbreth, *Laser Diagnostics for Combustion Temperature and Species, 2nd Ed.*, Gordon and Breach Publishers, Amsterdam B. V., Netherlands, **1996**.

62. J. Wulfrum, *Faraday Discuss.* **2001**, *119*, 1–26.

63. W. J. Hehre, L. Radom, P. v. R. Schleyer, and J. A. Pople, *Ab Initio Molecular Orbital Theory*, John Wiley & Sons, New York, **1986**.

64. (a) R. G. Parr and W. Yang, *Density Functional Theory in Atoms and Molecules*, Oxford University Press, New York, **1989**. (b) J. W. Labonowski and J. Andzelm, *Density Functional Methods in Chemistry*, Springer, New York, **1991**.

65. G. P. Smith, D. M. Golden, M. Frenklach, N. W. Moriarty, B. Eiteneer, M. Goldenberg, C. T. Bowman, R. K. Hanson, S. Song, W. C., Jr., Gardiner, V. V. Lissianski, and Z. Qin, http://www.me.berkeley.edu/gri_mech/. A listing of all of the species and the individual reactions is available on the web site.

66. R. H. Thring, *Homogeneous-Charge Compression-Ignition (HCCI) Engines*, Warrendale, Pa.: SAE, **1989**.

CHAPTER 7

Reactive Intermediates in Crystals: Form and Function

LUIS M. CAMPOS AND MIGUEL A. GARCIA-GARIBAY

Department of Chemistry and Biochemistry, University of California, Los Angeles, CA

7.1. INTRODUCTION

To date, the vast majority of organic transformations has been explored in solution media. Accordingly, most of our knowledge of the intermediates involved in

Reviews of Reactive Intermediate Chemistry. Edited by Matthew S. Platz, Robert A. Moss, Maitland Jones, Jr.
Copyright © 2007 John Wiley & Sons, Inc.

those transformations has been extracted from product analysis, stereochemical correlations, kinetic measurements, and spectroscopic studies carried out in solution phase. As the boundaries of our studies of chemical reactivity expand to the solid state, efforts have been directed to the study of reaction mechanisms and reactive intermediates in crystalline media. Interestingly, while giving little or no credit to basic organic chemistry textbooks, reactions in solids have played a remarkable role in the development of organic chemistry since the beginnings of the discipline. In fact, the "birth" of organic chemistry can be traced back to the thermal transformation of a crystalline "mineral," ammonium cyanate, into crystalline urea, serendipitously discovered by Wöhler in 1828.[1,2] A very strong impact can also be documented in events that took place in more recent times, as the main drive for semiconductor device industry is based on photoresist technology employing the photochemical transformation of diazonaphtoquinone in solid polymeric films.[3] Reactions *i – iv* illustrate these and other examples of solid state transformations that have revolutionized several aspects of organic chemistry. In addition to those already mentioned (reactions *i* and *iii*), one can include the solid state photo-degradation of santonin, the first drug formulated by Pfizer in the USA (reaction *ii*),[4] the first observation of a triplet molecule by Murray et al. (reaction *iv*),[5] a possible mechanism for the generation of optical activity under prebiotic conditions (reaction *v*),[6] and the formation of single crystalline conjugated polymers, which may have conductor and/or semiconductor applications.[7]

(i) Wöhler 1828: Synthesis of first "organic" compound

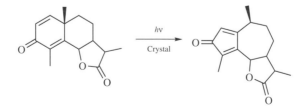

(ii) Pfizer 1849: First drug formulated in the USA

(iii) Süss 1949: First photoresist material

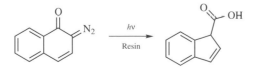

(iv) Murray et al. 1962: Observation of first triplet molecule

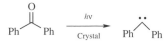

(v) Schmidt et al. 1971: "Spontaneous" generation of optical activity

(vi) Wegner et al. 1971: Single crystal conjugated polymers

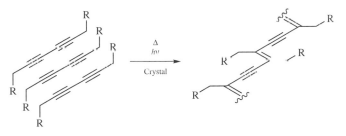

The study of reaction mechanisms and their intermediates in the solid state presents challenges and opportunities that are not available in other reaction media. Taking a "generic" stepwise reaction coordinate as a starting point for our analysis, we can identify three minima corresponding to the reactant R; a reactive intermediate RI, a product P, and two transition states that sequentially interconnect them. In general, RIs are characterized by their high-energy content, and by having one or more reaction pathways with low energy barriers, which make them exceedingly short-lived species. In order to isolate and characterize a reaction intermediate (RI), one can take advantage of two relatively simple variations to the reaction coordinate: *stabilization* and *persistence*.[8] *Stabilization* relies on lowering the energy content of the RIs (i.e., thermodynamic stabilization), usually by taking advantage of the electronic effects of substituents. *Persistence* is based on kinetic stabilization, which usually relies on steric effects, with either bulky groups or inert shielding groups such as fluorocarbons and other halogens. With sufficient stabilization and persistence, RIs can be manipulated, analyzed by standard spectroscopic techniques, and crystallized for single crystal X-ray diffraction.

While the design of RIs that can be crystallized is one of the most interesting challenges in physical organic chemistry, one can also study their structural properties and chemical behavior by selected chemical reactions in the solid state.[9] To that effect, strategies have been devised to "engineer" and control reactions in crystals by selecting crystalline precursors that generate RIs with very high energy content. Usually generated by photochemical processes, high energy species in crystals are able to break weak bonds to make stronger new ones despite having little or no molecular motion.[10]

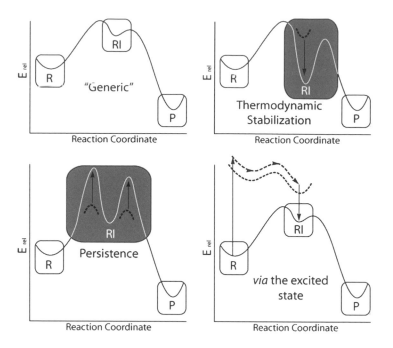

Strategies to determine the nature of reacting RIs in solids are often based on product analyses, stereochemical correlations, kinetics measurements, and so forth, just as they are generally applied in solution media.[9,11] While there are challenges associated with the implementation of spectroscopic methods in the solid state, the direct observation of short-lived RIs can be used to support or reject mechanistic models.

In this chapter, we will consider examples of RIs characterized by a hypervalent or valency-deficient carbon, such as carbocations, carbenes, carbanions, and carbon radicals. In the first part, we will consider examples that take advantage of stabilization and persistence to determine their structures by single crystal X-ray diffraction. In the second part we will describe several examples of transient reactive intermediates in crystals.[12]

7.2. REACTIVE INTERMEDIATES TRAPPED IN CRYSTALS: FORM

As a result of compelling three-dimensional models and remarkably high levels of precision, it is often assumed that structural elucidation by single crystal X-ray diffraction is the ultimate structural proof. Spatial information in the form of several thousands of X-ray reflection intensities are used to solve the position of a few dozen atoms so that the solution of a structure by X-ray diffraction methods is highly over-determined, with a statistically significant precision up to a few picometers. With precise atomic positions, structural parameters in the form of bond distances, bond angles, dihedral angles, and so forth, are generally analyzed in terms of hybridization models that reveal the electronic nature of the structure of interest. Having carbon as the center of attention in this chapter, it is important to have some benchmarks that define the idealized structure of stable carbon-containing molecules so that one can identify structural variations resulting from the unusual valency and high energy content of the RIs of interest. In the discussion that follows, we will highlight structural aspects that suggest changes in hybridization of carbon. For example, deviations in geometry from tetrahedral to planar trigonal to linear are interpreted as an increase in the "s" orbital character of the corresponding carbon center as it changes in character from sp^3 to sp^2 to sp. Variations in bond lengths are also interpreted in a similar manner, as bonds lengths are expected to decrease with an increase in s-orbital character of the two bonded atoms. For example, a sigma bond between two sp^3 carbons is longer than a sigma bond between sp^3 and sp^2 carbons. A set of typical bond length and bond angle values of particular relevance to our discussion is included in Table 7.1.[13]

7.2.1. Carbocations

Carbocations are ubiquitous reactive intermediates, with a formal charge +1 on carbon.[14] The recognition of their existence can be traced to the early 1900s, when carbocations were first the subject of much doubt, but then later were accepted as unstable, nonisolable species. Research efforts during the first 50 years after they

TABLE 7.1. Selected bond distances between various types of hybridized atoms.[13]

Bond type	Note	Distance*	Bond type	Note	Distance*
$C(sp^3)$–$C(sp^3)$		1.53	$C(sp^x)$–F		1.34
$C(sp^3)$–$C(sp^2)$		1.51	$C(sp^x)$–Cl		1.73
$C(sp^3)$–$C(sp^1)$		1.47	$C(sp^x)$–Br		1.97
$C(sp^3)$–$C(ar)$		1.51	$C(sp^x)$–I		2.16
$C(sp^2)$–$C(sp^2)$	Conjugated	1.46	$C(sp^3)$–$N(sp^3)$		1.47
	Nonconjugated	1.48	$C(sp^3)$–$N(sp^2)$	Planar N	1.46
$C(sp^2)$–$C(sp^1)$		1.43	$C(sp^2)$–$N(sp^2)$	Planar N	1.36
$C(sp^2)$–$C(ar)$	Conjugated	1.47	$C(sp^2)$–$N(sp^3)$	Pyramidal N	1.42
	Nonconjugated	1.49	$C(ar)$–$N(sp^y)$		1.38
$C(sp^1)$–$C(sp^1)$		1.38	$C(sp^3)$–O	Alcohols, ethers	1.43
$C(sp^2)$=$C(sp^2)$		1.32	$C(sp^2)$=O	Aldehydes, ketones	1.20
$C(sp^1)$≡$C(sp^1)$		1.18	C–Si	$C(sp^3)$, $C(ar)$	1.87
$C(ar)$–$C(ar)$	Aromatic	1.38	P=O		1.49
			C–P	type varies	>1.81

*Values in Angstroms, Å; ar = aromatic; x = 1, 2, or 3; y = 2 or 3.

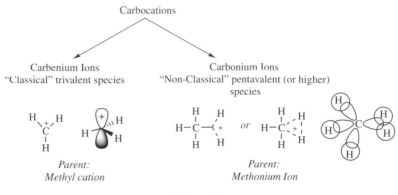

Scheme 7.1

were postulated were aimed toward their indirect characterization, and the later work yielded conditions that extended their lifetime in solution to the point that they could be crystallized, and their detailed structures corroborated by single crystal X-ray diffraction. Carbocations are the reactive intermediates with the most X-ray structures reported to date.

The International Union of Pure and Applied Chemistry recognizes two types of carbocations: carbenium and carbonium ions (Scheme 7.1).[15] Carbenium ions are the most common, also known as "classical" carbocations. They consist of a trivalent sp^2-hybridized carbon center and an empty p-orbital. As expected for such an electronic configuration, carbenium ions usually have a planar structure with the three groups attached to the carbenium center at angles of 120° from each other. Carbonium ions may be less common and are termed "nonclassical" due to the hypervalent nature of the carbon center. Carbonium ions are pentavalent carbon species characterized by a three-center, two-electron interaction, such as that shown in Scheme 7.1 (higher valency species have also been proposed).[14] The carbon center can be described as having sp^3 hybridization, and in the case of the methonium ion, CH_5^+, two hydrogen atoms share two electrons with carbon.[16] Thanks to the design of clever experiments, the X-ray structures of numerous carbocations have been solved and analyzed.[17]

Carbenium ions are key intermediates in many common organic processes such as the reaction of electrophiles with π-donors (e.g. protonation of alkenes), solvolytic substitution (S_N1) and elimination (E_1), among many others. Carbenium ions react readily with electron-rich species, which make them difficult to observe by standard spectroscopic methods. Efforts to generate them and increase their lifetime using superacid media were pioneered by George A. Olah (Nobel Prize, 1994) during the 1960s.[14b] Although the generation of long-lived carbocations allowed for their spectroscopic characterization in solution by ^{1}H NMR and infrared (IR) techniques, a low temperature X-ray structure of the simplest tertiary alkyl cation, *tert*-butyl carbocation, was not obtained until 1993.[18] Stressed in a more recent study with a different counterion, it has been shown that crystallization experiments must be cleverly designed to fulfill the following two main requirements:[19] (1) the counterion

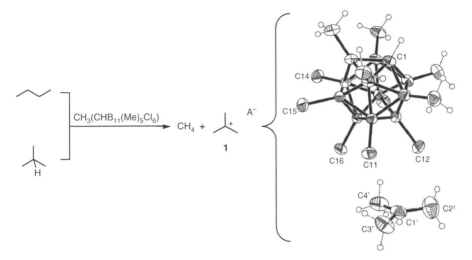

Figure 7.1. X-Ray structure of the *tert*-butyl cation and $(CHB_{11}Me_5Cl_6)^-$ obtained by the reaction of *n*-butane and isobutane with $CH_3(CHB_{11}Me_5Cl_6)$. (Adapted from reference 19.)

must be an extremely weak nucleophile to prevent covalent bond formation with the cation, and (2) it must also be an extremely weak base to avoid the loss of an α-proton to prevent the formation of an alkene.

The most recent generation and crystallization of the *tert*-butyl carbocation **1** were accomplished by reacting the noncrystalline methyl cation-carborane anion pair, $CH_3^+(CHB_{11}Me_5Cl_6)^-$, with butane, or *tert* butane by hydride transfer, leading to the formation of methane (Fig. 7.1). The removal of H^- from butane shows that the most stable tertiary carbenium ion is formed after a 1,2 methyl shift. Crystals grown under these conditions are indefinitely stable if kept in a dry environment and away from any source of nucleophiles. The X-ray structure of **1** reveals conclusive evidence of the structural parameters that had long been deduced by chemical inference and spectroscopic methods. The carbenium ion is indeed planar, and it has an average $C_1' - CH_3$ distance of 1.442 Å, which is shorter than a typical C−C distance of 1.53 Å between two sp^3 atoms. The angles $C2'−C1'−C3' = 120.3°$, $C2'−C1'−C4' = 120.0°$, $C3'−C1'−C4' = 119.7°$ are at, or very close to the expected value of 120°. These parameters clearly show that the carbocation is sp^2 hybridized, and the structure of the cation—anion pair does not exhibit any bonding interactions between the ions.

One factor that affects the relative stability in the decreasing order of tertiary, secondary, primary, and methyl cations is hyperconjugation. Hyperconjugation is an orbital interaction between the empty *p*-orbital at the carbenium center and the σ-orbital of an adjacent bond, aligned in a parallel manner to maximize overlapping (Scheme 7.2A). Although a formal resonance structure that delocalizes the positive charge at the neighboring hydrogen atom while forming a double bond is of significantly higher energy, its contribution is suggested by experimental observations

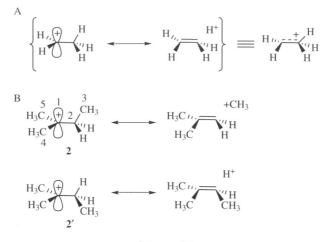

Scheme 7.2

showing that hyperconjugation plays a role along the bridging geometries of 1,2-group migrations.[20] In order to understand the extent to which C—C and C—H σ-bond hyperconjugation can stabilize substituted carbenium ions, computational studies have been carried out and compared to solution NMR studies of the 2-methyl-2-butyl cation (Scheme 7.2B).[21] Conformation **2**, with C—C σ-bond hyperconjugation, was found to be lower in energy than **2′** by 0.03 kcal/mol at a high level of theory (MP-4std/6-31G**//MP2(FU)6-31G*, zero-point energy inclusive). Accordingly, a comparison of experimental and computed NMR chemical shifts indicated that **2** is the preferred conformation (Scheme 7.2B, C labels correspond to Fig. 7.2). Noteworthy features of the calculated structure were an elongated C2—C3 distance of 1.58 Å and a C1—C2—C3 angle of 101.5°, which is smaller than the expected value of 109° and may be assigned to minimal bridging in the hyperconjugated structure.

More recently, single crystals of the 2-methyl-2-butyl cation were obtained in a similar manner as those of the *tert*-butyl cation, and their X-ray structure was obtained (Fig. 7.2).[19] The structure of the cation revealed no signs of the hyperconjugation expected in **2** (Scheme 7.2), but resembled that of **2′** with a dihedral angle D(C4′, C1′, C2′, C3′) of 25.8° and a C2′—C3′ distance of 1.485 Å, which is along the lines of that expected for a σ-bond between carbon atoms with sp³ and sp² character. An angle C1′—C2′—C3′ of 121.3° shows no signs of a bridging interaction, also indicating that the methyl group is not involved in hyperconjugation. However, as suggested by the authors, it may be possible that packing forces in the crystal may prevent the cation from experiencing the calculated interaction, especially given that the calculated energy difference is so small. Taken together, these results indicate that the thermochemical stability of conformation **2** is not drastically lower than that of **2′**. Although one would also expect the C—H bond distance of the C3 methyl group to increase and the C1′—C2′—H2′ angle to decrease with hyperconjugation, it is difficult to locate hydrogen atoms using X-ray methods.

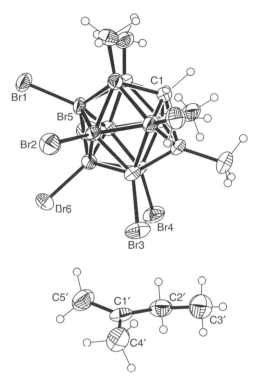

Figure 7.2. X-ray structure of the 2-methyl-2 butyl cation and $(CHB_{11}Me_5Br_6)^-$. (Taken from reference 19.)

Replacing an α-alkyl substituent by an α-aryl group is expected to stabilize the cationic center by the p-π resonance that characterizes the benzyl carbocations. In order to analyze such interaction in detail, the cumyl cation was crystallized with hexafluoroantimonate by Laube et al. (Fig. 7.3).[22] A simple analysis of cumyl cation suggests the potential contributions of aromatic delocalization (Scheme 7.3), which should be manifested in the X-ray structure in terms of a shortened cationic carbon—aromatic carbon bond distance (C^+-C_{Ar}). Similarly, one should also consider the potential role of σ-CH hyperconjugation, primarily observable in terms of shortened C^+-CH_3 distances. Notably, it was found experimentally that the C^+-C_{Ar} distance is indeed shortened to a value of 1.41 Å, which is between those of typical sp^2-sp^3 single bonds (1.51 Å) and sp^2-sp^2 double bonds (1.32 Å). In the meantime, a C^+-CH_3 distance of 1.49 Å is longer than that observed in the *tert*-butyl cation **1** (1.44 Å), and very close to the normal value for an sp^2-sp^3 single bond.

The structure of nonclassical carbocations, such as norbornenyl 3^+, has been the subject of debate since the 1950s when Saul Winstein published his milestone studies on the solvolysis of tosylated norbornenyl compounds.[23] It was proposed that the norbornenyl cation should be represented as the nonclassical structure 4^+, with a 3-center, 2-electron cyclic system (3c-2e), rather than as the classical equilibrium

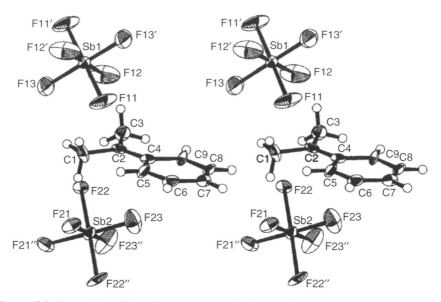

Figure 7.3. Stereoview of the X-ray structure of the cumyl cation and the hexafluoroantimonate anions. (Taken from reference 22.)

structures of **3**[+] (Scheme 7.4). Bartlett and Giddings followed with studies of the benzonorbornenyl cation **5**[+], for which solvolysis of the precursor was found to be slower than that of the precursor of **4**[+].[24]

Until recently, NMR[25] and computational[26] studies showed convincing evidence of the bridged (3c-2e) structure. However, Laube successfully obtained the X-ray structure of **Me-5**[+], with R = CH$_3$, by reacting SbF$_5$ and **6** (Fig. 7.4).[27] The most striking feature of the X-ray structure is the bending of the cationic center toward the aromatic system. The C4a-C9 (and C8a-C9) distance is 1.90 Å, surprisingly smaller than that of the calculated structure **H-5**[+] of 1.98 Å, and much smaller than that of the parent benzonorbornene (C8a-C9 = 2.35 Å). This result clearly shows that the cationic center is interacting strongly with the aromatic system by p-π-bonding. Another interesting feature is the puckering of the aromatic plane with respect to the plane formed by the C4a(C8a)-C4(C1) bonds, with a dihedral angle of 170.4° (see Fig. 7.4). This puckering causes a slight pyramidalization of C4a and

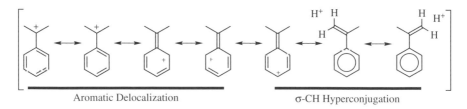

Aromatic Delocalization   σ-CH Hyperconjugation

Scheme 7.3

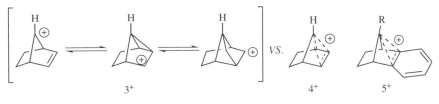

Scheme 7.4

C8a ($\Delta = 0.077°$) and an elongation of the 4a–8a bond (1.41 Å), indicating a weakened sp^2 bonding interaction between C4a and C8a in the aromatic system. It is clear that the structural features of **5**[+], and other bridging cations give convincing support to Winstein's 3-center, 2-electron non-classical structures.[28]

Laube has compiled an extensive list of X-ray structures of several types of carbocations to analyze how they interact with their counterions in a crystal.[29] As illustrated in Scheme 7.5, the analyzed structure set includes a range of coordination numbers for the C[+] centers, different neighboring groups, specific counterions, and structures involving interactions with α-hydrogens. In general terms, the cation–anion interactions are only electrostatic and the anions are usually found at a reasonable distance from the C[+] center. Following the elegant paradigm of Bürgi and Dunitz,[30] the structural information derived from the X-ray structures may be viewed as a snapshot during the early stage of a reaction trajectory for addition of a nucleophile to the cation, as illustrated in Scheme 7.5.

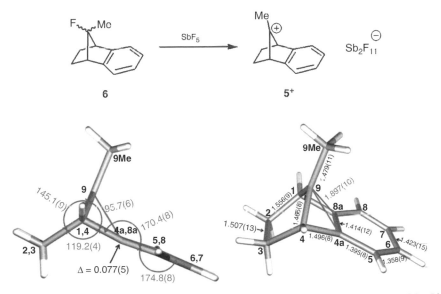

Figure 7.4. Top: Schematic representation of the reaction to form the cation-anion **Me-5**[+], $Sb_2F_{11}^-$. Bottom: Two different views of the X-ray structure of the **Me-5**[+], showing the important structural parameters. (Adapted from reference 27.) See color insert.

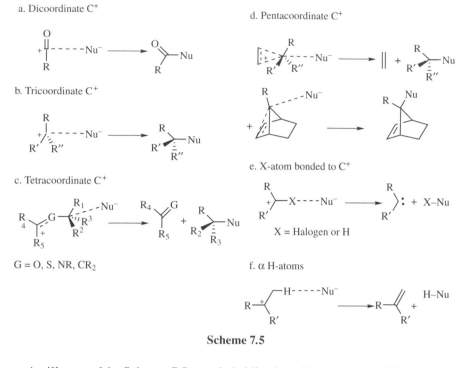

Scheme 7.5

As illustrated in Scheme 7.5a, sp hybridized acylium cations exhibit electrostatic interactions similar to that of an early approach by the nucleophile Nu^-. The angle $R-C^+(O)-Nu^-$ varied between 85° and 90°, as expected for the corresponding reaction trajectory. For the sp^2 hybridized cations represented in Scheme 7.5b, interaction with the Nu^- can be interpreted as attack to the empty p-orbital. The angle between the plane of the substituted C^+ atom and Nu^- is generally around 90°, but it shows some deviations when the C^+-Nu^- distance is shorter. Scheme 7.5c shows how the Nu^- species interacts with the cation when the C^+ center is bonded to the electron-rich substituent G. In such case, the Nu^- shows attraction to the polarized $C^{\delta+}$ center bonded to the electron-rich group G=O, S, NR, or CR_2, and not to the C^+ center. The angle $G-C^{\delta+}-Nu^-$ varied between 130° and 180°, lending support to the interpretation of an S_N2 type reaction with Walden inversion at the $C^{\delta+}$ center and a leaving-group product given by $G=C(R_4)R_5$. While only a few selected X-ray structures of carbonium ions have been reported, Scheme 7.5d shows the case where the Nu^- attacks the olefin-forming from the opposite side. Another type of interaction is that of a halogen or hydrogen (**X**) directly bonded both to the C^+ center and to the nucleophile, as shown in Scheme 7.5e. Such interaction can be interpreted as the heterolytic dissociation of the C^+ center and **X** to yield a carbene and Nu-X. Finally, nucleophiles were also found to interact with hydrogens attached in the α-position with respect to the cationic center (Scheme 7.5f). In this case, the heterolytic dissociation would yield an alkene along with the conjugate acid H-Nu.

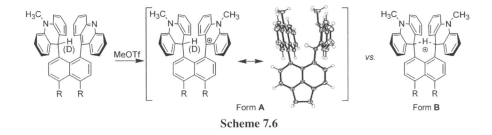

Form A vs. Form B

Scheme 7.6

The range of structural alternatives explored by valency-deficient carbon species and the subtle interplay of substituents is remarkable. Scheme 7.6 (ORTEP adapted from reference 31) illustrates an example of an X-ray structure clearly describing a localized [C−H C$^+$] carbenium ion (**A**) where a symmetric bridging structure [C− H− C]$^+$ (**B**) could have been assumed.[31] In this case it is proposed that a charge-transfer interaction between the resonance delocalized cation and the adjacent electron-rich carbazol moiety may be responsible for the stabilization of the localized form over the three-center, two-electron (3c-2e) bridging structure.

An interesting X-ray structure of the highly strained, sp hybridized vinyl cation **8** was recently reported by Müller et al.[32] The synthesis of **8** was accomplished by the reaction of alkynylsilane **7** and triphenylmethyl (trityl) cation.

Scheme 7.7, ORTEP adapted from reference 32), with [B(C$_6$F$_5$)$_4$]$^-$ or [CB$_{11}$H$_6$Br$_6$]$^-$ as the counterions. The X-ray structure clearly shows evidence of an sp hybridized C$^+$ center with a C$^\beta$−C$^\alpha$−C^4 angle of 178.8° and an abnormally short C$^\beta$ C$^\alpha$ double bond distance of 1.22 Å (compared to 1.32 Å, Table 7.1), and a nearly normal C$^\alpha$−C^4 (sp^3-sp) distance of 1.45 Å. The most striking feature of **8**, however, is the very long C$^\beta$−Si distance of 1.97 Å (compared to 1.87 Å, Table 7.1), which is attributed to hyperconjugation, as shown in the scheme. Elongation of the bonds α- to the cation center is a characteristic of hyperconjugation, also observed in the structure of the adamantyl cation.[28a,33]

While related to its carbon analogs, the existence of the R$_3$Si$^+$ species as a free ion in condensed phases had been doubted for a long time. However, NMR characterization using bulky aryl substituents has provided evidence for the triply coordinated silicon cation.[34] However, definitive evidence was recently reported by the groups of Reed and Lambert[35] with a silyl cation species bound to three mesityl groups and a carborane [HCB$_{11}$Me$_5$Br$_6$]$^-$ counterion (Fig. 7.5). It was suggested that

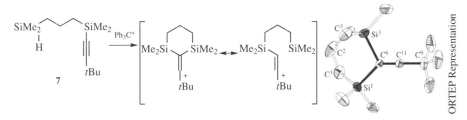

Scheme 7.7

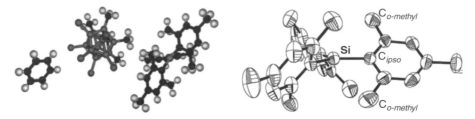

Figure 7.5. Left: Clathrate of benzene, $[HCB_{11}Me_5Br_6]^-$, and Si(Mesityl)$_3$. Right: ORTEP representation of the free silylium ion. (Adapted from reference 35.)

the structural arrangement, which included a benzene molecule in the crystal, is mainly determined by the packing efficiency, and not by electrostatic interactions. The trigonal silylium ion is planar and the Si−C bonds of 1.82 Å (on average) are relatively shorter than the 1.87 Å (Table 7.1) bond distances typically found in neutral molecules. Furthermore, the silylium ion center is completely shielded by the mesityl groups, allowing it to be free from interactions with the bromine atoms of the counterion or the aromatic system of the benzene molecule.

Finally, the structure of the first silaaromatic, a silacyclopropyl cation analogous to the cyclopropyl cation,[36,37,38] was recently characterized by single crystal X-ray diffraction methods (Scheme 7.8, ORTEP representation adapted from reference 39).[39] It was shown that the three-member ring is flat, with internal angles $Si^1-Si^2-Si^3$ ranging from 59.76° to 60.20°, and an average Si−Si distance of 2.22 Å, which is smaller than a single bond Si−Si distance of 2.37 Å, but larger than a double bond Si−Si distance of 2.16 Å.

7.2.2. Carbanions

The history of carbanions can be traced back to more than 100 years and several reviews detail many fascinating aspects of their history.[40] In general, carbanions are trigonal pyramidal and sp^3 hybridized, with $^-CH_3$ being isoelectronic with NH_3, having a similar structure and an inversion barrier of approximately 2 kcal/mol.[41] Carbanions with greater s-character have high electron binding energies

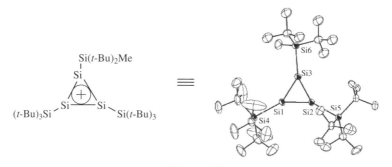

Scheme 7.8

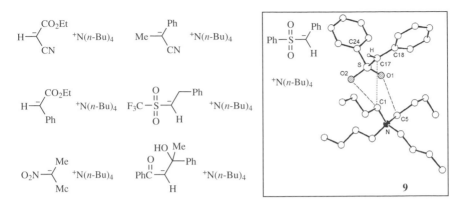

Figure 7.6. Structures of metal-free carbanions known to have a strong interaction with ammonium counterions. Shown in the box is the X-ray structure of salt **9**, where C17 is the negatively charged carbon, and C1–C17 = 3.70 Å, C1–O2 = 3.55 Å, and C5–O1 = 3.39 Å.[44] (Adapted from reference 44.)

and tend to be thermochemically more stable than those that are hybridized with more p-character (i.e. sp $>$ sp^2 $>$ sp^3). Carbanion stability in solution is usually evaluated by the pKa of the protonated species, so that the most stable carbanions come from the carbon acids with lower pKa values. Carbanions are also stabilized by α-groups that delocalize the charge,[42] such as carbonyl, sulfonyl, phosphonyl, phosphoryl, phosphonate, and cyanide, among others.[43] However, the best way to stabilize a carbanion is by salt formation with metals, as is the case for Grignard reagents and alkyllithium salts. While much is known about "metallorganic" carbanions, obtaining X-ray quality crystals of *free* carbanions has been exceedingly challenging. For example, Reetz et al.[44] have shown that carbanions formed with alkyl ammonium counterions are not free, as the carbanion center interacts strongly with the inductively charged alkyl groups of the ammonium ion, as illustrated in Figure 7.6.[45,46]

Recent work by Gais et al. has shown that free carbanions with sulfonyl stabilizing groups can be obtained if their metal counterions are complexed.[47,48,49] The X-ray structure of anion **10** in Figure 7.7 shows that there are no close intermolecular interactions of the carbanion center C1 with the lithium counterion, which is efficiently encapsulated by the cryptand host. Furthermore, the sp^3-hybridized carbanion is pyramidalized with an angle $\theta = 32.5°$, where θ is the dihedral angle defined by C2,C1,S1/C3,C1,S1 so that $\theta = 0°$ if the structure is flat, and $\theta = 60°$ if it is completely pyramidalized. A Newman projection in Figure 7.7 helps to visualize the structure and convey the fact that the anion would be chiral if two different groups were attached to C1. Though not available from the X-ray data, it is known that the barrier to rotation about the C–S bond is rather high, estimated at 14 kcal/mol for MeSO$_2$CH$_2^-$.[50] Thus, it is suggested that the $-$SO$_2$R group stabilizes the carbanion through electrostatic and d-orbital bonding interactions, which results in the formation of a pyramidalized carbanion center.[51,52]

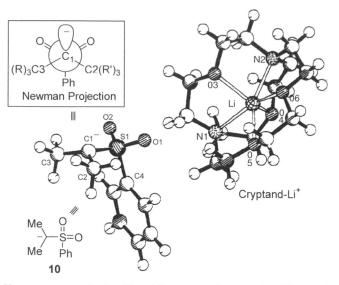

Figure 7.7. X-ray structure of anion **10** and the cryptand encapsulated lithium ion (Newman projection of the carbanion is shown in the box). (Adapted from reference 47.)

Phenyl rings[53] and/or heterocyclic[54] substituents have been used to stabilize carbanions and facilitate crystal growth to obtain their X-ray structures. However, they are usually flat at the C^- center, and in some cases they exhibit strong interactions with their counterions. Recent reports by Lawrence et al.[55] and Breher et al.[56] have described the use of pyrazole substituents with their π-systems kept out of plane from the anionic p-orbital by coordination with a metal ion (Fig. 7.8). The non-π-conjugated carbanion center adopts a pyramidalized sp^3 hybridization. The two examples in Figure 7.8 show how lithium and titanium atoms coordinate with the heterocycles, keeping the charge localized at the carbanion center. In carbanion **12**, the N–C^- distances ranged from 1.446 to 1.559 Å, in close agreement to the calculated distances for simple $N(sp^2)$–$C(sp^3)$ bonds, which range from 1.448 to 1.450 Å. Computational results showing a localized HOMO orbital on **11** also indicate the significant negative charge localization in these types of carbanions. In earlier work by Grim et al., the use of metal ions such as silver and mercury was shown to give analogous carbanions with the structure $^-C(P[Ph]_2S-)_3$–M^+, where the metal M^+ coordinates with the three sulfur atoms rather than to the negatively charged carbanion center.[57,58]

Anions are stabilized also by cyano groups. In the early 1960s, Bugg et al.[59] reported the nonplanar X-ray structure of $^-C(CN)_3$. However, the authors did not comment on any interactions with the counterion $^+NH_4$, which would most likely be strongly associated with the carbanion. Similarly, the pentacyano-*cyclo*-pentadienide ion $C_5(CN)_5^-$ is remarkably stabilized by the effect of the cyano substituents and the aromaticity of the ring. Recently, Richardson and Reed[60] challenged Webster's interesting suggestion[61] that $C_5(CN)_5^-$ is so stable that its conjugate acid cannot be prepared. Indeed, the mentioned authors were able to sililate and protonate

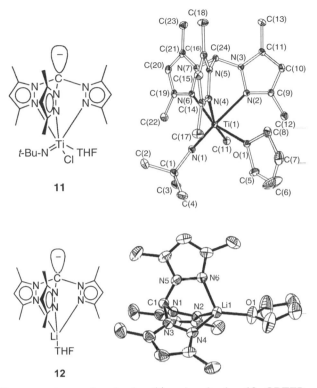

Figure 7.8. X-ray structures of carbanion **11** and carbanion **12**. ORTEP representations adapted from references 55 and 56.

the anion. They found that $^+Si(i\text{-}Pr)_3$ does not attack the aromatic system, but that it coordinates to a nitrogen in one of the cyano groups of the anion as shown in **13** (ORTEP representation adapted from reference 60). Reaction of **13** with triflic acid (HOTf) gave **14**, the conjugate acid of $C_5(CN)_5^-$, which unfortunately does not yield X-ray quality crystals and precipitates as a dark amorphous solid. The authors propose the polymeric structure **14** as the resulting product based on IR spectroscopy.

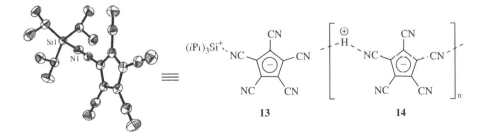

The X-ray structures of a few rather special dianions have also been obtained in recent years. The X-ray structure of a tribenzacepentalenediide[62] **15** in Figure 7.9

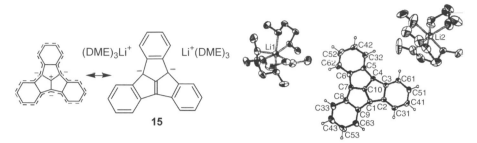

Figure 7.9. ORTEP representation of the X-ray structure of the dianion **15** complexed with two lithium ions chelated by dimethoxyethane (DME). (Adapted from ref. 62.)

suggests that the double negative charge is delocalized around the perimeter of the aromatic structure, as shown in the resonance structure in Figure 7.9 and proposed by the authors as a conjugated nonaromatic system. The dianion is not considered to be "free" since there are strong electrostatic interactions with the chelated lithium ions. The structure is bowl-shaped and the packing arrangements form layers of anions and cations. The three bonds in the center of the dianion structure are 1.40 Å on an average, which is relatively long as compared to the typical value of 1.32 Å for $C(sp^2)-C(sp^2)$ single bonds, leading the authors to suggest that the charge is delocalized around the perimeter. Furthermore, ^{13}C–NMR shows that the central carbon C10 is deshielded with $\delta = 177$ ppm, suggesting a partial positive charge at the center.

Triphosphonate **16** is an example of a 1,2-dianion stabilized both by the electronegative P=O groups[63] and by coordination of the phosphonate oxygens as shown in Figure 7.10.[64] The structure of **16** coordinates with two lithium ions, one THF molecule, and is solvated by three molecules of toluene. The Figure shows a complex of two dianions where the negative charges are located on carbons C1 and C2. Both centers are nearly orthogonal to each other to minimize the repulsion of the two negatively charged orbitals. The authors noted that the C1–C2 distance of 1.51 Å is

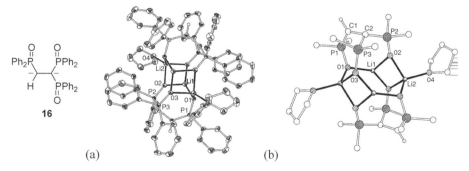

(a)   (b)

Figure 7.10. X-ray structure of dianion **16** shown in two different views: (a) ORTEP representation of two dianions complexed with four lithium ions and two molecules of THF, also excluding the toluene molecules found in the crystal. (b) View of the complex without the phenyl groups. (Adapted from ref. 64.)

short as compared to those determined for silicon substituted dicarbanions, which are 1.53–1.60 Å. This difference attributed to the stabilization of the P=O groups, which help delocalize the negative charges of the two carbanion centers. Indeed, they noticed an elongation of the P=O bonds of 1.54 Å, from a normal value of 1.49 Å (see Table 7.1).

7.2.3. Carbenes

Carbenes are divalent, neutral, highly reactive species with two nonbonding electrons in two orbitals. Though interest in carbenes can be traced back to more than a century ago,[65] it was not until the 1950s that the studies began to reveal more about the nature and reactivity of these intermediates.[66] The most prevalent electronic configurations are a closed shell singlet and an open shell triplet (Scheme 7.9). The closed shell singlet has the two electrons paired with opposite spin in a lower energy sp^2-hybridized, or σ-orbital, which leaves the higher energy pure p-orbital empty. This singlet state is expected to have a bent structure with an R−C−R angle of ca. 120°. During the early stages of electronic studies, triplet carbenes were believed to be linear, with an ideal R−C−R angle of 180°, resulting from the unpaired electrons occupying two pure p-orbitals. As computational chemistry and spectroscopic studies gained reliability, the structure of the parent carbene, :CH_2, was found to be bent, with an H−C−H angle of ca. 136° and the unpaired electrons occupying orbitals with high s-character. While electronic configurations with open shell singlets are also possible, they are higher in energy and are thought not to play a very important role. As computational chemistry gained momentum and proved reliability, a deeper insight into the structure of a large number of triplet carbenes uncovered the nonlinear nature of the ground state. In the case of the parent :CH_2 structure, the lowest energy state is a bent triplet with an H−C−H angle of approximately 136° with high s-character orbitals.[67,68,69,70] It has been shown that the structure and detailed electronic state of both singlet and triplet carbenes can be significantly affected by the nature of the R-substituents.

Studies have shown that carbene reactivity toward a wide variety of substrates is dramatically affected by the nature and multiplicity of the electronic state.[67,71] Similarly, the structure, electronic state, thermochemical stability, and reaction kinetics of both singlet and triplet carbenes can be significantly affected by the R-substituents.[72] If R provides steric hindrance, the carbene center can be shielded to slow down intermolecular reactions (kinetic stabilization). Additionally, bulky and/or geometrically

singlet or triplet

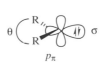

singlet

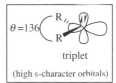

triplet
(high s-character orbitals)

Scheme 7.9

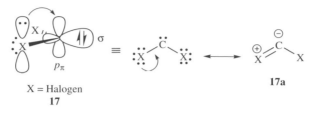

X = Halogen
17

17a

Scheme 7.10

constraining groups can affect the R−C−R angle, and hence influence the spin state of the system. The relative stability of the singlet and triplet states can also be affected by substituents that interact electronically with the carbene center.[73] Typical examples of carbenes with a lower triplet state include :CH_2, :CAr_2, :$C(CO_2R)_2$ and many (but not all) dialkyl carbenes. Singlet carbenes can be stabilized by groups R capable of "donating" electrons to the empty p_π-orbital of the singlet carbene such as R = -OR, -NR_2, halogen, etc. For example, the lone-pair electrons in dihalocarbenes **17** can be delocalized into the empty p_π-orbital (resonance structure **17a**, Scheme 7.10), thus stabilizing the singlet state structure as compared to that of the open shell triplet. The preparation of singlet and triplet carbenes suitable for single crystal growth and X-ray diffraction studies has been a remarkable challenge, which has met with more success in examples involving carbene singlet states.

7.2.3.1 Singlet Carbenes
Singlet carbenes have been the subject of many reviews.[65,72,74] It has been shown that electronic stabilization by neighboring groups can be achieved by electron donating groups **D**, such as –NR_2, which can delocalize their nonbonding electrons into the empty orbital as shown in structure **18a** in Figure 7.11. Alternatively, the lone pair of the carbene center can be delocalized into the available orbitals of electron withdrawing groups, **W**, or into the empty orbitals of substituents such as boron, as illustrated by **18b** in Figure 7.11. Finally, a combination of both **D** and **W** effects can also stabilize the singlet carbenes (**18c**) as represented by phosphinosilylcarbenes such as those shown in Figure 7.11. In this section, we will highlight a few examples of cyclic and acyclic carbenes stabilized by push-push and combined push-pull effects (Fig. 7.12).

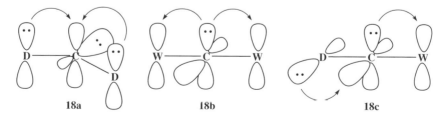

18a   **18b**   **18c**

Figure 7.11. Singlet carbene stabilizing substituents: **D** = electron donating groups, **W** = electron withdrawing groups.

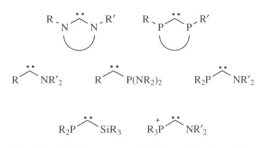

Figure 7.12. Generalized chemical structures of selected singlet carbenes with known X-ray structures.

The first X-ray structure of a singlet carbene was obtained in 1991 for carbene **19** by Arduengo et al.[75] The carbene center is a part of a five-membered ring with two adjacent nitrogen atoms linked by an ethylene bridge to complete a formal 6π aromatic system. Kinetic stabilization toward bimolecular reactions was accomplished by adding bulky adamantyl substituents to the nitrogen atoms. The resulting carbene displayed remarkable physical properties. The crystalline solid of **19** was stable in the absence of oxygen and moisture, and had a melting point of 240 °C. The X-ray structure revealed interesting geometric parameters. The N–:C–N angle of 102.2° was smaller than that of the imidazolium salt precursor by 7° and the: C–N distances of 1.37 Å are longer than those of the aromatic precursor, indicating a decrease in bond order from the delocalized aromatic system to the singlet carbene. The carbene is sterically and electronically stabilized, though it was later shown that bulky groups were not essential. Thus, carbene **20** was also obtained as a crystalline solid.[76] Furthermore, carbene **21** (Mes = Mesityl) demonstrated that aromaticity is not required.[77,78]

Changing from the cyclic 6π-aromatic system to a formal 7π-electrons in the pyrimidine core of naphthalene-substituted diamino, carbene **22** (ORTEP representation adapted from reference 79) resulted in a stable crystalline structure.[79] The X ray structure of **22** has N–:C–N angle of 115.5°, which is larger than those shown above; but the :C–N distance of 1.36 Å was very similar. Notably, ^{13}C-NMR spectroscopy showed a signal at δ 241.7 ppm corresponding to the deshielded carbene center, in accordance to the previously reported values.[75,76,77] These relatively subtle changes enabled the carbene to act as a strong σ-electron donor when used as a ligand for catalytic organometallic reactions.[80,81]

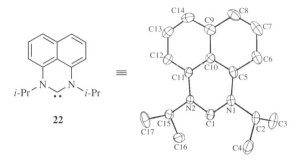

22

Bertrand and coworkers[82] have recently explored the effects of phosphorus in diphosphinocarbene **24** (ORTEP representation adapted from reference 82), which is analogous to the previously reported diamino carbene **23**.[83] Given that phosphorus is larger and more electron rich than nitrogen, the system was designed to test whether or not phosphorus would be a better donor than nitrogen toward the empty p-orbital of the carbene. Complete planarization of the phosphorus would suggest that it is a good donor with strong electron delocalization toward the vacant p-orbital. Indeed, the X-ray structure revealed a nearly planar structure near the phosphorus atoms, with slight puckering. The carbene exhibits interesting structural parameters. The :C–P distances are 1.692 Å on an average, which are short when compared to those of typical C–P single bonds (>1.80 Å). This shortening indicates a strong electronic interaction between the lone-pair electrons in phosphorus and the empty carbene orbital, as suggested by the authors. Interestingly, the P–:C–P angle of 98.2° is rather small, as compared to the 5-membered N-heterocyclic diaminocarbenes (NHCs) discussed above. Finally, the ^{13}C-NMR revealed a carbene signal at a higher field than that of the NHCs, which is thought to be because of the adjacent electron rich phosphorus. It has been suggested by Martin et al. that reactions requiring catalysts with electron-rich ligands are very likely to benefit from P-heterocyclic phosphinocarbenes (PHCs).[82]

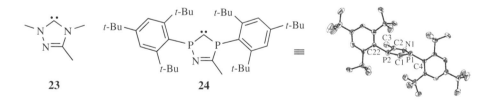

23   **24**

Insight into the effects of ring strain on the stability and structure of NHCs is available from the X-ray studies of the four-membered diamino carbene **25** (ORTEP representation adapted from reference 84), recently reported by Despagnet-Ayoub and Grubbs.[84] Unexpectedly, the highly strained carbene is not C_2 symmetric. The N1–C1–N2 angle of 96.7° is appreciably smaller than those of other NHCs, but only 1.5° smaller than the five-membered phosphinocarbene **24**. The C1–N1 and C1–N2 distances of 1.373 Å and 1.387 Å, respectively, are both short, indicating that the nitrogen lone-pair electrons are strongly interacting with the carbene center. Another striking feature of **25** is the slight pyramidalization around the N1 and N2 centers, which are generally flat in other diaminocarbenes. Finally, carbene **25** revealed a very deshielded carbene center with a ^{13}C-NMR signal at δ 285 ppm, which is 100 ppm downfield from that of phosphinocarbene **24**.

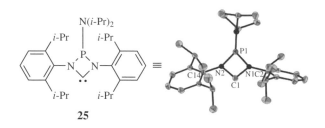

25

In another modification analyzed by single crystal X-ray diffraction, the carbene backbone of five- and six-membered ring NHC systems was modified with boron atoms (Fig. 7.13) [85,86] A simple qualitative analysis suggests that the empty p_π-orbital

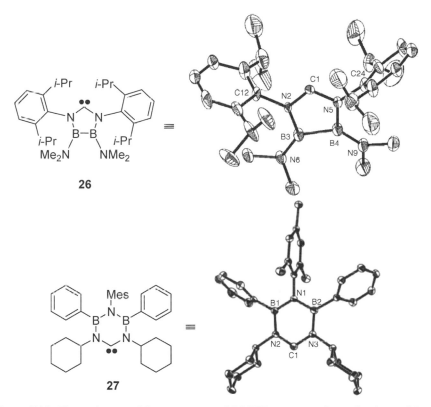

26

27

Figure 7.13. Top and center: Line structure and ORTEP representations of carbenes **26** and **27**. Bottom: N,B-heterocyclic carbenes (NBHCs) showing the competition between the N-C vs. N-B electronic interaction, that is mesomeric effect (the former is preferred). ORTEP representations adapted from references 85 and 86.

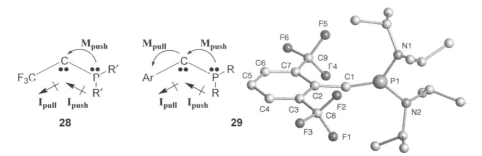

Figure 7.14. Mesomeric (M_{push}/ M_{pull}) and Inductive (I_{push}/ I_{pull}) effects in carbenes **28** and **29** (Ar = 2,6-bis[trifluoromethyl]phenyl, R = *iso*-propyl), and molecular representation of the X-ray structure of **29**. (Adapted from reference 87.)

of the carbene center should compete for π-lone pair donation from nitrogen with the empty *p*-orbital of trivalent boron. It was found that the N–:C–N angle of carbene **26** is 108.5° while that in **27** is slightly wider at 114.5°, while the: C-N distances in **26** and **27** are 1.38 Å (ave.) and 1.37 Å, respectively (the N-B distances are 1.47 Å and 1.44 Å, respectively). On account of these structural parameters in **26**, the authors suggested a weak electron donation interaction from nitrogen to boron, whereas in **27** it was suggested that there was a delocalized π-system. The[13]C-NMR shift of the carbene center **26** appears very downfield at δ 304 ppm, while that of **27** appears at δ 281 ppm, a value that is very similar to that of carbene **25**.

The effects of different push-pull substituents on the stability and structure of carbenes were recently analyzed by Buron et al. using carbenes **28** and **29** (Fig. 7.14).[87] While carbene **28** has inductive push–pull effects from the electron-poor trifluromethyl and electron-rich phosphine, and a donating mesomeric effect only from the latter, carbene **29** has both push–pull inductive and resonance effects from the fluoromethyl-aromatic and phosphine substituents. While carbene **28** is remarkably stable in solution at −30 °C, it exhibits the reactivity of a transient species (it should be noted that the CF_3 groups are inert toward insertion into the C–F bond). Carbene **29** was even more stable, yielding a crystalline solid with a melting point of 68–70 °C. The X-ray structure of **29** suggests that 2,6-bis(trifluoromethyl)phenyl, the "pull" substituent, withdraws electron density from the σ lone-pair of the carbene, as interpreted by the: C1–C2 distance of 1.390 Å, which clearly has a double bond character. The same effect may be responsible for the rather wide C2–C1–P1 angle of 162.1°. On the contrary, the groups around phosphorus are trigonal planar and orthogonal to the aromatic plane. The C1–P1 distance of 1.544 Å is shorter than a typical single C-P bond (i.e., 1.81 Å, Table 1), suggesting a strong delocalization from phosphorus to the carbene center. Carbene **29** was designed so that both groups significantly contribute to the overall stability.

Testing the limits of carbene stabilization by substituents, Bertrand and coworkers reported the synthesis and analysis of (amino)(aryl)carbenes where only one substituent can contribute to the stability of the carbene center.[88] Carbene **30** was designed to have the amino substituent as a π-donor and the 2,6-bis(trifluoromethyl)

(a)

(b)

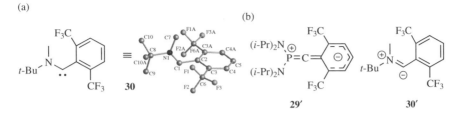

Figure 7.15. (a) Molecular representation of the X-ray structure of carbene **30**. (b) Better alternate representations of the electronic structures of carbenes **29** and **30**. (Adapted from ref. 88.)

phenyl group as a bulky spectator that may shield the carbene center from bimolecular and intramolecular reactions. While the aromatic system in the above carbene **29** acted as a "pull" substituent, the structure of carbene **30** was expected to have the aryl group π-system orthogonal to the doubly occupied carbene lone pair. The single crystal X-ray structure of **30** shown in Figure 7.15a confirmed this expectation. That the aryl group acts only as a "spectator" is suggested by the relatively long C1–C2 distance of 1.453 Å, and the very short C1–N1 distance of only 1.283 Å. The C1–C2 distance is close to that of an $C(sp^2)–C(sp^2)$ single bond (1.48 Å) and the C–N bond is significantly shorter than those of other carbene-nitrogen bonds (e.g., 1.37 Å in carbene **19**). Additionally, the carbene angle is smaller in **30** than in **29** (angle N1–C1–C2 = 121.0°). Comparing the structural parameters of **29** and **30** leads to the suggestion that structures **29′** and **30′** in Figure 7.15b are important contributors, with the aryl group in **30** not interacting with the carbene center.

"Push-spectator" carbenes of the type **31** (R, R′ = alkyl) were synthesized and reacted with various Lewis Acids to compare the reactivity of the phosphorus and carbene centers. Two such reactions are shown in Scheme 7.11.[89] From an X-ray structural analysis, the phosphorus substituent was shown to act as a spectator, leaving its lone pair available to react in a Lewis basic manner. When carbene **31** was reacted with BF_3, only the carbene adduct **32** was formed. By contrast, when **31** was reacted with the softer Lewis Acid BH_3, it was the phosphorus that reacted to yield adduct **33**. These types of carbenes exhibited ^{13}C-NMR shifts in the range of 320–348 ppm, a P–C–N angle of 116.5°, a short: C–N distance of 1.296 Å, and a long: C–P distance of 1.856 Å. The latter is very similar to that of a typical C–P single bond.

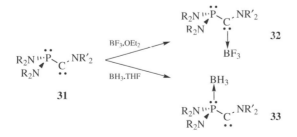

Scheme 7.11

In another study, Bertrand and co-workers analyzed the site selectivity of amino-phosphino-carbene **34** toward substitution reactions.[90] It was shown that **34** reacts with methyltriflate to give crystalline carbene **34a**, which has m.p. = 50°C and exhibits a ^{13}C-NMR shift of 302 ppm at the carbene center. This result suggests that the phosphorus group was primarily a spectator that did not interact with the carbene, in agreement to the long C1–P1 distance of 1.770 Å. The X-ray structure also suggests that the amino group interacts strongly with the carbene center, as indicated by a short C1–N1 distance of 1.287 Å (Scheme 7.10, ORTEP representation adapted from reference 90). Carbene **34** was subjected to the reactions shown in Scheme 7.12, where all the products were obtained quantitatively and were characterized fully.

7.2.3.2 Triplet Carbenes While the synthesis and modification of singlet carbenes continue yielding stable and isolable species that can be employed as ligands for catalysts,[91] triplet carbenes have proved to be far more difficult to isolate and crystallize.[92] Unlike singlet carbenes, which can be thermodynamically stabilized with electron donating and withdrawing substituents and kinetically stabilized with bulky substituents, persistent triplet carbenes can be obtained only by shielding the carbene center with bulky substituents in the adjacent α- and α'- positions or by one-electron delocalization. Bulky substituents tend to widen the carbene angle, simultaneously favoring the triplet state and providing the kinetic stability that extends its lifetime. Although the kinetic stabilization of triplet carbenes to lifetimes

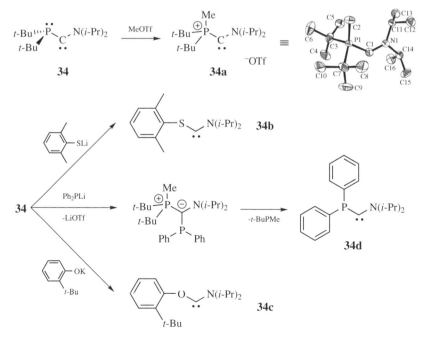

Scheme 7.12

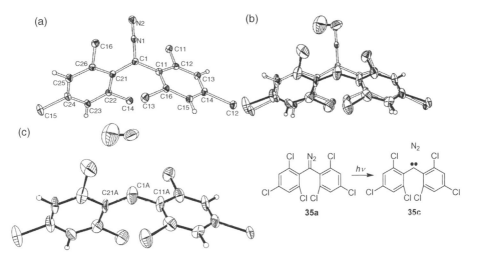

Figure 7.16. (a) ORTEP representation of the diazo compound **35a**; (b) disordered structure of both **35a** and **35c**; and (c) ORTEP representation of the carbene **35c** after removing that of **35a** in the disordered structure. (Adapted from ref. 93.)

of several days has been possible, to the best of our knowledge, the only X-ray structure obtained for a triplet carbene was reported in 2001 by Kawano et al. by generating carbene **35c** in situ upon UV irradiation of the diazo precursor **35a** at 80 K (Figure 7.16).[93] However, extended photolysis or crystal warming resulted in carbene dimerization and loss of crystallinity. A comparison of the X-ray structures of the pure diazo and the trapped carbene revealed that the carbene angle is significantly larger than that of its diazo precursor, 142° and 127°, respectively. A decrease in the C–C_{Ar} bond distance in going from 1.48 Å in the diazo compound to 1.43 Å in the carbene, indicates some electronic interaction between the aromatic system and the carbene center. Notably, the observations of this study are consistent with expectations from computational chemistry and other spectroscopic techniques such as EPR.[94,95,96,97]

7.2.4. Radicals and Biradicals

Carbon-centered organic radicals are highly reactive trivalent species with only one nonbonding electron. While most known radicals have their unpaired electron in a pure p- or a delocalized π-orbital, there are also examples of radicals centered in sp^n hybrid σ-orbitals, such as the well known phenyl and cyclopropyl radicals.[98] The first radical reported in the literature is credited to Gomberg's landmark paper in 1900 when he postulated the formation of triphenylmethyl radical **36**, also known as trityl.[99,100] The trityl radical is an example of a persistent radical that exists in equilibrium with the dimeric species **37**. Persistent carbon-centered radicals that are suitable for crystallization can be obtained by a combination of kinetic and thermodynamic stabilization, generally in the form of steric shielding of the radical center and resonance delocalization or substituent effects. Stable carbon-centered radicals in the solid state are very attractive due to their potential as organic magnetic materials.[101,102]

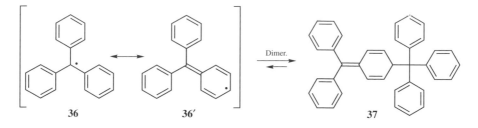

| 36 | 36′ | | 37 |

In a recent study, Armet et al.[103] carried out the synthesis, structural analysis, and spin-density determination of nine stable trityl radicals **38** with different degrees of chlorination. Six of the radicals gave X-ray quality crystals. Five were found to be trigonal planar about ·C the radical center, indicating that the unpaired electron is in a pure p-orbital. The sixth radical was slightly pyramidalized, suggesting that the singly occupied orbital has s-character. As expected for most triarylmethyl structures, the aromatic rings adopted a propeller conformation.[104] A bond distance of 1.72 Å between ·C and C_{Ar} is longer than the typical value of 1.48 Å for a $C(sp^2)$–$C(sp^2)$ single bond, but quite normal for these types of structures.

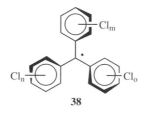

38

In a search of purely organic, carbon-centered, high-spin materials, Veciana and coworkers designed and synthesized the tri-radical **39**.[105] Interestingly, they obtained two types of solvent-containing single crystals with C_2 and D_3 symmetric structures by crystallization from n-alkanes (Fig. 7.17). The three radical centers in both **39-C_2** and **39-D_3** are close to trigonal planar and the ·C–$C_{Aromatic}$ bond lengths of 1.46 and 1.49 Å, respectively, which are much shorter than those in **38**, indicating that there is partial delocalization of the electron into the neighboring aromatic system. All the aromatic C_{Ar}–C_{Ar} distances had normal values of ca. 1.38 Å.

It is well known that delocalized stable radicals may have potential for the construction of solid state conducting materials.[106] The phenylalenyl radical **40** has been considered a good candidate with its spin density delocalized over 13 carbons in its π-conjugated system. Unfortunately, **40** exists in equilibrium with its dimer and it decomposes at modest temperatures. To overcome the dimerization problem, Goto et al.[107] and Koutentis et al.[108] synthesized substituted radicals **41** and **42**.

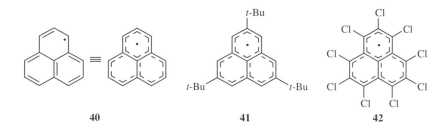

40   41   42

The X-ray structure of **41** revealed that the C–C distances vary from 1.374 to 1.421 Å in the cyclic system, with a nearly planar structure and a slightly distorted D_{3h} symmetry (Fig. 7.18a).[107] Radicals **41** pack in a herringbone pattern of π–π dimers separated by interplanar distances of 3.20 to 3.32 Å, which is significantly shorter than the sum of the van der Waals radii of 3.4 Å and result in antiferromagnetic order. A striking difference is observed in the X-ray structure and bulk properties of radical **42** (Fig. 7.18b).[108] First of all, the radical is no longer planar as the strong Cl$\cdots$Cl interactions with a distance of 3.5 Å result in a distorted propeller-like structure. The presence of bulky chlorine atoms and the nonplanarity of the radical increase the intermolecular π–π distances to 3.78 Å and allow for a columnar alignment of the π-systems with no significant spin–spin interactions above 100K.

Another example of π-delocalized radical is cyclopentadienyl **43**, which was reported by Sitzmann et al. to be rendered persistent and crystalline by derivatization with *iso*-propyl groups as in radical **44** (Fig. 7.19).[109,110] An interesting feature in the X-ray structure of **44** is that the *iso*-propyl groups adopt a paddlewheel-like conformation, also showing disorder in the crystal. The bond distances show that the radical is not D_{5h} symmetric, but that it is slightly distorted (C1 C2 = 1.412 Å, C2–C3 = 1.401 Å). Radical **44** also packs in a columnar motif with rings eclipsing each other

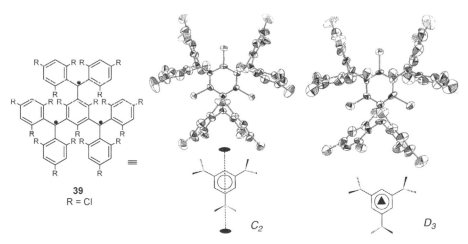

39
R = Cl

C_2 D_3

Figure 7.17. ORTEP representations of the triradical **39** in the two diastereomeric structures, C_2 and D_3. (Adapted from ref. 105.)

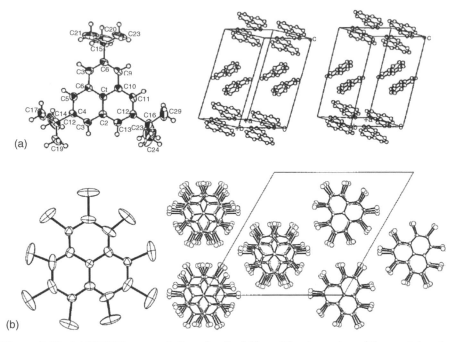

(a)

(b)

Figure 7.18. (a) ORTEP representation of radical **41**, and the stereoview of the crystal packing, with out the *t*-butyl groups. (b) ORTEP representation of radical **42**, and the packing arrangement viewed along the *c*-axis to show the trimeric interactions. (Adapted from refs. 107 and 108.)

at a rather long distance of 5.82 Å. While spin–spin interactions were not reported in this study they are likely to be weak, suggesting a paramagnetic material.

While structural evidence indicates that radical **44** is delocalized through the five carbon centers, it has been suggested that annelation of the cyclopentadienyl framework with rigid bicyclic systems[111,112] leads to localized structures, as in compound **45** (Fig. 7.20).[113] The X-ray structure of **45** indicates that the rather long distances between C1–C2 (1.445 Å), C1–C5 (1.442 Å), and C3–C4 (1.482 Å) correspond to

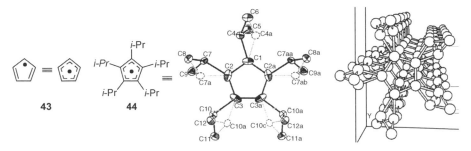

Figure 7.19. Cyclopentadienyl radical **43** and its per-*iso*-propyl analogue **44**, also showing the ORTEP representation of a columnar packing in the solid state of **44**. (Adapted from ref. 109.)

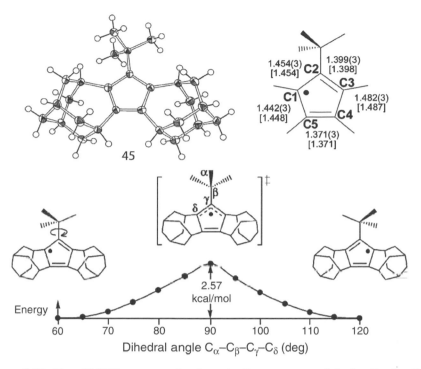

Figure 7.20. Top: ORTEP representation from the X-ray structure of the localized radical **45**, and selected bond parameters (the parenthesis show the values determined computationally at the UB3LYP/6-31G* level of theory). Bottom: Density functional theory calculations for the rotation of the *tert*-butyl group. (Taken from ref. 113.)

localized single-bonds while the shorter distances between C2–C3 (1.399 Å) and C4–C5 (1.371 Å) correspond to localized double bonds (Fig. 7.20). The conformation of the *tert*-butyl group is such that one methyl eclipses the plane of the five-membered ring, giving a C_s symmetric structure. Furthermore, calculations show that rotation of the *tert*-butyl group to achieve a C_{2v} symmetric structure leads to a transition structure (TS) that is 2.57 kcal/mol above the minima (Fig. 7.20, bottom). The TS corresponds to a delocalized allyl radical, favored by a Jahn-Teller distortion. As evidence of spin localization, the authors reported that radical **45** is easier to oxidize than radical **44**.

The short lifetimes of carbon-centered monoradicals are generally reduced in the case of diradicals due to their propensity to form covalent bonds. It has been suggested that stable diradicals may be observable from highly strained bicyclic molecules where the TS for inversion is a diradical.[114,115] Unfortunately, only persistent diradicals have been obtained in this way.[116] Akin to this approach, in a recent attempt to generate the oxyallyl diradical, Sorensen and co-workers synthesized two substituted bicyclobutanones hoping to stretch and homolytically break the central bond using bulky substituents, which would also stabilize the diradical.[117] Though the bicyclobutanones did not yield the desired oxyallyl derivative, the X-ray structures showed

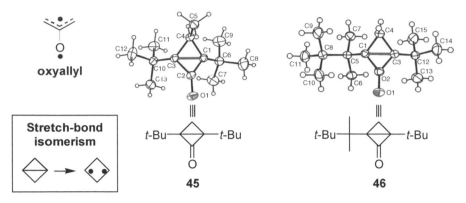

Figure 7.21. Structure of the oxyallyl radical, and the ORTEP representations from X-ray of the bicyclobutanones **45** and **46**. Box: Schematic representation of the stretch-bond isomerism of bicyclo[1.1.0]butane. (Adapted from ref. 117.)

contributions from such species (Fig. 7.21). In both **45** and **46**, the bridging bond and the α-carbonyl bonds are remarkably long, at 1.69 Å and ca. 1.43 Å, respectively. The C=O bond also showed a slight pyramidalization, suggesting some weakening of double bond as it would be expected for a species with allyl and alkoxy radicals.

Introducing heteroatoms into the bicyclic system has been fruitful, and spectacularly so for inorganic heterocycles that include phosphorus and boron.[118,119,120,121] However, the ring system containing only two phosphorus and two carbon atoms has been shown to exist in the ring-opened forms as in **47** and **48**.[122,123,124] The four-membered ring is planar for both diradicals and, in the case of **47**, both phosphorus and carbon radical centers are pyramidalized. The mesityl and chlorine groups are *syn-* to each other, leaving the mesityls *anti-*. In the modified structure **48**, both carbons are trigonal planar and the phosphorus pyramidalized (see Fig. 7.22). ^{13}C and ^{31}P NMR spectroscopies show strongly deshielded signals for the carbons and the phosphorus in the ring. Calculations showed that the most stable configuration for the diradicals is the singlet state, which is stabilized by conjugation with the nonbonding electrons of the phosphorus atoms,[122] and possibly by through-bond interactions.[118]

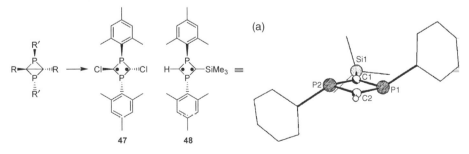

Figure 7.22. Diradical structures **47** and **48**, showing the model based on the X-ray structure of **48**. (Adapted from ref. 123.)

7.2.5. Conformational Analysis of Dioxocarbenium Ions

In the last example of this section, we describe a case where the X-ray structure of a reactive intermediate led to the convincing explanation of reaction selectivities. Woerpel and co-workers exploited the power of X-ray structural analysis and computations to design cyclic, stable dioxocarbenium ions, that were "trapped" by crystallization in their most stable conformers.[125–127] Since oxocarbenium ions are reactive, short-lived intermediates, dioxocarbenium ions were employed to increase stability by charge delocalization over two oxygen atoms.[127] Calculations showed that the preferred conformation of the R-group is pseudo-axial by 5.3 kcal/mol, when R is an alkoxy substituent. However, when R is an alkyl group, the pseudo-equatorial conformer is energetically preferred (Fig. 7.23a). Indeed, the above results were confirmed by the X-ray structures of carbenium ions **49**, R = methyl, and **50**, R = benzyl (Fig. 7.23b). The structures clearly show that a methyl substituent in the 4-position of the ring leads to the pseudo-equatorial conformation. In contrast, a benzyloxy group leads to the pseudo-axial conformer. On the basis of the calculations and X-ray analysis, the authors proposed that the conformational preference has to do more with through-bond electrostatic interactions than with neighboring group participation, also known as anchimeric assistance. This example elegantly shows how reactive intermediates

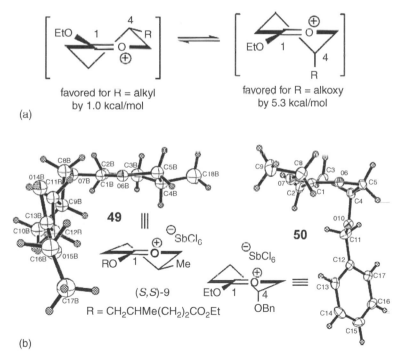

Figure 7.23. (a) Pseudo-equatorial and pseudo-axial conformers of the dioxocarbenium ions. (b) ORTEP representations of the X-ray structures obtained for carbenium ion **49** and **50**. (Taken from ref. 127.)

can be stabilized and crystallized to address specific structural questions, which may help to calibrate or validate a parallel computational study.

In conclusion, all the above examples highlight results that have led to the isolation of reactive intermediates, which at some point since their discovery were believed to be fugacious. At best, we have come to understand much information about their structure and reactivity. Through proper design, the crystallization properties of reactive intermediates have rendered them important not only for extracting structural parameters, but also for understanding reactivity in organic chemistry and designing unusual properties in the booming field of materials science.

7.3. REACTIONS IN CRYSTALLINE MEDIA: FUNCTION

7.3.1. Exerting Control of Reactive Intermediates in Crystalline Media: Concepts

Reactive intermediates in solution and in the gas phase tend to be indiscriminant and ineffective for synthetic applications, which require highly selective processes.[128] As reaction rates are often limited by bimolecular diffusion and conformational motion, it is not surprising that most strategies to control and exploit their reactivity are based on structural modifications that influence their conformational equilibrium, or by taking advantage of the microenvironment where their formation and reactions take place, including molecular crystals.[129]

In order to be useful in synthesis, reactions in crystals must be predictable, versatile, and efficient. As many solid state reactions have been shown to rival the selectivities observed in enzymatic processes, there has been an increased interest in their understanding and development.[130,131] Furthermore, as reactions in crystals require no solvents or reagents that may be harmful to the environment, they are particularly promising within the context of green chemistry.[132] With that in mind, we have recently suggested a very simple and general strategy to "engineer" reactions in crystals by taking advantage of high-energy species (Scheme 7.13). This suggestion is based on the fact that many high-energy species can release their energy through low barrier reactions with minimal atomic and molecular displacement, such as bond breaking and bond making processes (Scheme 7.13).[9,10,128] By appropriate design of crystalline precursors, the generation of high-energy species can be accomplished thermally or photochemically. Excited states themselves may lead directly to products in a concerted chemical process, or they may generate RIs in a stepwise manner under the structural control of the crystal to give potentially valuable products. Although the types of reactive intermediates that may be generated in the solid state is limited only by the availability of suitable precursors, and in principle they may be neutral or ionic, but to date only a few have been studied.

In order to appreciate reactions in crystals, it is of value to consider the topochemical postulate and the reaction cavity concepts as the starting point. Kohlshutter proposed in 1918 that the crystal lattice plays an important role on the outcome of chemical reactions as a result of its rigidity and topology, suggesting that *reactions*

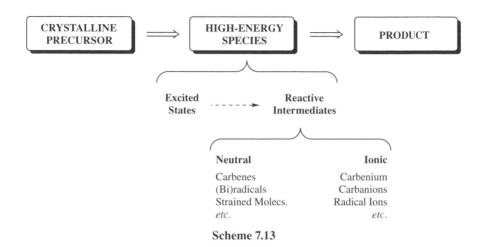

Scheme 7.13

in crystals can only occur with minimal atomic and molecular motion.[133] Refined by Schmidt and co-workers in the mid-1960s with the photodimerization of cinnamic acid derivatives as the main example, the "topochemical postulate" has been the most important source of insight into solid state reactions.[134,135] A few years later, Cohen complemented the topochemical postulate with the "reaction cavity" concept, which is used to describe *the size, shape, and volume of the space occupied by the reactant(s).* It was pointed out that the reaction cavity is determined by the boundary that results from the van der Waals contacts between the reactant and its neighboring molecules.[136] It is generally assumed that the reaction cavity can only withstand the most minimal deformations throughout the course of a reaction, so that only those transition states and products that closely resemble the reactant can be formed within the reaction cavity.

An example of a reaction cavity is illustrated by the van der Waals surface plots derived from the X-ray structures of *cis*-2,6-dihydroxy-2,6-diphenylcyclohexanone **51** and its photodecarbonylation products, the *cis-* and *trans*-2,6-diphenylcyclopentane-1,2-diols **52** and **53** (Scheme 7.14).[137] While the two products are formed in equal amounts in solution by the photochemically induced loss of CO,[128] the solid state reaction forms the *cis*-isomer **52** as the only product. The selectivity and least motion nature of the solid state reaction are determined by the reaction cavity (Scheme 7.15), while its topochemical nature is related to the order and rigidity of the crystal structure, and not to the influence of defect sites.

Elementary reactions in crystals may be unimolecular or bimolecular, and they may involve one or more components with properly designed complexes and mixed crystals. Photodimerizations and other bimolecular reactions[138] require the reactants to crystallize with precise distance, alignment, and orientation.[135a] In contrast, unimolecular reactions depend only on the intrinsic energetics of the reacting molecule and the rigidity and homogeneity of the crystalline environment. Given the high success rate of conformational predictions and the large amount of mechanistic

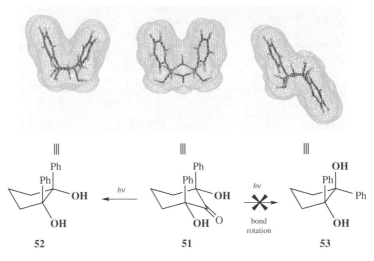

Scheme 7.14. See color insert.

information now available through quantum mechanical calculations, *chemists should be able to recognize (and perhaps invent) precursors for a wide range of RIs, predict their conformational preferences in the solid state, consider their most likely reaction pathways, calculate their energetics, and use this information to engineer new solid state reactions for mechanistic analysis or synthetic applications.*

It should be noted that reactions in crystals are nonlinear processes where product molecules accumulate as a function of reaction progress, so that the integrity of the lattice may be compromised. With azo compounds and diacylperoxides as precursors for radical pairs, and using EPR and FTIR as the main analytical tools, McBride and co-workers[139,140] have shown that the local stress caused by extrusion of N_2 or CO_2 has a major effect on the rates of radical recombination.[141] While very subtle changes in reaction kinetics and detailed reaction trajectories can be documented at very low conversions ($\ll 1\%$), there are many examples with product selectivities and conversions that are ideal from a synthetic perspective. Thus, it is possible that local stress may influence detailed reaction trajectories and elementary rate constants but it has a small effect on the final macroscopic outcome, which is ultimately determined by the reaction cavity and the tolerance of the reacting crystal to the product molecules. The experimental factors affecting the progress of solid-to-solid reactions have been recently reviewed.[11,142,143]

7.3.2. Radical Pairs and Biradicals as Transient Species

Two examples from ketone photochemistry that has been recently analyzed within the context of solid-to-solid transformations are the Norrish type I[128,143] and Norrish-Yang type II[144,145] reactions. In general terms, the type I reaction consists of a homolytic cleavage of bond α-to the carbonyl to generate an acyl-alkyl radical pair (**RP-A**) or an acyl alkyl biradical (**BR-A**) when the ketone is cyclic (Scheme 7.15).

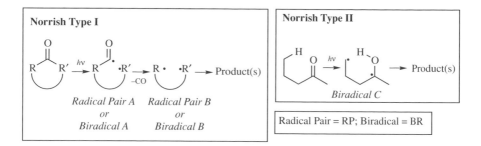

Scheme 7.15

So far, the solid state type I reaction has been reliable only when followed by the irreversible loss of CO to yield alkyl–alkyl radical species (**RP-B** or **BR-B**) in a net decarbonylation process.[128] The type II reaction relies on the presence of a γ-hydrogen that can be transferred to the carbonyl oxygen to generate the 1,4-hydroxy-biradical (**BR−C**). The type-I and type-II reactions are generally favored in the excited triplet state and they often compete with each other and with other excited state decay pathways. While the radical species generated in these reactions generate complex product mixtures in solution, they tend to be highly selective in the crystalline state.

7.3.2.1 Norrish Type I and Decarbonylation Reactions

The first examples of photochemical decarbonylations can be traced back to the early 1900s.[146] It was later shown by Norrish and Appleyard that reaction occurs in two separate steps, α-cleavage[147,148] and decarbonylation, via radical intermediates.[149] The first photodecarbonylations of crystalline ketones were reported by Quinkert et al.[150] in 1971 with samples of cis- and trans- diphenyl-2-indanones **54** and **55** (R=H, Me, Scheme 7.16). While reactions in solution yielded mixtures cis- and trans-benzocyclobutanes, reactions in crystals proceeded with very high stereocontrol (Scheme 7.14).[151] It was suggested that biradicals **BR1** and **BR2** lose their stereochemical information by conformational equilibration in solution, but they remain conformationally trapped within their corresponding reaction cavities in the crystalline state. Later studies showed that triplet bis-benzylic 1,4-biradicals correspond to the excited triplet states of the isomeric ortho-quinodimethanes. As indicated on the right side of Scheme 7.16, intersystem crossing in solution produces the observed benzocyclobutanes along with ground state ortho-quinodimethanes, which ultimately undergo a thermal cyclization to form the aromatic compounds. In crystals, however, the rigidity of the environment hinders bond rotations so that the two biradical intermediates, **BR1** and **BR2**, go on to form their corresponding products without losing their original stereochemistry.

While the examples in Scheme 7.16 hinted at the practicality of the solid state photodecarbonylation of ketones, the factors controlling this reaction remained unknown until very recently. As a starting point to understand and predict the photochemical behavior of ketones in terms of their molecular structures, we recall that most of the thermal (kinetic) energy of crystals is in the form of lattice vibrations,

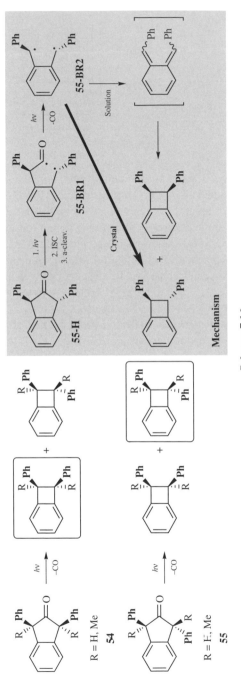

Scheme 7.16

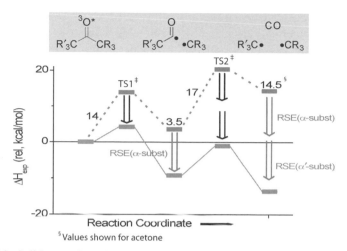

Figure 7.24. Solid-state photochemical decarbonylation model for ketones. The dashed path corresponds to the experimentally determined energies of acetone (in kcal/mol).[128] The effects of substituents with radical stabilizing energies (RSEs) are illustrated by the solid line in the reaction coordinate. See color insert.

rather than collisions, translations, rotations, or conformational motions. Since reactants can only explore a very limited region of their energy surface and require their environment to be structurally pre-organized, one may postulate that activation energies in crystals are determined by changes in enthalpy, which are relatively easy to obtain, rather than by changes in entropy, which are difficult to quantify.

To obtain a predictive reactivity model, we build a thermochemical diagram that considers the changes in enthalpy for the (high-energy) reactant, key intermediates, and transition states along the reaction coordinate (Fig. 7.24). On the basis of the Hammond's postulate,[152] one may expect that high-energy reactive intermediates will have highly exothermic reactions with low activation enthalpies and "reactant-like" transition sates. The diagram is first constructed for a "parent" compound, recognizing that one can make structural modifications to alter the enthalpies of "derivatives" according to known substituent effects. To implement this approach, one may rely on experimental enthalpies from thermochemical tables, on values derived from an increment approach, or on values obtained by quantum mechanical calculations.

As it pertains to the solid state photodecarbonylation reaction, the model assumes that most aliphatic ketones have similar excitation energies, that reactions are more likely along the longer-lived triplet excited state, and that each reaction step must be thermoneutral or exothermic to be viable in the solid state.[153,154,155] Using acetone and its decarbonylation intermediates as a reference reaction (dashed lines in Fig. 7.24), we can analyze the energetic requirements to predict the effects of substituents on the stability of the radical intermediates. The α-cleavage reaction of triplet acetone generates an acetyl–methyl radical pair in a process that is 3.5 kcal/mol endothermic and the further loss of CO from acetyl radical is endothermic by 11.0

kcal/mol. Recognizing that the triplet acyl–alkyl biradical formed by α-cleavage can go back to the ground state after intersystem crossing to the singlet, it was suggested that the irreversible loss of CO would be the product-determining step in solids. Not surprisingly, crystalline acetone is unable to overcome such unfavorable energetics within the limited lifetime of its excited triplet state and it is photochemically stable. Thus, for the overall reaction to be thermoneutral or exothermic, α-substituents with radical stabilizing energies RSE > 11 kcal/mol are required.

The radical stabilization energy RSE of a given α-substituent R (or group of substituents, R^1, R^2, etc.) is defined in terms of the difference in the C–H bond dissociation energy of the corresponding hydrocarbon, RH_2C–H, as compared to that in methane, H_3C–H.[156] The effects of one, two, or three different R groups can be readily estimated and experimentally tested with an appropriate crystalline ketone. It is known that the activation energies for the α-cleavage (TS1) and decarbonylation (TS2) steps are lowered as the radical pairs become more stable,[157] as indicated by the solid-line reaction coordinate in Figure 7.24. It should be noted that substituents with RSE > 11 kcal/mol are needed on the two α-carbons for the solid state reaction to take place since the alkyl–acyl biradical formed in the first step can intersystem cross to the singlet state and return to the starting ketone if the loss of CO is not sufficiently fast. Indeed, it has been shown experimentally that a wide variety of R-groups with RSE > 11 kcal/mol do enable the solid-state decarbonylation reaction when they are present on both α-carbons and that reaction is more efficient as the RSE values are larger.[155] Those results indicate that solid state reactivity can be predicted a priori for many types of substituents R, simply by taking advantage of well-documented[158] or appropriately calculated RSE values.

The first structure-reactivity correlation on the effects of phenyl substituents in the solid state was reported by Choi et al. in 1996.[159] Compounds **56–61** (Scheme 7.15) were selected because they form good crystals, have similar solution reactivities, and have different benzylic substitution patterns. The relative quantum efficiencies of **56–61**, determined in side-by-side photochemical experiments, met with the expectation that radical stabilizing groups are needed at the two α-positions for the two sequential bond-cleavage reactions to take place. The results are illustrated in Scheme 7.17 by ordering the structures with an increase in the efficiency of their α-cleavage and decarbonylation steps along the vertical and horizontal directions, respectively. While crystals of ketone **56** were photostable, crystals of ketones **57–61** yielded decarbonylation products with relative quantum efficiencies that varied by a factor of 100, as indicated in parenthesis below each structure. The photostability of **56** can be explained by the lack of a phenyl substituent at the 6-position, clearly showing that formation of a primary radical, by a ca. 6.5 kcal/mol endothermic decarbonylation is not favorable in the solid state. Indeed, with phenyl substituents at two α-positions, samples of *cis*-2,6-diphenylcyclohexanone **57** are photochemically active, although reaction times are about 100 times longer than those of *cis*-2,6-dimethyl-2,6-diphenylcyclohexanone **61**. An increase in solid state reactivity from **57** to **58** to **60**, and from **59** to **61** can be assigned to the increasing RSE effects of the substituents that facilitate the α-cleavage reaction. An increase in solid state reactivity from **56** to **57**, from **58** to **59**, and from **60** to **61** can be assigned to the increasing RSE effects of the

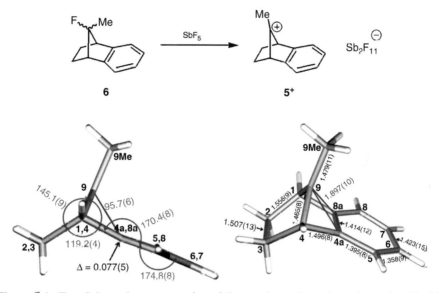

Figure 7.4. Top: Schematic representation of the reaction to form the cation-anion **Me-5**$^+$, $Sb_2F_{11}^-$. Bottom: Two different views of the X-ray structure of the **Me-5**$^+$, showing the important structural parameters. (Adapted from reference 27.)

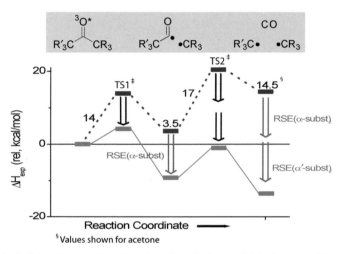

Figure 7.24. Solid-state photochemical decarbonylation model for ketones. The dashed path corresponds to the experimentally determined energies of acetone (in kcal/mol).[128] The effects of substituents with radical stabilizing energies (RSEs) are illustrated by the solid line in the reaction coordinate

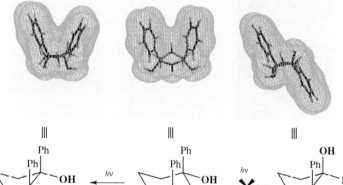

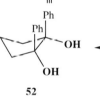

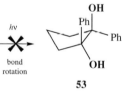

Scheme 7.14

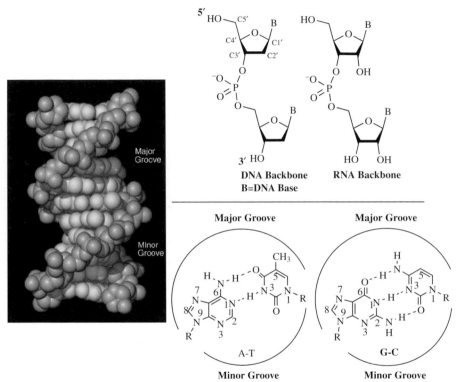

DNA Backbone
B=DNA Base

RNA Backbone

Major Groove

Major Groove

A-T

G-C

Minor Groove

Minor Groove

R=DNA Backbone

Figure 8.1.

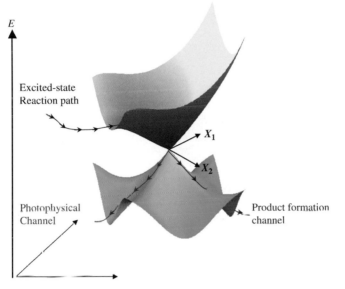

Figure 9.3. Cartoon of a "classic" double cone conical intersection, showing the excited state reaction path and two ground state reaction paths.

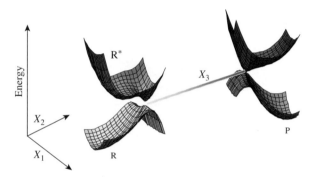

Figure 9.9. A cartoon showing the conical intersection hyperline traced out by a degeneracy-preserving coordinate X_3. The system remains degenerate as one traverses the coordinate X_3, but the energy and the shape of the double-cone must change in $X_1 X_2$.

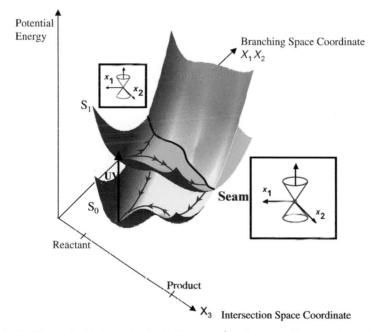

Figure 9.10. The conical intersection hyperline traced out by a coordinate X_3 plotted in a space containing the coordinate X_3 and one coordinate from the degeneracy-lifting space $X_1 X_2$.

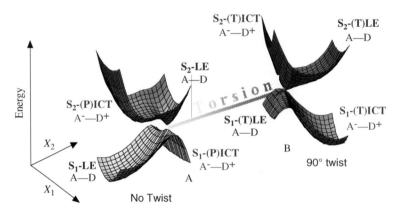

Figure 9.22. The geometries in Figure 9.21 located in the cone which changes shape along the conical intersection hyperline (adapted from reference 14).

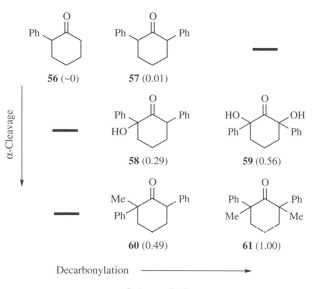

56 (~0)   **57** (0.01)

α-Cleavage

58 (0.29)   **59** (0.56)

60 (0.49)   **61** (1.00)

Decarbonylation ⟶

Scheme 7.17

substituents that facilitate the decarbonylation reaction. The RSE values for secondary benzylic [Ph(R)CH·], tertiary-hydroxy-benzylic [Ph(HO)(R)RC·], and tertiary benzylic [Ph(R)$_2$C·] are 19.6, 17.5 and 21.5 kcal/mol, respectively.[158]

An additional test of the reactivity model shown in Figure 24 was devised by analyzing the effects of alkyl and phenyl substituents in cyclopentanones **62-64** (Scheme 7.18).[160] The three alkyl groups with a combined RSE value of ca. 9.3 kcal/mol are not enough to facilitate the solid state reaction was shown by ketone **62**, which reacts well in solution but not in the solid state. Ketone **63** helped to confirm that radical stabilization in only one α-carbon is not enough for the solid state reaction to take place. While α-cleavage in **63** may be quite efficient, the loss of CO from **63-BR1** is unfavorable so that the biradical returns to the starting ketone **63** rather than going on to **63-BR2**. Finally, with tertiary-bis-benzylic radicals (RSE=22.2 kcal/mol) formed in each of the two bond-cleavage steps, ketone **64** reacted in the solid state as efficiently as in solution.[161] It is also worth noting that the intermediate 1,4-bradical **64-BR2** undergoes a radical–radical combination reaction exclusively to form cyclobutane **64-P1** in the solid state, while a third bond-cleavage reaction to form two equivalents of 1,1-diphenylethylene **63-P2** is the favored pathway in the liquid state.

The possibility of predicting solid state reactivity from calculated thermochemical data was first addressed with ketodiesters **65a-e**, which were substituted with methyl groups to vary the extent of the RSE in the radicals **65-BR1** – **65-BR3** involved along the photodecarbonylation pathway (Scheme 7.19).[154,155] All ketones reacted in solution to give complex product mixtures from radical combination (**66a-e**) and disproportionation processes. Calculations revealed RSEs of 8.9 kcal/mol, 15.1 kcal/mol, and 19.8 kcal/mol for radicals **65-BR1** (primary enol radical), **65-BR2** (secondary enol radical), and **65-BR3** (tertiary enol radical), respectively. In the

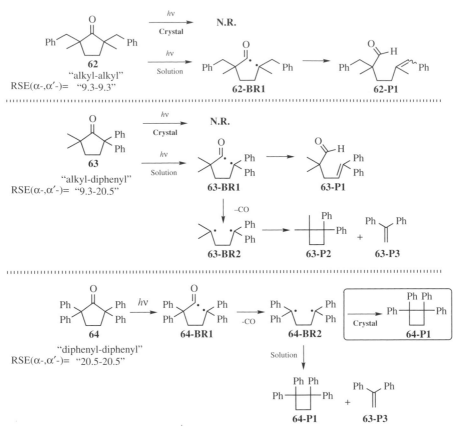

Scheme 7.18

solid state, only those ketones that reacted through secondary and/or tertiary enol radicals were reactive with relative efficiencies that increased in the order of increasing methyl substitution: **65e > 65d > 65c**. Ketones **65b** and **65a**, forming either one or two primary enol radicals along the way were photostable.

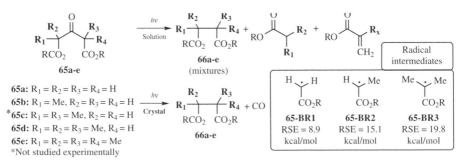

Scheme 7.19

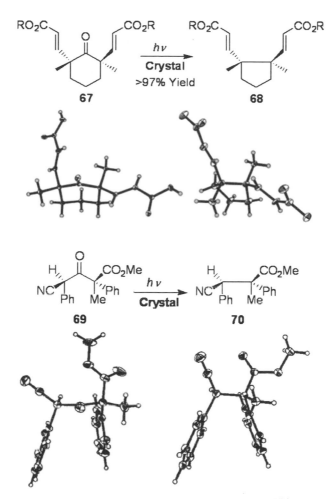

Figure 7.25. Scheme of the photochemical decarbonylation of **67** and **69** in the solid state, showing their respective ORTEP representation from X-ray data below the ChemDraw structure.

The high stereospecificity observed upon reaction of many cyclic and acyclic ketones suggests their application in solvent-free synthesis.[162] Knowing that the synthesis of compounds with adjacent chiral quaternary centers is a very challenging task,[163] recent efforts have aimed at exploring the synthetic potential of the solid-to-solid reactions. In a recent example, the trans-dialkenyl ketone **67** was shown to give trans cyclopentane **68** in almost quantitative yield with retention of stereochemistry (Figure 7.25).[164] A RSE value for the alkenyl substituents in **67** of ca. 25 kcal/mol can be obtained by comparing with the 3-methyl-1-buten-3-yl radical[158] (which exceeds the postulated requirement of RSE > 11 kcal/mol). Reactions can be designed with a wide range of substituents, as illustrated by the photochemical decarbonylation of the optically pure ketone **69**, which led to compound **70** in >98% yield and with nearly ideal >95% enantiomeric excess (*ee*).[165] The first example of a solid state decarbonylation reaction in total synthesis was reported recently in the case of natural product (±)-herbetenolide, which further illustrates the exquisite control that the solid state may exert on the chemical behavior of the otherwise highly promiscuous reactive intermediates.[166] As word or caution, it should be mentioned that intramolecular quenching effects known to act in solution can also affect that reaction in the solid state. Recently reported examples include the well-known intramolecular β-phenyl[167] and electron transfer quenching.[168]

7.3.2.2 Norrish-Yang Type II γ-Hydrogen Abstraction

Studies on the Norrish-Yang type II reactions of crystalline ketones have been extensively studied by Scheffer et al.[144] since the mid-1980s. Here we will present a few recent examples that highlight the use of X-ray structural parameters to address mechanistic questions regarding the geometry of the hydrogen transfer step and the selectivity of the 1,4-biradical intermediates. Key details of the reaction are summarized in Scheme 7.15 and Figure 7.26. It was Norrish who first showed that photolysis of 2-pentanone (Figure. 7.26, R_1=Me, R_2=R_3=H) generated acetone enol **74** (R_1=Me) and propene **75** (R_2=Me, R_3=H) by cleavage of the central 2,3-bond in biradical **72**.[147] A few years later, Yang uncovered the formation of cyclobutanols **73** by cyclization of the 1,4-biradical.[169,170]

In the spirit of the topochemical postulate and by correlating the presence or absence solid state reactivity with X-ray structural data, Scheffer proposed that reactions in crystals can provide a detailed insight into the structural parameters that make the reaction allowed or disallowed.[171] Assuming that the excited state structure is not very different from that of the ground state, he suggested that the ideal ground state parameters for the solid state reaction would be those that bear the closest resemblance to the postulated transition state (Fig. 7.26).[145,172] Four structural parameters, **d**, **Δ**, **Θ**, and **ω**, were used to describe the position of the γ-hydrogen with respect to the active *n*-orbital of the *n*,π* excited state. It was suggested that the distance between the carbonyl oxygen and the γ-hydrogen, **d**, should be less than the sum of their van der Waals radii of 2.72 Å; that the angle **Δ** (C=O⋅⋅⋅H) should be between 90° (sp) and 120° (sp²), ideally matching the hybridization of the oxygen atom; that the ideal angle θ (C–H⋅⋅⋅O)

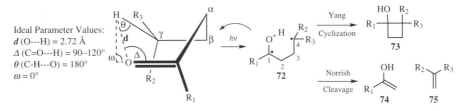

Figure 7.26. Photo-induced hydrogen abstraction from the γ-carbon leads to biradical **72**, which can (a) revert to the starting ketone, (b) cyclize, or (c) cleave the 2,3-CC bond. The structure for γ-H abstraction for the starting ketone is also shown and the "ideal" parameters defined and listed.

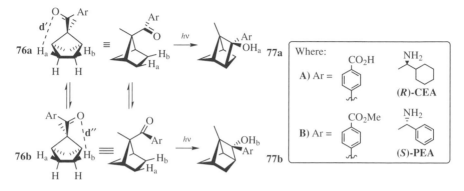

Scheme 7.20

would be 180°,[172] although smaller values should be expected for intra-molecular reactions;[171c] and finally, that the angle ω, formed between the plane of the carbonyl and the γ-hydrogen, should be 0°, since the nonbonding orbital lies on the same plane. [173]

While some variations around the ideal values are generally observed, the structural guidelines suggested by Scheffer have withstood many tests and have been the subject of numerous reviews.[144,145,170,171] Given that absolute rate constants in solids are difficult to obtain, the most valuable information has been obtained with structures that give two nonidentical products by abstraction of two nonequivalent γ hydrogens. The photochemistry of the bicyclic ketone **76** can be used as an example to illustrate this point.[174] In solution, carboxylic acid **76** (Ar=Ph-CO$_2$H, Scheme 7.20) exists as a 1:1 mixture of equilibrating conformers of **76a** and **76b**, which upon photolysis yield a racemic mixture of enantiomeric cyclobutanols **77a** and **77b**. However, an acid-base reaction of **76** with a chiral amine produces a chiral salt, which renders the equilibrating structures of **76a** and **76b** diastereomeric. As crystallization of one form is preferred over the other, the entire sample can crystallize as a single diastereomer[175] and the two γ-hydrogens, Ha and Hb, become diastereotopic and clearly nonequivalent.[176]

Thus, when acid **76** was crystallized as a salt with (S)-(−)-1-phenylethylamine ([S]-PEA), the X-ray structure showed that the conformational enantiomer **76a** was trapped in the crystal, displaying O···H$_a$ and O···H$_b$ distances of 2.47 Å and 3.41 Å, respectively. The conformation of **76a** placed the carbonyl oxygen and H$_a$ closer to the ideal values mentioned in Figure 7.26 as compared to H$_b$. A significant preference for H$_a$ was demonstrated after photolysis at 0 °C and diazomethane workup, when ester **77a** (**B**) was obtained in 65% ee after 90% conversion. Figure 7.27 illustrates the minimal atomic displacements required for reaction by comparing the X-ray structure of the reactant with that of the product, and with a structure obtained at 50% conversion. Better chemical results were obtained by photolysis of **76a** with (**R**)-**CEA**, which gave 90% ee of ester of **77a** (**B**) after diazomethane workup.

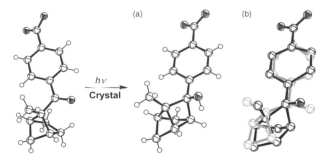

Figure 7.27. ORTEP representations from X-ray data of the salts **76a** with **(S)-PEA**, and (a) chiral product **76a-(S)-PEA**, and (b) the overlap of both starting material and product at 50% conversion of the crystal-to-crystal reaction. (Taken from reference 174.)

Having illustrated the use of chirality to document structure-reactivity correlations for γ-hydrogen abstraction, one may ask: *can one use a similar strategy to analyze the factors that determine cyclization vs. cleavage from the biradical intermediate 72* (Fig. 7.27). Based on arguments of orbital alignment that consider the formation of two double bonds in the ground state, it has been suggested that bond cleavage should be favored when there is good overlap between the C_2–C_3 sigma bond and the two singly occupied *p*-orbitals of the radical centers at C_1 and C_4 (Scheme 7.21).[177] Elegant evidence for this model, on the basis of a series of reactions that correlate X-ray structural parameters of desymmetrized chiral reactants with the selectivity for Yang cyclization and Norrish cleavage, is also available from recent work by Scheffer et al.[178,179]

In a very clever approach, bicyclic ketones **78** and **79** were designed as precursors for the solid-state photochemical generation of biradicals **78-BR** and **79-BR**, which are structurally rigid and with *only one* degree of rotational freedom about the C_1–C_2 bond.[179] Since those biradicals are transient species, the X-ray structures of their crystalline ketone precursors were used in search of a correlation between structural parameters and the preference for cyclization versus cleavage, and the preference for cleaving the C_2–C_3 versus the C_2–$C_{3'}$ bonds. Notably, the solution photochemistry of analogous ketones forming the analogous biradicals was already known.[179] The

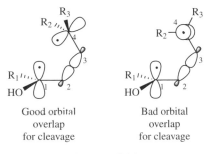

Good orbital
overlap
for cleavage

Bad orbital
overlap
for cleavage

Scheme 7.21

analogous **78-BR** had been shown to undergo 100% cyclization in solution, and the analogous **79-BR** to give 66% cleavage and 30% cyclization.

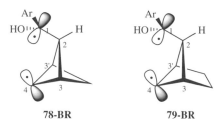

78-BR   **79-BR**

In analogy to the work described in the previous example,[174] desymmetrized structures were obtained in the homochiral crystalline salts **78s** and **79s**, simply prepared by mixing (S)-(−)-1-phenylethylamine ([S]-PEA) with ketones **78** and **79** (Scheme 7.22).[180] In principle, solid state photolysis of **78s** may yield the cyclization product **80** or the cleavage product **81**. If cleavage occurs, two possible enantiomers can be formed: (R)-**81** if the C_2–C_3 bond is broken, and (S) **81** if the C_2–$C_{3'}$ bond is cleaved. A similar outcome might be expected for the photolysis of **79s**. In practice, **78s** prefers *cyclization* to cleavage in 88:10 ratio, and **79s** prefers *cleavage* over cyclization in a 80:20 ratio, yielding (S)-**83** in a remarkable 98% ee.

An explanation for the differences in chemoselectivity was offered in terms of conformational parameters obtained from X-ray structural data of the crystallized salts (**78s** and **79s**, Fig. 7.28). in both cases, the $O \cdots H_4$ distance (hydrogen at carbon 4) was nearly the same, 2.67 Å in **78s** and 2.62 Å in **79s**, indicating that hydrogen abstraction should be equally likely. Similarly, the $C_1 \cdots C_4$ distance was also nearly identical, 2.81 Å in **78s** and 2.82 Å in **79s**, ruling out chemoselectivity arguments based on $O \cdots H_4$ and $C_1 \cdots C_4$ proximity. Another parameter considered was the angle β between the *p*-orbital on C_1 and the C_2–C_1 vector, which in the ideal case for *cyclization* would be 0°. The parameters more relevant for *cleavage* are the angles between the *p*-orbital on C_1 and C_2–C_3 (or C_2–$C_{3'}$ in parenthesis), φ_1, and similarly, the *p*-orbital on C_4 and C_2–C_3 (or C_2–$C_{3'}$ in parenthesis), φ_4. In the ideal case, $\varphi_1 = \varphi_4 = 0°$. For ketone **78s**, the following values were obtained: $\beta = \underline{\mathbf{22°}}$, $\varphi_1 = 22°$ (67°), and $\varphi_4 = 43°$ (43°). In ketone **79s**: $\beta = 32°$, $\varphi_1 = \underline{\mathbf{16°}}$ (78°), and $\varphi_4 = \underline{\mathbf{35°}}$ (35°).

From the parameters highlighted, it can be seen that the β value for **78s** is smaller or closer to 0°, than that in **79s** and thus preference for cyclization prevails over cleavage. On the contrary, for **79s** the values that would favor cleavage are also closer to 0° than in **78s**. Not only do these values indicate that **79s** is prone to cleavage, but comparing the φ_1 and φ_4 values to those in parenthesis, it is C_2–C_3 bond that cleaves preferentially over the C_2–$C_{3'}$. These parameters suggest that very small conformational changes may have drastic effects in the chemo-selectivity of biradicals.

7.3.4. Carbenes as Transient Species

Interest in carbenes in rigid media can be dated to the 1960s when Murray et al. reported diphenylcarbene as the first organic species with a triplet ground state.[181]

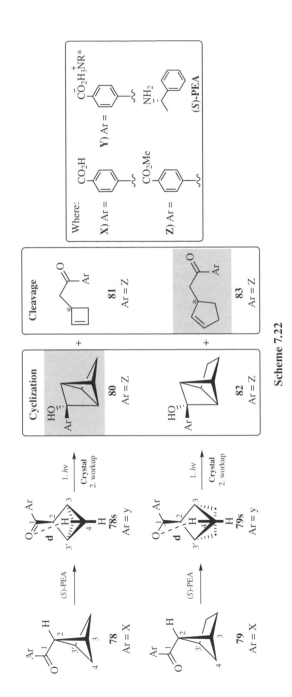

Scheme 7.22

(a)   (b)

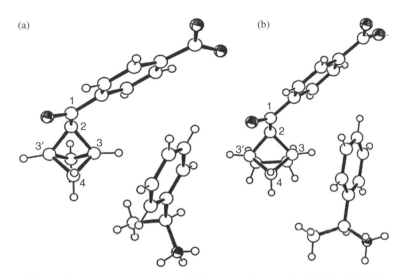

Figure 7.28. Molecular representations of the crystallized salts (a) **78s-(S)-PEA** and (b) **79s-(S)-PEA**, from the X-ray structure data. (Taken from reference 179.)

Shortly after, Doetschman and Hutchison reported the first example of a reactive carbene in the crystalline solid state, by preparing diphenylcarbene from diphenyldiazomethane in mixed crystals with 1,1-diphenylethylene **84** (Scheme 7.23).[182] When the mixed crystals were irradiated, carbene **85** was detected by electron paramagnetic resonance (EPR) and the disappearance of the signal was monitored to determine its kinetic behavior. Two reactions were shown to take place under topochemical

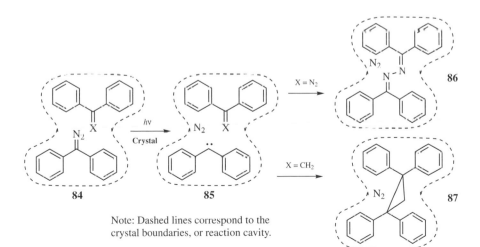

Note: Dashed lines correspond to the
crystal boundaries, or reaction cavity.

Scheme 7.23

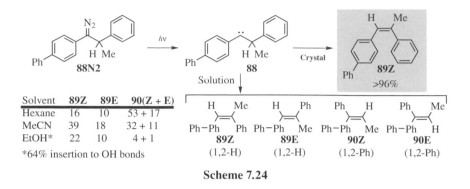

Solvent	89Z	89E	90(Z + E)
Hexane	16	10	53 + 17
MeCN	39	18	32 + 11
EtOH*	22	10	4 + 1

*64% insertion to OH bonds

Scheme 7.24

control, including the formation azine **86** by reaction of the singlet carbene with a nearby diazo precursor, and a cycloaddition of the triplet carbene to the alkene to form cyclopropane **87**, presumably through an intermediate triplet 1,3-biradical.

Subsequent research on the effect of rigid media on the selectivity of aryl-alkylcarbenes in polycrystalline media and in amorphous glasses at low temperature revealed the formation of a large number of products, highlighting the reactivity of these reactive intermediates.[183] More recently, solid-to-solid reactions with aryl-alkylcarbene **88** in the crystalline state revealed a remarkably high selectivity and an unprecedented reaction control (Scheme 7.24).[10] It was shown that crystals of the diazo **88N2** are remarkably stable, with a melting point of 94 °C,[184] and that the diazo compound readily extruded nitrogen upon photolysis. Experiments in solution gave complex mixtures of (E)- and (Z)- diastereomers of **89** and **90** along with other products arising from reactions with the solvents. It is well known that aryl–alkylcarbenes such as **88** exist in a very rapid equilibrium between their singlet and triplet states and that product formation in solution is determined by a complex interplay between singlet–triplet equilibration rates and the rates of spin-state specific reactions.[185] In the case of **89**, it was postulated that stilbenes **89** arise from 1,2-hydrogen shifts (1,2-H) from the singlet state and that 1,1-diphenylethenes **90** from a 1,2-phenyl migrations (1,2-Ph) in the triplet manifold.[186,187,188]

In contrast to the low selectivity observed in solution, photochemical generation of **88** in crystals resulted in one single product, **89Z**, formed by a highly selective 1,2-H shift. Comparing the lifetime of the carbene in different media near ambient temperature revealed that the reaction proceeds in crystals in a matter of seconds, while the lifetime of the carbene in solution is ca. 50 ns; a difference of 6-7 orders of magnitude.[189] In order to understand the remarkable selectivity of **88** toward product **89Z**, a structural comparison between crystals of the reacting diazo compound and crystals of the resulting alkene was carried out. Both reactant and product were found to have a close resemblance, suggesting that the reaction cavity was not deformed throughout the course of the reaction, and that the hydrogen shift proceeded through a least-motion path and with minimal heavy-atom displacements (Fig. 7.29). Such observations are consistent with the mechanism shown at the bottom of Figure 7.28, where only one carbene conformer of **88** is trapped in the crystal. It is known

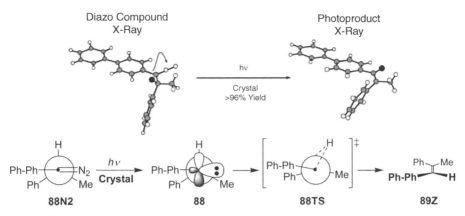

Figure 7.29. (Top) Molecular representations based on X-ray structural data of the diazo compound **88N2** and the alkene product **89Z** (the migrating hydrogen is shown in black in both reactant and product). (Bottom) Schematic reaction path showing the minimal structural changes in the transition from the diazo compound to the product, *via* the probable transition structure **88TS**.

that the preferred conformations of the carbene can be very similar to those of the precursor. In the case of **88**, the σ-bond of the migrating hydrogen is aligned with the empty *p*-orbital of the singlet carbene (**88**).[190,191] At the transition state **88TS**, the migrating hydrogen forms a bridge over the developing double bond, and the sp³ carbons proceed to planarize.

Knowing that carbene **88** has a triplet ground state and that the 1,2-H-shift occurs from the singlet state, one may explain the differences in solution and solid state kinetics by invoking a very inefficient intersystem crossing mechanism in the solid state.[10] It is well known that singlet and triplet carbenes tend to have very different R—C—R angles, so that the relative energetics of the singlet and triplet state may be different in crystals and their equilibrium dynamics may be slower. That being the case, one may ask whether a *ground state singlet carbene would have significantly faster [1,2]-H shifts in crystalline media?* As a first step to address this question, crystals of chlorodiazirine **91N2** were obtained and their reactivity investigated (Scheme 7.25).[192] When **91N2** was photolyzed in solution, only two products arising from [1,2]-H shifts were generated, **93E** and **93Z**. Unexpectedly, solid state photolysis also gave a mixture of products, arising from the same [1,2]-H shifts, and from reaction between the carbene **91** and the diazirine **91N2** to form the azine **92**. A close analysis of the X-ray structure of **91N2** revealed a conformationally disordered structure that helped to explain such unexpected behavior and confirmed the topochemical nature of the solid state reaction.

The conformational disorder and packing structure of diazirine **91N2** is shown on the left of Scheme 7.25. The formation of azine **92** can easily be explained by the head-to-head crystallization of the chloro-diazirine, which experiences close N···Cl interactions that ultimately bring the carbene carbon close to the diazirine nitrogen

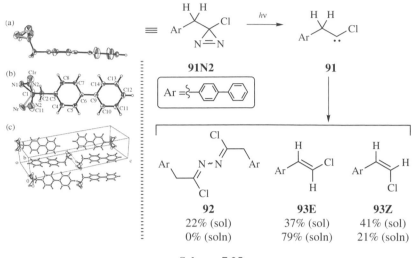

Scheme 7.25

of a neighboring molecule. The formation of E- and Z-isomers of chloroethylene **93** can be explained in terms of the conformational disorder that positions the chlorine atom on the same or opposite side of the aryl group (Scheme 7.26). Studies of the photolysis of **91N2** as a function of temperature down to 77K showed very small differences in product distribution, suggesting that a static disorder is more likely than having rapidly inter-converting structures. It is also interesting that all products are formed from the singlet state and that all reactions compete with similar efficiency. This suggests kinetic processes with similar activation energies. Unfortunately, absolute rate constants for these reactions are not available at this time.

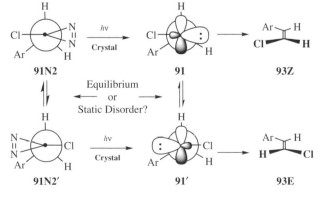

Scheme 7.26

7.4. CONCLUSIONS AND OUTLOOK

After a hundred years of studies in the gas phase and in solution, it seems only natural to advance our knowledge of reactive intermediates to the solid state. In this review, we have discussed the use of recent advances leading to stabilization and persistence of reactive intermediates to obtain single crystals for X-ray diffraction analysis. We have also stressed that transient species may be used to engineer reactions in crystals in a rational manner. It is very likely that many more reactive intermediates will be studied in solids in the next few years, and that methods will be developed to measure their absolute kinetics by standard laser flash photolysis and other transient methods. Having unusual electronic structures, persistent reactive intermediates will find uses as reagents and as materials. For example, stable high-spin carbenes and multi-radicals are attractive for the construction or organic magnetic materials and organic conductors. Within the scope of reactions in solids, it is clear that RIs are remarkably selective when constrained by the rigid and homogenous environment of the solid state. It is likely that we will see many examples of reactions in solids used for the total synthesis of natural products and biologically important molecules, and that solid state reactivity will be added to the many synthetic strategies that became available to chemists during the last century. Reactive intermediates have been tamed; now they will be domesticated.

7.5. ACKNOWLEDGEMENTS

LMC thanks the National Science Foundation, the Paul and Daisy Soros Fellowship, and the NSF IGERT: Materials Creation Training Program (MCTP) – (Grant number: DGE-0114443) for graduate support. We also thank NSF for support through grants CHE-0551938 and DMR-0605688.

7.6. SUGGESTED READING

a) Moss, R. A.; Platz, M. S.; Jones, Jr., M., Eds., *Reactive Intermediate Chemistry*, John Wiley & Sons: New Jersey, 2004, Chapters 1, 3, 4, 7, 8, 9.

b) Laube, T. *Chem. Rev.* **1998**, *98*, 1277.

c) Wentrup, C. *Science* **2002**, *295*, 1846.

d) Hollingsworth, M. D.; McBride, M. J., In: *Advances in Photochemistry*, Volman, D. H.; Hammond, G. S.; Gollnick, K., Eds.; John Wiley & Sons: New York, 1990, *Vol. 15*, p.279.

e) Desiraju, G. R. *Organic Solid State Chemistry*; Elsevier: Amsterdam, 1987.

f) Garcia-Garibay, M. A. *Acc. Chem. Res.* **2003**, *36*, 491.

g) Scheffer, J. R.; Garcia-Garibay, M. A.; Nalamasu, O., In *Organic Photochemistry*; Padwa, A., Ed.; Marcel Dekker: New York, 1987, *Vol. 8*, Ch. 4.

h) Horspool, W.; Lenci, F.; Eds., *CRC Handbook of Organic Photochemistry and Photobiology*, CRC Press, Boca Raton, 2nd Ed., 2004, Ch: 48, 54, 107.

7.7. REFERENCES

1. (a) Wöhler, F. *Pogg. Ann.* **1828**, *12*, 253; (b) For a historical account of Wöhler's discovery of Urea, incuding his initial failure to recognize it, please see: Kauffman, G.B; Chooljian, S.II. *Chem. Educator* **2001**, *6*, 121.

2. Dunitz, J. D., Harris, K. D. M., Johnston, R. L., Kariuki, B. M., MacLean, E. J., Psallidas, K., Schweizer, W. B., Tykwinski, R. R. *J. Am. Chem. Soc.* **1998**, *120*, 13274.

3. Reiser, A., Shih, H.-Y., Yeh, T.-F., Huang, J.-P. *Angew. Chem. Int. Ed.* **1996**, *35*, 2428.

4. http://www.pfizer.com/history/1849.htm

5. Murray, R. W., Trozzolo, A. M., Wasserman, E., Yager, W. A. *J. Am. Chem. Soc.* **1962**, *84*, 3213.

6. (a) Green, B. S., Lahav, M., Schmidt, G. J. M. *J. Chem. Soc. B.* **1971**, 1552; (b) Bonner, W. C. *Top. Stereochem.* **1988**, *18*, 1.

7. (a) Wegner, G. *Makromol. Chem.* **1971**, *134*, 219; (b) Wegner, G. *Pure Appl. Chem.* **1977**, *49*, 443.

8. Griller, D., Ingold, K. U. *Acc. Chem. Res.* **1976**, *9*, 13.

9. Garcia-Garibay, M. A. *Acc. Chem. Res.* **2003**, *36*, 491.

10. Garcia-Garibay, M. A., Shin, S. H., Sanrame, C. *Tetrahedron* **2000**, *56*, 6729.

11. Keating, A. E., Garcia-Garibay, M. A. In: *Organic and Inorganic Photochemistry.* Rammamurthy, V., Schanze, K. S., Eds. Marcel Dekker, Inc.: New York, 1998, 195.

12. See Section V: Suggested Reading.

13. Allen, F. H., Kennard, O., Watson, D. G., Brammer, L., Orpen, A. G. *J. Chem. Soc. Perkin Trans. II* **1987**, S1-S19.

14. (a) Olah, G. A., Prakash, G. K. S. *Carbocation Chemistry*; John Wiley & Sons: Hoboken, 2004; (b) Olah, G. A. *Angew. Chem. Int. Ed.* **1995**, *34*, 1393.

15. *IUPAC Compendium of Chemical Tehnology*, 2nd Edition, (Eds. McNaught, A. D., Wilkinson, A.), Eds., Blackwell Science, 1997.

16. White, E. T., Tang, J., Oka, T. *Science* **1999**, *284*, 135.

17. Laube, T. *Acc. Chem. Res.* **1995**, *28*, 399.

18. Hollenstein, S., Laube, T. *J. Am. Chem. Soc.* **1993**, *115*, 7240.

19. Kato, T., Reed, C. A. *Angew. Chem. Int. Ed.* **2004**, *43*, 2908.

20. Vorachek, J. H., Meisels, G. G., Geanangel, R. A., Emmel, R. H. *J. Am. Chem. Soc.* **1973**, *95*, 4078.

21. Schleyer, P. v. R., Carneiro, J. W. de M., Koch, W., Forsyth, D. A. *J. Am. Chem. Soc.* **1991**, *113*, 3990; and references therein.

22. Laube, T., Olah, G. A., Bau, R. *J. Am. Chem. Soc.* **1997**, *119*, 3087.

23. Winstein, S., Shatavsky, M., Norton, C., Woodward, R. B. *J. Am. Chem. Soc.* **1955**, *77*, 4183.

24. Bartlett, P. D., Giddings, W. P. *J. Am. Chem. Soc.* **1960**, *82*, 1240.

25. Volz, H., Shin, J. –H., Miess, R. *J. Chem. Soc., Chem. Commun.* **1993**, 543.

26. Tantillo, D. J., Hietbrink, B. N., Merlic, C. J., Houk, K. N. *J. Am. Chem. Soc.* **2000**, *122*, 399; **2001**, *123*, 5851.

27. Laube, T. *J. Am. Chem. Soc.* **2004**, *126*, 10904.

28. Laube, T., Lohse, C. *J. Am. Chem. Soc.* **1994**, *116*, 9001.

29. Laube, T. *Chem. Rev.* **1998**, *98*, 1277.

30. Bürgi, H. B., Dunitz, J. D. *Acc. Chem. Res.* **1983**, *16*, 153.

31. Kawai, H., Takeda, T., Fujiwara, K., Suzuki, T. *J. Am. Chem. Soc.* **2005**, *127*, 12172.

32. Müller, T., Juhasz, M., Reed, C. A. *Angew. Chem. Int. Ed.* **2004**, *43*, 1543.

33. Laube, T. *Angew. Chem. Int. Ed.* **1986**, *25*, 349.

34. (a) Nishinaga, T., Izukawa, Y., Komatsu, K. *J. Am. Chem. Soc.* **2000**, *122*, 9312; (b) Sekiguchi, A., Matsuno, T., Ichinohe, M. *J. Am. Chem. Soc.* **2000**, *122*, 11250.

35. Kim, K.-C., Reed, C. A., Elliott, D. W., Mueller, L. J., Tham, F., Lin, L., Lambert, J. B. *Science* **2002**, *297*, 825.

36. Ichinohe, M., Igarashi, M., Sanuki, K., Sekiguchi, A. *J. Am. Chem. Soc.* **2005**, *127*, 9978.

37. Breslow, R. *Pure Appl. Chem.* **1971**, *28*, 111.

38. Moss, R. A., Shen, S., Krogh-Jespersen, K., Potenza, J. A., Schugar, H. J., Munjal, R. C. *J. Am. Chem. Soc.* **1986**, *108*, 134.

39. Chance, J. M., Geiger, J. H., Okamoto, Y., Aburatani, R., Mislow, K. *J. Am. Chem. Soc.* **1990**, *112*, 3540.

40. Gronert, S. In: *Reactive Intermediate Chemistry* (Eds., Moss, R. A., Platz, M. S., Jones, Jr., M.), John Wiley & Sons: New Jersey, 2004, Chapter 1, and references therein.

41. Nobes, R. H., Poppinger, D., Li, W.-K., Radom, L., in *Comprehensive Carbanion Chemistry, Part C*, Buncel, E., Durst, T., Eds., Elsevier: New York, 1987; p. 1.

42. Forsyth, D. A., Yang, J.-R. *J. Am. Chem. Soc.* **1986**, *108*, 2157.

43. Crenshaw, M. D., Schmolka, S. J., Zimmer, H., Whittle, R., Elder, R. C. *J. Org. Chem.* **1982**, *47*, 101.

44. Reetz, M. T., Hütte, S., Goddard, R. *Eur. J. Org. Chem.* **1999**, 2475.

45. Reetz, M. T., Hütte, S., Goddard, R. *J. Am. Chem. Soc.* **1993**, *115*, 9339.

46. Gais, H.-J., Hellmann, G., Lindner, H. J. *Angew. Chem. Int. Ed.* **1990**, *1*, 100.

47. Gais, H.-J., Müller, J., Vollhardt, J. *J. Am. Chem. Soc.* **1991**, *113*, 4002.

48. Gais, H.-J., van Gumpel, M., Raabe, G., Müller, J., Braun, S., Lindner, H. J., Rohs, S., Runsink, J. *Eur. J. Org. Chem.* **1999**, 1627.

49. Gais, H.-J., van Gumpel, M., Schleusner, M., Raabe, G., Runsink, J., Vermeeren, C. *Eur. J. Org. Chem.* **2001**, 4275.

50. Bors, D. A., Streitweiser, Jr., A. *J. Am. Chem. Soc.* **1986**, *108*, 1397.

51. Cram, D. J., Nielsen, W. D., Rickborn, B. *J. Am. Chem. Soc.* **1960**, *82*, 6415.

52. Corey, E. J., Kaiser, E. T. *J. Am. Chem. Soc.* **1961**, *83*, 490.

53. Harder, S. *Chem. Eur. J.* **2002**, *8*, 3229, and references therein.

54. Pieper, U., Stalke, D. *Organometallics* **1993**, *12*, 1201.

55. Lawrence, S. C., Skinner, M. E. G., Green, J. C., Mountford, P. *Chem. Commun.* **2001**, 705.

56. Breher, F., Grunenberg, J., Lawrence, S. C., Mountford, P., Rüegger, H. *Angew. Chem. Int. Ed.* **2004**, *43*, 2521.

57. Grim, S. O., Smith, P. H., Nittolo, S., Ammon, H. L., Satek, L. C., Sangokoya, S. A., Khanna, R. K., Colquhoun, H. L., McFarlane, W., Holden, J. R. *Inorg. Chem.* **1985**, *24*, 2889.

58. Grim, S. O., Sangokoya, S. A., Rheingold, A. L., McFarlane, W., Colquhoun, H. L., Gilardi, R. D. *Inorg. Chem.* **1991**, *30*, 2519.

59. Bugg, C., Desiderato, R., Sass, R. L. *J. Am. Chem. Soc.* **1964**, *86*, 3157.

60. Richardson, C., Reed, C. A. *Chem. Commun.* **2004**, 706.

61. Webster, O. W. *J. Am. Chem. Soc.* **1966**, *88*, 4055.

62. Haag, R., Ohlhorst, B., Noltemeyer, M., Fleischer, R., Stalke, D., Schuster, A., Kuck, D., de Meijere, A. *J. Am. Chem. Soc.* **1995**, *117*, 10474.

63. Izod, K. *Coord. Chem. Rev.* **2002**, *227*, 153.

64. Izod, K., McFarlane, W., Clegg, W. *Chem. Commun.* **2002**, 2532, and references therein.

65. Bertrand, G. In: *Reactive Intermediate Chemistry*, (Eds., Moss, R. A., Platz, M. S., Jones, Jr., M.), John Wiley & Sons: New Jersey, 2004, Chapter 8, and references therein.

66. (a) Hine, J. *J. Am. Chem. Soc.* **1950**, *72*, 2438; (b) McClure, D. S., Blake, N. W., Hanst, P. L. *J. Phys. Chem.* **1954**, *22*, 255; (c) Skell, P. S. *J. Am. Chem. Soc.* **1956**, *78*, 4496; (d) Doering, W. von E., Buttery, R. G., Laughlin, R. G., Chaudhuri, N. *J. Am. Chem. Soc.* **1956**, *78*, 3224; (e) Breslow, R. *J. Am. Chem. Soc.* **1958**, *80*, 3719.

67. Tomioka, H. In: *Reactive Intermediate Chemistry*, (Eds., Moss, R. A., Platz, M. S., Jones, Jr., M.), John Wiley & Sons: New Jersey, 2004, Chapter 9, and references therein.

68. Schaefer III, H. F. *Science* **1986**, *231*, 1100, and references therein.

69. Wasserman, E., Schaefer III, H. F. *Science* **1986**, *233*, 829.

70. Wasserman, E., Yager, W. A., Kuck, V. J. *Chem. Phys. Lett.* **1970**, *92*, 4984.

71. Jones, Jr., M., Moss, R. A. In: *Reactive Intermediate Chemistry*, (Eds., Moss, R. A., Platz, M. S., Jones, Jr., M.), John Wiley & Sons: New Jersey, 2004, Chapter 7, and references therein.

72. Bourissou, D., Guerret, O., Gabbaï, F. P., Bertrand, G. *Chem. Rev.* **2000**, *100*, 39, and references therein.

73. Pauling, L. *Chem. Commun.* **1980**, 688.

74. Kirmse, W. *Angew. Chem. Int. Ed.* **2004**, *43*, 1767.

75. Arduengo III, A. J., Harlow, R. L., Kline, M. *J. Am. Chem. Soc.* **1991**, *113*, 361.

76. Arduengo III, A. J., Dias, H. V. R., Harlow, R. L., Kline, M. *J. Am. Chem. Soc.* **1992**, *114*, 5530.

77. Arduengo III, A. J., Goerlich, J. R., Marshall, W. J. *J. Am. Chem. Soc.* **1995**, *117*, 11027.

78. See also: Alder, R. W., Blake, M. E., Bortolotti, C., Bufalli, S., Butts, C. P., Linehan, E., Oliva, J. E., Orpen, A. G., Quayle, M. J. *Chem. Commun.* **1999**, 241.

79. Bazinet, P., Yap, G. P. A., Richeson, D. S. *J. Am. Chem. Soc.* **2003**, *125*, 13314.

80. Grubbs, R. H., Trnka, T. M. *Acc. Chem. Res.* **2001**, *34*, 18.

81. Herrmann, W. A. *Angew. Chem. Int. Ed.* **2002**, *41*, 1290.

82. Martin, D., Baceidero, A., Gornitzka, H., Schoeller, W. W., Bertrand, G. *Angew. Chem. Int. Ed.* **2005**, *44*, 1700, and references therein.

83. Enders, D., Breuer, K., Raabe, G., Runsink, J., Teles, J. H., Melder, J.-P., Ebel, K., Brode, S. *Angew. Chem. Int. Ed.* **1995**, *34*, 1021.

84. Despagnet-Ayoub, E., Grubbs, R. H. *J. Am. Chem. Soc.* **2004**, *126*, 10198.

85. Krahulic, K. E., Enright, G. D., Parvez, M., Roesler, R. *J. Am. Chem. Soc.* **2005**, *127*, 4142.

86. Präsang, C., Donnadieu, B., Bertrand, G. *J. Am. Chem. Soc.* **2005**, *127*, 10182.

87. Buron, C., Gornitzka, H., Romanenko, V., Bertrand, G. *Science* **2000**, *288*, 834.

88. Solé, S., Gornitzka, H., Schoeller, W. W., Bourissou, D., Bertrand, G. *Science* **2001**, *292*, 1901.

89. Merceron, N., Miqueu, K., Baceiredo, A., Bertrand, G. *J. Am. Chem. Soc.* **2002**, *124*, 6806.

90. Merceron-Saffon, N., Baceiredo, A., Gornitzka, H., Bertrand, G. *Science* **2003**, *301*, 1223.

91. Lavallo, V., Mafhouz, J., Canac, Y., Donnadieu, B., Schoeller, W. W., Bertrand, G. *J. Am. Chem. Soc.* **2004**, *126*, 8670.

92. Iikubo, T., Itoh, T., Hirai, K., Takahashi, Y., Kawano, M., Ohashi, Y., Tomioka, H. *Eur. J. Org. Chem.* **2004**, 3004.

93. Kawano, M., Hirai, K., Tomioka, H., Ohashi, Y. *J. Am. Chem. Soc.* **2001**, *123*, 6904.

94. Tomioka, H., Iwamoto, E., Itakura, H., Hirai, K. *Nature* **2001**, *412*, 626.

95. Tomioka, H. *Acc. Chem. Res.* **1997**, *30*, 315.

96. Itoh, T., Nakata, Y., Hirai, K., Tomioka, H. *J. Am. Chem. Soc.* **2006**, *128*, 957.

97. Kirmse, W. *Angew. Chem. Int. Ed.* **2004**, *42*, 2117.

98. Newcomb, M. in *Reactive Intermediate Chemistry*, (Eds., Moss, R. A., Platz, M. S., Jones, Jr., M.), John Wiley & Sons: New Jersey, 2004, Chapter 4, and references therein.

99. Gomberg, M. *J. Am. Chem. Soc.* **1900**, *22*, 757.

100. McBride, J. M. *Tetrahedron* **1974**, *30*, 2009.

101. Chi, X., Itkis, M. E., Patrick, B. O., Barclay, T. M., Reed, R. W., Oakley, R. T., Cordes, A. W., Haddon, R. C. *J. Am. Chem. Soc.* **1999**, *121*, 10395.

102. Pal, S. K., Itkis, M. E., Tham, F. S., Reed, R. W., Oakley, R. T., Haddon, R. C. *Science* **2005**, *309*, 281.

103. Armet, O., Veciana, J., Rovira, C., Riera, J., Castaner, J., Molins, E., Rius, J., Miravitlles, C., Olivella, S., Brichfeus, J. *J. Phys. Chem.* **1987**, *91*, 5608.

104. Hayes, K., Nagumo, M., Blount, J. F., Mislow, K. *J. Am. Chem. Soc.* **1980**, *102*, 2773.

105. Sedó, J., Ventosa, N., Ruiz-Molina, D., Mas, M., Molins, E., Rovira, C., Veciana, J. *Angew. Chem. Int. Ed.* **1998**, *37*, 330.

106. Haddon, R. C. *Nature* **1975**, *256*, 394.

107. Goto, K., Kubo, T., Yamamoto, K., Nakasuji, K., Sato, K., Shiomi, D., Takui, T., Kubota, M., Kobayashi, T., Yakusi, K., Ouyang, J. *J. Am. Chem. Soc.* **1999**, *121*, 1619.

108. Koutentis, P. A., Chen, Y., Cao, Y., Best, T. P., Itkis, M. E., Beer, L., Oakley, R. T., Corder, A. W., Brock, C. P., Haddon, R. C. *J. Am. Chem. Soc.* **2001**, *123*, 3864.

109. Sitzmann, H., Bock, H., Boese, R., Dezember, T., Havlas, Z., Kaim, W., Moscherosch, M., Zanathy, L. *J. Am. Chem. Soc.* **1993**, *115*, 12003.

110. Sitzmann, H., Boese, R. *Angew. Chem. Int. Ed.* **1991**, *30*, 971.

111. Bürgi, H.-B., Baldrige, K. K., Hardcastel, K., Frank, N., Siegel, J. *Angew. Chem. Int. Ed.* **1995**, *34*, 1454.

112. Nishinaga, T., Wakamiya, A., Yamakazi, D., Komatsu, K. *J. Am. Chem. Soc.* **2004**, *126*, 3163.

113. Kitagawa, T., Ogawa, K., Komatsu, K. *J. Am. Chem. Soc.* **2004**, *126*, 9930.

114. Rohmer, M.-M., Bénard, M. *Chem. Soc. Rev.* **2001**, *30*, 340.

115. Nguyen, K. A., Gordon, M. S., Boatz, J. A. *J. Am. Chem. Soc.* **1994**, *116*, 9241.

116. Wentrup, C. *Science* **2002**, *295*, 1846.

117. Bhargava, S., Hou, J., Parvez, M., Sorensen, T. S. *J. Am. Chem. Soc.* **2005**, *127*, 3704.

118. Scheschkewitz, D., Amii, H., Gornitzka, H., Schoeller, W. W., Bourissou, D., Bertrand, G. *Science* **2002**, *295*, 1880.

119. Rodriguez, A., Tham, F. S., Schoeller, W. W., Bertrand, G. *Angew. Chem. Int. Ed.* **2004**, *43*, 4876.

120. Rodriguez, A., Olsen, R. A., Ghaderi, N., Scheschkewitz, D., Tham, F. S., Mueller, L. J., Bertrand, G. *Angew. Chem. Int. Ed.* **2004**, *43*, 4880.

121. Amii, H., Vranicar, L., Gornitzka, H., Bourissou, D., Bertrand, G. *J. Am. Chem. Soc.* **2004**, *126*, 1344.

122. Niecke, E., Fuchs, A., Baumeister, F., Nieger, M., Schoeller, W. W. *Angew. Chem. Int. Ed.* **1995**, *34*, 555.

123. Niecke, E., Fuchs, A., Nieger, M., *Angew. Chem. Int. Ed.* **1999**, *38*, 3028.

124. Sebastien, M., Nieger, M., Szieberth, D., Nyulászi, L., Niecke, E. *Angew. Chem. Int. Ed.* **2004**, *43*, 637.

125. Larsen, C. H., Ridgway, B. H., Shaw, J. T., Smith, D. M., Woerpel, K. A. *J. Am. Chem. Soc.* **2005**, *127*, 10879.

126. Ayala, L., Lucero, C. G., Romero, J. A. C., Tabacco, S. A., Woerpel, K. A. *J. Am. Chem. Soc.* **2003**, *125*, 15521.

127. Chamberland, S., Ziller, J. W., Woerpel, K. A. *J. Am. Chem. Soc.* **2005**, *127*, 5322, and references therein.

128. Garcia-Garibay, M. A., Campos, L. M. In: *CRC Handbook of Organic Photochemistry and Photobiology*, Horspool, W., Lenci, F., Eds., CRC Press, Boca Raton, 2nd Ed., 2004, Ch. 48, and references therein.

129. (a) Toda, F. *Acc. Chem. Res.* **1995**, *28*, 480; (b) Desiraju, G. R. *Organic Solid State Chemistry*; Elsevier: Amsterdam, 1987.

130. Ellison, M. E., Ng, D., Dang, H., Garcia-Garibay, M. A. *Org. Lett.* **2003**, *5*, 2531.

131. Ng, D., Yang, Z., Garcia-Garibay, M. A. *Org. Lett.* **2004**, *6*, 645.

132. Anastas, P. T., Warner, J. C. *Green Chemistry: Theory and Practice.* 1998: Oxford University Press. p. 1.

133. Kohlshutter, H. W. *Anorg. Allg. Chem.* **1918**, *105*, 121.

134. (a) Cohen, M. D., Schmidt, G. M. J. *J. Chem. Soc.* **1964**, 1996; (b) Schmidt, G. M. J. *Pure Appl. Chem.* **1971**, *27*, 647.

135. (a) Schmidt, G. M. J. *Solid State Photochemistry.* Verlag Chemie: New York, 1976; (b) Hollingsworth, M. D. *Science*, **2002**, *295*, 2410; (c) Desiraju, G. R. *Current Science*, **2001**, *81*, 1038; (d) Desiraju, G. R. *Crystal Engineering: The Design of Organic Solids.* Elsevier: Amsterdam, 1989; (e) Thomas, J. M. *Nature* **1981**, *289*, 633.

136. Recent computational studies also support the concept of the reaction cavity: (a) Keating, A. E., Shin, S., Houk, K. N., Garcia-Garibay, M. A. *J. Am. Chem. Soc.* **1997**, *119*, 1474; (b) Keating, A. E., Shin, S.H., Huang, F., Garrell, R. L., Garcia-Garibay, M. A. *Tetrahedron Lett.* **1999**, *40*, 261; (c) Garcia-Garibay, M. A., Houk, K. N., Keating, A. E., Cheer, C. J., Leibovitch, M., Scheffer, J. R., Wu, L.-C. *Org. Lett.* **1999**, *8*, 1279; (d) Zimmerman, H. E., Zhu, Z. *J. Am. Chem. Soc.* **1994**, *116*, 9757; (e) Zimmerman, H. E.,

Zuraw, M. J. *J. Am. Chem. Soc.* 1989, *111*, 2358; (f) Gavezotti, A., Simonetta, M. *Chem. Rev.* **1982**, *82*, 5220.

137. Choi, T., Cizmeciyan, D., Khan, S. I., Garcia-Garibay, M. A., *J. Am. Chem. Soc.* **1995**, *118*, 12893.

138. An example of a bimolecular reaction can be found at: Koshima, H., Ding, K., Chisaka, Y., Matsura, T. *J. Am. Chem. Soc.* **1996**, *118*, 12059.

139. Hollingsworth, M. D., Swift, J. A., Kahr, B. *Cryst. Growth & Design* **2005**, *5*, 2022.

140. Hollingsworth, M. D., McBride, M. J. In: *Advances in Photochemistry*, Volman, D. H., Hammond, G. S., Gollnick, K., Eds., John Wiley & Sons: New York, 1990, *Vol. 15*, p. 279.

141. McBride, M. J. *Acc. Chem. Res.* **1983**, *16*, 304.

142. Garcia-Garibay, M. A., Constable, A. E., Jernelius, J., Choi, T., Cizmeciyan, D., Shin, S. H. *Physical Supramolecular Chemistry*; Kluwer Academic Publishers: Dordrecht, 1996.

143. Mortko, C. J., Garcia-Garibay, M. A. *Top. Stereochem.* **2006**, *24*, Ch. 7.

144. (a) Scheffer, J. R., Xia, W. *Top. Curr. Chem.* **2005**, *254*, 233; (b) Scheffer, J. R., Scott, C. In *CRC Handbook of Organic Photochemistry and Photobiology*, (Eds., Horspool, W., Lenci, F.), CRC Press, Boca Raton, 2nd Ed., 2004, Chapter 54, and references therein.

145. Ihmels, H., Scheffer, J. R. *Tetrahedron* **1999**, *55*, 885.

146. Bowen, E. J., Watts, H. G. *J. Chem. Soc.* **1926**, 1607, and references therein.

147. Norrish, R. G. W., Appleyard, M. E. S. *J. Chem. Soc.* **1934**, 874.

148. (a) Norrish, R. W. G. *Trans. Faraday Soc.* **1937**, *33*, 1521; (b) Norrish, R. G. W. *Trans. Faraday Soc.* **1934**, *30*, 103; (c) Norrish, R. G. W., Crone, H. G., Saltmarsh, O. D. *J. Chem. Soc.* **1934**, 1456; (d) Bamford, C. H., Norrish, R. G. W. *J. Chem. Soc.* **1935**, 1504.

149. (a) Chatgilialoglu, C., Crich, D., Komatsu, M., Ryu, I. *Chem. Rev.* 1999, *99*, 1991; (b) Vinogradov, M. G., Nikishin, G. I. *Russ. Chem. Rev.* **1971**, *40*, 916.

150. Quinkert, G., Tbata, T., Hickmann, E.A.J., Dobrat, W. *Angew. Chem. Int. Ed.* **1971**, *10*, 199.

151. Quinkert, G., Palmowski, J., Lorenz, H. P., Wiersdorff, W. W., Finke, M. *Angew. Chem. Int. Ed.* 1971, *10*, 198.

152. Hammond, G. S. *J. Am. Chem. Soc.* **1955**, *77*, 334.

153. Yang, Z., Garcia-Garibay, M. A. *Org. Lett.* **2000**, *2*, 1963.

154. Yang, Z., Ng, D., Garcia-Garibay, M. A. *J. Org. Chem.* **2001**, *66*, 4468.

155. Campos, L. M., Ng, D., Yang, Z., Dang, H., Martinez, H. L., Garcia Garibay, M. A. *J. Org. Chem.* **2002**, *67*, 3749.

156. Parkinson, C. J., Mayer, P. M., Radom, L. *J. Chem. Soc. Perkin Trans. 2* **1999**, *16*, 2305.

157. Fisher, H., Paul, H., *Acc. Chem. Res.* **1987**, *20*, 200.

158. Luo, Y.-R. *Handbook of Dissociation Energies in Organic Compounds*. CRC Press: Boca Raton, 2003.

159. Choi, T., Peterfy, K., Khan, S. I., Garcia-Garibay, M. A. *J. Am. Chem. Soc.* **1996**, *118*, 12477.

160. Peterfy, K., Garcia-Garibay, M. A. *J. Am. Chem. Soc.* **1998**, *120*, 4540.

161. Effects of other substituted-aryl groups have been discussed: Resendiz, M. J. E., Garcia-Garibay, M. A. *Org. Lett.* **2005**, *7*, 371.

162. Mortko, C. J., Garcia-Garibay, M. A. *Topics in Stereochemistry.* (Eds., Green, M. M., Nolte, R. J. M., Meijer, E. W.), John Wiley & Sons, **2006**, Chapter 7.

163. Kodanko, J. J., Overman, L. E. *Angew. Chem. Int. Ed.* **2003**, *42*, 2528.

164. Mortko, C. J., Garcia-Garibay, M. A. *J. Am. Chem. Soc.* **2005**, *127*, 7994.

165. Ellison, M. E., Ng, D., Dang, H., Garcia-Garibay, M. A. *Org. Lett.* **2003**, *5*, 2531.

166. Ng, D., Yang, Z. Garcia-Garibay, M. A. *Org. Lett.* **2004**, *6*, 645.

167. Ng, D., Yang, Z. Garcia-Garibay, M. A. *Tetrahedron Lett.* **2001**, *42*, 9113.

168. Ng, D., Yang, Z., Garcia-Garibay, M. A., *Tetrahedron Lett.* **2002**, *43*, 7063.

169. Yang, N. C., Yang, D. H. *J. Am. Chem. Soc.* **1958**, *80*, 2913.

170. For a recent review, see: Wagner, P. J. In *CRC Handbook of Organic Photochemistry and Photobiology*, Horspool, W., Lenci, F., Eds., CRC Press, Boca Raton, 2nd Ed., 2004, Chapter 58, and references therein.

171. (a) Scheffer, J. R., Garcia-Garibay, M. A., Nalamasu, O., In *Organic Photochemistry*; (Ed., Padwa, A.), Marcel Dekker: New York, 1987, *Vol. 8*, Chapter 4; (b) Scheffer, J. R., In *Organic Solid State Chemistry*; Desiraju, G. R., Ed.; Elsevier: Amsterdam, 1987; 1; (c) Gudmondsdottir, A. D., Lewis, T. J., Randall, L. H., Scheffer, J. R., Rettig, S. J., Trotter, J., Wu, C.-H. *J. Am. Chem. Soc.* **1996**, *118*, 6167; (d) Scheffer, J. R. *Acc. Chem. Res.* **1980**, *13*, 283.

172. Dorigo, A. E., Houk, K. N. *J. Am. Chem. Soc.* **1987**, *109*, 2195.

173. Wagner, P. J. *Top. Curr. Chem.* **1976**, *66*, 1.

174. Xia, W., Scheffer, J. R., Patrick, O. B. *CrystEngComm.* **2005**, *7*, 728, and references therin.

175. If one of two equilibrating diastereomers crystallizes out of solution, shifting the equilibrium in one direction, the process is referred to as an asymmetric transformation of the second kind: *IUPAC Compendium of Chemical Technology*, 2nd Edition, McNaught, A. D., Wilkinson, A., Eds.; *Blackwell Science*, 1997.

176. Patrick, B. O., Scheffer, J. R., Scott, C. *Angew. Chem. Int. Ed.* **2003**, *42*, 3775.

177. (a) Wagner, P. J., Kemppainen, A. E. *J. Am. Chem. Soc.* **1968**, *90*, 5896; (b) Hoffmann, R., Swaminathan, S., Odell, B. G., Gleiter, R. *J. Am. Chem. Soc.* **1970**, *92*, 7091.

178. Braga, D., Chen, S., Filson, H., Maini, L., Netherton, M. R., Patrick, B. O., Scheffer, J. R., Scott, C., Xia, W. *J. Am. Chem. Soc.* **2004**, *126*, 3511.

179. Yang, C., Xia, W., Scheffer, J. R., Botoshansky, M., Kaftory, M. *Angew. Chem. Int. Ed.* **2005**, *44*, 5087, and references therein.

180. Photolysis of the salts was followed by diazomethane workup to yield the corresponding methyl esters.

181. Murray, R. W., Trozzolo, A. M., Wasserman, E., Yager, W. A. *J. Am. Chem. Soc.* **1962**, *84*, 3213.

182. Doetschman, D. C., Hutchison, C. A. *J. Chem. Phys.* **1972**, *56*, 3964.

183. Tomioka, H., Hayashi, N., Izawa, Y., Senthilnathan, V. P., Platz, M. S. *J. Am. Chem. Soc.* **1983**, *105*, 5053.

184. Shin, S. H., Cizmeciyan, D., Keating, A. E., Khan, S. I., Garcia-Garibay, M. A. *J. Am. Chem. Soc.* **1997**, *119*, 1859.

185. Platz, M. S. *Kinetics and Spectroscopy of Carbenes and Biradicals*; (Ed., Platz, M. S.), Plennum Press: New York, 1990, pp 143 212.

186. Shin, S. H., Keating, A. E., Garcia-Garibay, M. A. *J. Am. Chem. Soc.* **1996**, *118*, 7626.

187. Motschiedler, K. R., Toscano, J. P., Garcia-Garibay, M. A. *Tetrahedron Lett.* **1997**, *38*, 949.

188. Keating, A. E., Garcia-Garibay, M. A., Houk, K. N. *J. Phys. Chem.* **1998**, *102*, 8467.

189. Motschiedler, K. R., Gudmundsdottir, A., Toscano, J. P., Platz, M. S., Garcia-Garibay, M. A. *J. Org. Chem.* **1999**, 5139

190. (a) Liu, M. T. H. *Acc. Chem. Res.* **1994**, *27*, 287; (b) Moss, R. A. *Pure Appl. Chem.* **1995**, *67*, 741.

191. Evanseck, J. D., Houk, K. N. *J. Am. Chem. Soc.* **1990**, *112*, 9148.

192. Sanrame, C. N., Suhrada, C. P., Dang, H., Garcia-Garibay, M. A. *J. Phys. Chem. A* **2003**, *107*, 3287.

█████ CHAPTER 8

The Chemical Reactions of DNA Damage and Degradation

KENT S. GATES

Department of Chemistry, University of Missouri, Columbia, MO

Reviews of Reactive Intermediate Chemistry. Edited by Matthew S. Platz, Robert A. Moss, Maitland Jones, Jr.
Copyright © 2007 John Wiley & Sons, Inc.

8.1. INTRODUCTION AND HISTORICAL PERSPECTIVE

8.1.1. The Importance of DNA Damage

The sequence of heterocyclic bases in double-helical deoxyribonucleic acid (DNA) forms the genetic code that serves as the blueprint for all cellular operations.[1,2] Accurate readout of genes during transcription of cellular DNA is required for production of functional proteins.[2] In addition, faithful replication of DNA must take place during cell division to yield daughter cells containing exact copies of the genetic code.[2,3] With this said, it is not surprising that chemical modification of cellular DNA can have profound biological consequences including induction of DNA repair proteins, inhibition of cell growth (cell cycle arrest), or cell death via either necrotic or apoptotic mechanisms.[4–7] Several cellular systems repair damaged DNA.[8–11] Nonetheless, in attempts to replicate damaged DNA, polymerases may introduce errors into the genetic code (mutagenesis).[12–19] Accordingly, the study of agents that damage DNA is of both practical and fundamental importance to diverse fields including medicinal chemistry, carcinogenesis, toxicology, and biotechnology.

8.1.2. Chemical Agents that Damage DNA and the Chemistry of DNA Degradation

There is rather a small number of functional groups or structural motifs that have the ability to carry out efficient modification of cellular DNA.[20–24] These molecules are of chemical and biological interest because they successfully execute a difficult balancing act—they possess sufficient *reactivity* to make and break covalent bonds

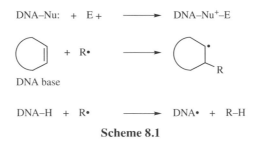

$$\text{DNA–Nu:} \;\; + \;\; E+ \;\; \longrightarrow \;\; \text{DNA–Nu}^+\text{–E}$$

DNA base

$$\text{DNA–H} \;\; + \;\; R\bullet \;\; \longrightarrow \;\; \text{DNA}\bullet \;\; + \;\; \text{R–H}$$

Scheme 8.1

within DNA, yet possess sufficient *stability* to survive in the aqueous environment of the cell. Almost all of the cellular DNA damage reactions carried out by drugs, toxins, and mutagens fall into just two general categories: (1) the reaction of a DNA nucleophile with an electrophile or (2) the reaction of a DNA pi bond or C–H bond with a radical (Scheme 8.1). In the following sections of this chapter, we will examine the degradation of DNA that is initiated by its reaction with electrophiles and radicals. In addition, we will consider examples that illustrate chemical strategies by which organic molecules can deliver these highly reactive intermediates to DNA under physiological conditions. It should be noted that many of the reaction mechanisms shown in this review are schematic in nature. For example, protonation, deprotonation, and proton transfer steps may not be explicitly depicted or may not distinguish specific acid–base catalysis from general acid–base catalysis. In some cases, arrows indicating electron movement are used to direct the reader's attention to sites where the "action" is occurring rather than to illustrate a complete reaction mechanism.

8.2. REACTIONS OF ELECTROPHILES WITH DNA

8.2.1 Nucleophilic Sites in DNA and the Factors that Determine the Sites at Which Electrophiles React with DNA

Virtually, all of the heteroatoms in DNA (Fig. 8.1) have the potential to act as nucleophiles in reaction with electrophiles.[25,26] As one might expect, access to some sites is limited in double-stranded DNA relative to single-stranded DNA.[25,26] Interestingly, however, reactions are not completely precluded even at locations on Watson–Crick hydrogen bonding surfaces of the bases that reside near the helical axis of the duplex. The factors that determine the atom site selectivity for a given DNA-alkylating agent are complex.[25–29] A recent detailed study of alkylation by diazonium ions led to the conclusion that atom site selectivities seen in duplex DNA do *not* reflect intrinsic nucleophilicities of the heteroatoms in the nucleobases.[29] Rather, placement of the nucleobases into the environment of the double helix substantially alters the nucleophilicity of base heteroatoms. Factors that alter the nucleophilicities of various heteroatoms in the DNA bases, when placed within the context of double helix, include proximity of the polyanionic sugar-phosphate backbone, lower dielectric constants in the DNA grooves relative to bulk water, and interaction of the inherent dipoles of the nucleobases with the electrostatic environment of the double helix (e.g., charges of the backbone and neighboring bases).[29]

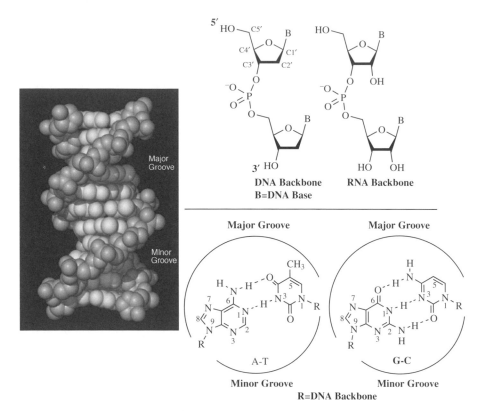

Figure 8.1. See color insert.

8.2.2. The Sequence Context of a DNA Base Affects its Nucleophilicity

The sequence context in which a given DNA base resides can also affect its nucleophilicity. For example, many studies have documented the fact that flanking bases have a marked effect on the reaction of the N7-position of guanine residues with positively charged electrophiles such as aziridinium ions and diazaonium ions.[30–34] The observed sequence specificity for DNA alkylation by these agents often correlates well with the calculated sequence-dependent variations in the molecular electrostatic potential at the N7-position of the guanine alkylation site.[30–35] For example, for many agents, alkylation at 5′-GGG sequences is markedly favored over that at 5′-CGC sequences (where the G is alkylated).

8.2.3. The Nature of the Reactive Intermediate Involved in the Alkylation Reaction Influences Atom Site Selectivity

The favored sites for alkylation of duplex DNA vary depending upon the nature of the alkylating agent. For example, in the alkylation of double-stranded DNA by dimethylsulfate, the preferred sites of attack follow the order: N7G $\gg$ N3A $\gg$ N1A~N3A~N3G $\gg$ O^{6}G.[25,26] On the contrary, when methyldiazonium ion is the alkylating agent, the preferred sites of attack are: phosphate oxygen $>$ N7G $>$ O^{2}T $>$ O^{6}G $>$ N3A.[25,26] Such trends have often been rationalized by noting that hard, S$_N$1-type alkylating agents such as methyldiazonium ion display increased reactivity with hard oxygen nucleophiles in DNA.[25–28] This analysis is of some practical value in predicting and rationalizing alkylation site preferences; however, in some cases the traditional classifications of alkylating agents as "S$_N$1 or S$_N$2" may be inaccurate and, therefore, confusing. For example, primary diazonium ions that have typically been classified as S$_N$1-type alkylating agents,[28] in fact, are known to react strictly via bimolecular S$_N$2 mechanisms.[36,37] Accordingly, Fishbein and co-workers sharply refined the analysis of alkylation site specificity when they pointed out that "the low degree of covalency in the transition state for S$_N$2 substitution on primary diazonium ions means that electrostatic stabilization of the transition state by the nucleophile will become a dominant interaction and this gives rise to enhanced oxygen atom alkylation by primary diazonium ions."[29]

8.2.4. Noncovalent Association of the Alkylating Agent can Direct Reaction to Selected Atoms and Selected Sequences in DNA

Alkylation atom-site specificity can be strongly influenced by noncovalent association of an agent with the DNA duplex. Localization of an electrophilic functional group near a particular DNA nucleophile can drive alkylation exclusively to a single site.[38] For example, small episulfonium ions alkylate a variety of sites in duplex DNA, including N7G, N3A, and O^{6}G.[39] In contrast, noncovalent DNA binding directs the episulfonium ion derived from natural product leinamycin exclusively toward reaction at the N7-position of guanine residues.[34,40] Similarly, Gold and co-workers designed nitrosamine-intercalator conjugates that alter the atom-site selectivity of nitrosamines.[41] In addition to modulating atom-site selectivity, noncovalent binding can also afford sequence selectivity in DNA reactions. For example, attachment of DNA-damaging agents to sequence-selective major groove[42] or minor groove binders[43,44] can direct covalent DNA modification to a small number of targeted sites in genomic DNA.

8.3. POSSIBLE FATES OF ALKYLATED DNA

8.3.1. Alkylation at Some Sites in DNA Yields Stable Lesions

Alkylation at several sites on the DNA bases affords an attachment that is chemically stable. Nucleophilic sites that typically yield stable alkyl attachments include the exocyclic nitrogen atoms N^{2}G, N^{6}A, N^{6}C, the amide-type nitrogens N1G, N1T, O^{6}G, and O^{4}T.[45–52] Alkyl groups on phosphate residues in DNA are typically stable under physiological conditions,[53,54] although the charge-neutralized phosphotriester groups are somewhat more prone to hydrolysis under basic conditions, compared with the native phosphodiesters.[55,56]

8.3.2. Alkylation at Some Sites in DNA Yields Unstable (Labile) Lesions

Alkylation at many of the endocyclic nitrogen atoms in the DNA bases places a formal positive charge on the nucleobase. As a result, alkylation at N7G, N7A, N3G, N3A, N1A, and N3C leads to destabilization of the bases. The decomposition reactions of these alkylated bases are driven by the "desire" to neutralize the formal charge imparted by alkylation. There are three principle types of reactions involved in the decomposition of these alkylated bases: (1) deglycosylation (also known as depurination or depyrimidination) — hydrolytic loss of the alkylated base from the DNA backbone; (2) ring opening; and (3) loss of the alkyl group from the base. These reactions are perhaps best characterized for N7-alkylguanine lesions[57] and will be discussed here primarily in this context.

8.3.3. Deglycosylation Reactions of Alkylated Bases

Hydrolytic cleavage of the glycosidic bond holding the DNA bases to the sugar-phosphate backbone is typically a very slow process under physiological conditions (pH 7.4; 37°C).[58] Loss of the pyrimidine bases cytosine and thymine occurs with a rate constant of 1.5×10^{-12} s^{-1} ($t_{1/2}$ = 14,700 years), while loss of the purine bases guanine and adenine proceeds slightly faster, with a rate constant of 3.0×10^{-11} s^{-1} ($t_{1/2}$ = 730 years). Alkylation of the DNA bases can vastly increase the rate of deglycosylation (Scheme 8.2), with the degree of destabilization depending upon the exact site of alkylation. For example, the half-lives for deglycosylaton of bases modified with simple alkyl groups are as follows: N7dA (3 h) > N3dA (24 h) > N7dG (150 h) > N3dG (greater than 150 h) > O^{2}dC (750 h) > O^{2}dT (6300 h) > N3dC (7700 h).[52,57,59–61]

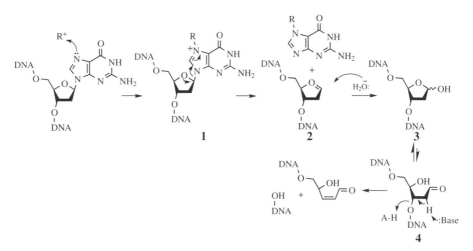

Scheme 8.2

Analogous to the well-studied acid-catalyzed deglycosylation of guanine residues from DNA,[62–64] the loss of alkylated bases from the sugar-phosphate backbone likely proceeds via an oxygen-assisted S_N1 hydrolysis reaction, formally classified as a $D_N + A_N$ reaction (Scheme 8.2).[65] In the case of N7-alkylguanine residues (**1**, and probably for other alkylated bases as well), electron-withdrawing groups on the alkyl substitutents increase the leaving group ability of the alkylated base, thus increasing the rate of deglycosylation.[57,66,67] Loss of alkylated bases is slower in duplex DNA than in single-stranded DNA and monomeric nucleosides. For example, the reaction is 50–100 times slower in duplex DNA than it is in nucleosides.[57] Addition of electron-withdrawing substituents to the sugar residue also slows the deglycosylation of the alkylated bases, presumably through destabilization of the carbocation ion intermediate **2** (Scheme 8.2).[57,63,68,69] For example, the 2'-OH groups found in RNA significantly slow depurination of alkylguanine residues.

The abasic sites (**3**, Scheme 8.2) resulting from the loss of alkylated bases from DNA are both cytotoxic and mutagenic.[70–72] The cyclic acetal (**3**) exists in equilibrium with small amounts (~1%) of the open chain aldehyde (**4**).[73] The acidic nature of α-proton in the aldehyde form of the abasic lesion facilitates β-elimination of the 3'-phosphate residue to yield a strand break.[74,75] This reaction occurs with a half-life of about 200 h under physiological conditions (pH 7.4, 37°C), but can be accelerated by heat, basic conditions, or the presence of various amines.[76–80]

8.3.4. Ring-Opening Reactions of Alkylated Bases

The formal positive charge imposed on the DNA bases by alkylation at certain endocyclic nitrogens makes the ring systems susceptible to hydrolytic attack that may ultimately lead to ring-opening reactions. For example, attack of hydroxide at the C8-position of N7-alkylguanine residues leads to fragmentation of the imidazole ring to produce the 5-(alkyl)formamidopyrimidine (FAPy) derivatives **5** (Scheme 8.3).[57,81–88] The glycosidic bond that joins alkyl-FAPy derivatives to the deoxyribose sugar is hydrolytically stable, and the half-life for release of this type of ring-opened base has been estimated[57] from the data of Greenberg's group[89] to be about 1500 h in a single-stranded DNA at pH 7.5 and at 37°C. FAPy derivatives undergo anomerization via an imine intermediate **6** (Scheme 8.3). The half-life for this process in monomeric FAPy nucleosides is approximately 3.5 h.[89,90] Either acidic (0.1 N HCl) or basic (0.2 N NaOH) conditions lead to the loss of the formyl group from (alkyl)-FAPy-guanine lesions.[86,91,92]

The conversion of N7-alkylguanine residues into the corresponding FAPy derivative is facilitated by electron-withdrawing groups on the alkyl substituent or the sugar residues, but is slowed by neighboring phosphate residues in nucleotides or polynucleotides, presumably due to electrostatic repulsion of the hydroxide nucleophile.[66,67] In general, the ring-opening reaction is quite slow relative to the competing depurination reaction (Scheme 8.2) under physiological conditions.[57,93] However, basic conditions (e.g., pH 10, 37°C, 4 h) can be employed to prepare the FAPy lesion from N7-alkylguanine lesions in oligonucleotides.[88,94] Alkyl-FAPy lesions typically may not be formed in high yields under physiological conditions; but, once formed

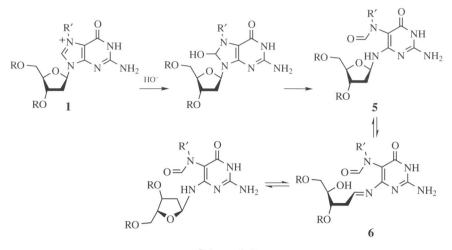

Scheme 8.3

in cellular DNA, they persist[95–97] and are likely to be mutagenic or cytotoxic.[98–100] Thus, even low yields of these DNA-damage products could have significant biological effects.

Similar to the ring opening of N7-alkylguanine residues described above, adenine residues alkylated at the N1-position are prone to hydrolytic ring opening. In this case, however, ring opening initiates a rearrangement reaction that leads to an apparent migration of the alkyl group from the N1-position to the exocyclic N^6-position of the nucleobase.[101] The overall process is known as a Dimroth rearrangement (Scheme 8.4). From a practical perspective, this reaction makes it challenging to determine whether N^6-alkyladenine adducts observed in DNA arise via direct reaction at this position or via rearrangement of an initial N1-adduct. The half-life for the Dimroth rearrangement of the 1-methyl-2′-deoxyadenosine-5′-phosphate nucleotide is approximately 150 h at pH 7.[59] One might suspect that the C5-C6 bond rotation required for the Dimroth rearrangement would be hindered by hydrogen bonding and stacking of the adducted base within the double helix; however, it has been shown that the rearrangement can occur to a significant extent in duplex DNA.[102]

N3-Alkyladenine nucleosides can also undergo ring opening under basic conditions. At pH 8.98, the ring-opening reaction of 3-methyl-2′-deoxyadenosine is

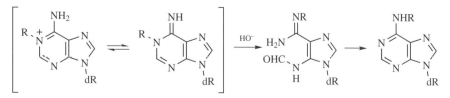

Scheme 8.4

observed alongside hydrolytic cleavage of the glycosidic bond (depurination). At neutral pH values, this ring-opening reaction may not compete effectively with the relatively fast depurination of 3-alkyladenine residues in duplex DNA.[103]

8.3.5. Alkylation of Adenine and Cytosine Residues can Accelerate Deamination

Deamination, the hydrolytic loss of exocyclic amino groups on the DNA bases, is typically a very slow reaction. For example, deamination of cytosine residues in duplex DNA occurs with a half-life of about 30,000 years under physiological conditions, and the deamination of adenine residues is still more sluggish.[104,105] Alkylation at the N3-position of cytosine (Scheme 8.5) greatly increases the rate of deamination ($t_{1/2}$ = 406 h). Deamination of 3-methyl-2'-deoxycytidine proceeds 4000 times faster than the same reaction in the unalkylated nucleoside.[61] Alkylation of the N3-position in cytosine residues also facilitates deglycosylation ($t_{1/2}$ = 7700 h, lower pathway in Scheme 8.5), but the deamination reaction is 20 times faster and, therefore, predominates.[61]

The analogous deamination reaction is not observed in 1-methyl-2'-deoxyadenosine nucleosides.[106] Rather, in the adenine series, the Dimroth rearrangement occurs (Scheme 8.4). On the contrary, in styrene adducts of 2'-deoxyadenosine, the hydroxyl residue of the adduct undergoes intramolecular reaction with the base to initiate deamination (Scheme 8.6).[107,108] Similarly, cytosine residues bearing styrene adducts at the N3-position undergo rapid deamination (nearly complete deamination is seen within 75 h).[109]

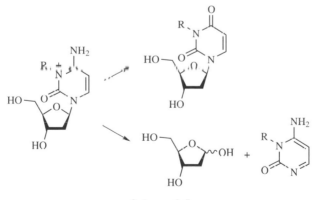

Scheme 8.5

8.3.6. Reactions Where the DNA Base Serves as a Leaving Group: Thiolysis, Hydrolysis, Elimination, and Reversible DNA Alkylation

In some DNA adducts, the nucleobase can serve as a leaving group in decomposition reactions. For example, guanine is expected to be a reasonably good leaving group

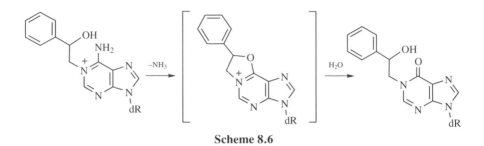

Scheme 8.6

in N7-alkyl guanine adducts, as judged by the pK_a of its conjugate acid (the pK_a of an N7-protonated guanine residue is 3.5).[110] Accordingly, attack of *tert*-butylthiolate on 7-methylguanosine regenerates the unmodified nucleoside (Scheme 8.7).[111] The nitrosamine-derived guanine adduct shown in Scheme 8.8 undergoes hydrolysis

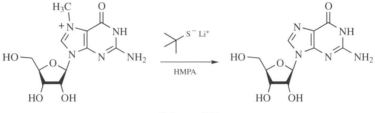

Scheme 8.7

to regenerate the unmodified DNA base.[112] In one example (Scheme 8.9), an N7-guanine adduct is thought to decompose via elimination of the nucleobase.[113]

In a few interesting cases, adducts decompose via reactions in which DNA serves as a leaving group and the original DNA-alkylating electrophile is regenerated. Reversible

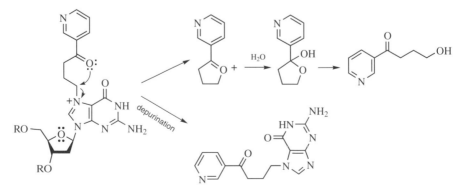

Scheme 8.8

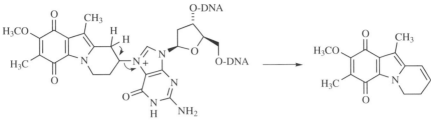

Scheme 8.9

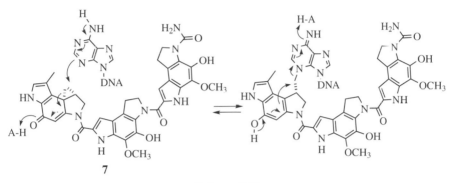

7

Scheme 8.10

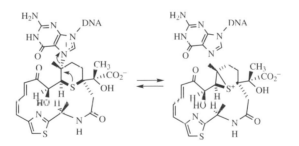

Scheme 8.11

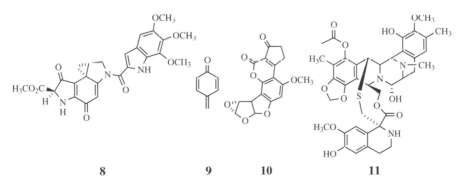

8　　　　　9　　　10　　　　　11

DNA alkylation has the potential to yield a time-dependent spectrum of adducts, in which initially formed kinetically favored lesions give way to the thermodynamically favored adducts over time.[114] Reversible alkylation has been observed at several of the nucleophilic sites in DNA, including N3A (CC-1065, **7**, Scheme 8.10, duocarmycin, **8**),[115–117] N1A (quinone methide, **9**),[114] N7G (leinamycin, Scheme 8.11, aflatoxin B₁ epoxide, **10** and quinone methide, **9**),[57,114,118] N3C (quinone methide, **9**),[114] and N²G (ecteinascidin 743, **11**).[119] The bidentate N1/N²G adduct of malondialdehyde also forms reversibly.[120]

8.4. SELECTED EXAMPLES OF DNA-ALKYLATING AGENTS

A large number of DNA-alkylating agents are known and we will not attempt a comprehensive survey here. A number of excellent reviews provide an overview of this area.[20–22,24,25,121,122] Here we will review DNA alkylation by three types of reactive intermediates that are important in medicinal chemistry and toxicology—episulfonium ions, aziridinium ions, and carbocations.

8.4.1. Episulfonium Ions

Sulfur mustards (e.g., **12**, Scheme 8.12) decompose to generate episulfonium ions (**13**).[123–125] This reaction involves rate-limiting intramolecular displacement of chloride by the attack of sulfur on the β-carbon followed by the attack of nucleophiles on the resulting episulfonium ion (Scheme 8.12). Episulfonium ions are typically fleeting intermediates that can be observed only under special conditions.[124] Mustard-derived episulfonium ions alkylate DNA at a variety of positions, including N7G, N3A, and O⁶G (Scheme 8.12).[39] The major products are the N7-guanine adducts. Alkylation of DNA by sulfur mustards is not particularly efficient because these agents react with a variety of biological macromolecules and undergo significant amounts of hydrolytic decomposition.[123] Nonetheless, reactions of mustard-derived

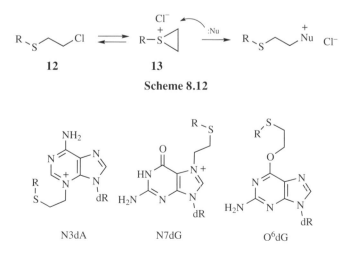

Scheme 8.12

N3dA   N7dG   O⁶dG

episulfonium ions with DNA are central to the biological activities of war gases and several environmental mutagens.[126,127]

In the antibiotic leinamycin, Nature has revealed novel strategies for the efficient delivery of an episulfonium ion alkylating intermediate to cellular DNA.[128] Leinamycin (**14**, Scheme 8.13) is a *Streptomyces*-derived metabolite that displays potent antitumor properties (IC_{50} of 27 nM against HeLa cells) stemming from its ability to damage cellular DNA.[129] The DNA-damaging properties of leinamycin are triggered by the reaction of the natural product with thiolate. This interesting thiol-dependent chemistry is probably central to the biological activity of leinamycin because it represents a strategy for the selective generation of a cytotoxic reactive intermediate when the compound enters the thiol-rich environment of the cell. Mammalian cells contain millimolar concentrations of thiols (primarily glutathione, **15**).[130]

Reaction of thiolates with the 1,2-dithiolan-3-one 1-oxide heterocycle of leinamycin generates the sulfenate intermediate **16**.[131,132] Cyclization of this sulfenate onto the neighboring carbonyl group yields two unusual reactive intermediates: (1) a persulfide (also known as a hydrodisulfide, **17**) residue derived from the attacking thiol and (2) the 1,2-oxathiolan-5-one derivative of leinamycin (**18**).[131,132] The persulfide residue is thought to contribute to the cytotoxicity of leinamycin through its ability to generate oxidative stress in the cell.[133–135] The electrophilic sulfur atom in the 1,2-oxathiolan-5-one (**18**) reacts with the C6–C7 alkene of leinamycin to generate the episulfonium ion alkylating agent (**19**).[40] This episulfonium ion exists in equilibrium with an epoxide isomer **20**. In fact, the epoxide is the only product observed in NMR experiments, following thiol activation of leinamycin; however, product analysis shows that all subsequent alkylation reactions proceed via equilibrium amounts of the episulfonium **19**.[40,136] The half-life of activated leinamycin (consisting of the **19/20** mixture) in neutral aqueous buffer is approximately 3 h.[136,137] The episulfonium-epoxide equilibrium, involving a thia-Payne rearrangement,[138,139] may allow the epoxide isomer to serve as a semistable "carrier" that limits unproductive hydrolysis, thus increasing the yield of DNA alkylation.

Leinamycin alkylates the N7-position of guanine residues in duplex DNA. The reaction occurs selectively at 5′-GG sequences.[34] This sequence selectivity, along with the inability to alkylate single-stranded DNA or free nucleosides efficiently,[34] led to the suspicion that leinamycin associates noncovalently in the major groove of DNA. Indeed, recent work shows that the alkylation of DNA by leinamycin is dependent upon initial noncovalent binding.[140] Both biophysical and kinetic measurements reveal a modest binding constant of about 1000–2000 M^{-1} for the association of leinamycin (**10**) or activated leinamycin (**19/20**) with duplex DNA.[140,141] Kinetic analysis shows that noncovalent binding provides a 150,000-fold rate acceleration for the alkylation of guanine residues in duplex DNA as against the free 2′-deoxyguanosine nucleoside.[140] This corresponds to an effective concentration of 74 M leinamycin in the major groove of DNA.[140] In addition, noncovalent DNA binding helps leinamycin to achieve an apparent second-order rate constant for DNA alkylation that is 62 million times larger than that for hydrolytic decomposition.[140] The thiol-triggered rearrangement and DNA-alkylation by leinamycin is a remarkably efficient process. The DNA-alkylation reaction is absolutely dependent upon

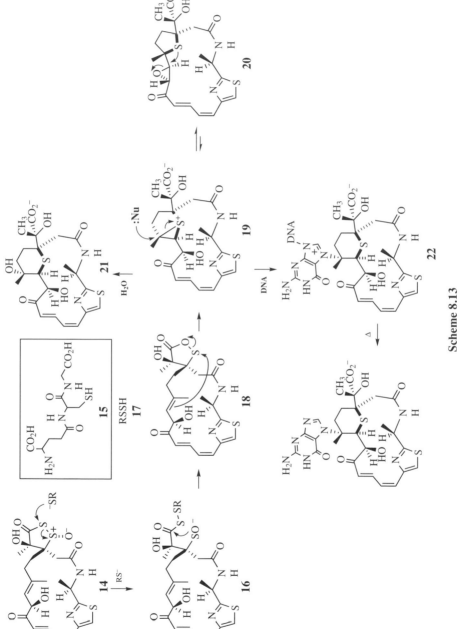

Scheme 8.13

noncovalent association of leinamycin with the duplex. At physiologically relevant DNA concentrations (≥ 10 mM base pairs), >90% of the input leinamycin goes on to form the guanine adduct. Interestingly, if leinamycin *lacked* its modest DNA-binding abilities, it is estimated that the yields of DNA alkylation would be less than 1% and the natural product would largely undergo hydrolytic decomposition to yield **21**. Overall, these studies with leinamycin provide a stark indication of the important role that noncovalent binding can play in the covalent modification of DNA by reactive intermediates. Finally, the guanine adduct **22**, produced by leinamycin, undergoes unusually rapid depurination ($t_{1/2} = 3$ h) to yield a "burst" of abasic sites in duplex DNA that may present a special challenge to the cell.[118,142]

8.4.2. Aziridinium Ions

Nitrogen mustards such as **23** decompose to yield aziridinium ions **24** (Scheme 8.14).[143] As in the case of sulfur mustards, the reaction involves rate-limiting intramolecular displacement of β-chlorine by nitrogen. The resulting aziridinium ions

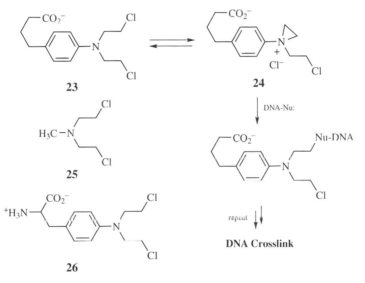

Scheme 8.14

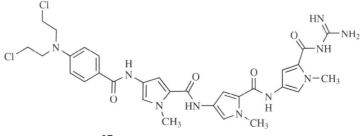

27

alkylate DNA at a variety of positions including N7G, N3A, N1A, N^6A, and the phosphate groups.[144-146] Several N,N-bis(2-chloroethyl) bisalkylating agents such as chlorambucil (**23**), mechlorethamine (**24**), and melphalan (**25**) see widespread clinical use for the treatment of various cancers.[144] These agents form interstrand cross-links that involve N7G-to-N7G attachments at 5'GNC sites.[147] In addition, Tretyakova and co-workers have identified N7G-N3A, N7G-N1A, and N7G-N^6A cross-links formed by these agents in duplex DNA.[146] Although cross-links are formed in relatively low yields as compared with monoadducts, they are thought to account for much of the cytotoxic activity of these drugs.[144,147] Bisalkylating nitrogen mustards have been conjugated with a variety of noncovalent DNA-binding units in an effort to improve the yields of DNA alkylation.[44,148,149] Many of these agents do, indeed, show improved potency or selectivity against cancer cell lines, and the minor groove-binder-mustard conjugate tallimustine (**27**) has reached phase II clinical trials.[150]

Fasicularin (**28**) is the first natural product observed to generate a DNA-alkylating aziridinium ion via a chemical mechanism analogous to the clinically used anticancer drugs mechlorethamine and chlorambucil.[151] Fasicularin is a thiocyanate-containing alkaloid that was isolated in 1997 from the sea squirt, *Nephteis fasicularis*.[152] This compound displays an IC_{50} of 14 μg/mL against Vero cells. Furthermore, experiments showing that DNA repair-deficient cell lines are hypersensitive to fasicularin provided an early indication that DNA is an important cellular target for this agent. Recent studies employed gel electrophoretic analysis along with LC/MS/MS to demonstrate that fasicularin, at biologically relevant concentrations near its IC_{50} value, alkylates guanine residues in duplex DNA.[151] After thermal workup of the DNA to depurinate the alkylguanine residue (**30**), the adduct **31** was identified by HPLC/MS/MS. A simple secondary thiocyanate, isopropyl thiocyanate, did not cause DNA damage in these assays, thus supporting the idea that the thiocyanate group and the nitrogen atom in fasicularin work in concert to generate the DNA-alkylating aziridinium ion **29** (Scheme 8.15). Interestingly, this result represents for the first time that an aziridinium ion has been generated under physiological conditions by intramolecular displacement of a thiocyanate anion. The thiocyanate leaving group (pK_a of HSCN = −1.9) may strike an appropriate balance of reactivity and stability in this molecular framework. For stereoelectronic reasons, aziridinium ion formation is expected to occur only from certain conformations of fasicularin. Thus, it is possible that the azatricyclic ring system introduces conformational rigidity that controls the rate of aziridinium ion formation in this natural product. Finally, the structurally related natural products cylindricines A and B (**34** and **32**) interconvert via a presumed aziridinium ion intermediate (**33**, Scheme 8.16).[153] It will be interesting to examine the DNA-damaging chemistry and antitumor biology of these agents.

8.4.3. Alkenylbenzenes: DNA Alkylation by Resonance-Stabilized Carbocations

Estragole (**35**), methyleugenol (**36**), and safrole (**37**) occur naturally in various herbs and spices and are approved favor additives. [154,155] These compounds are

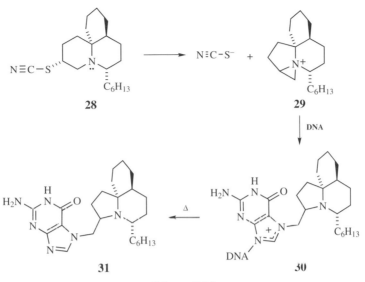

Scheme 8.15

procarcinogens,[154] that is, they are not themselves carcinogenic but are metabolically converted to carcinogens in vivo.[156] Cytochrome P450-mediated oxidation at the benzylic position of these compounds can yield the hydroxylated
analog (**39**), which is further converted by phase II metabolism to the sulfate
derivative **40** (Scheme 8.17).[155] This sulfate derivative is thought to decompose

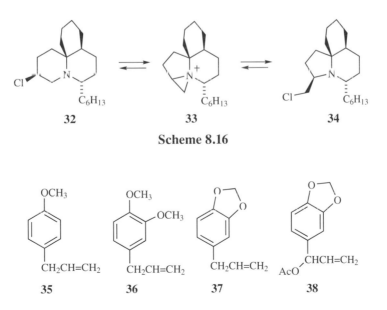

Scheme 8.16

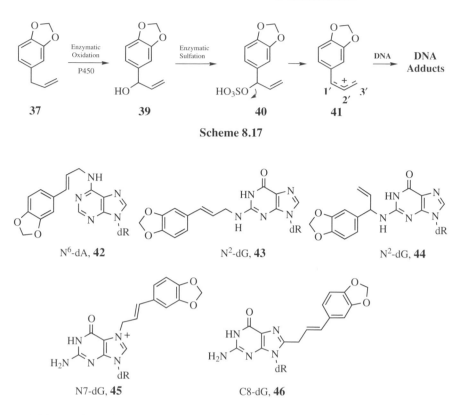

Scheme 8.17

to yield a carbocation that is responsible for the observed DNA alkylation by these compounds.[157] Acetoxy derivatives such as **38** have been used to model DNA alkylation by the sulfate **40**.[158] In vitro reactions of the acetoxy compound **38** with nucleosides produced adducts with attachment to DNA at both the benzylic and terminal carbon of the propene side chain. Adducts connected at N^6dA (**42**), N^2dG (**43**, **44**), N7dG (**45**), and C8dG (**46**), were obtained.[158,159] These same adducts were formed in vivo, in the livers of mice treated with 1'-hydroxysafrole.[158,159]

Formation of the C8dG adduct in these reactions is striking. As described in subsequent sections, the C8-position of guanine is a common site for the addition of radicals. However, the C8-position of guanine residues is not strongly nucleophilic and, consequently, does not typically react with electrophiles (although reactions of nitrenium ions with 2'-deoxyguanosine may present an exception to this rule).[160] In the case of these metabolically activated alkenylbenzenes, it is possible that the C8-adducts arise via rearrangement of an initially formed N7-guanine adduct (**47**, Scheme 8.18). The acidity of the C8-proton in guanine residues is vastly increased upon N7-alkylation of base.[161,162] Thus, equilibrium amounts of the resulting C8-anion (**48**) may carry out an intramolecular S_N2' displacement of the N7 atom to yield the C8-adduct (**46**).

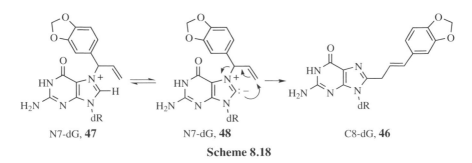

N7-dG, **47**   N7-dG, **48**   C8-dG, **46**

Scheme 8.18

8.5. REACTION OF RADICAL INTERMEDIATES WITH DNA

8.5.1. General Considerations Regarding Hydrogen Atom Abstraction from the 2′-Deoxyribose Sugar

Abstraction of hydrogen atoms from the sugar-phosphate backbone (Fig. 8.1, Scheme 8.1) typically leads to breakage of the DNA chain by complex reaction cascades.[163–167] Reaction barriers of hydrogen abstraction reactions typically correlate with the C–H bond strength.[168] The C–H bond enthalpies calculated for abstraction of hydrogen from the 1′, 2′, 3′, and 4′ positions of deoxyribose reveal similar bond strengths for the 1′, 4′, and 3′ C–H bonds. In these cases, the resulting carbon-centered radicals are stabilized by an adjacent oxygen.[168] The bond energy of the 2′ C–H bond is computed to be slightly higher. Bond enthalpies calculated at the 6–31G* level range from 87.3 to 91.2 kcal/mol.[168] This computation compares reasonably well with the experimentally measured C–H bond strength of 92 kcal/mol for the α hydrogen in tetrahydrofuran.[169] In practice, highly reactive radicals such as the hydroxyl radical (HO•) can abstract hydrogen atoms from every position on the DNA backbone, with the relative amount of reaction at each position dictated by steric accessibility, rather than C–H bond strength.[170] The accessibility of the deoxyribose hydrogens in duplex DNA follows the trend: H5′ > H4′ >> H3′ ~ H2′ ~ H1′.[170] There are many excellent, comprehensive reviews that discuss the chemistry stemming from hydrogen atom abstraction from the deoxyribose sugars of DNA.[163–167] Here, we will discuss two of the better characterized strand cleavage pathways involving hydrogen atom abstraction from the C1′ and C4′ positions of deoxyribose.

8.5.2. Strand Cleavage Stemming from Abstraction of the C1′-Hydrogen Atom

Although the C1′-hydrogen is the least accessible deoxyribose hydrogen,[170] deeply buried in the minor groove of duplex DNA, abstraction of this hydrogen atom is an important reaction for a variety of agents including Cu(o-phenanthroline)$_2$, neocarzinostatin, esperamicin, and dynemicin.[164] Work of Greenberg and co-workers, employing photogeneration of discrete C1′-DNA radicals, has provided a wealth

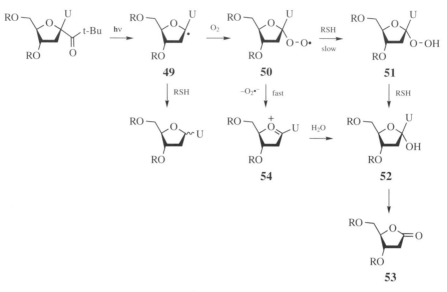

Scheme 8.19

of information regarding the reactions of the C1′-radical under physiological conditions.[171] Cells contain two agents, molecular oxygen (~100 μM) and thiols (primarily glutathione, **15**, at 1–10 mM) that react readily with the C1′-deoxyribose radical. The C1′-radical abstracts a hydrogen atom from thiols with a rate constant of about $2 \times 10^6 M^{-1} s^{-1}$ in duplex DNA.[172] This reaction can generate either the natural β-anomer (a "chemical repair" reaction) or the mutagenic α-anomer. Measurements from Greenberg's group suggest that, in duplex DNA under physiologically relevant conditions (5 mM thiol and 200 μM O_2), only about 0.4% of the C1′ radical goes on to generate the mutagenic α-anomer.[172]

Reaction of the C1′ radical **49** with O_2 to yield the peroxyl radical **50** (Scheme 8.19) is a very fast reaction, occurring with a rate constant of $1 \times 10^9 M^{-1} s^{-1}$.[173] Two potential fates have been recognized for this peroxyl radical. This intermediate can abstract hydrogen atoms from thiol with a rate constant $\leq 200 M^{-1} s^{-1}$ to yield the hydroperoxide **51**.[174] Under physiological conditions, this hydroperoxide could be reduced to the alcohol **52** that, in turn, decomposes to the 2-deoxyribonolactone abasic site (**53**) (peroxides are readily reduced by the biological thiol glutathione at pH 7, with a rate constant of about $1 M^{-1} s^{-1}$, thus allowing us to estimate a half-life of about 1 min for the reduction of a hydroperoxide like **51**, in the presence of 10 mM glutathione).[175] The peroxyl radical can also eliminate superoxide radical anion to yield the carbocation **54**. Superoxide radical anion is a reasonably good leaving group (pK_a of the conjugate acid HOO• is 5.1). Competition kinetic analysis provided a rate constant of $1.2 s^{-1}$ for this reaction,[176] while laser flash photolysis studies yield a value of $2 \times 10^4 s^{-1}$.[173] Regardless of the discrepancy in these rate constants, it seems clear that, under physiological conditions (5–10 mM RSH), loss

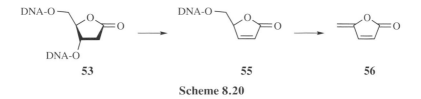

Scheme 8.20

of superoxide radical from the peroxyl radical **50** is likely to predominate over its reaction with glutathione. The resulting carbocation **54** undergoes hydrolysis and loss of the DNA base to yield the 2-deoxyribonolactone abasic site **53**.

The 2-deoxyribonolactone lesion (**53**) undergoes α,β-elimination of the 3′-phosphate residue to cause DNA strand cleavage (Scheme 8.20).[177–179] This strand-cleavage reaction proceeds with a half-life of 38–54 h in double-stranded DNA, depending on the identity of the base that opposes the lactone abasic site.[177] The resulting 3′-ene-lactone intermediate **55** undergoes a subsequent rapid γ,δ-elimination to give the methylene lactone sugar by-product (**56**) and an oligonucleotide with a 3′-phosphate residue.[177,178] The transient ene-lactone (**55**) has been trapped by a variety of chemical reagents.[179] In addition, the 2-ribonolactone and the ene-lactone generate covalent cross-links with DNA-binding proteins.[180,181]

8.5.3. Strand Cleavage Initiated by Abstraction of the 4′-Hydrogen Atom

The 4′-hydrogen atom of deoxyribose is an important reaction site for a variety of oxidative DNA-damaging agents including hydroxyl radical (HO•), bleomycin, and several enediynes.[166] Studies involving photogeneration of discrete C4′-sugar radicals have elucidated details of the reactions stemming from hydrogen atom abstraction at this position of the DNA backbone. The C4′ radical **57** resulting from hydrogen

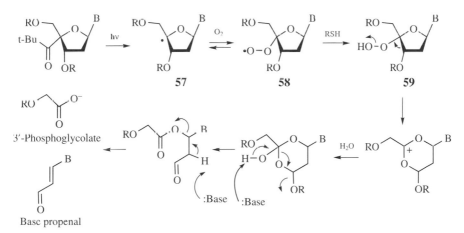

Scheme 8.21

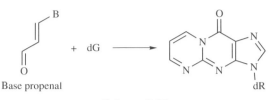

Base propenal   + dG   dR

Scheme 8.22

atom abstraction readily reacts with biological thiols, with a rate constant of about $2 \times 10^6 \, M^{-1} \, s^{-1}$ in single-stranded DNA.[182] This reaction in duplex DNA yields a 10:1 ratio of natural:unnatural stereochemistry at the C4′ center.[183] The C4′ radical **57** also reacts rapidly with molecular oxygen (presumed $k_{O_2} = 2 \times 10^9 \, M^{-1} \, s^{-1}$) to yield the peroxyl radical **58** (Scheme 8.21).[182] Reaction of the peroxyl radical with thiol (estimated $k = 200 \, M^{-1} \, s^{-1}$) yields hydroperoxide **59**. This intermediate has been observed directly by MALDI mass spectroscopy in an experiment in which a single C4′ radical was generated at a defined site in a single-stranded 2′-deoxyribonucleotide.[184] The hydroperoxide (**59**) is thought to yield strand cleavage via Criegee rearrangement, followed by hydrolysis and fragmentation that produces the 5′-phosphate, 3′-phosphoglycolate, and base propenal end products (Scheme 8.21).[184] Interestingly, the base propenal products of oxidative DNA damage can further damage DNA via the formation of covalent adducts at guanine residues, as shown in Scheme 8.22.[120,185]

Early γ-radiolysis studies by Schulte–Frohlinde,[186,187] followed by more recent experiments[184,188] employing photogeneration of discrete C4′ radicals, identified an alternative reaction pathway that yields the well-known 5′-phosphate, 3′-phosphoglycolate, and base propenal end products. The C4′ radical can eliminate phosphate to yield a radical cation intermediate **60** that undergoes hydrolysis to afford the neutral radical **61**.[184,188] Reaction with molecular oxygen followed by reduction of the peroxyl radical gives the hydroperoxide **62** (Scheme 8.23).[184] Giese's group has shown that this peroxide

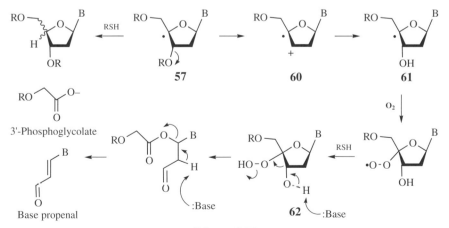

Scheme 8.23

undergoes a Grob fragmentation followed by elimination to afford the final 3'-phosphoglycolate and base propenal products.[184,189] The Schulte–Frohlinde fragmentation of the C4'-radical to afford **60** (Scheme 8.23) occurs with a rate constant of $100\,s^{-1}$ in duplex DNA.[183,189] Thus, under typical physiological conditions (e.g., $100\,\mu M\ O_2$, 10 mM GSH), this pathway is not likely to compete favorably either against trapping of the radical **57** by thiol or the trapping by molecular oxygen that precedes the Criegee rearrangement shown in Scheme 8.21. Interestingly, addition of O_2 to the C4' radical is reversible (Scheme 8.21), with loss of molecular oxygen from the peroxyl radical **58** occurring with a rate constant of $1\,s^{-1}$.[182] The reversible nature of the reaction with molecular oxygen means that, at low thiol concentrations where irreversible trapping of the peroxyl radical **58** by thiol to produce **59** is slow, the damage reaction may instead channel through the Schulte–Frohlinde cleavage pathway shown in Scheme 8.23.

8.5.4. Oxidative RNA Damage

Interestingly, RNA is less susceptible to oxidative strand cleavage than DNA.[190] This observation can be understood broadly in terms of the oxidative cleavage pathways shown above. The 2'-hydroxyl group that distinguishes RNA from DNA (Fig. 8.1) has the potential to destabilize cationic intermediates such as those generated following abstraction of both the 1'- and 4'-hydrogen atoms (Schemes 8.19, 8.21, and 8.23).[191]

8.5.5. General Considerations Regarding Reactions of Radicals with the DNA Bases

Radicals can react with bases via hydrogen atom abstraction or, more commonly, by addition to the pi bonds in the heterocyclic nucleobases (Scheme 8.1). These reactions have been extensively studied in the context of hydroxyl radical (HO•), which is generated by γ-radiolysis of water. When DNA is exposed to the hydroxyl radical, approximately 80% of the reactions occur at the bases.[163,192,193] Many base damage products arising from the reaction hydroxyl radical with DNA have been characterized (Fig. 8.2).[193–198]

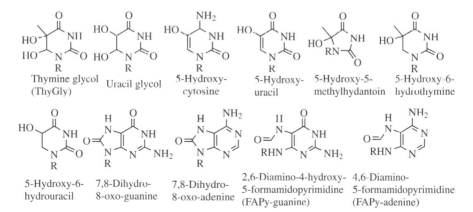

Figure 8.2

Radical attack yields nucleobase radical adducts that must undergo either oxidation or reduction to yield a stable final product. The cellular oxidant in these reactions may be molecular oxygen or high-valent transition metal ions (e.g., Fe^{3+}), while the reductant may be either thiols, superoxide radical, or low-valent transition metal ions (e.g., Fe^{2+}). In many cases, the base remains largely intact and the sequence of chemical events can be readily inferred. In some other cases, more extensive base decomposition occurs. Here, we will consider a set of representative examples that provide a framework for understanding virtually all radical-mediated base damage reactions.

8.5.6. Hydrogen Atom Abstraction from the C5-Methyl Group of Thymine

Highly reactive radicals such as hydroxyl radical (HO•) can abstract hydrogen atoms from the methyl group at the C5-position of thymine residues.[199] Trapping of the resulting radical **63** (Scheme 8.24) by molecular oxygen yields an intermediate peroxyl radical **64** that ultimately produces the 5-hydroxymethylthymidine lesion (**66**) in DNA.[199] Two mechanisms have been considered for this process. First, elimination of superoxide radical anion, followed by addition of water to the resulting methide intermediate has been suggested.[163] However, the precedent cited for this chemistry involves reactions of the free base (Scheme 8.25) which may not serve as a good model for the DNA nucleotides.[200] Indeed, recent isotopic labeling studies from Greenberg's group show that the oxygen in **66** is derived from O_2 and not from H_2O.[201] This observation strongly argues against the methide intermediate shown in Scheme 8.25. Rather, the reaction of the peroxyl radical **64** with a reducing agent (e.g., thiol or $O_2\bullet^-$) will yield the hydroperoxide **65** (Scheme 8.24). This species decomposes to **66** with a half-life of 35 h in water.[202] Importantly, in the presence of biologically relevant thiol concentrations, reduction of the hydroperoxide **65** to the hydroxymethyl nucleobase **66** may be quite rapid. The notion that hydroperoxide reduction by thiols is facile

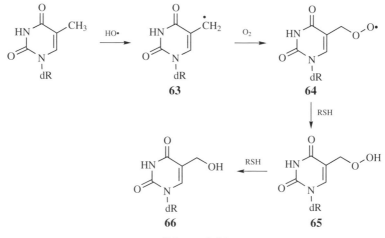

Scheme 8.24

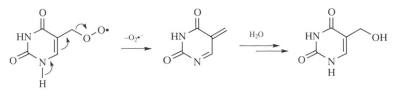

Scheme 8.25

is supported by the fact that hydrogen peroxide is readily reduced by the biological thiol glutathione at pH 7, with a rate constant of about 1 M^{-1} s^{-1}.[175] The reactivities of organic hydroperoxides and hydrogen peroxide with thiols are similar.[203] Together, these facts allow us to estimate a half-life of about 1 min for thiol-mediated reduction

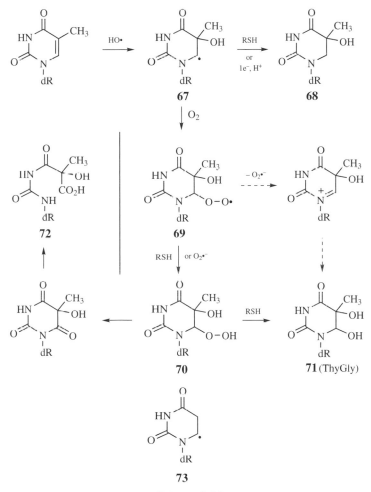

Scheme 8.26

of a DNA hydroperoxide in the presence of 10 mM glutathione. Analogous oxidation of the C5-methyl group is observed for 5-methylcytosine residues in DNA.[204]

8.5.7. Addition of Radicals to C5 of Thymine Residues

Radicals such as HO• readily add to the carbon–carbon pi bond of thymine residues.[199] Addition to the C5-position yields a C6-thymidine radical (**67**, Scheme 8.26). Reaction of this radical with reducing agents such as a thiol or superoxide radical anion yields the 5-hydroxy-6-hydrothymine residue (**68**, Scheme 8.26). Alternatively, reaction with molecular oxygen affords the peroxyl radical **69**. It has been suggested that elimination of superoxide radical anion from **69**, followed by subsequent addition of water, yields thymine glycol (**71**).[163] This sort of heteroatom-assisted elimination of superoxide has been seen in the context of the C1′-sugar radical (Scheme 8.19) and for the uracil free base (Scheme 8.25).[176,200] Importantly, the uracil free base may not serve as a good model for the DNA nucleotides (Schemes 8.24 and 8.25). In fact, recent studies did not detect ejection of superoxide from the peroxyl radical derived from 5,6-dihydro-2′-deoxyuridin-6-yl (**73**), indicating that the rate for superoxide elimination is $<10^{-3}$ s^{-1}.[205] Therefore, in the presence of biological thiols, it is likely that the peroxyl radical **69** undergoes reduction to the hydroperoxide **70**. In the presence of biological thiols, the hydroperoxide **70** may be reduced to yield thymine glycol **71**, which is a major product stemming from oxidative damage of DNA.[193–197] In the absence of thiol, the hydroperoxide **70** undergoes decomposition with a half-life of 1.5–10 h (this value is for the nucleoside in water, with the *trans*-isomer decomposing more rapidly than the *cis*-isomer).[202] In the thiol-independent decomposition process, formation of the ring-opened lesion **72** predominates, while thymine glycol **71** is formed as a minor product (Scheme 8.26).[202]

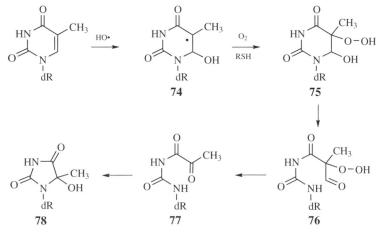

Scheme 8.27

8.5.8. Addition of Radicals to C6 of Thymine Residues

Addition of a radical to the C6-position of thymine residues in DNA generates the C5-thymine radical **74** (Scheme 8.27).[199] Reaction with molecular oxygen, followed by reduction, yields the hydroperoxide **75**. Decomposition of the hydroperoxide ultimately yields the hydantoin nucleobase **78** via the ring-opened derivatives **76** and **77** ($t_{1/2}$ for the decomposition of **75** in aqueous solution is slow, with a 9 h for the *trans*-isomer and 20 h for the *cis*-isomer of the nucleoside).[202]

8.5.9. Addition of Radicals to Guanine Residues

Guanine is the most easily oxidized DNA base[206,207] and, accordingly, is a major target for oxidative damage.[193] A common reaction for radicals involves addition to the C8-position of guanine residues in DNA (Scheme 8.28).[193,208,209] The resulting nucleobase radical **79** can undergo either one-electron reduction or oxidation to yield closed-shell products.[193,209] Reduction leads to a ring-opened formamidopyrimidine (FAPy) lesion (**80**). As discussed above in the context of alkyl-FAPy lesions, these ring-opened lesions are mutagenic and are chemically stable in DNA under physiological conditions.[100,210–212] Oxidation of the nucleobase radical intermediate **79** yields 8-oxo-7,8-dihydroguanosine (**81**), often referred to simply as 8-oxo-G. The 8-oxo-G lesion is sufficiently stable to allow its incorporation into synthetic 2′-deoxyoligonucleotides,[213] but is prone to oxidative decomposition under biologically relevant conditions.[193,214] Specifically, oxidation of 8-oxo-G (**81**) yields 5-hydroxy-8-oxo-7,8-dihydroguanosine

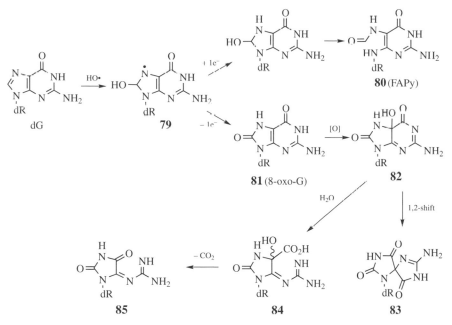

Scheme 8.28

(**82**).[215] This compound can decompose to yield the spiroaminodihydantoin (**83**) via a 1,2-shift.[215] In nucleosides, this reaction is favored at pH values ≥7. Alternatively, hydrolysis of **82** can yield the ureidoimidazoline (**84**), which undergoes decarboxylation to the guanidinohydantoin **85**.[215] In nucleosides, guanidinohydantoin formation is favored at pH values <7. In the context of duplex DNA, oxidative degradation of 8-oxo-G residues yields the guanidinohydantoin (**85**) residue as the major product, though this lesion may be prone to deglycosylation and further oxidative decomposition.[214]

8-Oxo-G (**81**), spiroaminodihydantoin (**83**), and guanidinohydantoin (**85**) are mutagenic lesions in duplex DNA.[216–220] Interestingly, the electrophilic intermediate generated upon oxidation 8-oxo-G can yield protein–DNA cross-links.[221–223]

8.5.10. Damage Amplification: Reaction of DNA Radical Intermediates with Proximal DNA Bases and Sugars

In some circumstances, DNA radical lesions can react with an adjacent base or the sugar residues. In these cases, a single radical "hit" can be transformed into two adjacent damage sites on the DNA. The resulting "tandem lesions" may present special challenges to DNA replication and repair systems.[224,225]

Nucleobase radicals can lead to damage at adjacent sugar residues. Greenberg's group observed an oxygen-dependent damage amplification involving the 5,6-dihydrothymidine radical **86** generated in single-stranded oligonucleotides.[226] Tallman and Greenberg subsequently studied this process in a minimal, dinucleotide system where the products could be clearly characterized.[227] In these studies, the 5,6-dihydrothymidine radical (**86**, Scheme 8.29) was generated from a phenyl selenide precursor. In the presence of molecular oxygen, the peroxyl radical (**87**) abstracts a C1′-hydrogen atom from the sugar

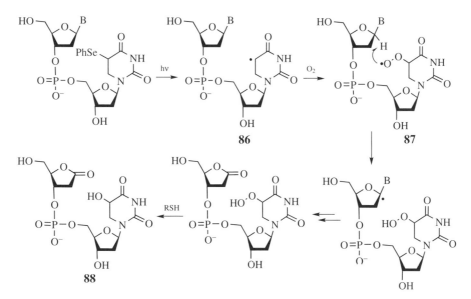

Scheme 8.29

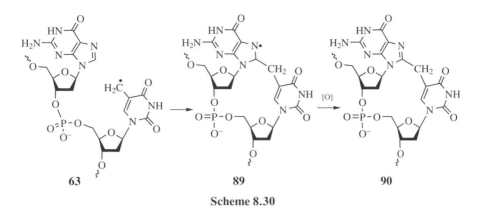

Scheme 8.30

residue on the 5'-side. This overall process produces a tandem lesion **88** consisting of a damaged pyrimidine base adjacent to a 2-deoxyribonolactone residue.

There are examples in which base radicals undergo reaction with adjacent base residues. The 5-(2'-deoxyuridinyl)methyl radical (**63**, Scheme 8.30) can forge an intrastrand cross-link with adjacent purine residues.[228] Cross-link formation is favored with a guanine residue on the 5'-side of the pyrimidine radical and occurs under low-oxygen conditions.[229] A mechanism was not proposed for this process, but presumably the reaction involves addition of the nucleobase alkyl radical to the C8-position of the adjacent purine residue. Molecular oxygen likely inhibits cross-link formation by trapping the radical **63**, as shown in Scheme 8.24. The radical intermediate **89** must undergo oxidation to yield the final cross-linked product **90**,

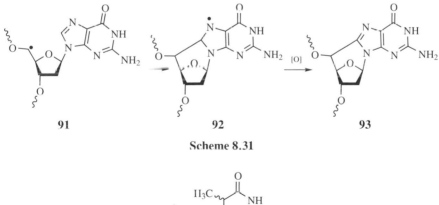

Scheme 8.31

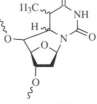

94

although it is unclear what serves an oxidant under the anaerobic conditions required for the formation of this lesion.

Under low oxygen conditions, C5'-sugar radicals can react with the base residue on the same nucleotide. In purine nucleotides, the carbon-centered radical **91** can add to the C8-position of the nucleobase (Scheme 8.31). Oxidation of the intermediate nucleobase radical **92** yields the 8,5'-cyclo-2'-deoxypurine lesion **93**.[197,224,225,230–233] Similarly, in pyrimidine nucleotides, the C5'-radical can add to the C6-position of nucleobase. Reduction of the resulting radical intermediate yields the 5',6-cyclo-5,6-dihydro-2'-deoxypyrimidine lesion **94**.[234–236]

8.6. EXAMPLES OF DNA-DAMAGING RADICALS

Many compounds that damage DNA via radical intermediates have been identified. Some of the agents, such as bleomycin and the enediynes, damage DNA primarily through abstraction of hydrogen atoms.[237–240] In these cases, chemical reactions are directed to certain positions on the DNA backbone by noncovalent binding that places the reactive intermediates in close proximity to particular deoxyribose sugar residues.[38] Similar to the reactions of HO• described above, small radicals, such as •CH$_3$,[241–243] Ph•,[244] RO•,[245,246] and RS•,[247–249] typically undergo reactions at both the sugars and the nucleobases of DNA. Several reviews provide thorough overviews of agents that damage DNA via radical intermediates.[20,22,38,163] Here, we will briefly discuss agents that generate hydroxyl radical (HO•) under physiological conditions.

8.6.1. γ-Radiolysis

Hydroxyl radical is the key biologically active species generated by γ-radiolysis of water.[163,167,250] Accordingly, due to the long-standing medical importance of radiation therapy, the biological reactions of hydroxyl radical have been extensively studied for many years. Hydroxyl radical (HO•) reacts at near diffusion-controlled rates with almost any organic substance that it encounters.[251] For example, the rate constant for reaction with glutathione is $1.4 \times 10^{10} \, M^{-1} s^{-1}$, that for reaction with alanine is $7.5 \times 10^{8} \, M^{-1} s^{-1}$, and that for reaction with adenine is $6.1 \times 10^{9} \, M^{-1} s^{-1}$. Thus, the reactions of hydroxyl radical generated inside cells are not highly selective for any particular biomolecule. Nonetheless, the reactions of hydroxyl radical with DNA have received special attention because hydroxyl radical-mediated modification of DNA is central to the cell-killing properties of radiation.[163]

8.6.2. Tirapazamine

A few interesting organic molecules may possess the ability to release hydroxyl radical under physiological conditions. The most extensively studied of these is 3-amino-1,2,4-benzotriazine 1,4-dioxide (tirapazamine, **95**, Scheme 8.32). Tirapazamine is a bioreductively activated DNA-damaging agent that selectively kills the oxygen-poor (hypoxic) cells found in solid tumors.[252–255] The biological activity of

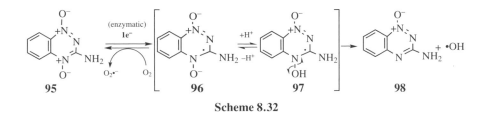

Scheme 8.32

tirapazamine stems from its DNA-damaging properties.[256,257] This compound is currently undergoing phase I, II, and III clinical trials for the treatment of various cancers.[258] The drug is reduced by cellular enzymes such as NADPH:cytochrome P450 reductase to yield a crucial radical intermediate (**96**, Scheme 8.32).[259] The radical anion exists in equilibrium with the protonated, neutral radical (**97**, pK_a = 6).[260,261] In normally oxygenated cells, the activated, radical anion of tirapazamine (**96**) is converted back into starting material[260] by reaction with O_2 (k_{O_2} = 6.2 × $10^6 M^{-1} s^{-1}$) but, in the hypoxic environment of tumors (oxygen concentrations around 2 µM), the neutral radical partitions forward to cause DNA damage.[260,262–266] The mono-N-oxide **98** is the major drug metabolite generated in this process.[260,267,268] This compound is nontoxic.[262] It is important to note that the so-called futile cycling tirapazamine involving enzymatic reduction-O_2 oxidation under aerobic conditions generates superoxide radical. Although superoxide radical itself can be cytotoxic (as described in the next section); enzymes such as superoxide dismutase, catalase, and glutathione peroxidase work together to mitigate the toxicity of superoxide and its breakdown products.[269] In the context of tirapazamine, it is clear that futile redox-cycling of tirapazamine is relatively nontoxic compared with the reactions that proceed under anaerobic conditions.

The exact nature of the DNA-damaging intermediate generated by one-electron reduction of tirapazamine remains a matter of ongoing study, but several lines of evidence suggest that this drug may deliver the active agent of radiation therapy, hydroxyl radical, selectively to hypoxic tumor cells (Scheme 8.32). Tirapazamine causes sequence-independent strand cleavage of duplex DNA, consistent with the involvement of a small, highly reactive species like hydroxyl radical.[265] Further evidence for the involvement of a small, highly reactive oxidant was obtained through analysis of the DNA "end products" resulting from oxidative strand cleavage.[270] Strand cleavage mediated by the drug yields the methylene lactone by-product (**56**, Scheme 8.20) along with base propenals and 3'-phosphoglycolate termini (Scheme 8.21). These products indicate that tirapazamine-mediated cleavage of duplex DNA is initiated by hydrogen atom abstraction at the most hindered sites in the duplex (C1'–H) as well as the more accessible (C4'–H). In addition, furfural, a characteristic product generated by abstraction of a C5'–hydrogen atom, is observed. In fact, the ratio of products (**56** and furfural) stemming from abstraction at the C1' and C5' positions mirrors that seen for hydroxyl radical.

In addition to causing DNA strand breaks, tirapazamine damages the heterocyclic base residues of double-stranded DNA.[264,266] Similar to authentic, radiolytically generated hydroxyl radical, the drug causes approximately four times more base

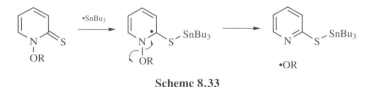

Scheme 8.33

damage than direct strand breaks.[266] Gas chromatography/mass spectrometry and liquid chromatography/mass spectrometry were used to characterize and quantify tirapazamine-mediated oxidative DNA base damage, under hypoxic conditions.[264] A variety of hydroxylated DNA bases were observed (Fig. 8.2), including 5-hydroxy-6-hydrothymine, 5-hydroxy-6-hydrouracil, 8-hydroxyguanine, 8-hydroxyadenine, 2-hydroxyadenine, thymine glycol, 5-hydroxycytosine, 5-hydroxyuracil, isodialuric acid, and 5-(hydroxymethyl)uracil. In addition, 2,6-diamino-4-hydroxy-5-formamidopyrimidine, 4,6-diamino-5-formamidopyrimidine, and 8,5'-cyclopurine-2'-deoxynucleoside lesions (e.g., **93**) were detected. This is the first time that any agent other than radiation has been found to generate the lethal 8,5'-cyclopurine lesions. More generally, tirapazamine-mediated production of various hydroxylated nucleobases is consistent with the hypothesis that activated tirapazamine generates hydroxyl radical under anaerobic conditions.

Fragmentations analogous to that proposed for tirapazamine in Scheme 8.32, involving aromatization of a nitrogen heterocycle, are well precedented in the literature and are expected to be thermodynamically favorable. For example, the well known fragmentation of Barton's N-hydroxypyridinethione esters (Scheme 8.33) is directly analogous.[271-275] In addition, the groups of Schuster and Gould have studied similar fragmentations of N-(alkoxy)pyridinium salts (Scheme 8.34).[276,277] In the case of tirapazamine, neutralization-reionization mass spectroscopy revealed that the neutral tirapazamine radical (**97**) decomposes in the gas phase to yield (**98**).[278] On the basis of a 5 µs "flight time" in the spectrometer, an activation energy of approximately 14 kcal/mol was estimated for the unimolecular, gas phase fragmentation of **97**.

The bioreductively-activated, hypoxia-selective DNA-damaging properties found in tirapazamine are not unique to this nitrogen heterocycle. Analogous properties have been observed for quinoxaline di-N-oxides such as **99** and **100**.[279,280] In addition, early work by Hecht and co-workers invoked similar mechanisms to explain redox-activated DNA damage by phenazine di-N-oxides **101**.[281]

Much of the data related to redox-activated DNA damage by tirapazamine seems to be well explained by a mechanism involving release of hydroxyl radical (Scheme 8.32), but other mechanisms have been proposed. Early reports often invoked direct hydrogen atom abstraction by the neutral radical **97** to explain DNA strand cleavage by the drug.[282] More recently, it has been proposed that the neutral radical undergoes dehydration to yield the benzotriazinyl radical **102** that is suggested to cause DNA damage.[283] This proposal is based upon UV-vis spectroscopic data from pulse radiolysis experiments suggesting the existence of an intermediate between **97** and **98**. These mechanisms deserve further consideration, and it is important to recognize that activated tirapazamine may decompose by more than one reaction pathway.

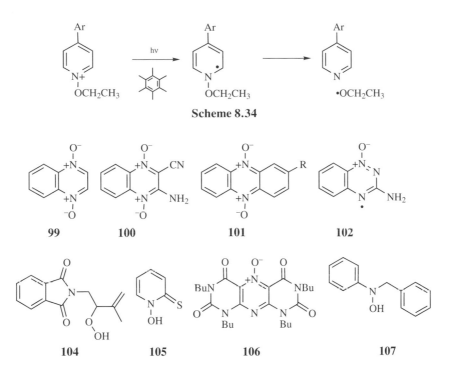

Scheme 8.34

99   **100**   **101**   **102**

104   **105**   **106**   **107**

Careful consideration of the mechanisms involved in the hypoxia-selective DNA damage by tirapazamine reveals a paradox. That is, tirapazamine generates DNA-damaging radicals only in the *absence* of molecular oxygen, yet it is well known that efficient radical-mediated DNA damage requires the *presence* of molecular oxygen.[163,165,166] This apparent paradox was resolved by the observation that tirapazamine not only initiates the formation of DNA radicals under low oxygen conditions, but the drug and its metabolites can also substitute for molecular oxygen in reactions that convert deoxyribose radicals into strand breaks.[284–286] The mono-N-oxide **98** traps a C1′ radical in duplex DNA with a rate constant of $1.4 \times 10^7 \ M^{-1} \ s^{-1}$ and tirapazamine reacts with a rate constant of $4.6 \times 10^6 \ M^{-1} \ s^{-1}$.[285] Further studies showed that these N-oxides convert the C1′ radical in duplex DNA into the 2-deoxyribonolactone lesion (**53**, Scheme 8.20).[270] Model studies with a C1′-nucleoside radical provide insight regarding the mechanism(s) of this reaction. When the reaction was conducted using ^{16}O-containing tirapazamine in ^{18}O-labeled water, a 70:30 ratio of $^{16}O{:}^{18}O$ was observed in the lactone product.[285] This result suggests that the major pathway for this reaction involves oxygenation of the C1′ radical via a covalent N-oxide intermediate (**102**, upper pathway of Scheme 8.35). A smaller, but significant, portion of the reaction proceeds with incorporation of oxygen from water, presumably via the oxygen-stabilized carbocation intermediate **103** (Scheme 8.35). The realization that these N-oxides can oxygenate DNA radicals has led researchers to employ the nontoxic mono-N-oxide **68**, as an oxygen-mimetic for sensitizing hypoxic cells to killing by tirapazamine.[287]

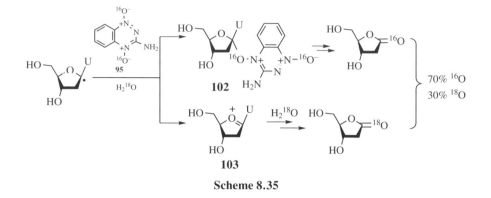

Scheme 8.35

8.6.3. Photolytic Generation of Hydroxyl Radical

Photolysis of a variety of substances including hydrogen peroxide, phthalimide hydro-peroxides (**104**), N-hydroxypyridinethiones (**105**), pyrimido[5,4-g]pteridinetetrone N-oxides (**106**), and N-arylalkyl-N-phenylhydroxylamines (**107**) has been reported to generate hydroxyl radical in aqueous solutions.[273,275,288–291]

8.6.4. Agents that Reduce Molecular Oxygen to Superoxide Radical ($O_2 \bullet^-$)

Generation of superoxide radical under physiological conditions ultimately leads to the production of hydroxyl radical through a cascade of redox reactions. Initially, superoxide disproportionates to generate hydrogen peroxide (Eq. 3, Scheme 8.36).[292–297] Superoxide radical exists in equilibrium with its protonated form ($HO_2 \bullet$, $pK_a = 5$). Under physiological conditions, the disproportionation of superoxide radical is very fast and involves reaction of the radical anion $O_2 \bullet^-$ with a molecule of the neutral radical, $HO_2 \bullet$ (Eq. 2, Scheme 8.36). Disproportionation yields one molecule of molecular oxygen and one molecule of hydrogen peroxide.

The resulting hydrogen peroxide undergoes metal-mediated decomposition to yield hydroxyl radical (Eq. 4, Scheme 8.36). Metals such as Cu(I) and Fe(II) can

$$O_2 + e^- \longrightarrow O_2 \bullet^- \tag{1}$$

$$O_2 \bullet^- + H^+ \rightleftharpoons HO_2 \bullet \qquad (pK_a = 5) \tag{2}$$

$$O_2 \bullet^- + HO_2 \bullet \longrightarrow H_2O_2 + O_2 \qquad (1 \times 10^5 \text{ M}^{-1} \text{ s}^{-1}) \tag{3}$$

$$H_2O_2 + M^{n+} \longrightarrow HO \bullet + HO^- + M^{(n+1)+} \qquad (76 \text{ M}^{-1} \text{ s}^{-1}) \tag{4}$$

$$M^{(n+1)+} + O_2 \bullet^- \longrightarrow M^{n+} + O_2 \quad (M = Fe, 3.1 \times 10^5 \text{ M}^{-1} \text{ s}^{-1}) \tag{5}$$

Scheme 8.36

participate in this Fenton-type reaction. There has been much discussion regarding the exact nature of the reactive intermediates produced in the reaction of metal salts with hydrogen peroxide;[298–300] however, in the context of DNA damage, it has been established that the reactive intermediate behaves in a manner identical to authentic, radiolytically generated hydroxyl radical.[301] The concentrations of free metal ions present inside the cell are undoubtedly low, but it is clear that sources of metals capable of participating in the Fenton-type reactions are available.[293, 297] When low concentrations of metal salts are present, efficient generation of hydroxyl radical from hydrogen peroxide via the Fenton reaction requires the action of some reducing agent that can return the metal ion catalyst to the active, reduced form (e.g., Fe^{2+}, Eq. 5, Scheme 8.36). Depending on the reaction conditions, this role may be performed by superoxide anion or other reducing agents, such as thiols or ascorbate.[293,296,302]

Many organic agents are able to carry out the net reduction of molecular oxygen to superoxide radical under physiological conditions. In many cases, the reaction is stoichiometric, with the organic reducing agent consumed in the reaction. On the other hand, some agents can serve as catalysts for the net transfer of electrons from biologically plentiful reducing agents such as thiols or NAD(P)H to molecular oxygen. Compounds with this capability may show potent biological activity because small amounts of the agent have the potential to generate significant quantities of superoxide radical while consuming cellular stores reducing agents. Simultaneous generation of reactive oxygen species ($O_2^{\bullet-}$, H_2O_2, and $HO\bullet$), along with the depletion of cellular thiols, constitutes a condition generally described as "oxidative stress."[293,297,303–305] Organic agents capable of generating superoxide radical under physiologically relevant conditions include: thiols,[306–309] ascorbate,[310] vitamin A,[311,312] tecapentaene-12,[311] hydrazines,[313] hydroxylamines,[314] 4-hydroxy-5-methyl-3-(2H)-furanones,[315] quinocarcin,[316] dihyropyrazines,[317] quinones,[318] pyrimido[5,4-3][1,2,4]triazine-5,7-diamines,[319] iminoquinones,[320] nitro compounds,[321,322] heterocyclic-N-oxides,[265,281] 1,2-dithiole-3-thiones,[323] persulfides,[135] and polysulfides.[133,134,324]

8.7. CONCLUSION AND OUTLOOK

The reactions described above provide a foundation for understanding the chemical events underlying the biological activity of a wide variety of DNA-damaging drugs, carcinogens, and toxins. The chemistry of DNA is rich and ongoing studies in this field will continue to reveal reactions that are both chemically interesting and biologically important. Ultimately, the nature of the biological response elicited by a given DNA-damaging agent is dictated by the chemical structure of the damaged DNA. For the most part, however, the relationships between the chemical structure of damaged DNA and the resulting biological response are not well understood. Thus, careful characterization of the DNA lesions generated by various drugs, toxins, and mutagens represents a mainstay of efforts directed toward reaching a comprehensive understanding of the chemical and biological mechanisms by which DNA-damaging agents elicit diverse biological responses.

SUGGESTED READING

A good overview of DNA structure and function. E. Pennisi, *Science* **2003**, *300*, 282.

Reviews of nucleic acid chemistry:

(a) B. Singer, D. Grunberger, *Molecular biology of mutagens and carcinogens*, Plenum, New York, 1983.

(b) P. S. Miller, *Bioconjugate Chem.* **1990**, *1*, 187.

(c) L. J. Marnett, P. C. Burcham, *Chem. Res. Toxicol.* **1993**, *6*, 771.

(d) T. Lindahl, *Nature* **1993**, *362*, 709

Surveys of organic molecules that covalently modify DNA:

(a) K. S. Gates, Covalent modification of DNA by natural products. in *Comprehensive Natural Products Chemistry, Vol. 7* (Ed. E. T. Kool), Pergamon, New York, **1999**; 491.

(b) L. H. Hurley, *Nature Rev. Cancer* **2002**, *2*, 188.

(c) S. E. Wolkenberg, D. L. Boger, *Chem. Rev.* **2002**, *102*, 2477.

A comprehensive review of N7-alkylguanine chemistry. N7-alkylguanine lesions may be the most common product formed in the reaction of DNA with electrophiles:

K. S. Gates, T. Nooner, S. Dutta, *Chem. Res. Toxicol.* **2004**, *17*, 839.

Overviews of DNA repair:

(a) T. Lindahl, R. D. Wood, *Science* **1999**, *286*, 1897.

(b) O. D. Schärer, *Angew. Chem. Int. Ed.* **2003**, *42*, 2946.

Measurement, chemistry, and biological importance of oxidative DNA base damage:

(a) M. D. Evans, M. Dizdaroglu, M. S. Cooke, *Mutation Res.* **2004**, *567*, 1.

(b) C. J. Burrows, J. G. Muller, *Chem. Rev.* **1998**, *98*, 1109.

Comprehensive discussions of radical reactions at the deoxyribose residues in DNA:

(a) G. Pratviel, J. Bernadou, B. Meunier, *Angew. Chem. Int. Ed. Engl.* **1995**, *34*, 746

(b) A. P. Breen, J. A. Murphy, *Free Rad. Biol. Med.* **1995**, *18*, 1033

(c) W. K. Pogozelski, T. D. Tullius, *Chem. Rev.* **1998**, *98*, 1089.

(d) M. M. Greenberg, *Chem. Res. Toxicol.* **1998**, *11*, 1235.

Regarding oxidative stress:

(a) B. Halliwell, J. M. C. Gutteridge, *Methods Enzymol.* **1990**, *186*, 1.

(b) T. Finkel, *Curr. Opin. Cell Biol.* **2003**, *15*, 247.

REFERENCES

1. J. D. Watson, F. H. C. Crick, *Nature* **1953**, *171*, 737.
2. B. Alberts, A. Johnson, J. Lewis, M. Raff, K. Roberts, P. Walter, *Molecular Biology of the Cell*. 4th edition, Garland Science: New York, **2002**.
3. B. Alberts, *Nature* **2003**, *421*, 431.
4. B.-B. S. Zhou, S. J. Elledge, *Nature* **2000**, *408*, 433.
5. J. Rouse, S. P. Jackson, *Science* **2002**, *297*, 547.
6. D. R. Green, *Cell* **2005**, *121*, 671.

7. C. J. Norbury, I. D. Hickson, *Ann. Rev. Pharmacol. Toxicol.* **2001**, *41*, 367.

8. O. D. Schärer, *Angew. Chem. Int. Ed.* **2003**, *42*, 2946.

9. E. C. Friedberg, *Nature* **2003**, *421*, 436.

10. J. H. J. Hoeijmakers, *Nature* **2001**, *411*, 366.

11. T. Lindahl, R. D. Wood, *Science* **1999**, *286*, 1897.

12. K. Hemminki, *Carcinogenesis* **1993**, *14*, 2007.

13. J. P. O'Neill, *Proc. Nat. Acad. Sci. USA* **2000**, *97*, 11137.

14. A. Luch, *Nature Rev. Cancer* **2005**, *5*, 113.

15. C. H. McGowan, *Mutat. Res.* **2003**, *532*, 75.

16. B. Tippin, P. Pham, M. F. Goodman, *Trends in Microbiology* **2004**, *12*, 288.

17. A. K. Showalter, B. J. Lamarche, M. Bakhtina, M.-I. Su, K.-H. Tang, M.-D. Tsai, *Chem. Rev.* **2006**, *106*, 340.

18. B. A. Bridges, *DNA Repair* **2005**, *4*, 725.

19. F. P. Guengerich, *Chem. Rev.* **2006**, *106*, 420.

20. K. S. Gates, Covalent modification of DNA by natural products, in *Comprehensive Natural Products Chemistry Vol. 7*, (Ed. E. T. Kool), Pergamon, New York, **1999**, 491.

21. L. H. Hurley, *Nature Rev. Cancer* **2002**, *2*, 188.

22. S. E. Wolkenberg, D. L. Boger, *Chem. Rev.* **2002**, *102*, 2477.

23. W. A. Remers, Textbook of organic, medicinal and pharmaceutical chemistry, in *Antineoplastic Agents* (Eds. J. W. a. R., W. A. Delgado), Ed. Lippincott, Philadelphia, PA., **1991**, 313.

24. P. D. Lawley, D. H. Phillips, *Mutation Res.* **1996**, *355*, 13.

25. D. T. Beranek, *Mutation Res.* **1990**, *231*, 11.

26. B. Singer, D. Grunberger, *Molecular Biology of Mutagens and Carcinogens.* Plenum: New York, **1983**.

27. R. C. Moschel, W. R. Hudgins, A. Dipple, *J. Org. Chem.* **1979**, *44*, 3324.

28. E. L. Loechler, *Chem. Res. Toxicol.* **1994**, *7*, 277.

29. X. Lu, J. M. Heilman, P. Blans, J. C. Fishbein, *Chem. Res. Toxicol.* **2005**, *18*, 1462.

30. J. A. Hartley, Selectivity in alkylating agent-DNA interactions. in *Molecular Aspects of Anticancer Drug-DNA Interactions Vol. 1* (Eds. S. Neidle, M. Waring), CRC Press, Boca Raton, FL, **1993**, 1.

31. K. W. Kohn, J. A. Hartley, W. B. Mattes, *Nucleic Acids Res.* **1987**, *15*, 10531.

32. M. A. Warpehoski, L. H. Hurley, *Chem. Res. Toxicol.* **1988**, *1*, 315.

33. R. L. Wurdeman, B. Gold, *Chem. Res. Toxicol.* **1988**, *1*, 146.

34. H. Zang, K. S. Gates, *Chem. Res. Toxicol.* **2003**, *16*, 1539.

35. A. Pullman, B. Pullman, *Q. Rev. Biophys.* **1981**, *14*, 289.

36. D. Brosch, W. Kirmse, *J. Org. Chem.* **1991**, *56*, 907.

37. T. E. Agnew, H.-J. Kim, J. C. Fishbein, *J. Phys. Org. Chem.* **2005**, *17*, 483.

38. X.-L. Yang, A. H.-J. Wang, *Pharm. Ther.* **1999**, *83*, 181.

39. A. Fidder, G. W. H. Moes, A. G. Scheffer, G. P. ver der Schans, R. A. Baan, L. P. A. de Jong, H. P. Benschop, *Chem. Res. Toxicol.* **1994**, *7*, 199.

40. A. Asai, M. Hara, S. Kakita, Y. Kanda, M. Yoshida, H. Saito, Y. Saitoh, *J. Am. Chem. Soc.* **1996**, *118*, 6802.

41. P. Mehta, K. Church, J. Williams, F.-X. Chen, L. Encell, D. E. G. Shuker, B. Gold, *Chem. Res. Toxicol.* **1996**, *9*, 939.

42. K. M. Vasquez, L. Narayanan, P. M. Glazer, *Science* **2000**, *290*, 530.

43. Y.-D. Wang, J. Dziegielewski, N. R. Wurtz, B. Dziegielewska, P. B. Dervan, T. A. Beerman, *Nucleic Acids Res.* **2003**, *31*, 1208.

44. L. A. Dickinson, R. Burnett, C. Melander, B. S. Edelson, P. S. Arora, P. B. Dervan, J. M. Gottesfeld, *Chem. Biol.* **2004**, *11*, 1583.

45. F. W. Perrino, P. Blans, S. Harvey, S. L. Gelhaus, C. McGrath, S. A. Akman, S. Jenkins, W. R. LaCourse, J. C. Fishbein, *Chem. Res. Toxicol.* **2003**, *16*, 1616.

46. A. Guy, D. Molko, L. Wagrez, R. Teoule, *Helv. Chim. Acta* **1986**, *69*, 1034.

47. T. Horn, M. S. Urdea, *Nucleic Acids Res.* **1989**, *17*, 6959.

48. J. Bernadou, M. Blandin, B. Meunier, *Nucleosides & Nucleotides* **1983**, *2*, 459.

49. P. L. Fischhaber, A. S. Gall, J. A. Duncan, P. B. Hopkins, *Cancer Res.* **1999**, *17*, 4363.

50. C. J. Wilds, A. M. Noronha, S. Robidoux, P. S. Miller, *J. Am. Chem. Soc.* **2004**, *126*, 9257.

51. T. E. Spratt, C. R. Campbell, *Biochemistry* **1994**, *33*, 11364.

52. B. Singer, M. Kroger, M. Carrano, *Biochemistry* **1978**, *17*, 1246.

53. Y. Guichard, G. D. D. Jones, P. B. Farmer, *Cancer Res.* **2000**, *60*, 1276.

54. R. c. Le Pla, Y. Guichard, K. J. Bowman, M. Gaskell, P. B. Farmer, G. D. D. Jones, *Chem. Res. Toxicol.* **2004**, *17*, 1491.

55. U. Siebenlist, W. Gilbert, *Proc. Nat. Acad. Sci. USA* **1980**, *77*, 122.

56. O. Nyanguile, G. L. Verdine, *Org. Lett.* **2001**, *3*, 71.

57. K. S. Gates, T. Nooner, S. Dutta, *Chem. Res. Toxicol.* **2004**, *17*, 839.

58. T. Lindahl, B. Nyberg, *Biochemistry* **1972**, *11*, 3610.

59. P. D. Lawley, P. Brookes, *Biochem. J.* **1963**, *89*, 127.

60. P. D. Lawley, W. Warren, *Chem.-Biol. Interact.* **1976**, *12*, 211.

61. L. C. Sowers, W. D. Sedwick, B. Ramsay Shaw, *Mutation Res.* **1989**, *215*, 131.

62. J. A. Zoltewicz, D. F. Clark, *J. Org. Chem.* **1972**, *37*, 1193.

63. J. A. Zoltewicz, D. F. Clark, T. W. Sharpless, G. Grahe, *J. Am. Chem. Soc.* **1970**, *92*, 1741.

64. L. Hevesi, E. Wolfson-Davidson, J. B. Nagy, O. B. Nagy, A. Bruylants, *J. Am. Chem. Soc.* **1972**, *94*, 4715.

65. R. D. Guthrie, W. P. Jencks, *Acc. Chem. Res.* **1989**, *22*, 343.

66. R. C. Moschel, W. R. Hudgins, A. Dipple, *J. Org. Chem.* **1984**, *49*, 363.

67. N. Muller, G. Eisenbrand, *Chem.-Biol. Int.* **1985**, *53*, 173.

68. E. R. Garrett, P. J. Mehta, *J. Am. Chem. Soc.* **1972**, *94*, 8542.

69. J. L. York, *J. Org. Chem.* **1981**, *46*, 2171.

70. A. R. Evans, M. Limp-Foster, M. R. Kelley, *Mutation Res.* **2000**, *461*, 83.

71. J. Lhomme, J.-F. Constant, M. Demeunynck, *Biopolymers* **2000**, *52*, 65.

72. L. A. Loeb, B. D. Preston, *Ann. Rev. Genet.* **1986**, *20*, 201.

73. J. A. Wilde, P. H. Bolton, A. Mazumdar, M. Manoharan, J. A. Gerlt, *J. Am. Chem. Soc.* **1989**, *111*, 1894.

74. P. Crine, W. G. Verly, *Biochim. Biophys. Acta* **1976**, *442*, 50.

75. T. Lindahl, A. Andersson, *Biochemistry* **1972**, *11*, 3618.

76. A. Fkyerat, M. Demeunynck, J.-F. Constant, P. Michon, J. Lhomme, *J. Am. Chem. Soc.* **1993**, *115*, 9952.

77. M. Lefrancois, J.-R. Bertrand, C. Malvy, *Mutation Res.* **1990**, *236*, 9.

78. A. M. Maxam, W. Gilbert, *Methods Enzymol.* **1980**, *65*, 499.

79. P. J. McHugh, J. Knowland, *Nucleic Acids Res.* **1995**, *23*, 1664.

80. H. Sugiyama, T. Fujiwara, A. Ura, T. Tashiro, K. Yamamoto, S. Kawanishi, I. Saito, *Chem. Res. Toxicol.* **1994**, *7*, 673.

81. P. D. Lawley, P. Brookes, *Nature* **1961**, *192*, 1081.

82. J. A. Haines, C. B. Reese, L. Todd, *J. Chem. Soc.* **1962**, 5281.

83. L. B. Townsend, R. K. Robins, *J. Am. Chem. Soc.* **1963**, *85*, 242.

84. H. C. Box, K. T. Lilgam, J. B. French, G. Potienko, J. L. Alderfer, *Carbohydrates, Nucleosides, Nucleotides* **1981**, *8*, 189.

85. C. J. Chetsanga, B. Bearie, C. Makaroff, *Chem.-Biol. Int.* **1982**, *41*, 217.

86. C. J. Chetsanga, B. Bearie, C. Makaroff, *Chem.-Biol. Interact.* **1982**, *41*, 235.

87. S. M. Hecht, B. L. Adams, J. W. Kozarich, *J. Org. Chem.* **1976**, *13*, 2303.

88. W. G. Humphreys, F. P. Guengerich, *Chem. Res. Toxicol.* **1991**, *4*, 632.

89. M. M. Greenberg, Z. Hantosi, C. J. Wiederholt, C. D. Rithner, *Biochemistry* **2001**, *40*, 15856.

90. L. T. Bergdorf, T. Carrel, *Chem. Eur. J.* **2002**, *8*, 293.

91. R. Barak, A. Vincze, P. Bel, S. P. Dutta, G. B. Chedda, *Chem.-Biol. Interact.* **1993**, *86*, 29.

92. S. Boiteux, J. Belleney, B. P. Roques, J. Laval, *Nucleic Acids Res.* **1984**, *12*, 5429.

93. S. Hendler, E. Furer, P. R. Srinivasan, *Biochemistry* **1970**, *9*, 4141.

94. H. Mao, Z. Deng, F. Wang, T. M. Harris, M. P. Stone, *Biochemistry* **1998**, *37*, 4374.

95. D. T. Beranek, C. C. Weiss, F. E. Evans, C. J. Chetsanga, F. F. Kadlubar, *Biochem. Biophys. Res. Commun.* **1983**, *110*, 625.

96. R. G. Croy, G. N. Wogan, *J. Natl. Cancer Inst.* **1981**, *66*, 761.

97. R. G. Croy, G. N. Wogan, *Cancer Res.* **1981**, *41*, 197.

98. M. E. Smela, S. S. Currier, E. A. Bailey, J. M. Essigmann, *Carcinogenesis* **2001**, *22*, 535.

99. B. Tudek, *J. Biochem. Mol. Biol.* **2003**, *36*, 12.

100. C. J. Wiederholt, M. M. Greenberg, *J. Am. Chem. Soc.* **2002**, *124*, 7278.

101. T. Fujii, T. Itaya, *Heterocycles* **1998**, *48*, 359.

102. T. Barlow, J. Takeshita, A. Dipple, *Chem. Res. Toxicol.* **1998**, *11*, 838.

103. T. Fujii, T. Saito, T. Nakasaka, *Chem. Pharm. Bull.* **1989**, *37*, 2601.

104. T. Lindahl, B. Nyberg, *Biochemistry* **1974**, *13*, 3405.

105. P. Karran, T. Lindahl, *Biochemistry* **1980**, *19*, 6005.

106. J. B. Macon, R. Wolfenden, *Biochemistry* **1968**, *7*, 3453.

107. T. Barlow, J. Ding, P. Vouros, A. Dipple, *Chem. Res. Toxicol.* **1997**, *10*, 1247.

108. M. Koskinnen, K. Hemminki, *Org. Lett.* **1999**, *1*, 1233.

109. T. Barlow, A. Dipple, *Chem. Res. Toxicol.* **1999**, *12*, 883.

110. G. Kampf, L. E. Kapinos, R. Griesser, B. Lippert, H. Sigel, *J. Chem. Soc. Perkin Trans. 2* **2002**, 1320.

111. S. M. Hecht, J. W. Kozarich, *JCS Chem. Commun.* **1973**, 387.

112. P. Upadhyaya, S. J. Sturla, N. Tretyakova, R. Ziegel, P. W. Villalta, M. Wang, S. S. Hecht, *Chem. Res. Toxicol.* **2003**, *16*, 180.

113. A. Ouyang, E. B. Skibo, *J. Org. Chem.* **1998**, *63*, 1893.

114. E. E. Weinert, K. N. Frankenfield, S. E. Rokita, *Chem. Res. Toxicol.* **2005**, *18*, 1364.

115. D. L. Boger, W. Yun, *J. Am. Chem. Soc.* **1993**, *115*, 9872.

116. A. Asai, S. Nagamura, H. Saito, I. Takahashi, H. Nakano, *Nucleic Acids Res.* **1994**, *22*, 88.

117. M. A. Warpehoski, D. E. Harper, M. A. Mitchell, T. J. Monroe, *Biochemistry* **1992**, *31*, 2502.

118. T. Nooner, S. Dutta, K. S. Gates, *Chem. Res. Toxicol.* **2004**, *17*, 942.

119. M. Zewail-Foote, L. H. Hurley, *J. Am. Chem. Soc.* **2001**, *123*, 6485.

120. J. P. Plastaras, J. N. Riggins, M. Otteneder, L. J. Marnett, *Chem. Res. Toxicol.* **2000**, *13*, 1235.

121. W. A. Remers, B. S. Iyengar, Cancer chemotherapeutic agents, in *Antitumor Antibiotics* (Ed. W. O. Foye), American Chemical Society, Washington, DC, **1995**; 577.

122. B. J. S. Sanderson, A. J. Shield, *Mutation Res.* **1996**, *355*, 41.

123. A. G. Ogston, E. R. Holiday, J. S. L. Philpot, L. A. Stocken, *Trans. Faraday Soc.* **1948**, *44*, 45.

124. W. A. Smit, N. S. Zefirov, I. V. Bodrikov, M. Z. Krimer, *Acc. Chem. Res.* **1979**, *12*, 282.

125. Y.-C. Yang, L. L. Szafraniec, W. T. Beaudry, J. R. Ward, *J. Org. Chem.* **1988**, *53*, 3293.

126. M. Fox, D. Scott, *Mutation Res.* **1980**, *75*, 131.

127. J. G. Valadez, F. P. Guengerich, *J. Biol. Chem.* **2004**, *279*, 13435.

128. K. S. Gates, *Chem. Res. Toxicol.* **2000**, *13*, 953.

129. M. Hara, Y. Saitoh, H. Nakano, *Biochemistry* **1990**, *29*, 5676.

130. A. Meister, M. E. Anderson, *Ann. Rev. Biochem.* **1983**, *52*, 711.

131. S. B. Behroozi, W. Kim, K. S. Gates, *J. Org. Chem.* **1995**, *60*, 3964.

132. L. Breydo, K. S. Gates, *J. Org. Chem.* **2002**, *67*, 9054.

133. T. Chatterji, K. S. Gates, *Bioorg. Med. Chem. Lett.* **1998**, *8*, 535.

134. T. Chatterji, K. S. Gates, *Bioorg. Med. Chem. Lett.* **2003**, *13*, 1349.

135. T. Chatterji, K. Keerthi, K. S. Gates, *Bioorg. Med. Chem. Lett.* **2005**, *15*, 3921.

136. A. Asai, H. Saito, Y. Saitoh, *Bioorg. Med. Chem.* **1997**, *5*, 723.

137. L. Breydo, H. Zang, K. Mitra, K. S. Gates, *J. Am. Chem. Soc.* **2001**, *123*, 2060.

138. D. M. Gill, N. A. Pegg, C. M. Rayner, *Tetrahedron* **1996**, *52*, 3609.

139. G. B. Payne, *J. Org. Chem.* **1962**, 27.

140. J. Szekely, S. Dutta, L. Breydo, T. Nooner, K. S. Gates, *Manuscript in preparation* **2006**.

141. L. Breydo, H. Zang, K. S. Gates, *Tetrahedron Lett.* **2004**, *45*, 5711.

142. K. Shipova, K. S. Gates, *Bioorg. Med. Chem. Lett.* **2005**, *15*, 2111.

143. P. M. Cullis, R. E. Green, M. E. Malone, *J. Chem. Soc. Perkin 2* **1995**, 1503.

144. L. F. Povirk, D. E. Shuker, *Mutation Res.* **1994**, *318*, 205.

145. E. Haapala, K. Hakala, E. Jokipelto, J. Vilpo, J. Hovinen, *Chem. Res. Toxicol.* **2001**, *14*, 988.

146. S. Balcome, S. Park, D. R. Quirk Dorr, L. Hafner, L. Phillips, N. Tretyakova, *Chem. Res. Toxicol.* **2004**, *17*, 950.

147. J. T. Millard, S. Raucher, P. B. Hopkins, *J. Am. Chem. Soc.* **1990**, *112*, 2459.

148. P. R. Turner, W. A. Denny, L. R. Ferguson, *Anti-Cancer Drug Design* **2000**, *15*, 245.

149. J. C. Marquis, S. M. Hillier, A. N. Dinaut, D. Rodrigues, K. Mitra, J. M. Essigmann, R. G. Croy, *Chem. Biol.* **2005**, *12*, 779.

150. G. Sunavala-Dossabhoy, M. W. Van Dyke, *Biochemistry* **2005**, *44*, 2510.

151. S. Dutta, H. Abe, S. Aoyagi, C. Kibayashi, K. S. Gates, *J. Am. Chem. Soc.* **2005**, *127*, 15004.

152. A. D. Patil, A. J. Freyer, R. Reichwein, B. Carte, L. B. Killmer, L. Faucette, R. K. Johnson, D. J. Faulkner, *Tetrahedron Lett.* **1997**, *38*, 363.

153. A. J. Blackman, C. Li, D. C. R. Hockless, B. W. Skelton, A. H. White, *Tetrahedron* **1993**, *49*, 8645.

154. E. C. Miller, A. B. Swanson, D. H. Phillips, T. L. Fletcher, A. Liem, J. A. Miller, *Cancer Res.* **1983**, *43*, 1124.

155. S. M. F. Jeurissen, J. J. P. Bogaards, M. G. Boersma, J. P. F. ter Horst, H. M. Awad, Y. C. Fiamegos, T. A. van Beek, G. M. Alink, E. J. R. Sudhölter, N. H. P. Cnubben, I. M. C. M. Rietjens, *Chem. Res. Toxicol.* **2006**, *19*, 111.

156. F. P. Guengerich, *Carcinogenesis* **2000**, *21*, 345.

157. J. L. Burkey, J. M. Sauer, C. A. McQueen, I. G. Sipes, *Mutation Res.* **2000**, *453*, 25.

158. R. W. Wiseman, T. R. Fennell, J. A. Miller, E. C. Miller, *Cancer Res.* **1985**, *45*, 3096.

159. D. H. Phillips, DNA adducts: Identification and biological significance, in *DNA Adducts Derived From Safrole, Estragole and Related Compounds, and From Benzene and Its Metabolites*, Vol. 125 (Eds. K. Hemminki, A. Dipple, D. E. G. Shuker, F. F. Kadlubar, D. Segerback, H. Bartsch). IARC Scientific Publications, Lyon, **1994**, P 131.

160. R. A. McClelland, A. Ahmad, A. P. Dicks, V. E. Licence, *J. Am. Chem. Soc.* **1999**, *121*, 3303.

161. M. Tomasz, *Biochim. Biophys. Acta* **1970**, *199*, 18.

162. M. Tomasz, J. Olson, C. M. Mercado, *Biochemistry* **1972**, *11*, 1235.

163. A. P. Breen, J. A. Murphy, *Free Rad. Biol. Med.* **1995**, *18*, 1033.

164. M. M. Greenberg, *Chem. Res. Toxicol.* **1998**, *11*, 1235.

165. W. K. Pogozelski, T. D. Tullius, *Chem. Rev.* **1998**, *98*, 1089.

166. G. Pratviel, J. Bernadou, B. Meunier, *Angew. Chem. Int. Ed. Engl.* **1995**, *34*, 746.

167. C. von Sonntag, U. Hagen, A. Schon-Bopp, D. Schulte-Frohlinde, *Adv. Radiat. Biol.* **1981**, *9*, 109.

168. K. Miaskiewicz, R. Osman, *J. Am. Chem. Soc.* **1994**, *116*, 232.

169. D. M. Golden, S. W. Benson, *Chem. Rev.* **1969**, *69*, 125.

170. B. Balasubramanian, W. K. Pogozelski, T. D. Tullius, *Proc. Nat. Acad. Sci. USA* **1998**, *95*, 9738.

171. B. K. Goodman, M. M. Greenberg, *J. Org. Chem.* **1996**, *61*, 2.

172. J.-T. Hwang, M. M. Greenberg, *J. Am. Chem. Soc.* **1999**, *121*, 4311.

173. C. J. Emmanuel, M. Newcomb, C. Ferreri, C. Chagilialoglu, *J. Am. Chem. Soc.* **1999**, *121*, 2927.

174. K. Hildebrand, D. Schulte-Frohlinde, *Int. J. Radiat. Biol.* **1997**, *71*, 377.

175. C. C. Winterbourn, D. Metodiewa, *Free Rad. Biol. Med.* **1999**, *27*, 322.

176. K. A. Tallman, C. Tronche, D. J. Yoo, M. M. Greenberg, *J. Am. Chem. Soc.* **1998**, *120*, 4903.

177. Y. Zheng, T. L. Sheppard, *Chem. Res. Toxicol.* **2004**, *17*, 197.

178. Y. Roupioz, J. Lhomme, M. Kotera, *J. Am. Chem. Soc.* **2002**, *124*, 9129.

179. J.-T. Hwang, K. A. Tallman, M. M. Greenberg, *Nucleic Acids Res.* **1999**, *27*, 3805.

180. K. M. Kroeger, M. Hashimoto, Y. W. Kow, M. M. Greenberg, *Biochemistry* **2003**, *42*, 2449.

181. M. Hashimoto, M. M. Greenberg, Y. W. Kow, J.-T. Hwang, R. P. Cunningham, *J. Am. Chem. Soc.* **2001**, *123*, 3161.

182. A. Dussy, E. Meggers, B. Giese, *J. Am. Chem. Soc.* **1998**, *120*, 7399.

183. B. Giese, A. Dussy, E. Meggers, M. Petretta, U. Schwitter, *J. Am. Chem. Soc.* **1997**, *119*, 11130.

184. B. Giese, X. Beyrich-Graf, P. Erdmann, M. Petretta, U. Schwitter, *Chem. Biol.* **1995**, *2*, 367.

185. P. C. Dedon, J. P. Plastaras, C. A. Rouzer, L. J. Marnett, *Proc. Nat. Acad. Sci. USA* **1998**, *95*, 11113.

186. G. Behrens, G. Koltzenburg, D. Schulte-Frohlinde, *Z. Naturforsch* **1982**, *37c*, 1205.

187. G. Koltzenburg, G. Behrens, D. Schulte-Frohlinde, *J. Am. Chem. Soc.* **1982**, *104*, 7311.

188. B. Giese, X. Beyrich-Graf, J. Burger, C. Kesselheim, M. Senn, T. Schäfer, *Angew. Chem. Int. Ed. Engl.* **1993**, *32*, 1742.

189. B. Giese, X. Beyrich-Graf, P. Erdmann, L. Giraud, P. Imwinkelried, S. N. Müller, U. Schwitter, *J. Am. Chem. Soc.* **1995**, *117*, 6146.

190. H. H. Thorpe, *Chem. Biol.* **2000**, *7*, R33.

191. D. Crich, X.-S. Mo, *J. Am. Chem. Soc.* **1997**, *119*, 249.

192. G. Scholes, J. F. Ward, J. Weiss, *J. Mol. Biol.* **1960**, *2*, 379.

193. C. J. Burrows, J. G. Muller, *Chem. Rev.* **1998**, *98*, 1109.

194. E. Gajewski, G. Rao, Z. Nackerdien, M. Dizdaroglu, *Biochemistry* **1990**, *29*, 7876.

195. S. S. Wallace, *Free Rad. Biol. Med.* **2002**, *33*, 1.

196. M. Dizdaroglu, *Methods Enzymol.* **1994**, *234*, 3.

197. M. Dizdaroglu, P. Jaruga, M. Birincioglu, H. Rodriguez, *Free Rad. Biol. Med.* **2002**, *32*, 1102.

198. M. D. Evans, M. Dizdaroglu, M. S. Cooke, *Mutation Res.* **2004**, *567*, 1.

199. S. V. Jovanovic, M. G. Simic, *J. Am. Chem. Soc.* **1986**, 108.

200. M. N. Schuchmann, C. von Sonntag, *J. Chem. Soc. Perkin 2* **1983**.

201. I. S. Hong, H. Ding, M. M. Greenberg, *J. Am. Chem. Soc.* **2006**, *128*, 485.

202. J. R. Wagner, J. E. van Lier, M. Berger, J. Cadet, *J. Am. Chem. Soc.* **1994**, *116*, 2235.

203. I. M. Bell, D. Hilvert, *Biochemistry* **1993**, *32*, 13969.

204. S. Tardy-Planechaud, J. Fujimoto, S. S. Lin, L. C. Sowers, *Nucleic Acids Res.* **1997**, *25*, 553.

205. K. N. Carter, M. M. Greenberg, *J. Org. Chem.* **2003**, *68*, 4275.

206. S. Steenken, S. V. Jovanovic, *J. Am. Chem. Soc.* **1997**, *119*, 617.

207. F. D. Lewis, R. L. Letsinger, M. R. Wasielewski, *Acc. Chem. Res.* **2001**, *34*, 159.

208. L. P. Candeias, S. Steenken, *Chem. Eur. J.* **2000**, *6*, 475.

209. T. Douki, S. Spinelli, J.-L. Ravanat, J. Cadet, *J. Chem. Soc. Perkin 2* **1999**, 1875.

210. K. Haraguchi, M. O. Delaney, C. J. Wiederholt, A. Sambandam, Z. Hantosi, M. M. Greenberg, *J. Am. Chem. Soc.* **2002**, *124*, 3263.

211. K. Haraguchi, M. M. Greenberg, *J. Am. Chem. Soc.* **2001**, *123*, 8636.

212. M. O. Delaney, C. J. Weiderholt, M. M. Greenberg, *Angew. Chem.* **2002**, *41*, 771.

213. V. Bodepudi, S. Shibutani, F. Johnson, *Chem. Res. Toxicol.* **1992**, *5*, 608.

214. C. J. Burrows, J. G. Muller, O. Kornyushyna, W. Luo, V. Duarte, M. D. Leipold, S. S. David, *Environ. Health Perspect. Supp.* **2002**, *110*, 713.

215. W. Luo, J. G. Muller, E. M. Rachlin, C. J. Burrows, *Chem. Res. Toxicol.* **2001**, *14*, 927.

216. W. L. Neeley, J. M. Essigmann, *Chem. Res. Toxicol.* **2006**, *19*, 491.

217. Y. Kuchino, F. Mori, H. Kasai, H. Inoue, S. Iwai, K. Miura, E. Ohtsuka, S. Nishimura, *Nature* **1987**, *327*, 77.

218. O. Kornyushyna, A. M. Berges, J. G. Muller, C. J. Burrows, *Biochemistry* **2002**, *41*, 15304.

219. P. T. Henderson, J. C. Delaney, J. G. Muller, W. L. Neeley, S. R. Tannenbaum, C. J. Burrows, J. M. Essigmann, *Biochemistry* **2003**, *42*, 9257.

220. S. Shibutani, M. Takeshita, A. P. Grollman, *Nature* **1991**, *349*, 431.

221. R. P. Hickerson, C. L. Chepanoske, S. D. Williams, S. S. David, C. J. Burrows, *J. Am. Chem. Soc.* **1999**, *121*, 9901.

222. M. E. Hosford, J. G. Muller, C. J. Burrows, *J. Am. Chem. Soc.* **2004**, *126*, 9540.

223. M. E. Johansen, J. G. Muller, X. Xu, C. J. Burrows, *Biochemistry* **2005**, *44*, 5660.

224. P. J. Brooks, D. S. Wise, D. A. Berry, J. V. Kosmoski, M. J. Smerdon, R. L. Somers, H. Mackie, A. Y. Spoonde, E. J. Ackerman, K. Coleman, R. E. Tarone, J. H. Robbins, *J. Biol. Chem.* **2000**, *275*, 22355.

225. I. Kuraoka, C. Bender, A. Romieu, J. Cadet, R. Wood, T. Lindahl, *Proc. Nat. Acad. Sci. USA* **2000**, *97*, 3832.

226. M. M. Greenberg, M. R. Barvian, G. P. Cook, B. K. Goodman, T. J. Matray, C. Tronche, H. Venkatesan, *J. Am. Chem. Soc.* **1997**, *119*, 1828.

227. K. A. Tallman, M. M. Greenberg, *J. Am. Chem. Soc.* **2001**, *123*, 5181.

228. H. C. Box, E. E. Budzinski, J. B. Dawdzik, J. C. Wallace, H. Iijima, *Radiat. Res.* **1998**, *149*, 433.

229. S. Bellon, J.-L. Ravanat, D. Gasparutto, J. Cadet, *Chem. Res. Toxicol.* **2002**, *15*, 598.

230. M. Dizdaroglu, P. Jaruga, H. Rodriguez, *Free Rad. Biol. Med.* **2001**, *30*, 774.

231. A. Romieu, D. Gasparutto, D. Molko, J. Cadet, *J. Org. Chem.* **1998**, *63*, 5245.

232. P. Jaruga, M. Birincioglu, H. Rodriguez, M. Dizdaroglu, *Biochemistry* **2002**, *41*, 3703.

233. I. Kuraoka, P. Robins, C. Masutani, F. Hanaoka, D. Gasparutto, J. Cadet, R. D. Wood, T. Lindahl, *J. Biol. Chem.* **2001**, *276*, 49283.

234. E. Muller, D. Gasparutto, M. Jaquinod, A. Romieu, J. Cadet, *Tetrahedron* **2000**, *56*, 8689.

235. E. Muller, D. Gasparutto, J. Cadet, *Chem. Bio. Chem.* **2002**, *3*, 534.

236. A. Romieu, D. Gasparutto, J. Cadet, *J. Chem. Soc. Perkin 1* **1999**.

237. D. L. Boger, H. Cai, *Angew. Chem. Int. Ed.* **1999**, *38*, 448.

238. J. Stubbe, J. W. Kozarich, W. Wu, D. E. Vanderwall, *Acc. Chem. Res.* **1996**, *29*, 322.

239. A. L. Smith, K. C. Nicolaou, *J. Med. Chem.* **1996**, *39*, 2103.

240. Z. Xi, I. H. Goldberg, DNA-damaging enediyne compounds, in *Comprehensive Natural Products Chemistry Vol. 7*, (Ed. E. T. Kool), Pergammon, Oxford, **1999**, 553.

241. C. G. Riordan, P. Wei, *J. Am. Chem. Soc.* **1994**, *116*, 2189.

242. M. F. Zady, J. L. Wong, *J. Am. Chem. Soc.* **1977**, *99*, 5096.

243. M. F. Zady, J. L. Wong, *J. Org. Chem.* **1980**, *45*, 2373.

244. J. Liu, C. J. Petzold, L. E. Ramirez Arizmendi, J. Perez, H. Kenttämaa, *J. Am. Chem. Soc.* **2005**, *127*, 12758.

245. W. Adam, J. Hartung, H. Okamoto, S. Marquardt, W. M. Nau, U. Pischel, C. R. Saha-Moeller, K. Spehar, *J. Org. Chem.* **2002**, *67*, 6041.

246. J. Termini, *Critical Reviews of Oxidative Stress and Aging* **2003**, *1*, 39.

247. M. J. Robins, G. J. Ewing, *J. Am. Chem. Soc.* **1999**, *121*, 5823.

248. D. Pogocki, C. Schöneich, *Free Rad. Biol. Med.* **2001**, *31*, 98.

249. K. N. Carter, T. Taverner, C. H. Schlesser, M. M. Greenberg, *J. Org. Chem.* **2000**, *65*, 8375.

250. C. von Sonntag, *The Chemical Basis of Radiation Biology.* Taylor and Francis, London, 1987.

251. G. V. Buxton, C. L. Greenstock, W. Phillip, A. B. Ross, *J. Phys. Chem. Ref. Data* **1988**, *17*, 513.

252. J. M. Brown, W. R. Wilson, *Nature Rev. Cancer* **2004**, *4*, 437.

253. W. A. Denny, *The Lancet Oncol.* **2000**, *1*, 25.

254. J. M. Brown, *Cancer Res.* **1999**, *59*, 5863.

255. D. R. Gandara, P. N. Lara, Z. Goldberg, Q. T. Le, P. C. Mack, D. H. M. Lau, P. H. Gumerlock, *Sem. Oncol.* **2002**, *29*, 102.

256. K. A. Biedermann, J. Wang, R. P. Graham, *Br. J. Cancer* **1991**, *63*, 358.

257. B. G. Siim, P. L. van Zijl, J. M. Brown, *Br. J. Cancer* **1996**, *73*, 952.

258. L. Marcu, I. Olver, *Curr. Clin. Oncol.* **2006**, *1*, 71.

259. A. V. Patterson, M. P. Saunders, E. C. Chinje, L. H. Patterson, I. J. Stratford, *Anti-Cancer Drug Des.* **1998**, *13*, 541.

260. K. L. Laderoute, P. Wardman, M. Rauth, *Biochem. Pharmacol.* **1988**, *37*, 1487.

261. K. I. Priyadarsini, M. Tracy, P. Wardman, *Free Rad. Res.* **1996**, *25*, 393.

262. M. A. Baker, E. M. Zeman, V. K. Hirst, J. M. Brown, *Cancer Res.* **1988**, *48*, 5947.

263. J. Wang, K. A. Biedermann, M. J. Brown, *Cancer Res.* **1992**, *52*, 4473.

264. M. Birincioglu, P. Jaruga, G. Chowdhury, H. Rodriguez, M. Dizdaroglu, K. S. Gates, *J. Am. Chem. Soc.* **2003**, *125*, 11607.

265. J. S. Daniels, K. S. Gates, *J. Am. Chem. Soc.* **1996**, *118*, 3380.

266. D. Kotandeniya, B. Ganley, K. S. Gates, *Bioorg. Med. Chem. Lett.* **2002**, *12*, 2325.

267. T. Fuchs, G. Chowdhary, C. L. Barnes, K. S. Gates, *J. Org. Chem.* **2001**, *66*, 107.

268. K. Laderoute, A. M. Rauth, *Biochem. Pharmacol.* **1986**, *35*, 3417.

269. H. Sies, *Angew. Chem. Int. Ed. Engl.* **1986**, *25*, 1058.

270. G. Chowdhury, V. Junnutula, J.-T. Hwang, M. M. Greenberg, K. S. Gates, *Manuscript in preparation* **2006**.

271. D. H. R. Barton, D. Crich, W. B. Motherwell, *Tetrahedron* **1985**, *41*, 3901.

272. J. Boivin, E. Crepon, S. Z. Zard, *Tetrahedron Lett.* **1990**, *31*, 6869.

273. W. Adam, D. Ballmaier, B. Epe, G. N. Grimm, C. R. Saha-Moller, *Angew. Chem. Int. Ed. Engl.* **1995**, *34*, 2156.

274. D. H. R. Barton, J. C. Jasberenyi, A. I. Morrell, *Tetrahedron Lett.* **1991**, *32*, 311.

275. B. M. Aveline, I. E. Kochevar, R. W. Redmond, *J. Am. Chem. Soc.* **1996**, *118*, 289.

276. I. Wölfle, J. Lodays, B. Sauerwein, G. B. Schuster, *J. Am. Chem. Soc.* **1992**, *114*, 9304.

277. E. D. Lorance, W. H. Kramer, I. R. Gould, *J. Am. Chem. Soc.* **2002**, *124*, 15225.

278. D. Zagorevski, Y. Yuan, T. Fuchs, K. S. Gates, M. Song, C. Breneman, C. M. Greenlief, *J. Am. Soc. Mass Spec.* **2003**, *14*, 881.

279. G. Chowdhury, D. Kotandeniya, C. L. Barnes, K. S. Gates, *Chem. Res. Toxicol.* **2004**, *17*, 1399.

280. B. Ganley, G. Chowdhury, J. Bhansali, J. S. Daniels, K. S. Gates, *Bioorg. Med. Chem.* **2001**, *9*, 2395.

281. K. Nagai, B. J. Carter, J. Xu, S. M. Hecht, *J. Am. Chem. Soc.* **1991**, *113*, 5099.

282. J. M. Brown, *Br. J. Cancer* **1993**, *67*, 1163.

283. R. F. Anderson, S. S. Shinde, M. P. Hay, S. A. Gamage, W. A. Denny, *J. Am. Chem. Soc.* **2003**, *125*, 748.

284. J. S. Daniels, K. S. Gates, C. Tronche, M. M. Greenberg, *Chem. Res. Toxicol.* **1998**, *11*, 1254.

285. J.-T. Hwang, M. M. Greenberg, T. Fuchs, K. S. Gates, *Biochemistry* **1999**, *38*, 14248.

286. G. D. D. Jones, M. Weinfeld, *Cancer Res.* **1996**, *56*, 1584.

287. B. G. Siim, F. B. Pruijn, J. R. Sturman, A Hogg, M. P. Hay, J. M. Brown, W. R. Wilson, *Cancer Res.* **2004**, *64*, 736.

288. R. B. J. MacGregor, *Anal. Biochem.* **1992**, *204*, 66.

289. I. Saito, M. Takayama, T. Matsuura, S. Matsugo, S. Kawanishi, *J. Am. Chem. Soc.* **1990**, *112*, 883.

290. M. Sako, K. Nagai, Y. Maki, *J. Chem. Soc., Chem. Commun.* **1993**, 750.

291. J. R. Hwu, S.-C. Tsay, B.-L. Chen, H. V. Patel, C.-T. Chou, *J. Chem. Sec. Chem. Commun.* **1994**, 1427.

292. T. A. Dix, K. M. Hess, M. A. Medina, R. W. Sullivan, S. L. Tilly, ; T. L. L. Webb, *Biochemistry* **1996**, *35*, 4578.

293. B. Halliwell, J. M. C. Gutteridge, *Methods Enzymol.* **1990**, *186*, 1.

294. S. A. Lesko, R. J. Lorentzen, P. O. P. Ts'o, *Biochemistry* **1980**, *19*, 3023.

295. J. Wilshire, D. T. Sawyer, *Acc. Chem. Res.* **1979**, *12*, 105.

296. J. A. Imlay, S. Linn, *Science* **1988**, *240*, 1302.

297. B. Halliwell, J. M. C. Gutteridge, *Biochem. J.* **1984**, *219*, 1.

298. A. U. Khan, M. Kasha, *Proc. Nat. Acad. Sci. USA* **1994**, *91*, 12365.

299. Y. Luo, Z. Han, S. M. Chin, S. Linn, *Proc. Nat. Acad. Sci. USA* **1994**, *91*, 12438.

300. D. A. Wink, R. W. Nims, J. E. Saavedra, W. E. J. Utermahlen, P. C. Ford, *Proc. Nat. Acad. Sci. USA* **1994**, *91*, 6604.

301. W. K. Pogozelski, T. J. McNeese, T. D. Tullius, *J. Am. Chem. Soc.* **1995**, *117*, 6428.

302. E. S. Henle, Y. Luo, S. Linn, *Biochemistry* **1996**, *35*, 12212.

303. T. Finkel, *Curr. Opin. Cell Biol.* **2003**, *15*, 247.

304. R. S. Sohal, R. Weindruch, *Science* **1996**, *273*, 59.

305. J. E. Klaunig, L. M. Kamendulis, *Ann. Rev. Pharm. Tox.* **2004**, 44.

306. V. C. Bode, *J. Mol. Biol.* **1967**, *26*, 125.

307. H. S. Rosenkrantz, S. Rosenkrantz, *Arch. Biochem. Biophys.* **1971**, *146*, 483.

308. H. P. Misra, *J. Biol. Chem.* **1974**, *249*, 2151.

309. C. Giulivi, E. Cadenas, *Biochim. Biophys. Acta* **1998**, *1366*, 265.

310. R. A. Morgan, R. L. Cone, T. M. Elgert, *Nucleic Acids Res.* **1976**, *3*, 1139.

311. J. Szekely, K. S. Gates, *Chem. Res. Toxicol.* **2006**, *19*, 117.

312. M. Murata, S. Kawanishi, *J. Biol. Chem.* **2000**, *275*, 2003.

313. S. Jeon, D. T. Sawyer, *Inorg. Chem.* **1990**, *29*, 4612.

314. H.-J. Rhaese, E. Freese, M. S. Melzer, *Biochim. Biophys. Acta* **1968**, *155*, 491.

315. K. Hiramoto, R. Aso-o, H. Ni-iyama, S. Hikage, T. Kato, K. Kikugawa, *Mutation. Res.* **1996**, *359*, 17.

316. R. M. Williams, T. Glinka, M. E. Flanagan, R. Gallegos, H. Coffman, D. Pei, *J. Am. Chem. Soc.* **1992**, *114*, 733.

317. N. Kashige, T. Takeuchi, S. Matsumoto, S. Takechi, F. Miake, T. Yamaguchi, *Biol. Pharm. Bull.* **2005**, *28*, 419.

318. J. L. Bolton, M. A. Trush, T. M. Penning, G. Dryhurst, T. J. Monks, *Chem. Res. Toxicol.* **2000**, *13*, 135.

319. K. R. Guertin, L. Setti, L. Qi, R. M. Dunsdon, B. W. Dymock, P. S. Jones, H. Overton, M. Taylor, G. Williams, J. A. Sergi, K. Wang, Y. Peng, M. Renzetti, R. Boyce, F. Falcioni, R. Garippa, A. R. Olivier, *Bioorg. Med. Chem. Lett.* **2003**, *13*, 2895.

320. S. S. Matsumoto, J. Biggs, B. R. Copp, J. A. Holden, L. R. Barrows, *Chem. Res. Toxicol.* **2003**, *16*, 113.

321. W. A. Denny, W. R. Wilson, *J. Med. Chem.* **1986**, *29*, 879.

322. M. P. Hay, H. H. Lee, W. R. Wilson, P. B. Roberts, W. A. Denny, *J. Med. Chem.* **1995**, *38*, 1928.

323. W. Kim, K. S. Gates, *Chem. Res. Toxicol.* **1997**, *10*, 296

324. K. Mitra, W. Kim, J. S. Daniels, K. S. Gates, *J. Am. Chem. Soc.* **1997**, *119*, 11691.

■■■■■ **CHAPTER 9**

Conical Intersection Species as Reactive Intermediates

MICHAEL J. BEARPARK AND MICHAEL A. ROBB

Chemistry Department, Imperial College London, London SW7 2AZ UK

9.1. INTRODUCTION AND HISTORICAL PERSPECTIVE

The textbook definition of a reactive intermediate is a short-lived, high-energy, highly reactive molecule that determines the outcome of a chemical reaction. Well-known examples are radicals and carbenes: such species cannot be isolated in general, but are usually postulated as part of a reaction mechanism, and evidence for their existence is usually indirect. In thermal reactivity, for example, the Wheland intermediate (Scheme 9.1) is a key intermediate in aromatic substitution.

Reaction mechanisms in chemistry are often written as a sequence of chemical structures, and such structures are referred to as reactive intermediates. Indeed, these intermediates often correspond to local minima on the potential energy curve—as shown in Figure 9.1—but need not do so, and the definition of a reactive intermediate

Reviews of Reactive Intermediate Chemistry. Edited by Matthew S. Platz, Robert A. Moss, Maitland Jones, Jr.
Copyright © 2007 John Wiley & Sons, Inc.

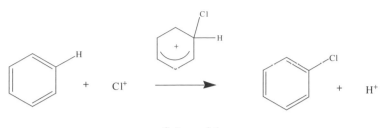

Scheme 9.1

is more general than that. Thus a reactive intermediate is usually postulated in terms of a nuclear geometry and valence bond (VB) electronic structure, but only a detailed computation or indirect experimental measurement can determine whether the postulated structure is a critical point on the potential surface (local minimum or transition state) or not. For our purposes, we take the definition of a reactive intermediate to include any point on the reaction path that can be characterized as a critical point on the potential surface, or that has a well-defined VB structure without necessarily being a critical point.

Thus we adopt the view (see suggested reading 1) that any structure along the reaction path is an "intermediate." With modern laser spectroscopy (suggested reading 2), one can now directly probe such structures experimentally. Thus it is a reactive intermediate, because it can be "seen" spectroscopically, not because it can be put in a bottle for a short time. As we shall see, this revised view of an intermediate is particularly relevant for photochemical and other nonadiabatic processes.

A photochemical reaction coordinate has two branches: an excited state branch and a ground state branch that is reached after decay at a conical intersection. Thus a conical intersection between ground and excited states of a molecule is a precursor to ground state reactivity, and conforms to the above definition of a reactive intermediate. The main focus of our article will be to develop this idea. In Figure 9.1b, we show the energy profile for a photochemical reaction with a conical intersection

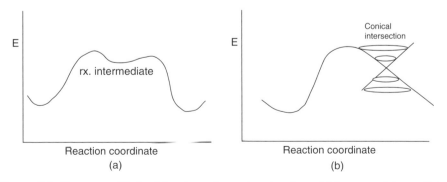

Figure 9.1. Energy profile for (a) a thermal reaction showing a high-energy intermediate, and (b) a photochemical reaction showing a conical intersection.

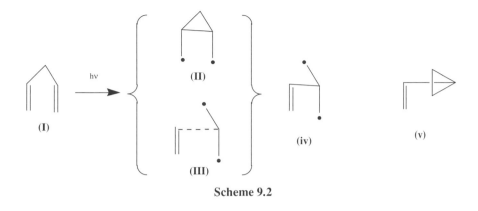

Scheme 9.2

along the reaction path. Obviously, there are many important differences between the energy profiles shown in Figures 9.1a and 9.1b, the main one being that the photochemical reaction is nonadiabatic and involves two potential energy surfaces.

A simple example serves to illustrate the similarities between a reaction mechanism with a "conventional" intermediate and a reaction mechanism with a conical intersection. Consider Scheme 9.2 for the photochemical di-π-methane rearrangement. Chemical intuition suggests two possible key intermediate structures, **II** and **III**. Computations confirm[1] that, for the singlet photochemical di-π-methane rearrangement, structure **III** is a conical intersection that divides the excited-state branch of the reaction coordinate from the ground state branch. In contrast, structure **II** is a conventional biradical intermediate for the triplet reaction.

Conical intersections are involved in other types of chemistry in addition to photochemistry. Photochemical reactions are nonadiabatic because they involve at least two potential energy surfaces, and decay from the excited state to the ground state takes place as shown, for example, in Figure 9.2a. However, there are also other types of nonadiabatic chemistry, which start on the ground state, followed by an excursion upward onto the excited state (Fig. 9.2b). Electron transfer problems belong to this class of nonadiabatic chemistry, and we have documented conical intersection

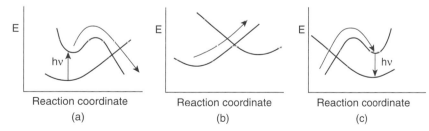

Figure 9.2. Reaction profiles involving a conical intersection: (a) a photochemical reaction; (b) an upwards excursion via a conical intersection in a nonadiabatic reaction; (c) a chemiluminescent reaction.

processes in electron transfer in references 2–5. Similarly, excited states can be generated thermally rather than by photo excitation. Thus, in chemiluminescence, the nonadiabatic process also involves moving from the ground state to the excited state (Fig. 9.2c). We have characterized conical intersections in chemiluminescent reactions in reference 6, and we shall discuss one example of thermal nonadiabatic chemistry in a subsequent section.

Quantum chemistry can routinely provide quantitative information about certain classes of ground state reactive intermediates, particularly those that correspond to a local minimum or transition state. More recently, it has been possible to study excited state photochemical reaction paths and conical intersections with the same degree of confidence as ground state reactivity. In fact, nonadiabatic chemistry in general is now amenable to theoretical study. We can determine the molecular geometries at which conical intersections occur, and subsequently analyze the electronic structure. Many generalities have emerged that allow one to understand the role of such species and to make predictions about chemical reactivity. In this article we shall discuss some of these generalities, and demonstrate that they serve to unify many diverse mechanistic problems in photochemistry and other nonadiabatic processes such as electron transfer.

Mechanistic photochemistry has made tremendous strides in the last 15 years, in large part due to the application of computational methods. Some of this progress is charted in reviews and review collections (see suggested reading 3–8) as well as in textbooks (suggested reading 9–13). As a result of computational studies on many photochemical problems, it is now clear that the "photochemical funnel" (suggested reading 11) first described by Zimmerman and Michl manifests itself in a conical intersection. (A useful historical presentation can be found in the first chapter of suggested reading 6.) The present article is aimed at a broad audience, including junior research workers with an experimental rather than a theoretical background. Thus, we will not assume a high level of mathematical sophistication (the reader is referred to suggested reading 5 for more mathematical detail). Instead, we will try to present a self-contained account of the role played by "conical intersection intermediates" in the mechanisms of nonadiabatic chemistry. We aim to illustrate these ideas with recent examples, largely drawn from our own work. Thus this article is more like a broadly based "Accounts of Chemical Research." We begin by trying to establish some quite general but simple principles that relate the nature of a conical intersection to mechanistic photochemistry. We shall return to nonadiabatic chemistry in general and electron transfer (suggested reading 14) in particular later on.

9.2. UNDERSTANDING CONICAL INTERSECTIONS

Part of the explanation that follows concerns the reasons why a conical intersection may be found in a particular system. This part could be skipped over in a first reading, as it is more mathematical. Nevertheless, it is closely connected with an explanation of the "shape" of a conical intersection, which in turn determines the crossing's accessibility on the excited state, and the subsequent reaction paths on the ground

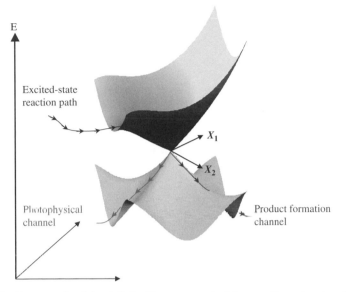

Figure 9.3. Cartoon of a "classic" double cone conical intersection, showing the excited state reaction path and two ground state reaction paths. See color insert.

state. Although these can all be appreciated without the mathematical background, our experience is that the mathematical models have been useful.

We begin with a very simple "sand in a funnel" picture for a photochemical mechanism involving excited and ground state branches (with two ground state reaction pathways) and a conical intersection (Fig. 9.3). We shall use cartoons of this form to illustrate many aspects of nonadiabatic chemistry in this article. We begin by offering a few comments on how such cartoons should be interpreted. We have plotted the energy in two geometrical coordinates, X_1 and X_2. In general, these two coordinates will be combinations of changes in the bond lengths and bond angles of the molecular species under investigation. We are limited in such cartoons to using two or three combinations of molecular variables. However, we must emphasize that these are just cartoons used to illustrate a mechanistic idea. All the computations we shall discuss are done in the full space of molecular geometries without any constraint.

The coordinates X_1 and X_2 in Figure 9.3 correspond to the space of molecular geometrical deformations that lift the degeneracy. These coordinates are precisely defined quantities that can be computed explicitly. Similarly, the apex of the cone corresponds in general to an optimized molecular geometry. The shape or topology in the region of the apex of the double cone will change from one photochemical system to another, and it is the generalities associated with the shape that form part of the mechanistic scenario that we will discuss.

Of course the shape of the potential energy surface is a reflection of the electronic structure, which differs between one electronic state and another. We now introduce

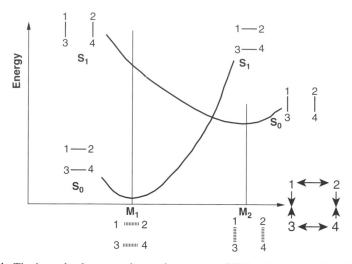

Figure 9.4. The interplay between electronic structure (VB bonding pattern) and geometrical change on the potential energy surface.

this concept with a simple example. In Figure 9.4 we show potential energy curves for a hypothetical photochemical bond exchange reaction (for example, [2+2] cycloaddition) in which the 1,2 and 3,4 bonds become 1,3 and 2,4 bonds. In general, in photochemistry, the excited state is a valence bond (VB) isomer of the ground state. Looking at the left-hand side of Figure 9.4, when the ground state S_0 at the geometry M_1 is excited vertically to S_1, one can see that the corresponding excited state valence bond structure has single bonds in new positions, 1,3 and 2,4. The electronic (VB) structure on S_1 drives geometry relaxation to move the nuclei and place the geometrical bonds in positions 1,3 and 2,4. (Note that we use the term "VB structure" in a very general way to indicate the electronic structure in terms of "arrangement of bonds.") Thus the geometric structure M_1 on state S_1 will relax toward the new structure M_2 on state S_0, which has a geometric structure compatible with the electronic (VB) structure. (i.e., the nuclei have "caught up" with the new electron distribution). However, at the same time it is also clear that, as one changes the molecular structure to relax the energy on S_1 (from the geometry M_1), the energy associated with S_0 must rise, because the S_0 electronic structure is associated with bonds 1,2 and 3,4 on S_0. At some point one must have a curve crossing as shown in the Figure 9.4. The geometry where this crossing occurs must be the geometry for which the energy of the VB structure 1,2 3,4 is equal to the energy of the VB structure 1,3 2,4.

There is another aspect of this picture (Figure 9.4) that is very important for future discussions. If we trace the curve labeled S_1 from the geometry M_1 through the crossing from left to right to the geometry M_2, we see that the VB structure does not change. A similar observation can be made for the curve S_0. Thus in Figure 9.4, the VB structure 1,3 2,4 lies on the excited state at the initial geometry M_1 but lies on the ground state at the final structure M_2. Thus in our future discussions, the states

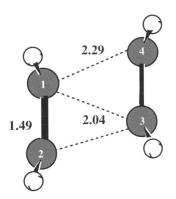

Figure 9.5. Geometry (Å) of the MECI for the [2+2] photocycloaddition of two ethylenes.

must always carry two labels: S_0 or S_1 to indicate ground or excited state, and a VB label to indicate the nature of the excited state. (A brief word about terminology: "S_0" or "S_1" label *adiabatic* electronic states that can change character; a particular VB electronic structure corresponds to a *diabatic* state, having a particular character, that can be either S_0 or S_1 depending on the geometry of the molecule; and *nonadiabatic* refers to a process involving more than one adiabatic state, which may correspond to involving one diabatic state.)

The conical intersection structure (see reference 7) for the real [2+2] photochemical cycloaddition of two ethylenes is shown in Figure 9.5. On the one hand, it looks something like a transition structure, with stretched ethylenic C–C double bonds and partly formed new C C single bonds. But the trapezoidal distortion at first sight seems rather strange. It turns out that we can understand a model state crossing problem of four electrons in four s orbitals quite easily: we now digress to discuss this, and then show that when one replaces the s orbitals of the model by p^π orbitals, the origin of the trapezoidal structure in Figure 9.5 becomes clear. But going further, our objective is also to explain the nature of the essential geometrical coordinates X_1 and X_2 that lift the degeneracy at the apex of the cone. We will develop the idea that the point of degeneracy in Figure 9.3 or 9.4 is not a single point in higher dimensions, but rather an extended hyperline, and the conical intersection for the [2+2] problem just discussed is a point on such a line.

In Figure 9.6 we show some examples of abstract geometries (for four electrons in four 1s orbitals) where a real crossing (viz. Fig. 9.3) occurs. If the bonding arrangement is 1,2 and 3,4 we have one electronic state. If the bonding arrangement is 1,3 and 2,4 we have the other electronic state (as shown in Fig. 9.4). In chemistry, the energy of a bond is associated with orbital overlap. Thus in Figure 9.6a, the bonding arrangements 1,2 3,4 and 1,3 2,4 would appear to have the same energy (i.e., a conical intersection) because the overlaps are equal (cubic geometry). However, this intuitive idea is not quite correct and we must develop it more precisely. (For more detailed discussion see suggested reading 13 and reference 8.) Thus we need an expression for the VB bond energy involving

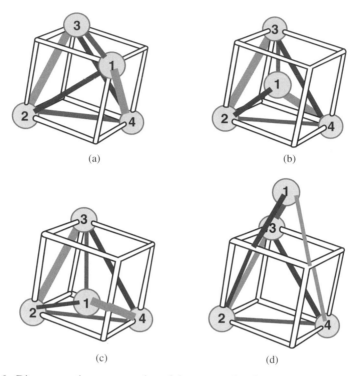

(a)   (b)

(c)   (d)

Figure 9.6. Diagrammatic representation of the geometries that correspond to $T = 0$ in the London formula for four electrons in four 1s orbitals (e.g., H_4).

the overlap and an expression for the total energy, which we shall see needs additional "cross-terms."

The energy K_{ij} of an isolated VB bond is given in Eq. 9.1. (This is the stabilization energy relative to a fully homolytically cleaved bond in H_2 for example.) The first term is the exchange repulsion, which is small and positive, while the second is the exchange attraction, which depends upon the overlap and is negative. Thus the exchange K_{ij} is in general negative and depends upon the overlap. The total energy for the model system shown in Figure 9.6 is given by the London formula in Eq. 9.2 in terms of the individual K_{ij}. We use K_R for the bond energy of the bonding arrangement 1,2 3,4 (which could be the ground state) and K_P for the 1,3 2,4 bonding arrangement (which could be the excited state). The term K_X is a "cross-term." The total energy given in Eq. 9.2a is the same for ground and excited state if $T = 0$, which implies $K_P = K_R$, $K_X = K_P$ and $K_X = K_R$. Q is the coulomb energy, which is the same for both ground and excited states.

$$K_{ij} = [ij \,|\, ij] + S_{ij} h_{ij} \tag{9.1}$$

$$E = Q \pm T \tag{9.2a}$$

$$T = \sqrt{(K_R - K_P)^2 + (K_P - K_X)^2 + (K_X - K_R)^2} \qquad (9.2b)$$

$$K_R = K_{12} + K_{34} \quad K_P = K_{13} + K_{24} \quad K_X = K_{14} + K_{23} \qquad (9.2c)$$

The K_{ij} depends on orbital overlap, and the overlap between s orbitals depends simply on the distance between the orbital sites. Thus we can represent K_{ij} by the length of the lines 1,2 1,3 and so forth, and the geometries where $T = 0$ in the London formula in Eq. 9.2 can be illustrated geometrically as in Figure 9.6. Each solid line in Figure 9.6 corresponds to an exchange integral K_{ij} associated with centers i and j in Eq. 9.2. The pairs of exchange integrals corresponding to K_R, K_P, and K_X correspond to the pairs of equally-shaded lines. It should be clear that T in Eq. 9.2 will be zero (and consequently the ground and excited states will have the same energy), whenever the sum of the lengths of the blue lines is the same as the sum of the lengths of the red lines and also the same as sum of the lengths of the green lines, that is, $K_P = K_R$, $K_X = K_P$, and $K_X = K_R$. Thus in Figure 9.6 we give examples of such geometries where these conditions are satisfied exactly.

The curve crossing of the sort discussed in Figure 9.4 in one dimension, and in Figure 9.3 in two dimensions is a conical intersection. As shown in Figure 9.6, there are many geometries that satisfy the geometrical/mathematical constraint which corresponds to $T = 0$ (Eq. 9.2b). Thus a conical intersection is a family of geometries or a "hyperline" in the space of the 3N-6 internal coordinates where the energies of the ground and excited states are equal. Of course, the energy associated with this hyperline will change (i.e., Q changes in Eq. 9.2a as one traces out geometries on the conical intersection), and thus one may have many relative maxima and minima on the conical intersection hyperline. The lowest minimum energy point is often referred to as *the* "conical intersection point" or more correctly the minimum energy conical intersection (MECI). A photochemical reaction path must cross this hyperline at the nonadiabatic transition point that connects the excited state reaction path and the ground state reaction path. However, it needs not cross at the minimum energy point.

At this stage, we wish to emphasize that a point (molecular geometry) on a conical intersection hyperline has a well-defined electronic structure (illustrated in Figure 9.6 or Eq. 9.2 with $T = 0$) and a well-defined geometry. Of course, the four electrons in four 1s orbitals shown in Figure 9.6 is a very simple example, but we believe it is useful in order to be able to appreciate the generality of the conical intersection construct. In more complex systems, the conical intersection hyperline concept persists, but the rationalization may be less obvious.

Now let us return to our discussion of the conical intersection structure for the [2+2] photochemical cycloaddition of two ethylenes and photochemical di-π-methane rearrangement. They are both similar to the 4 orbital 4 electron model just discussed, except that we have p^π and p^σ overlaps rather than 1s orbital overlaps. In Figure 9.5 it is clear that the conical intersection geometry is associated with $T = 0$ in Eq. 9.2b. Thus (inspecting Figure 9.5) we can deduce that

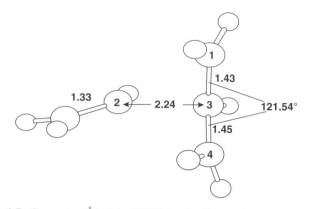

Figure 9.7. Geometry (Å) of the MECI for the di-π-methane rearrangement.

(a) $K_R = K_P$ since the p$^\sigma$ overlap is larger than the p$^\pi$ overlap,
(b) $K_{24} = 0$ and $K_X = K_{13}$ thus with an appropriate trapezoidal distortion $K_X = K_R = K_P$.

Similarly for the photochemical di-π-methane rearrangement conical intersection (Fig. 9.7), one has $T = 0$ under similar conditions:

(a) $K_R = K_{34}$ with $K_{12} = 0$,
(b) $K_P = K_{13}$ with $K_{24} = 0$, so that
(c) $K_X = K_{23} = K_R = K_P$.

Again we assume that p$^\sigma$ overlap is larger than the p$^\pi$ overlap.

In Figure 9.8a, we show a cartoon of the potential energy surface for the [2+2] photochemical cycloaddition of two ethylenes (reference 7). The branching space $(X_1 \, X_2)$ coordinates, correspond to the distance between the two ethylenes R (X_1) and a trapezoidal distortion (X_2). As we have just discussed, our analysis in terms of exchange integrals K_{ij} rationalizes the trapezoidal structure of the conical intersection at point E. The cross-section in Figure 9.8b through points A and A' along the coordinate R (X_1) corresponds to the classic Woodward–Hoffmann [2$_s$+2$_s$] cycloaddition reaction coordinate. Notice the avoided crossing along this coordinate with a transition state at A' and an excited state minimum at A. However, this avoided crossing topology is just an artifact of the use of one "distinguished" coordinate R (X_1). When one adds the trapezoidal coordinate corresponding to the other branching space coordinate X_2, we see that we have both the avoided crossing A A' and the conical intersection at geometry E. However, the excited state minimum A is not a real minimum at all but rather a transition state that connects the two conical intersection points. We shall discuss the avoided crossing concept in more detail subsequently (Fig. 9.14). However, the point we wish to emphasize is "avoided" crossings are an artifact of a low-dimensional treatment. In higher dimensions, the avoided crossing will always become a conical intersection.

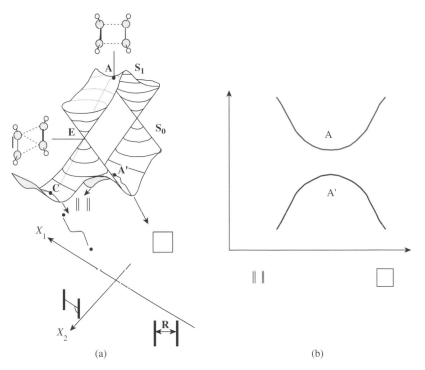

(a)   (b)

Figure 9.8. A cartoon showing (a) the conical intersection for the [2+2] photocycloaddition of two ethylenes, drawn in the branching space corresponding to the distance between the two ethylenes R (X_1) and a trapezoidal distortion (X_2), and (b) an avoided crossing in a cross-section R (X_1).

Let us summarize briefly at this stage. We have seen that the point of degeneracy forms an extended hyperline which we have illustrated in detail for a four electrons in four 1s orbitals model. The geometries that lie on the hyperline are predictable for the 4 orbital 4 electron case using the VB bond energy (Eq. 9.1) and the London formula (Eq. 9.2). This concept can be used to provide useful qualitative information in other problems. Thus we were able to rationalize the conical intersection geometry for a [2+2] photochemical cycloaddition and the di-π-methane rearrangement.

We now want to discuss the hyperline concept and the nature of the branching space $X_1 X_2$ in more depth using the four electrons in four 1s orbitals model, and show how one might relate the hyperline traced out by the geometries a–d in Figure 9.6 with the conical intersection "point" shown in Figure 9.3. Also, referring to Figure 9.3, we must now ask about the directions X_1 and X_2.

In Figure 9.9, we have presented another cartoon that attempts to show the potential energy surfaces in the space of the degeneracy lifting coordinates as one traverses a third coordinate X_3. Of course, this is really a four-dimensional picture, which is not easy to assimilate. Nevertheless, it should be clear that the degeneracy

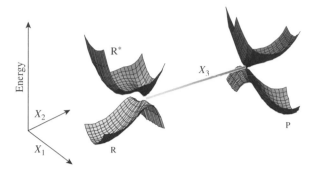

Figure 9.9. A cartoon showing the conical intersection hyperline traced out by a degeneracy-preserving coordinate X_3. The system remains degenerate as one traverses the coordinate X_3, but the energy and the shape of the double-cone must change in $X_1 X_2$. See color insert.

persists along the coordinate X_3. However, the energy will change and so will the shape of the double-cone near the apex. Thus one may think about the geometries a–d in Figure 9.6 as points on the coordinate X_3 in Figure 9.9.

It is useful to consider the information contained in Figure 9.9 in a different way. In Figure 9.10, we show the conical intersection hyperline traced out by a coordinate

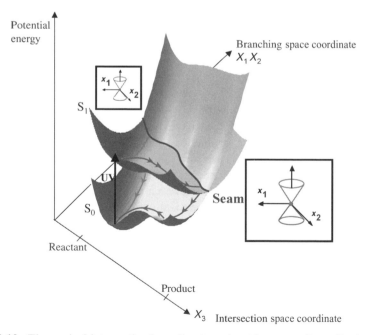

Figure 9.10. The conical intersection hyperline traced out by a coordinate X_3 plotted in a space containing the coordinate X_3 and one coordinate from the degeneracy-lifting space X_1X_2. See color insert.

X_3 plotted, this time, in a plane containing the intersection space coordinate X_3 and one coordinate from the degeneracy-lifting (or branching) space $X_1 X_2$. In this figure, the conical intersection line now appears as a seam. In contrast to the "sand in the funnel" model shown in Figure 9.3, it is clear that the reaction path could be almost parallel to this seam. Of course this figure can be misunderstood, because each point on the seam line lies in the space of the double-cone $X_1 X_2$ as suggested by the insert. Thus, it is essential to appreciate that decay at the conical intersection is associated with three coordinates, the branching space $X_1 X_2$, and the reaction path X_3 which may or may not lie in the space $X_1 X_2$.

Figures 9.9 and 9.10 establish two important mechanistic points:

1. The important points on a conical intersection hyperline are those where the reaction path meets with the seam (see Fig. 9.10)
2. Radiationless decay takes place in the coordinates $X_1 X_2$ as one passes through the conical intersection diabatically (the VB structure does not change, see, for example, Fig. 9.4).

In Figure 9.10, this second principle appears to be violated since the reaction path appears to pass through the hyperline adiabatically. However, we emphasize—as indicated by the double-cone insert—that as one passes through the hyperline, decay takes place in the coordinates $X_1 X_2$ and in general their VB structure does not change. This idea is obviously easier to appreciate in Figure 9.9. We shall use both Figures 9.9 and 9.10 as models in subsequent discussions; but the reader needs to remember the conceptual limitations.

It now remains to discuss the nature of the two geometrical coordinates, corresponding to the branching space $X_1 X_2$, that lift the degeneracy at a conical intersection. For illustrative purposes, let us consider the geometry at point (a) shown in Figure 9.6. In Figure 9.11, we show two geometrical coordinates that lie in the space $X_1 X_2$. It should be obvious that moving in these coordinates will break the condition $T = 0$ in Eq. 9.2b. In general, such directions are computed as part of calculations to optimize the geometries of conical intersections. However, in this case, because of the high symmetry, these two directions are obvious.

In order to understand more general situations, it is helpful to discuss the origin of the space $X_1 X_2$. The structure (a) in Figure 9.6 belongs to the T_d point group and the ground and excited state wave functions (corresponding to the couplings 1,2 3,4 and 1,3 2,4) belong to the degenerate E irreducible representation of that group. The degeneracy is lifted on lowering the symmetry to the subgroup D_{2d} where the remaining C_2 principal axis is taken to be vertical to the 1,2 plane. The degenerate E irreducible representation then splits into two nondegenerate representations, one of which must be symmetric, and the other antisymmetric with respect to the principal axis. The two vectors shown in Figure 9.11 span symmetric and antisymmetric representations and clearly lift the degeneracy. Furthermore, referring to Eq. 9.2, it is clear that such distortions will break the equalities $K_R = K_P$ along X_1 and $K_X = K_R$ and so forth along X_2. The general point to appreciate is that one can rationalize the directions spanning the branching space $X_1 X_2$ in the

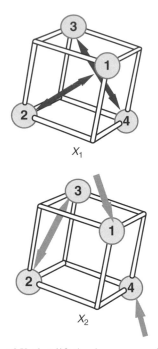

Figure 9.11. Two vectors X_1 and X_2 that lift the degeneracy of the conical intersection seam at point (a) in Figure 9.6.

same intuitive way that we rationalize the driving force for structural change using VB diagrams.

At this stage it is helpful to consider an additional example to illustrate the extended conical intersection hyperline concept with a very simple molecule. We now discuss the conical intersection occurring in fulvene (for details see references 9 and 10). In Figure 9.12, we show a plot of the potential energy along the stretching motion associated with the C−C bond between the methylene group and the five-membered ring. This is one of the coordinates X_1 that lifts the degeneracy and corresponds to the gradient difference vector which we shall discuss subsequently. The antisymmetric skeletal deformation coordinate X_2 (in Figure 9.12) is the second coordinate that lifts the degeneracy and together they form the branching space. The resulting conical intersection corresponds to a planar geometry illustrated in the figure. As we shall now discuss, this geometry corresponds to a local maximum on the conical intersection line, as shown in Figure 9.13.

Figure 9.13 shows a cartoon of the potential energy surfaces for fulvene in the space X_1 and X_3 = torsion. It is clear that the degeneracy persists along the "curved" extended seam topology, analogous to the one illustrated in Figure 9.10. However, it is clear from Figure 9.13 that a lower energy point on the conical intersection hyperline exists, with the methylene group twisted 90 degrees. In fact, there are even lower energy points on the conical intersection line along coordinates that involve the

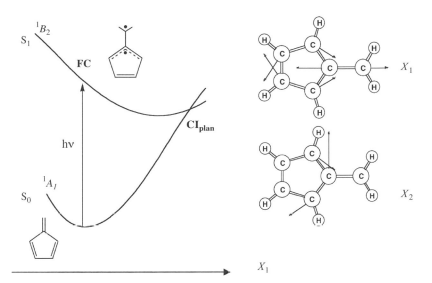

Figure 9.12. Potential energy profile along X_1 (adapted from reference 10) near the fulvene conical intersection. The branching space consists of stretching and skeletal deformation of the five-membered ring.

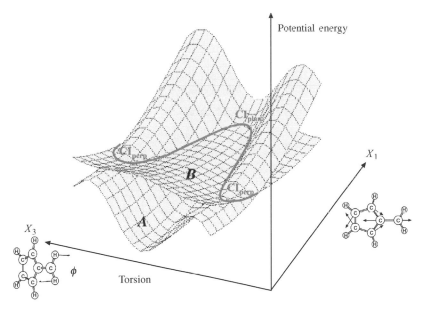

Figure 9.13. Potential energy profile (adapted from reference 10) for fulvene in the space spanned by X_1 and the X_3 coordinate (torsion).

pyramidalization of the methylene group and the reader is referred to references 9 or 10 for more details. Perhaps the most important point to appreciate from this figure is that the branching space $X_1 X_2$ does not contain *any* component of torsion. Thus the reaction path X_3 (torsional coordinate) is rigorously orthogonal to the branching space. In general, of course, the angle between the reaction path X_3 and the branching space $X_1 X_2$ will vary between zero, in which case one has the "sand in the funnel model" shown in Figure 9.3, and 90 degrees, in which case one has the extended hyperline model as shown in Figure 9.9 or 9.10 and illustrated in Figure 9.13 for fulvene.

To conclude this section, it could be helpful to make a connection between the pictorial discussion we have just given and the type of computation that one can carry out in quantum chemistry. The double cone topology shown in Figure 9.3 can be represented mathematically by Eqs 9.3a and 9.3b. Q_{x_1}, Q_{x_2} are the branching space coordinates. This equation is valid close to the apex of the cone. (A full discussion of the analytical representation of conical intersections can be found in references 9 and 10.)

$$U_A = E^0_{AB} + \frac{\lambda_1}{2}\bar{Q}_{x_1} + \frac{\lambda_2}{2}\bar{Q}_{x_2} - \frac{1}{2}\sqrt{(\delta\kappa\bar{Q}_{x_1})^2 + 4(\kappa^{AB}\bar{Q}_{x_2})^2} \qquad (9.3a)$$

$$U_B = E^0_{AB} + \frac{\lambda_1}{2}\bar{Q}_{x_1} + \frac{\lambda_2}{2}\bar{Q}_{x_2} + \frac{1}{2}\sqrt{(\delta\kappa\bar{Q}_{x_1})^2 + 4(\kappa^{AB}\bar{Q}_{x_2})^2} \qquad (9.3b)$$

$$H_{AB} = \int \phi_A^* \hat{H}_e \phi_B d\tau \qquad (9.3c)$$

$$S = H_{BB} + H_{AA} \qquad (9.3d)$$

$$\Delta H = H_{BB} - H_{AA} \qquad (9.3e)$$

$$\lambda_i = \frac{\partial S}{\partial \bar{Q}_i}\Big|_0 \qquad (9.3f)$$

$$\delta\kappa_i \equiv \frac{\partial \Delta H}{\partial \bar{Q}_i}\Big|_0 \qquad (9.3g)$$

$$\kappa_i^{AB} \equiv \frac{\partial H_{AB}}{\partial \bar{Q}_i}\Big|_0 \qquad (9.3h)$$

Equations 9.3a and 9.3b give the energy of the upper and lower part of the cone (U_A and U_B). In Eqs 9.3a and 9.3b, the first term represents Q in Eq. 9.2, while the expression under the square root sign corresponds to T in Eq. 9.2. E_{AB} is the reference energy at the apex of the cone. The remaining quantities in these two equations are energy derivatives. The quantity in Eq. 9.3g is the gradient difference vector, while the quantity in Eq. 9.3h

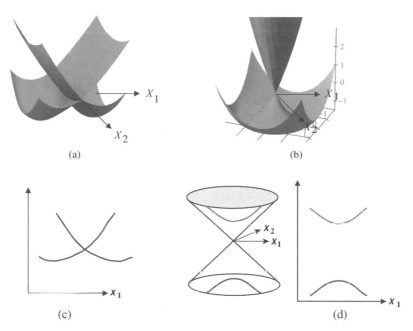

Figure 9.14. (a) Diabatic potential energy surfaces (the interstate coupling Eq. 9.3h is zero); (b) The double-cone that occurs when the interstate coupling, Eq. 9.3h, is finite; (c) Cross-section along X_1 through Figure 9.14a showing an "unavoided" crossing; (d) Cross-section along X_1 through Figure 9.14b, not passing through the apex of the cone, showing an "avoided" crossing.

is proportional to the derivative coupling. The directions $X_1 X_2$ are parallel to these two directions, 3g and 3h. The important point is that the directions can be obtained from any theoretical calculation using gradient-type technology

In Figure 9.14, we show the pictorial representation of Eqs 9.3a and 9.3b where $\lambda_1 = \lambda_2$, that is, a circular cone. (In the figure we have used quadratic terms in the potential as well, so the figure represents Eqs 9.3a, 9.3b only in the immediate apex of the cone.) In Figure 9.14a we show the diabatic potential energy surfaces corresponding to the case where the interstate coupling (Eq. 9.3h) is zero. In this case, the branching space has dimension one: only X_1 is defined. This situation is important in electron transfer problems and we shall return to it later. In Figure 9.14b, we show the double-cone that occurs when the interstate coupling is finite.

In Figures 9.14c and 9.14d we show complementary one-dimensional cross-sections through the potential energy surfaces of Figures 9.14a and 9.14b. Figures 9.14c and 9.14d allow us to discuss the concepts of an unavoided and avoided crossing. These are terms that are in common usage but have meaning only in the context of such one-dimensional cross-sections. In Figure 9.14a, any cross-section must result in a (unavoided) crossing of the two potential energy surfaces (as shown in Fig. 9.14c). This situation is rare and we will discuss this subsequently. In Figure 9.14d, we show a cross-section through Figure 9.14b in a plane that does not pass through the apex of the cone. In this case the crossing is said to be "avoided." But this is an artifact of the use of only one distinguished coordinate. It is clear that if one has an avoided crossing of the sort discussed, there must always be a "real" crossing (at the apex of the cone), if one adds the dimension X_2. Finally, we should point out that the double cone can have different shape than the circular cone shown in the figure and thus the excited state minimum on the cross-section may not always have the same energetic relationship to the apex of the cone. However, the central point is that an avoided crossing is an artifact of a low dimensional description.

If we examine the conical intersection hyperline shown in Figure 9.10 (or Fig. 9.13), it is clear that one can distinguish local minima and maxima. One could in principle optimize such critical points on a conical intersection hyperline in the same way that one might optimize a transition structure for a thermal reaction. In a transition structure optimization, one minimizes with respect to all coordinates except the direction of the transition vector. One then maximizes along the transition vector or reaction path direction. If one examines the conical intersection hyperline shown in either Figure 9.9 or Figure 9.10, it is clear that one first needs to minimize the energy difference between the two states to reach a point on a crossing hyperline. Then one wishes to remain on the hyperline, minimizing the energy on this line (i.e., minimizing the energy in the space spanned by X_3 to X_{3N-6}). The mathematical equations have been discussed in several places (see references 11–13). Minimization of the energy difference between the two states yields the conditions shown in Eq. 9.4a, while minimization along a hyperline requires vanishing of the gradient orthogonal to the branching space X_1X_2 given in Eq. 9.4b. The corresponding gradients are given in Eqs 9.4c and 9.4d. The projector **P** projects the gradient orthogonal to the space X_1X_2. One can then use usual gradient optimization methods, and techniques of the form discussed in references 9 and 10 to determine the curvature.

$$U_B - U_A = 0 \qquad (9.4a)$$

$$\frac{\partial U}{\partial \bar{Q}_3} = \frac{\partial U}{\partial \bar{Q}_4} = \ldots = \frac{\partial U}{\partial \bar{Q}_{3N-6}} = 0 \qquad (9.4b)$$

$$\frac{\partial (U_B - U_A)}{\partial Q_i} = -2|U_B - U_A|\delta\kappa_i = 0 \qquad (9.4c)$$

$$\mathbf{g} = \mathbf{P}\frac{\partial U_B}{\partial \mathbf{Q}} \qquad (9.4d)$$

Our hypothesis for discussion in this section has been that the conical intersection can be characterized like any other reactive intermediate. On examining Figure 9.3 or 9.10, it is clear that a conical intersection divides the excited-state branch of the reaction path from the ground-state branch in a photochemical transformation. (We shall look at other types of nonadiabatic chemistry subsequently.) Thus the conical inter-section is indeed the precursor to ground state reactivity. However, the essential dif-ference with other types of reactive intermediate is that a conical intersection, rather than being best described as a single geometry, is in fact a family of geometries (Fig. 9.6), associated with a hyperline (sketched in Fig. 9/Fig. 10.) As we have just discussed, one can distinguish certain points on a hyperline associated with maxima and minima, where the gradient vanishes. However, one needs to know the relation-ship between the reaction path and a conical intersection geometry associated with the hyperline. Thus the correct analogy with other reactive intermediates of organic chemistry probably corresponds to the point where the reaction path intersects the conical intersection seam. However, looking at Figure 9.10, there may be several relevant intersection points that require study here. Thus in Figure 9.10 one could decay in the reactant region, or one could pass over the adiabatic reaction barrier on the excited state before decaying in the product region. Thus, one may be able to find critical points on the conical intersection hyperline in the region of the reactants and in the region of their products. Either of these could be thought of as a distinct reactive intermediate.

The other aspect of a conical intersection that we have tried to emphasize is that there is a relationship between the valence bond structures associated with the ground state or the excited state and the position of the surface crossing. In any mechanistic study this is also very interesting because it provides information that can be used to think intuitively about mechanisms. We will try to emphasize this point of view in the rationalization of all the examples we will look at.

We now proceed to look at three examples from recent work in some depth. In the first example, we wish to illustrate that a knowledge of the VB structure or of the states involved in photophysics and photochemistry rationalize the potential surface topology in an intuitively appealing way. We then proceed to look at an example where the extended hyperline concept has interesting mechanistic implications. Finally, we shall look at an example of how conical intersections can control electron transfer problems.

9.3. INTERPRETATION OF CONICAL INTERSECTIONS USING VB STRUCTURES (TICT PROCESSES)

In this section we would like to consider an example which illustrates that one can understand the occurrence of conical intersections—as well as the direc-tions $X_1 X_2$ corresponding to the branching space—if one has an understanding of the electronic structure of the two states involved. We address the following two questions:

1. What is the connection between the molecular geometry and the electronic (VB) structure?
2. Can the nature of the adiabatic and nonadiabatic pathways (and the position of the conical intersection) be predicted from VB structures?

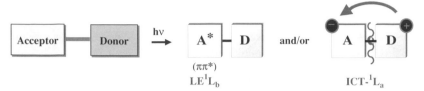

Figure 9.15. Schematic representation of the TICT process and the generation of LE and ICT states.

The examples we will use are TICT (twisted intermolecular charge transfer) aminobenzonitrile (ABN) compounds (Figs. 9.15 and 9.16). We have recently completed theoretical work on this class of compound and the reader is referred to reference 14 for a complete bibliography.

In ABNs there are two low-lying excited states: a locally excited (LE) state where the excitation is localized on the phenyl ring, and the intramolecular charge transfer (ICT) state, where there is a transfer of charge from the amino group to the benzene ring (see Figs. 9.15 and 9.16). The ICT state is thus similar (electronically) to a benzene radical anion. In spectroscopy, with suitable substitution R and in the appropriate solvent, one can see emission from each state (LE or ICT diabatically, but both on S_1). The lowest energy equilibrium geometry of the ICT state is usually assumed to be twisted; hence the acronym TICT. Since one observes dual fluorescence, there must be two S_1 minima, associated with the LE and ICT electronic structures. An adiabatic reaction path must therefore connect these two electronic structures on S_1. Thus, there is an adiabatic reaction coordinate associated with the electron transfer process. However, the absorption from the ground state to the LE state in the Franck–Condon region will be forbidden. Rather the absorption takes place to S_2, which is the ICT state at the Franck–Condon geometry. Thus, there is also a nonadiabatic ICT process associated with the radiationless decay from S_2 (ICT) to S_1 (LE).

The ideas just discussed can be summarized in the potential energy diagram shown in Figure 9.17. From the figure it is clear that the (adiabatic) state labels S_1 and S_2 and the (diabatic) VB structure labels ICT or LE are independent. The adiabatic reaction path (solid arrow), involving a transition state (i.e., avoided crossing) appears to be associated with the real crossing. The TICT coordinate (amino group torsion) is assumed to be the reaction path. The transition state on this reaction path is associated with a state change from LE to ICT. This state change can also be associated with the nonadiabatic process via the real crossing. However, the real crossing

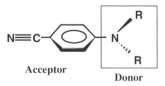

Figure 9.16. The TICT coordinate in aminobenzonitrile (ABN) compounds.

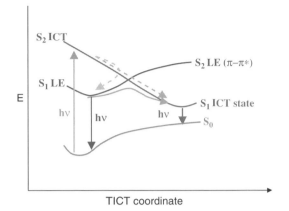

TICT coordinate

Figure 9.17. Adiabatic and nonadiabatic reaction profiles for the TICT process.

and the nature of the branching space and its relationship to the adiabatic reaction path can be understood only by moving to higher dimensions and by consideration of the relationship between the branching space coordinates and the twisting coordinate. In other words, we need to consider a potential energy surface model of the form suggested in Figure 9.9 or 9.10.

Let us begin with a discussion of the VB states involved in Figure 9.17. In the ABN problem, there are four VB structures that are relevant and these are shown in Figure 9.18. There are two "dot-dot" (covalent) configurations **I** and **II** and two zwitterionic configurations **III** and **IV**. Structures **I** and **II** are just the Kekule and

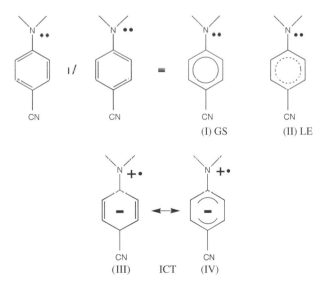

Figure 9.18. VB states involved in the ICT LE surface in the TICT process (adapted from reference 14).

anti-Kekule structures of benzene. The LE structure corresponds to the anti-Kekule electronic structure, where excitation takes place in the phenyl ring. The zwitterionic structures **III** and **IV** are the ICT states. The ICT state has a positive charge on the amino group and an extra electron on the phenyl group and we expect similarities to the benzene radical anion. Thus there will be a quinoid (**III**) and an anti-quinoid structure (**IV**). In a theoretical calculation on ABN species, an inspection of the wave function will yield the information about which resonance structure dominates. We have recently developed a method for analyzing CASSCF calculations in this way (see reference 15) and the VB analysis has been carried out using this approach.

Now we must return to a concept that we introduced earlier (see Fig. 9.4): as one passes through a conical intersection, the electronic nature (i.e., VB structure) of the state does not change. If we trace the curve labeled S_2-S_1 through the crossing from left to right in Figure 9.19, we see that the VB structure does not change. Thus in Figure 9.19, the ICT VB structure lies on the excited state at the geometry A, but lies on the lower S_1 state at geometry B, so that the states should always carry two labels: (a) S_2 S_1 to indicate the first or second excited state, and (b) the VB label (LE/ICT) to indicate the VB nature of the excited state. Referring to Figure 9.17, it should be clear that the ICT VB structure exists on both S_2 and S_1. Thus passage through a conical intersection is essentially diabatic unless the system decays before the surface crossing.

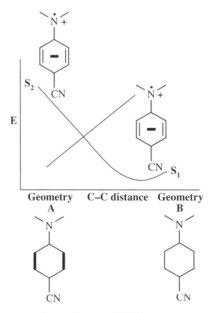

Figure 9.19. Geometry versus electronic state (S_1/S_2) as one passes through the TICT conical intersection line (adapted from reference 14). The ICT electronic structure exists on both S_2 and S_1.

We are now in a position to discuss the reaction profile outlined in Figure 9.17 in the full space of coordinates corresponding to the branching space $X_1 X_2$ of a conical intersection and the torsional coordinate X_3. This discussion will be focused on four related concepts:

1. The S_2 to S_1 radiationless decay,
2. The geometry and electronic structure of the two S_1 minima,
3. The geometry of the S_1/S_2 conical intersection together with the nature of the $X_1 X_2$ branching space, and
4. The nature of reaction path X_3 connecting the LE and ICT regions of S_1. Our objective is to rationalize all this data using the four VB structures in Figure 9.18 and to illustrate the overall surface topology according to the models or cartoons given in Figures 9.3, 9.9, and 9.10.

In Figure 9.20a, we show the geometry of the S_2/S_1 ICT/LE crossing MECI, together with the directions X_1 and X_2. The crossing occurs between the LE and ICT (**III** quinoid) VB structures. The most important point about the geometry is that the amino group is not twisted. The directions X_1 and X_2 are mainly the skeletal deformations of the phenyl ring and do not involve torsion. This is completely consistent with the fact that the LE and ICT (**III** quinoid) VB structures differ essentially only in the phenyl ring. Thus we have established that the nonadiabatic

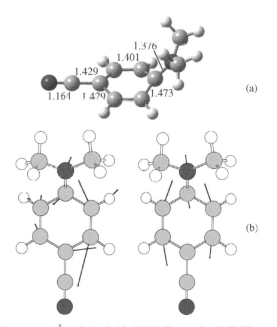

Figure 9.20. (a) Geometry (Å) of the S_2/S_1 ICT/LE crossing MECI, together with (b) the directions X_1 and X_2 for ABN systems (adapted from reference 14).

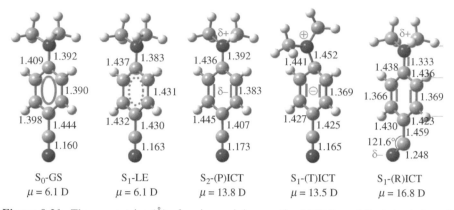

Figure 9.21. The geometries (Å) of various minima on the ABN S_2 and S_1 states (adapted from reference 14).

decay does not involve the amino group twist, since the directions X_1 and X_2 exclude this coordinate. This in turn follows from the bonding pattern of the two VB states.

Let us now consider the geometries of the various minima on the S_2 and S_1 states shown in Figure 9.21. Notice that we are careful to include both the adiabatic label S_2 or S_1 and the diabatic VB state label LE or ICT. One can see that the main difference in the "dot-dot" covalent VB structures associated with S_0-GS and S_1-LE geometries occurs in the C–C bond lengths of the phenyl ring, which are lengthened in S_1-LE because of the anti-Kekule nature of the VB structure. If we examine the CT structures, we see that we have a planar (P)ICT structure on S_2 and a twisted (T)ICT structure on S_1. (There is also a high-energy (R)ICT structure that has a bent cyano group.) The reaction pathways that connect these structures must include (i) an adiabatic reaction path that connects the S_1-LE and (T)ICT structures on S_1 along a torsional coordinate and (ii) a nonadiabatic reaction path that connects the S_2 planar (P)ICT structure with the S_1-LE structure and the S_1 (T)ICT structure via an extended conical intersection seam that lies along a torsional coordinate. We now discuss this.

The optimized geometries of the various minima on S_2 and S_1 (Fig. 9.21), together with the nature of the branching space vectors X_1 X_2 (Fig. 9.20b), suggests that the topology of the potential surface has the form shown in the model surface in Figure 9.9. Thus we have a conical intersection seam along X_3 = NR_2 torsion with the branching space X_1 X_2 spanning the phenyl group skeletal deformations shown in Figure 9.20b. We collect together all this information in Figure 9.22, corresponding to the general model given in Figure 9.9. At the left-hand side, corresponding to untwisted geometries, one can see the S_1-LE minimum and the planar S_2 (P)ICT state minima. Because the branching space excludes X_3 = NR_2 torsion, the conical intersection seam can persist as an extended seam along this coordinate. The double

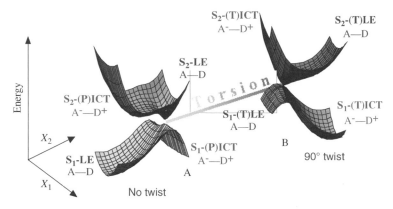

Figure 9.22. The geometries in Figure 9.21 located in the cone which changes shape along the conical intersection hyperline (adapted from reference 14). See color insert.

cone at the twisted geometry is shown on the right-hand side of the figure. Here the double cone shape changes and the twisted S_1 (T)ICT state minima develops. Thus, we have added the two branching space dimensions $X_1 X_2$ to Figure 9.17 to yield Figure 9.22. The origin of the nonadiabatic S_2 to S_1 process is now clear. The initially created state at the Franck–Condon geometry is near the S_2 (P)ICT state minimum. This state can decay to either the S_1-LE minimum or the S_1 (T)ICT minimum along the extended seam. The S_1 adiabatic process can occur following S_2 (P)ICT to S_1-LE decay via a path on S_1 involving the $X_3 = NR_2$ torsion.

9.4. THE EXTENDED CONICAL INTERSECTION SEAM: APPLICATIONS IN PHOTOCHEMISTRY

We now move to another example with an extended conical intersection seam. We shall use o-hydroxyphenyl-(1,3,5)-triazine species (Fig. 9.23) as an example because such compounds are effective photostabilizers and it is the extended conical intersection seam that is one of the contributing factors to the efficiency in such species. The enol form Figure 9.23a absorbs light and decays to the keto form (Fig. 9.23b) on the ground state. The ground state keto form is metastable and interconverts back to the enol form over a small barrier. Thus, we have light absorption followed by no net chemical change and a photostabilizing cycle. The low-lying excited states of such species are π–π^*, yet the hydrogen transfer involves the σ-electrons. Thus the reaction coordinate X_3, since it involves these σ-electrons, must be completely independent from the electronic state changes, since the latter clearly involve only the π-electrons. This is, therefore, an example where, *a priori*, the branching space coordinates must be completely different and independent from the reaction path, and one knows from the outset that the surfaces must involve the extended seam

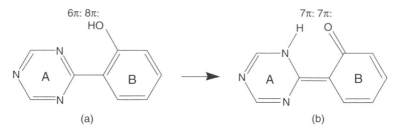

Figure 9.23. Enol (a) - keto (b) tautomerism in o-hydroxyphenyl-(1,3,5)-triazine, indicating the number of π-electrons in the ground state for each ring A and B (adapted from reference 16).

topology shown in Figure 9.10. It is possible to classify the ground state and the two types of π–π* excited state according to the number of π-electrons associated with the two rings A and B (as indicated in Fig. 9.23).

As we shall now discuss, the extended seam of conical intersection, which is parallel to the reaction path, allows for radiationless decay at any point along the proton transfer reaction path, even on the enol side. This topology explains the experimental observation that the proton transfer is in competition with a temperature-dependent deactivation process. For photostability, this paradigm is ideal, since the seam has everywhere a sloped topology and the ground state enol form is regenerated on an ultrafast timescale. These mechanistic features are independent of the ordering of the locally excited versus charge-transfer configurations. The notion of a seam of intersection explains the high photostability of the o-hydroxyphenyl-triazine class of photostabilizers in particular, but more generally highlights an important photochemical feature that should be considered when designing a photostabilizer.

We begin with a VB analysis of ground and excited states at the enol and keto geometries. In Figure 9.24, we show a valence bond correlation diagram for the lowest excited states along a proton transfer coordinate. This correlation diagram was elucidated by analysis of the excited states in recent theoretical calculations (reference 16). However the main ideas can be seen from more elementary considerations. At a given geometry (keto or enol) the locally excited states preserve the number of π-electrons in each ring, while the CT states change this population. (Notice that LE and CT, as we use them in this context, are relative to the ground state electronic configuration at a given geometry.) Thus, the state with the configuration 6π–8π is locally excited at the enol geometry but formally CT at the keto geometry because of the migration of the proton. To avoid ambiguity, we will be classifying the excited states according to the number of π-electrons in each ring. Only the ordering of the various states has to be determined from theoretical computations. However, it will be the vertical excitation to the CT state that will be observed experimentally, because of its larger oscillator strength.

If we look at the correlation (Figure 9.24) between the enol ground state electronic configuration and the keto ground state configuration, we observe a change

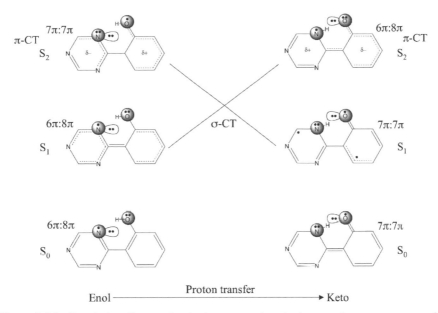

Figure 9.24. Correlation diagram for the lowest π–π* excited states along a proton transfer coordinate (adapted from reference 16).

in the number of ring A and B electrons. One might expect an activation barrier due to the change in electronic configuration. In fact, computations (reference 16) sug gest that the barrier to back formation of the enol form from the keto form is small (4 kcal/mol). Thus if the keto form is generated photochemically, the enol form will be rapidly regenerated thermally over a small barrier. It only remains to discuss the photochemical proton transfer to generate a ground state keto form.

The excited state proton transfer can be understood using Figure 9.25, where we have labeled the various excited state potential energy surfaces consistent with Figure 9.24. In Figure 9.25, we show potential energy surfaces in a cartoon involv ing the proton transfer coordinate and one coordinate from the branching space of the extended conical intersection seam. In reference 16 we have optimized a point on the conical intersection at four geometries as indicated by the four points/stars in Figure 9.25. In each case the branching space coordinates X_1 X_2 involve the skeletal deformations of the two rings and do not include a component along the proton transfer coordinate. Thus, in this case, the branching space is rigorously distinct from the reaction coordinate corresponding to proton transfer. Of course, along an adiabatic reaction path from the enol S_1 6π–8π minimum to the keto S_1 7π–7π minimum, the real crossing will become avoided and generates a transition state.

However, the initial excitation is to the enol S_2 7π–7π state. It is clear from Figure 9.25 that there is an extended conical intersection seam between the 7π–$7\pi/6\pi$–8π excited states and the ground state. Thus, the system can decay

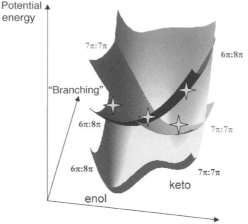

Figure 9.25. A cartoon showing the lowest π–π^* excited states along a proton transfer coordinate and a skeletal deformation coordinate from the branching space (adapted from Figure 4 of reference 16).

efficiently after photoexcitation at any point along the seam. Since the ground state barrier between the keto and enol form is negligible, the regeneration of the ground state enol form, following photoexcitation, must be exceedingly efficient. Thus, the presence of a conical intersection seam along the reaction path, where the branching space coordinates are rigorously orthogonal to the reaction path, can be identified as a desirable design feature for efficient photostabilizers.

9.5. THERMAL NONADIABATIC CHEMISTRY INVOLVING CONICAL INTERSECTIONS (ELECTRON TRANSFER)

We now turn to an example of nonadiabatic chemistry where the nonadiabatic process starts on the ground state, and is followed by an excursion upward onto the excited state: electron transfer (see references 2–5).

Let us begin by looking at a one-dimensional reaction coordinate for a photochemical process and an electron transfer process, as illustrated in Figure 9.26. If we take a "slice" near a ground-state transition state for a photochemical problem, we expect an avoided crossing (see Fig. 9.8 or 9.14 and the associated discussion). Of course, when we add another coordinate and look at the picture in three dimensions, the avoided crossing becomes a real crossing at a conical intersection (see, for example, Fig. 9.8). On the right-hand side of Figure 9.26, one has the analogous picture for electron transfer. The transition state here is also an avoided crossing, so there should be a real crossing when one adds an additional coordinate, just like a photochemical problem. If such a "real" conical intersection occurs, then, of course, there is the possibility of going up through the intersection, taking an excursion on the excited

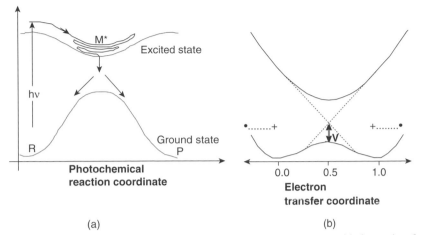

(a)   (b)

Figure 9.26. Energy profile along (a) the reaction coordinate at an avoided crossing for a photochemical reaction and (b) an electron transfer process.

state, before emerging again on the ground state. This may seem unlikely, since it appears to be obviously easier to go around the conical intersection. However, this relies on the coupling between the ground and excited states being large. As we will show, there are situations where the branching space becomes one dimensional (i.e., X_1 is well defined and has finite length but X_2 is not well defined because it has zero length: see Fig. 9.14 and the associated discussion).

In Figure 9.27, we show a two-dimensional picture with a cutting plane that corresponds to the one-dimensional cross-section shown in Figure 9.26b. This figure illustrates the existence of a conical intersection, with an avoided crossing in an adjacent cutting plane. We have left out the upper sheet of the double-cone for clarity.

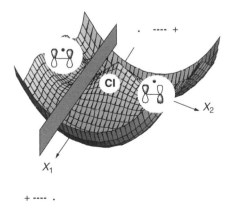

Figure 9.27. A cartoon of a conical intersection, with an avoided crossing in an adjacent cutting plane that corresponds to the one-dimensional cross-section shown in Figure 9.26b.

It is not obvious that electron transfer in a radical cation can be associated with a conical intersection in this way. So it will be helpful to develop this figure a little bit further with some additional one-dimensional cuts. We imagine that the charge transfer is taking place along a coordinate X_1. The reactant has an unpaired electron at one end and the product has an unpaired electron at the other end. Thus, the positive charge and the unpaired electron are localized in reactant and product, but in opposite regions. It should also be clear that at the transition state, both the charge and the electron are delocalized. Accordingly, if the electron transfer is taking place between two localized p-orbitals, then the transition state, and obviously the CI, have delocalized orbitals, as shown in Figure 9.27.

In Figure 9.28 we show additional one-dimensional cuts through Figure 9.27. The plane X_1 with $X_2 = 0$ is the reaction coordinate (Fig. 9.28b) with the charge and the electron localized, and the plane $X_1 = 0$, X_2 is orthogonal to the reaction coordinate (Fig. 9.28a), where both the electron and the charge are delocalized. Thus, the charge transfer is assumed to take place between two localized orbitals, and the coordinate X_1 is the usual reaction coordinate in electron transfer theory. Similarly, the ground and excited states have the form usually used in electron transfer theory.

The coordinate X_2 may at first sight be unfamiliar in electron transfer problems. If we look at the left-hand side of Figure 9.28, on the lower energy profile one sees the hole-pair associated with an in phase combination of the two p-orbitals, that is, a π-orbital. Obviously, the excited state involves the corresponding out of phase combination π^*. If one then passes through the surface crossing from left to right, then the lowest energy surface, to the right of the surface crossing, is associated diabatically with the π^*-orbital. Looking back at Figure 9.27, we see that there are two possible reaction paths going around opposite sides of the cone. Thus, in general, in electron transfer theory, if the excited state is low enough in energy, there should be two reaction pathways around and avoiding a conical intersection.

In Figure 9.29, we show three surface topologies that we have been able to document in electron transfer processes in radical cations. Two views are presented in each case. At the top, we show a cross-section along the coordinate orthogonal to the reaction path (analogous to that shown for X_2 in Fig. 9.28a). Then, at the bottom, we

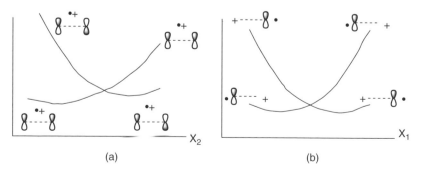

(a)   (b)

Figure 9.28. One-dimensional cuts through Figure 9.27: (a) X_2 (b) X_1 is the reaction coordinate with the charge and the electron localized.

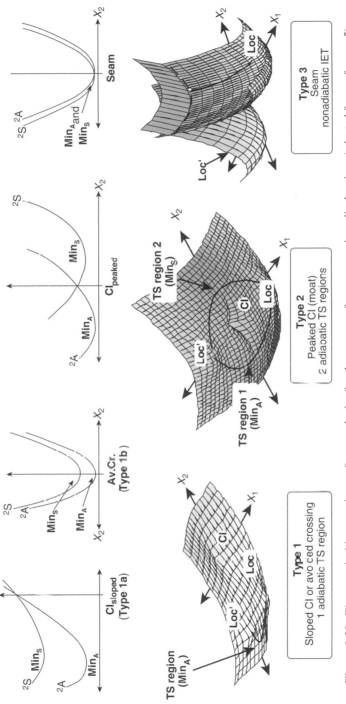

Figure 9.29. Three corical intersection surface topologies for electron transfer processes in radical cations (adapted from reference 5).

show three corresponding perspective pictures of the lower part of the conical intersection surface. We have already discussed type 2, and it remains to discuss the other two possibilities. In type 1, there is only one ground state adiabatic reaction path. Two possible variants are shown in the top line of figure. Clearly, in this case, and in type 2, nonadiabatic effects (involving an excursion on an excited state surface) will not be important, because the conical intersection can be avoided by following a lower energy path. Type 3 is the most interesting possibility (see Fig. 9.14a), because in this case, there is no ground state adiabatic path for electron transfer: the product can only be reached by an excursion on the excited state. However, the most probable trajectory will simply move to the excited state through the crossing seam and revert on a reverse trajectory back to the starting point. It should be clear that type 3 is a special case of type 2 where the two minima occur at virtually the same geometry.

The conical intersection seam, which we see for type 3, appears to be similar to the extended conical intersection seam that we have been discussing previously for photochemistry, except that the dimensionality is lower. This type of topology occurs when the vector X_2 has zero length (see Fig. 9.14a). At first sight this is confusing. However, another way of describing this situation is to say that the branching space is spanned by one vector X_1 rather than by two. In this interpretation, the vector shown as X_2, in fact, lies in the intersection space (i.e., it should be labeled as X_3). When one plots the energy in the space containing one vector from the branching space and one vector from the intersection space one sees a seam. The crucial difference in the present case is that since there is only one branching space direction X_1, there is no way to avoid the conical intersection. The magnitude of the vector X_2 is given in Eq. 9.3h. Thus, the interaction matrix element in the case corresponding to type 3 is always zero and the two states do not interact. Thus, the difference between type 2 and type 3 has its origins in the magnitude of the coupling matrix element given in Eq. 9.3h.

We have established an important principle in electron transfer theory that is not present in conventional one-dimensional models. The reaction coordinate is always localizing and corresponds to coordinate X_1. The coordinate X_2 corresponds to the direction in which the matrix element between ground and excited states is "switched on." If this coordinate has zero length then the branching space becomes one dimensional and an adiabatic reaction path does not exist. We now consider two examples.

In reference 3, we have discussed electron transfer in bis(hydrazine) radical cations. We find type 2 reaction pathways and a type 3 reaction pathway indicated as "chemical 1, 2" or "nonadiabatic" in Figure 9.30. The potential energy surface for the type 2 reactions is illustrated schematically in Figure 9.31. In each case, one has an unpaired electron on the phenyl ring corresponding to a benzene radical anion with the quinoid and antiquinoid structures (shown in the "chemical 1, 2 structures" in Fig. 9.30). The direct path, which does not involve the phenyl ring, is type 3. However, as can be seen from Figure 9.30, this nonadiabatic pathway is much higher in energy. Thus, we are able to document the existence of both type 2 and type 3 electron transfer pathways in the same simple chemical electron transfer reaction.

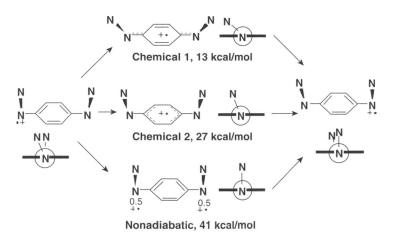

Figure 9.30. Electron transfer pathways in bis(hydrazine) radical cations (adapted from reference 3).

Our recent study of the dynamics for electron transfer in the bis(methylene) adamantyl radical cation (reference 5) provides an interesting example of type 3 electron transfer (see Fig. 9.32). Two ethylene groups are held in orthogonal orientation by the adamantane cage. Electron transfer takes place directly (i.e., there is no intermediate geometry with an electron on the adamantane species). The crossing occurs at a geometry where the formal double bonds at each end of the molecule are symmetric, with half an electron each. The reaction coordinate starts with a double bond at one end with two π-electrons, with the odd electron at the other end. The direction X_2 is vanishingly small in actual calculations. Thus, X_2 actually belongs to the intersection space (X_3) and there is almost no adiabatic pathway. Although, the direction X_2 is not

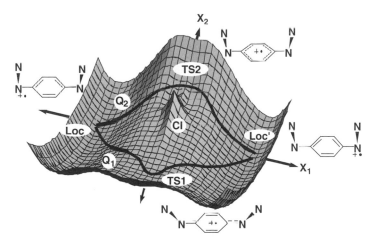

Figure 9.31. Type 2 electron transfer pathways in bis(hydrazine) radical cations.

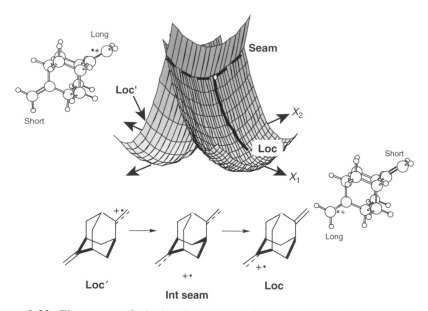

Figure 9.32. Electron transfer in the adamantane radical cation (bis(methylene) adamantyl radical cation).

very well defined, dynamics computations showed some trajectories that followed an adiabatic path. These involved anti-symmetric deformations of the adamantane cage that increase the coupling corresponding to Eq. 9.3h.

Thus, we have shown that nonadiabatic effects can be important in problems such as electron transfer where excited and ground states may be close together. We believe that future investigations in this area will be fruitful.

9.6. CONCLUSIONS AND OUTLOOK

We hope that the preceding discussions have developed the concept of a conical intersection as being as "real" as many other reactive intermediates. The major difference compared with other types of reactive intermediate is that a conical intersection is really a family of structures, rather than an individual structure. However, the molecular structures corresponding to conical intersections are completely amenable to computation, even if their existence can only be inferred from experimental information. They have a well-defined geometry. Like the transition state, the crucial directions governing dynamics can be determined (X_1 and X_2) even if there are now two such directions rather than one. As for a transition structure, the nature of optimized geometries on the conical intersection hyperline can be determined from second derivative analysis.

Work over the last 15 years has demonstrated that photochemical reactions certainly involve conical intersections as part of the mechanism for rapid radiationless decay. In the future, dynamics calculations will reveal the importance of the extended seam. Calculations have already been performed on biological chromophores, and the effect of the environment (solvent and the field of the protein) on the topology of the conical intersection is already being studied. The next frontier will undoubtedly be the study of the role of conical intersections in ground-state chemistry such as electron transfer and radical reactions where low-lying excited states can play a role.

ACKNOWLEDGEMENTS

Our work on photochemistry was started in collaboration with Massimo Olivucci, and the study of electron transfer problems was initiated with Lluis Blancafort. However, our study of conical intersections has involved many other collaborators, postdocs, and students. Only some of their work has been cited explicitly here. However, articles such as this, which collect many ideas and thoughts, are possible only through the dedicated hard work of our co-workers over many years who helped to develop them.

SUGGESTED READING

1. A. H. Zewail, *Science* **1988**, *242*, 1645.

2. A. H. Zewail, "Femtochemistry: Atomic-scale dynamics of the chemical bond using ultrafast lasers - (Nobel lecture)," *Angew. Chem. Int. Ed.* **2000**, *39*, 2587.

3. F. Bernardi, M. A. Robb, and M. Olivucci, *Chem. Soc. Rev.* **1996**, *25*, 321.

4. M. A. Robb, M. Garavelli, M. Olivucci, and F. Bernardi, A Computational Strategy for Organic Photochemistry, in *Reviews in Computational Chemistry*, (Eds. K. B. Lipkowitz and D. B. Boyd), Wiley-VCH Publishers, New York, **2000**, *Vol. 15*, p. 87.

5. W. Domcke, D. R. Yarkony, H. Koppel Ed. *Conical intersections: Electronic structure dynamics and spectroscopy*, World Scientific Singapore **2005**.

6. *Computational Photochemistry* (Ed. M. Olivucci), Elsevier **2006**.

7. L. Blancafort, F. Ogliaro, M. Olivucci, M. A. Robb, M. J. Bearpark, A. Sinicropi, Computational Investigation of Photochemical Reaction Mechanisms, *Computational Methods in Photochemistry* (Ed. A. Kutateladze), CRC Press, Boca Raton **2005**, p. 31.

8. D. R. Yarkony, *Acct. Chem. Res.* **1998**, *31*, 511.

9. N. J. Turro, *Modern Molecular Photochemistry*, Benjamin, Menlo Park **1978**.

10. A. Gilbert and J. Baggott, *Essentials of Molecular Photochemistry*, Blackwell Scientific, Oxford, **1991**.

11. J. Michl and V. Bonacic-Koutecky, *Electronic Aspects of Organic Photochemistry*, Wiley, New York, **1990**.

12. M. Klessinger and J. Michl, *Excited States and Photochemistry of Organic Molecules*, VCH Publishers, New York, **1994**.

13. L. Salem, *Electrons in Chemical Reactions: First Principles*, Wiley, New York, **1982**.

14. R. A. Marcus, "Electron-transfer reactions in chemistry - theory and experiment (Nobel Lecture)," *Angew. Chem. Int. Ed.* **1993**, *32*, 1111.

REFERENCES

1. M. Reguero, F. Bernardi, H. Jones, M. Olivucci, I. N. Ragazos, and M. A. Robb, *J. Amer. Chem. Soc.* **1993**, *115*, 2073.

2. L. Blancafort, W. Adam, D. González, M. Olivucci, T. Vreven, and M. A. Robb, *J. Amer. Chem. Soc.* **1999**, *121*, 10583.

3. E. Fernández, L. Blancafort, M. Olivucci, and M. A. Robb, *J. Amer. Chem. Soc.* **2000**, *122*, 7528.

4. L. Blancafort, F. Jolibois, M. Olivucci, M. A. Robb, *J. Amer. Chem. Soc.* **2001**, *123*, 722.

5. L. Blancafort, P. Hunt, and M. A. Robb, *J. Amer. Chem. Soc.* **2005**, *127*, 3391.

6. S. Wilsey, F. Bernardi, M. Olivucci, M. A. Robb, S. Murphy, and W. Adam, *J. Phys. Chem. A.* **1999**, *103*, 1669.

7. F. Bernardi, S. De, M. Olivucci, and M. A. Robb, *J. Amer. Chem. Soc.* **1990**, *112*, 1737.

8. F. Bernardi, M. Olivucci, M. A. Robb, and G. Tonachini, *J. Amer. Chem. Soc.* **1992**, *114*, 5805.

9. M. J. Paterson, M. J. Bearpark, M. A. Robb, L. Blancafort, and G. A. Worth, *Phys. Chem. Chem. Phys.* **2005**, *7*, 2100.

10. M. J. Paterson, M. J. Bearpark, M. A. Robb, and L. Blancafort, *J. Chem. Phys.* **2004**, *121*, 11562.

11. I. N. Ragazos, M. A. Robb, F. Bernardi, and M. Olivucci, *Chem. Phys. Lett.* **1992**, *197*, 217.

12. D. R. Yarkony, *J. Phys. Chem.* **1993**, *97*, 4407.

13. M. J. Bearpark, M. A. Robb, and H. B. Schlegel, *Chem. Phys. Lett.* **1994**, *223*, 269.

14. I. Gómez, M. Reguero, M. Boggio-Pasqua, and M. A. Robb, *J. Amer. Chem. Soc.* **2005**, *127*, 7119.

15. L. Blancafort, P. Celani, M. J. Bearpark, and M. A. Robb, *Theo. Chem. Acc.* **2003**, *110*, 92.

16. M. J. Paterson, M. A. Robb, L. Blancafort, and A. D. DeBellis, *J. Phys. Chem. A* **2005**, *109*, 7527.

CHAPTER 10

Quantum Mechanical Tunneling in Organic Reactive Intermediates

ROBERT S. SHERIDAN

Department of Chemistry, University of Nevada, Reno, Nevada

Reviews of Reactive Intermediate Chemistry. Edited by Matthew S. Platz, Robert A. Moss, Maitland Jones, Jr.
Copyright © 2007 John Wiley & Sons, Inc.

10.1. INTRODUCTION

Transition state theory has long served the organic community with great success. The concept of two-dimensional reaction coordinate diagrams depicting minimum energy pathways is deeply ingrained in the way that we view reaction kinetics. It is useful to visualize reactions, in simple-minded terms, as occurring through thermal activation of molecules. Those molecules that are endowed with sufficient internal energy can overcome the transition state barriers separating reactants from products. Molecules with energies less than that of the transition state cannot pass the barriers. Most differences in the relative rates of related reactions can be nicely rationalized by considering perturbations on transition state and/or reactant energies caused by structural and/or electronic influences. The reaction coordinate diagram also provides a handy framework for conceptualizing the temperature dependence of reaction rates. Higher temperatures correspond to greater percentages of reactants with enough energy to traverse the transition state, and hence to higher reaction rates. As temperatures are lowered, the statistical number of molecules which possess the necessary energy decreases, and reaction slows and eventually ceases.

It is becoming clear, however, that many reactions of highly reactive molecules do not quite follow this simple picture. In some instances, reactions do not slow as much as expected when temperature is lowered, or proceed at rates that are impossibly rapid for the existing reaction barriers. Certain reactions are found to continue at measurable rates even at temperatures near absolute zero. Moreover, the structure-reactivity concepts that have proved so valuable for understanding and predicting the reactions of "ordinary" organic molecules are sometimes found to be incomplete when applied to highly reactive intermediates. Factors such as the extent of geometry change that is required, or the masses of the nuclei that move during a reaction, can assume disproportionate roles.

Many of these nonclassical reaction effects can be traced to the quantum nature of the particles involved. On the atomic level, molecules need not necessarily surmount an activation barrier to transform to products. The quantum uncertainties in the location of the nuclei make it possible, in certain cases with sufficiently "narrow" barriers, for reactants to penetrate the barriers rather than pass over them. This phenomenon has been termed quantum mechanical tunneling (QMT).[1,2] Many organic reactive intermediates have very small activation barriers protecting them from reaction, and require minimal geometry changes in their transformation to products. Thus, it is not surprising that a wide variety of reactive intermediates have shown dramatic effects of QMT in their chemistry. In this chapter, we will first provide a brief theoretical background of tunneling in organic reactions, with particular emphasis on understanding the experimentally observable ramifications of quantum effects. Then, we will consider in detail a number of case studies that illustrate the application of these principles in specific reactive intermediate reactions.

It is important to recognize at the outset that it has long been recognized, even in the early days of the development of quantum mechanics, that tunneling plays a role in a very broad range of chemical and spectroscopic properties of molecules. Our narrow focus here on organic reactive intermediates compels us to forgo discussion of such fascinating and important areas as, for example, enzymology,[3] interstellar chemistry,[4] electron transfer chemistry,[5] spectroscopy,[6] catalysis,[7] and photochemistry,[8] where the critical involvement of QMT is still being unraveled. It must also be noted that tunneling corrections have been found to be important in precise modeling of isotope effects in various "ordinary" organic reactions,[9] even involving heavy atoms, but these examples are also outside the scope of this chapter.

10.2. THEORETICAL BACKGROUND

It is worthwhile to first review several elementary concepts of reaction rates and transition state theory, since deviations from such "classical" behavior often signal tunneling in reactions.[10] For a simple unimolecular reaction, A→B, the rate of decrease of reactant concentration (equal to rate of product formation) can be described by the first-order rate equation (Eq. 10.1).

$$-d[A]/dt = k[A] \quad \text{or} \quad \ln([A]/[A_0]) = -kt \tag{10.1}$$

The rate constant k is generally determined from the slope of a plot of the integrated form of the rate equation where $[A_0]$ is the starting concentration of the reactant, and t is the time.

Experimentally, rate constants of most organic reactions depend on temperature exponentially according to the Arrhenius equation (Eq. 10.2),

$$k = Ae^{(-E_a/RT)} \tag{10.2}$$

where A, the so-called preexponential term, is a constant in units of s^{-1}, E_a is the Arrhenius activation energy, R is the gas constant, and T is absolute temperature. Commonly, plotting ln k versus 1/T affords a straight line from which values for A and E_a can be derived.

Various statistical treatments of reaction kinetics provide a physical picture for the underlying molecular basis for Arrhenius temperature dependence.[10] One of the most common approaches is Eyring transition state theory, which postulates a thermal equilibrium between reactants and the transition state. Applying statistical mechanical methods to this equilibrium and to the inherent rate of activated molecules transiting the barrier leads to the Eyring equation (Eq. 10.3), where k is the Boltzmann constant, h is the Planck's constant, and $\Delta G^{\ddagger}$ is the relative free energy of the transition state [note: Eq. (10.3) ignores a transmission factor, which is normally ≈ 1, in the preexponential term].

$$k = (kT/h)e^{(-\Delta G^{\ddagger}/RT)} \tag{10.3}$$

$$A = (ekT/h)e^{(\Delta S^{\ddagger}/R)} \tag{10.4}$$

$$E_a = \Delta H^{\ddagger} + RT \tag{10.5}$$

It is sometimes informative to separate $\Delta G^{\ddagger}$ into hypothetical enthalpic and entropic terms, and then the Arrhenius factors may be related to the transition state activation parameters by Eqs 10.4 and 10.5. Thus, the Arrhenius activation energy can be approximately related to the potential energy of a transition state, and the preexponential A value includes probability factors.

Transition state theory, as embodied in Eq. 10.3, or implicitly in Arrhenius theory, is inherently "semiclassical." Quantum mechanics plays a role only in consideration of the quantized nature of molecular vibrations, etc., in a statistical fashion. But, a critical assumption is that only those molecules with energies exceeding that of the transition state barrier may undergo reaction. In reality, however, the quantum nature of the nuclei themselves permits reaction by some fraction of molecules possessing less than the energy required to surmount the barrier. This phenomenon forms the basis for QMT.[1,2]

Tunneling may be conceptualized in several ways. For example, the wave-particle duality of quantum mechanics indicates that particles of mass m and velocity v have an associated de Broglie wavelength given in Eq. 10.6, where h is Planck's constant.[11]

$$\lambda = h/mv \qquad (10.6)$$

If the wavelengths of the reacting nuclei become comparable to barrier widths, that is, the distance nuclei must move to go from reactant "well" to product "well," then there is some probability that the nuclear wave functions extend to the other side of the barrier. Thus, the quantum nature of the nuclei allows the possibility that molecules tunnel through, rather than pass over, a barrier.

It is useful to reformulate Eq. 10.6 in kinetic energy (KE) terms, giving Eq. 10.7.

$$\lambda = h/(2\,m[\text{KE}])^{1/2} \qquad (10.7)$$

For a proton in thermal equilibrium at 295 K, assuming mean translational kinetic energy of $3kT/2$, the de Broglie wavelength is ca. 1.5 Å. Thus, the wavelength is of the order of molecular dimensions. Equation 10.7 illustrates that the nuclear wavelengths are inversely proportional to $(\text{mass})^{1/2}$ of different nuclei, suggesting that tunneling of heavier atoms should decrease significantly compared with H (e.g., $\lambda \approx 0.4$ Å for carbon at the same temperature and kinetic energy).

Alternatively, it can be shown that the Heisenberg uncertainty in a particle's position is proportional to its de Broglie wavelength by Eq. 10.8. This underscores the point that when reactant nuclear de Broglie wavelengths become comparable with the width of an energy barrier, there is a significant probability of finding nuclei on the product side of the barrier. Again, the uncertainty in the position of a particle depends on the square root of its mass.[2,11]

$$|\Delta x| \approx \lambda/4\pi \qquad (10.8)$$

Finally, it is a well-known result of quantum mechanics[11] that the wavefunctions of harmonic oscillators extend outside of the bounds dictated by classical energy barriers, as shown schematically in Figure 10.1. Thus, in situations with narrow barriers it can

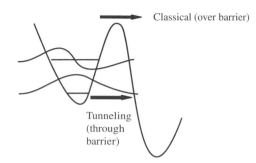

Figure 10.1. Exaggerated reaction coordinate diagram showing approximate wavefunctions for the zero-point and first vibrational levels of the reactant.

be seen that molecular vibrations can lead to finite penetration of the barriers. "Inside" the barrier, the total energy of the system is less than the potential energy, so the kinetic energy is formally negative during the tunneling process. The higher the barrier energy, the more negative the kinetic energy must become, and hence the lower probability of tunneling. On the contrary, note that the barriers will effectively be narrower for higher vibrational wave functions, translating to higher tunneling probabilities.

A number of theoretical models have been developed to treat tunneling through differently shaped energy barriers.[1,2] One of the most commonly used and mathematically convenient models was developed by Bell[1,2,12] for tunneling through a hypothetical parabolic energy barrier. Although this model only very crudely represents a physically realistic potential surface, it is useful for considering the relative influence of various molecular factors on QMT rates. Here, the Arrhenius rate expression is modified to include a tunneling correction Q, as shown in Eq. 10.9. Q is defined in Eq. 10.10, where m is the mass of the "tunneling particle," $2a$ is the base width of the truncated parabolic barrier, $\alpha = E/kT$, and $\beta = 2\pi^2 a(2mE)^{1/2}/h$.

$$k = QAe^{(-E/RT)} \tag{10.9}$$

$$Q = e^{\alpha}(\beta e^{-\alpha} - \alpha e^{-\beta})/(\beta - \alpha) \tag{10.10}$$

Caldin[2] has pointed out that in systems of chemical interest, $\beta > \alpha$, and Eq. 10.10 can be reduced approximately to Eqs 10.11 and 10.12. The effect of Q is to increase the rate of reaction at a given temperature relative to that expected based on A and E_a. Therefore, Eq. 10.11 indicates that the greater the α/β, the greater will be the apparent effect of tunneling "acceleration."

$$Q = 1/(1 - \alpha/\beta) \tag{10.11}$$

$$\alpha/\beta = hE^{1/2}/2\pi^2 akT(2m)^{1/2} \tag{10.12}$$

Equations 10.11 and 10.12 confirm our qualitative predictions. The degree of tunneling depends inversely both on the square root of the mass and on the barrier width. Moreover, it turns out that Q increases as temperature is lowered. The importance of all these factors will be seen in the various examples described later in this chapter where QMT plays a role in reactive intermediate chemistry. But first, we will consider common kinetic anomalies that signal QMT in general.

10.2.1. Experimental Signatures of Tunneling

10.2.1.1. Nonlinear Arrhenius Plots   For most organic reactions, plots of ln k versus $1/T$ are linear, and afford E_a and A values in accord with the Arrhenius equation.[11] However, for systems where QMT is involved, rate constants fall off less steeply than expected as temperatures are lowered, which often leads to upwardly curved Arrhenius plots as illustrated in Figure 10.2.[1,2]

In contrast to classical overbarrier reactions, QMT can occur from the lowest vibrational quantum levels without thermal activation. Under these circumstances at the lowest temperatures, the degree of tunneling, and hence the reaction rate, is independent of temperature. At some point as the temperature is raised, higher vibrational levels become populated. As illustrated in Figure 10.1, the effective barrier is narrower for excited vibrational levels, and hence tunneling becomes more facile, leading to an increase in rate. Finally, as temperatures are raised further, classical reaction begins to compete, and usually dominates at room temperature (but, not always).

It is important to dispel two common misconceptions regarding Arrhenius plots and tunneling. First, it is sometimes assumed that the onset of temperature dependence, as in Figure 10.2, signals emerging contributions from classical reaction. As noted above, tunneling is also a thermally activated process, and population of excited vibrational states dominates curvature at low temperatures. Second, in a practical sense, because of the temperature dependencies of both classical and tunneling components of reactions, Arrhenius plots may only begin to show obvious curvature at very low temperatures. Linear Arrhenius plots are, therefore, no guarantee of the absence of tunneling in a reaction. But, on the contrary, temperature independence at very low temperatures is a strong indicator of QMT.

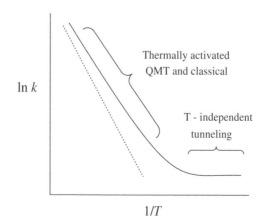

Figure 10.2. Schematic Arrhenius plots with (solid line) and without (dotted line) tunneling.

In practice, fittings of nonlinear Arrhenius plots to kinetic models based on differently shaped potential energy barriers have been used to estimate the barrier heights and widths for tunneling reactions. Commonly, the Bell truncated parabolic model (Eq. 10.10) provides a good starting point. In cases where good rate data over a broad temperature range is available, numerical fitting of more sophisticated and realistic energy barriers, such as the Eckart-type, has yielded more accurate estimates of tunneling parameters.[1,2]

10.2.1.2. Reaction Rates Faster than "Expected"

Modern calculational methods have made it convenient and routine to estimate transition state barriers very accurately.[13] It is easy to predict a reasonable approximate rate for a classical organic reaction. However, QMT permits reactions to occur at rates that can be considerably higher than predicted by calculation or by extrapolation from rates measured at room temperature with rapid spectroscopic methods.

The effects of QMT at cryogenic temperatures can be quite spectacular. At extremely low temperatures, even very small energy barriers can be prohibitive for classical overbarrier reactions. For example, if $E_a = 1$ kcal/mol and A has a conventional value of 10^{12} s^{-1} for a unimolecular reaction of a molecule, Arrhenius theory would predict $k = 2 \times 10^{-10}$ s^{-1}, or a half-life of 114 years at 10 K. But, many tunneling reactions of reactive intermediates have been observed to occur at measurable rates at this and lower temperatures, even when energy barriers are considerably higher. Reactive intermediates can, thus, still be quite elusive at extremely low temperatures if protected only by small and narrow energy barriers.

10.2.1.3. Anomalous Kinetic Isotope Effects

Because of the (mass)$^{1/2}$ dependence of the rates of tunneling, isotope effects on rates of reaction can be more dramatic than predicted, based on zero-point vibrational energies. For example, transition state theory predicts a maximum k_H/k_D ratio of ca. 7 at 25°C for a linear H-transfer reaction, taking into account only C–H versus C–D stretching vibrations.[1,2,10] This "classical" ratio may rarely increase to as high as 17 if bending vibrations also become important. However, isotope effects in tunneling-assisted transfers can greatly exceed these values. These anomalously large isotope effects can become especially pronounced at low temperatures, where tunneling dominates.

The greater facility of tunneling in H-transfer reactions versus D-transfer comes from two effects. First, the longer de Broglie wavelengths of the lighter isotope give greater barrier penetration. But second, C–D bonds have lower zero-point vibrational energies, sitting lower in the energy wells, and hence experience effectively broader barriers. These effects can manifest themselves in apparent $E_a(D) - E_a(H)$ differences that are larger than the theoretical value of ca. 1.2 kcal/mol.[10]

Transition state statistical treatments predict that for classical reactions, A_H and A_D should be comparable in magnitude, with $A_H/A_D > 0.7$ approximately.[14] Ratios of A values smaller than this are also indicative of QMT. The relatively smaller than expected measured A_H values can be envisioned to arise from the greater deviation of H-transfers from linear Arrhenius plots at lower temperatures; extrapolation to $1/T = 0$ then gives a lower intercept corresponding to the apparent A value.[2]

10.2.2. Matrix Effects

A final note must be made about a common problem that has plagued many kinetic treatments of reactive intermediate chemistry at low temperatures. Most observations of QMT in reactive intermediates have been in solid matrices at cryogenic temperatures.[15] Routinely, reactive intermediates are prepared for spectroscopy by photolyses of precursors imbedded in glassy organic or noble gas (or N_2) solids. The low temperatures and inert surroundings generally inhibit inter- and intramolecular reactions sufficiently to allow spectroscopic measurements on conventional and convenient timescales. It is under such conditions, where overbarrier reactions are diminished, that QMT effects become most pronounced.

Invariably, measurements of decay of reactive molecules in solid glasses are found to be nonexponential, that is, first-order plots of ln[intensity] versus time are upwardly curved, as shown in Figure 10.3.

It is generally understood that the reactive intermediates are generated in a random distribution of different microenvironments, each with its own energy barrier.[16,17] The complex decay of this dispersion of rates leads to the nonexponential kinetics. Thus, disappearance plots are dominated at early times by reaction of those species in "fast sites", which have lower energy barriers. As these sites are cleared, the distribution of rates over time becomes more reflective of sites with higher barriers. Finally, at longer times, the decay curves are dominated by the slowest sites. It is often observed that plots of ln[intensity] versus $t^{1/2}$ or $t^{1/3}$ are approximately linear. It has been shown that this is an outcome of a statistical distribution of matrix rates.

Determination of QMT effects often rests upon the temperature or isotope dependence of rates, as described above. Thus, the matrix site dispersity presents an immediate dilemma: Which matrix sites should be compared at different temperatures or for different isotopes? There have been different approaches to this problem. The most simple has been to compare the first 10–20% of the decay curves after irradiation is shut off. First-order plots are generally linear in those time frames. However,

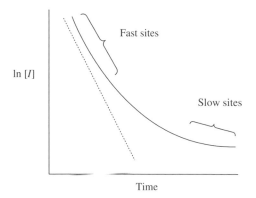

Figure 10.3. "First-order" plots of ln[Intensity] versus time, showing idealized exponential decay (dotted line), and nonexponential decay from statistical distribution of matrix sites (solid line).

the difficulty is that significant reaction of the fastest sites may have already occurred during precursor irradiation. Moreover, at different temperatures it is possible that different populations of matrix sites are being detected, namely those whose rates fall conveniently into the observation time period. Likewise, kinetics observed for different isotopes may correspond to very dissimilar matrix sites, or even to entirely different reactions. These complications are particularly troublesome with certain spectroscopic methods that detect only reactive intermediate decays and not corresponding product formation. Nevertheless, despite these weaknesses, evidence for QMT based on initial decay rates has repeatedly been validated, at least qualitatively.

It has been shown that the "rate constants" obtained from the slopes of ln[intensity] versus $t^{1/2}$ plots approximate the rates of the highest-probability matrix sites.[16] Hence, workers have utilized the temperature dependence of these "$k_{1/2}$" values, or other empirically derived "stretched exponential" time dependencies, to estimate low temperature Arrhenius plots. The validity of such methods, however, depends critically on obtaining accurate time-dependence data on the fastest matrix sites, which is increasingly difficult as temperatures are raised.

Dougherty and co-workers[17] have conducted an especially detailed treatment of matrix effects in inhomogeneous solids, and have developed several valuable tools for modeling the kinetics of such systems. The details and applications of these methods on specific systems will be described later in this chapter.

10.3. EXAMPLES OF REACTIVE INTERMEDIATE TUNNELING

10.3.1. Intramolecular Hydrogen Abstraction by Radicals

Hydrogen is the lightest element, with correspondingly the longest de Broglie wavelengths or greatest positional uncertainty. Quantum mechanical tunneling (QMT) is thus most prevalent in reactions that involve transfers of hydrogen, either H·, H⁺, or H⁻, from one center to another.[1,2] Indeed, one of the early groundbreaking reports of nonclassical behavior in organic reactive intermediate chemistry was the observation of tunneling in the low temperature rearrangements of aryl radicals involving intramolecular hydrogen transfer.

Although tunneling in H-abstractions by methyl radical in low temperature glasses had been well established,[16] one of the earliest examples where tunneling was clearly demonstrated in the rearrangement of an organic reactive intermediate was reported in 1973.[18] Phenyl radical (**1**) had been characterized by gas-phase electronic spectroscopy and by EPR spectroscopy in cryogenic matrices. However, attempts to detect this highly reactive σ radical in the solution phase were unsuccessful. Borrowing on previous successes with similarly reactive radicals, Ingold and co-workers[18] attempted to utilize steric blocking to kinetically stabilize an aryl radical. The strategy worked, and it was found that photolysis of 1-bromo-2,4,6-tri-*tert*-butylbenzene and hexamethylditin in cyclopropane solvent produced an EPR signal attributable to the very hindered aryl radical **2**.

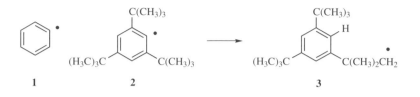

The first indication of something unusual in the reactivity of this species was that the EPR signal of **2** was found to decay via a first-order process to produce a new radical. The product was shown to be the neophyl radical **3**, whose EPR spectrum was identical with an independently prepared authentic sample. Over the temperature range −30 to −90°C, Arrhenius plots indicated an unusually low preexponential log A (s^{-1}) value of only 5.3, and a strikingly large k_H/k_D ratio of ca. 50 observed at −30°C (based on comparison of reaction rates of **2** versus the tri-d$_9$-*tert*-butyl analog) for the rearrangement.

Subsequently, very thorough work by these investigators[18,19] demonstrated conclusively that these intramolecular H-abstractions occurred via quantum mechanical tunneling. Multiple criteria were invoked to support this proposal as follows:

(1) Based on C–H versus C–D zero point vibrational differences, the authors estimated maximum classical kinetic isotope effects of 17, 53, and 260 for k_H/k_D at −30, −100, and −150°C, respectively. In contrast, ratios of 80, 1400, and 13,000 were measured experimentally at those temperatures. Based on the temperature dependence of the atom transfers, the difference in activation energies for H- versus D-abstraction was found to be significantly greater than the theoretical difference of 1.3 kcal/mol. These results clearly reflected the smaller tunneling probability of the heavier deuterium atom.

(2) Arrhenius plots of both $\ln(k_H)$ and $\ln(k_D)$ versus $1/T$ were found to be curved significantly, exhibiting decreasing dependence on temperature at lower temperatures. As described in the Introduction, such nonlinear Arrhenius plots are telling indicators of QMT; as temperature is lowered, the classical over-barrier reaction slows more significantly than does one that proceeds via tunneling. Later work[19] showed that the abstraction reaction of **2** to **3** persisted at a measurable rate at least down to 28 K in frozen media, where the decay became nearly temperature independent.

(3) In general, as temperatures are increased, theoretically to infinite temperature, rate constants for similar H- and D-abstraction reactions should become comparable. For conventional linear Arrhenius kinetics, this suggests that $A_H \approx A_D$. In the case of **2** and its deuterated analog, however, "least squares" lines through all of the data points taken in the Arrhenius plots gave $A_H = 10^{3.1}$ and $A_D = 10^{5.1}$ s^{-1}. Alternatively, tangents to the curved plots at −30°C gave $A_H = 10^{6.3}$ and $A_D = 10^{7.5}$ s^{-1}. These low A values, and the small A_H/A_D ratios, were also taken as evidence of tunneling.

The temperature dependence data[18,19] were fit to Eckart, Gaussian, and truncated parabolic barriers based on theoretical methods of LeRoy.[16] It was found that an Eckart-shaped barrier fitted the experimental data best, and predicted "uncorrected" parameters $A = 10^{11}$ s^{-1}, a classical barrier height of 14.5 kcal/mol, and a tunneling barrier "width" of 0.330 Å (corresponding approximately to half-width at half-height of the barrier). When these numbers were applied to estimate the corresponding tunneling rearrangement rates for C–D transfer in d_9-**2** at different temperatures, excellent agreement with experiment was also found. Several key points were noted from this modeling. The classical barrier and A values were considered to be reasonable for H-transfer by a phenyl-type radical. With the obtained parameters, the corresponding overbarrier H-transfer rate constant was estimated to be $k = 10^{-14.8}$ s^{-1} at −150°C. Thus, the actual rate constant of $10^{-1.2}$ s^{-1} corresponded to a "staggering" rate acceleration of $10^{13.6}$ attributable to tunneling. Finally, the estimated width of the one-dimensional Eckart barrier near its "base" was suggested to be consistent with the 1.34 Å distance over which the H must be transferred in **2**.

In subsequent experiments,[19] it was found that rearrangement of **2** persisted in solid glasses at temperatures as low as 28 K, where the reaction rates became nearly temperature independent, as expected. The fact that similar kinetics were measured in solution and in low temperature solids indicated that matrix effects were minimal, at least in these intramolecular hydrogen abstractions.

It should not be assumed that evidence for tunneling in radical hydrogen abstraction reactions rests only on low-temperature reactions by particularly reactive radicals. In fact, there is abundant evidence for QMT in a wide variety of hydrogen transfer reactions under a broad range of conditions.[1,2,16] For example, for H-transfer from tetralin **4** to benzyl radical **5** at temperatures ranging from ca. 100–170°C, workers found $E_a(D) - E_a(H) = 3.0$ kcal/mol, well exceeding the difference of 1.2 kcal/mol predicted from C–H versus C–D zero-point vibrational energies.[20] Likewise, the ratio of preexponential factors, $A_H/A_D = 0.24$ is significantly smaller than the $\geqslant 0.7$ limit indicated for classical reactions. Both observations are good indicators of tunneling in this hydrogen abstraction.

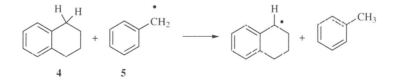

Finally, we note the informative work of Garcia–Garibay and co-workers,[9] who have extensively studied QMT in hydrogen transfer reactions in the excited triplet states of *ortho*-alkylarylketones, for example, **6** → **7**, which are electronically similar to radical rearrangements.

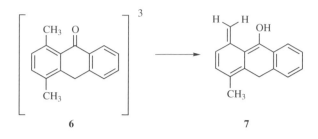

10.3.2. Tunneling in Reactions of Biradicals

There are few organic reactions as conceptually simple as coupling of two radicals to form a carbon–carbon sigma bond. Although in principle, reaction of two radicals might occur without an activation barrier, the actual kinetic details depend on the relative electronic spins on the two centers, as well as other potential stabilizing factors such as delocalization, strain in the potential products, etc. Molecules possessing two radical centers, termed biradicals or diradicals, have long been postulated as fleeting intermediates in a wide variety of thermal and photochemical organic reactions. However, until the mid-1970s, no short-chain 1,3- or 1,4-biradicals had been spectroscopically detected, although several delocalized triplet biradicals had been characterized by EPR spectroscopy. Indeed, there was a considerable discussion whether many singlet biradicals represent true chemical intermediates, rather than simply transition states.[21]

It was thus a landmark when Buchwalter and Closs[22] reported the first spectroscopic observation of a localized 1,3-biradical in 1975. Irradiation of azo compound **8**, frozen in cyclohexane at 5.5 K, was found to produce a clear EPR signal that could be assigned to triplet 1,3-cyclopentanediyl (**9**). The EPR spectrum exhibited appropriate splitting for a species with two electron spins separated by the expected distance in **9**, and could be satisfactorily modeled by semiempirical calculations. Interestingly, when irradiation was stopped, the EPR signal decayed rapidly with a half-life of approximately 30 min. The disappearance of the triplet signal was non-exponential, signifying a random distribution of differently reacting matrix sites. The rates of disappearance also varied in different matrix materials. Significantly, the rate of decay was found to be independent of temperature over the temperature range 1.3 K to ca. 20 K, and then rose rapidly between 20 and 40 K. Based primarily on this temperature dependence, it was suggested that the disappearance was due to a tunneling reaction.

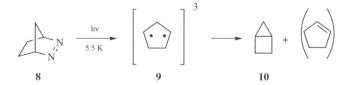

The evidence strongly suggested that the triplet was in fact the ground state of biradical **9**. The observation of an EPR signal even at 1.3 K made it unlikely that a thermally populated triplet state was being observed. Moreover, the irreversible decay of the triplet signal was many orders of magnitude slower than expected for intersystem crossing to a lower lying singlet state. Finally, solution CIDNP NMR experiments showed product polarizations that were most consistent with a ground state triplet biradical.

Various possibilities were considered for the underlying reaction of the biradical. No radical signals grew when the biradical decayed, so H-abstraction from the matrix did not appear to be occurring. Analysis of products formed from irradiations of **8** at 5.5 K showed both bicyclopentane **10** and cyclopentene, in a ratio of 30:1. Very similar ratios, ca. 25:1, were observed in solution irradiations at room temperature. It was noted that if the major tunneling reaction was H-shift to produce cyclopentene, this product should be enhanced as temperatures were lowered, in contrast to the experimental observations. Hence, it was concluded that the observed decay of the EPR spectrum of **9** was due to ring closure to give **10**.

At this time, awareness of the involvement of quantum tunneling in organic reactions, predominantly in H-transfer reactions, was just awakening among the scientific community.[2] Hence, it was a bold proposal that tunneling was occurring in a reaction dominated by heavy atom movement. It was found that the decay of the d_8-biradical was considerably slower at 5.5 K under similar conditions. The implication was that this rate deceleration was due to the requirement for out-of-plane movement of the heavier CD_2 group in the deuterated isomer. The authors were able to fit the temperature data to a Bell-type parabolic barrier with a height of 2.3 kcal/mol, width of 0.64 Å, and a "mass of tunneling particle" of 14 mass units, corresponding to a CH_2 group. Importantly, it was noted that it was impossible to fit the experimental data with a model involving movement of a single H-atom. To reproduce the observed temperature dependence, a barrier width of 2.4 Å for H-shift would have been required, much larger than physically reasonable.

The authors[22] suggested that the results were consistent with the simple potential surfaces illustrated qualitatively in Figure 10.4. Reaction of triplet **9** was proposed

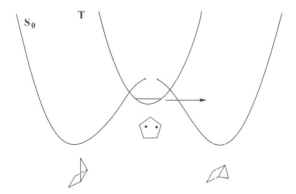

Figure 10.4. Proposed energy surfaces showing tunneling from triplet **9** to singlet **10**.

to occur via tunneling through the barrier imposed by intersection of the triplet and singlet surfaces, facilitated by out-of-plane ring bending. It was noted that the "classical" preexponential A value for this reaction would also be lowered substantially by the requirement for simultaneous intersystem crossing, and $A = 10^8$ s^{-1} was estimated for this spin-forbidden process.

Michl and co-workers reported evidence for tunneling reactions in a related, but somewhat more delocalized, 1,3-biradical. It was found[23] that irradiation of hydrocarbon **11** in low-temperature organic glasses, polyethylene, or inert matrices produced 1,3-perinaphthadiyl (**12**), which could be characterized by UV/VIS, fluorescence, IR and EPR spectroscopy. It was initially proposed that the biradical possessed a singlet ground state which was responsible for the electronic spectrum, and that the EPR spectrum arose from a thermally populated triplet state slightly higher in energy. However, these conclusions were shown later to be in error, and it was determined that, in fact, biradical **12** has a triplet ground state.[24,25] Part of the original misassignment resulted from the thermal lability of **12**, which compromised measurement of the temperature dependence of the EPR signals.[23]

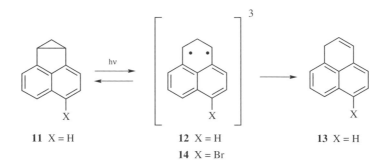

$$
\mathbf{11} \ X = H \qquad\qquad \mathbf{12} \ X = H \qquad\qquad \mathbf{13} \ X = H
$$
$$
\mathbf{14} \ X = Br
$$

When irradiations of **11** were discontinued, the various spectra of **12** were found to decay at low temperatures.[23,25,26] However, in contrast to cyclopentadiyl **9**, perinaphthadiyl **12** primarily underwent H-shift to give **13** at low temperatures. The alkene could be directly observed to grow concurrently with disappearance of **12**. The rearrangement of **12** to **13** in solid polyethylene could be followed spectroscopically over a broad range of temperature. Arrhenius plots for the hydrogen shift exhibited two distinct regions: (a) a linear segment from 120 to 160 K, with $E_a = 5.3$ kcal/mol, $A = 10^{5.1}$ s^{-1}, and (b) a temperature independent region below ca. 100 K, with $E_a = 0.0$ kcal/mol and $A = k = 10^{-6.1}$ s^{-1}. These results suggested that tunneling from triplet **12** to the ground state singlet **13** dominated the reaction. The low-temperature rate constant for 2,2-d_2-**12** was found to be $10^{-9.2}$ s^{-1}, signifying a kinetic isotope effect of 1300.

Interestingly, it was possible to probe the spin-forbidden component of the tunneling reaction with internal and external heavy atom effects.[26] Such effects are well known to enhance the rates of intersystem crossing of electronically excited triplets to ground singlet states, where the presence of heavier nuclei increases spin-orbit coupling.[27] Relative rates for the low-temperature rearrangements of **12** to **13** were found to be 1.0, 2.3, 6.3, and 23 in polyethylene, Ar, Kr, and Xe matrices, respectively. Moreover, 6-bromo-1,3-perinaphthadiyl (**14**) reacted more rapidly than the parent, with $E_a = 5.5$ kcal/mol, and $A = 10^{6.2}$ s^{-1} at temperatures above ca. 120 K, and with a temperature independent rate constant of $10^{-5.4}$ s^{-1} below ca. 100 K. It was noted that the heavy atom effect logically increased the A values for the process, where spin-orbit effects increased the probability of multiplicity change. It was suggested that the results illustrated the absence of a difference between the "photophysical" process of intersystem crossing in electronic excited states and the "chemical" rearrangement of a ground-state triplet molecule to the singlet state of its isomer.[26]

Citing unpublished work,[26] it was reported that under certain conditions, depending on temperature and isotopic substitution, it was possible to also observe thermal ring closure of **12** back to the starting material **11**. However, although the H-shift reactions to give alkene **13** were found to be linearly first-order over at least several half-lives, the rates of formation of **11** were nonexponential. The ring-closure reaction requires greater geometric deformation, and therefore might be expected to be more sensitive to matrix site influence.

Dougherty and co-workers[17] have described exceedingly detailed studies of the ring-closure reactions of triplet 1,3-cyclobutanediyls. Irradiations of a variety of disubstituted bicyclo[2.1.1]diazenes **15** in organic glasses at 3.8 K produced the corresponding triplet biradicals **16–20**, which could be characterized by EPR spectroscopy. Depending on substitution, two different kinetic behaviors were observed. Biradicals with one or more resonance stabilizing substitutents (**16, 17,** and **18**) were persistent at 3.8 K, but began to disappear irreversibly at temperatures around 20 K. In contrast, EPR signals of those systems with only methyl or ethyl substitution, **19** and **20**, decayed even at 3.8 K.

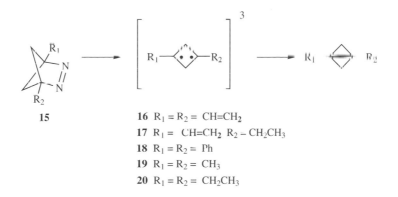

15

16 $R_1 = R_2 = CH{=}CH_2$
17 $R_1 = CH{=}CH_2$ $R_2 = CH_2CH_3$
18 $R_1 = R_2 = Ph$
19 $R_1 = R_2 = CH_3$
20 $R_1 = R_2 = CH_2CH_3$

The disappearance of the spectra of the biradicals was attributed to ring closure to form the corresponding bicyclobutanes. H-abstraction from the surrounding matrices was excluded because (a) rates of decay were much too fast below 65 K compared with known radical H-abstraction reactions, (b) no radical signals were observed to grow in the EPR spectra, and (c) no rate differences were observed in deuterated compared with protio matrices. NMR and GC analyses of the EPR samples showed only bicyclobutane products, and no traces of cyclobutenes, ruling out hydrogen shifts in the biradicals.

Kinetic analyses of the decay rates of the biradicals were complicated by severely nonexponential disappearance in the low temperature glasses. As usual in low temperature glasses, the decay kinetics reflected random distributions of rate constants due to differently reacting matrix sites. To address this dispersive kinetics problem, two new analysis methods were developed. The first was termed "distribution slicing." Irradiations were conducted at 3.8 K, spectra were recorded, the matrices were warmed to varying temperatures, and then samples were recooled to 3.8 K so that a spectral "snap shot" could be recorded of the remaining triplet species. In this fashion, it was possible to estimate the relative populations of biradicals in matrix sites having differing activation barriers. Hence, a qualitative picture of the distribution of activation barriers could be derived, at least in the well-behaved resonance stabilized systems. It was noted that these results confirmed the origins of the matrix-site effect, and that nonexponential decays arose from a distribution of E_a values and not A values. In the cases of partially resonance stabilized biradicals **16**, **17**, and **18**, Gaussian distributions of E_a values, with constant A values, satisfactorily fit the distribution slicing results.

To compare rate data at different temperatures, "most probable" rate constants for each biradical were determined at various temperatures. The distribution slicing information was utilized to help fit the EPR decay curves at different temperatures through a numerical Laplace transform method. Arrhenius plots of the resulting ln k values corresponding to the most probable matrix sites versus $1/T$ were found to be linear for all three resonance stabilized biradicals, giving activation parameters E_a=1.7, 1.4, and 2.3 kcal/mol for **16**, **17**, and **18**, respectively, in 2-methyltetrahydrofuran matrices; log A (s^{-1}) for all three ranged from 6 to 8. The linear Arrhenius behavior, together with the lack of reactivity below ca. 20 K, led to the conclusion that these systems were not undergoing tunneling.

The fully localized biradicals **19** and **20** showed substantially different behavior, as did the cyclopenta-1,3-diyl (**9**), which was reinvestigated for comparison. Because all three systems underwent considerable amounts of rearrangement at lowest temperatures, it was not possible to apply a distribution-slicing method to their kinetics. Fitting the nonexponential decay curves was thus less precise, but it was still possible to derive approximate most-probable rate constants at various temperatures. Arrhenius treatment of these rate constants gave nonlinear plots, displaying approximately temperature independent rates for all three biradicals between ca. 20 and 4 K. Decay rates rose rapidly above ca. 20 K. It was noted that there was no physically reasonable distribution of matrix-site dependent E_a values that could recreate the apparent lack of temperature dependence over such a large temperature range. For example, assuming log A (s^{-1}) = 8.0, the rate of closure of **19** at 4 K implied E_a = 170 kcal/mol. The predicted rate constant at 42 K, for example, would then be 4.1×10^6 s^{-1}, compared with the observed value of only 0.082 s^{-1}.

Thus, these results indicated the involvement of heavy atom tunneling in the localized biradicals. The rates of decay for **19**, **20**, and **9** could be fitted with Bell's simple model of tunneling through a parabolic barrier. Assuming log A (s^{-1}) = 8.0, and a tunneling mass corresponding to the entire mass of the biradicals for simplicity's sake, the Bell model gave barriers of 0.75, 0.825, and 1.60 kcal/mol, and barrier widths of 0.38, 0.325, and 0.29 Å, for **19**, **20**, and **9**, respectively. It was suggested that the lack of stabilizing substituents in these localized biradicals led to narrower barrier widths for the geometric distortions required for crossing from the triplet to the singlet product surface, affording a higher probability of tunneling.[17]

10.3.3. Cyclobutadiene—Heavy Atom Tunneling Without Intersystem Crossing

Cram and co-workers[28] termed 1,3-cyclobutadiene (**21**), "…the Mona Lisa of organic chemistry in its ability to elicit wonder, stimulate the imagination, and challenge interpretive instincts." Hence, depending on one's point of view, the fact that cyclobutadiene has turned out to be one of the most widely accepted examples of heavy atom tunneling might be considered either ironic or perfectly fitting. Although QMT in reactions involving primarily translation of hydrogen has become broadly appreciated, confirmed tunneling in reactions dominated by movement of carbon or other heavier nuclei is still quite rare. Moreover, in contrast to biradical examples described in the previous section, carbon tunneling in cyclobutadiene is uncomplicated by additional requirements of change in multiplicity.

21a	**21b**	**21a**	**21b**

A focal point of interest with cyclobutadiene, or [4]-annulene, is whether **21a** and **21b** are equilibrating valence isomers, or instead represent resonance structures analogous to benzene's. Although the antiaromaticity of cyclobutadiene has long been recognized and understood theoretically, the fact that **21** is rectangular (D_{2h}) rather than square (D_{4h}) was only conclusively established in the late 1970s. Following many years of study, and considerable controversy, the bulk of evidence for the nonsquare geometry came to rest mainly on theoretical and spectroscopic evidence.[29,30] Calculations suggested that the two D_{2h} forms were separated by a barrier of somewhere between 8.3 and 14 kcal/mol, corresponding to the fully delocalized square structure. By 1980, however, there was still little experimental information on the rates of interconversion of the two valence tautomeric forms. Variable temperature NMR studies on highly substituted derivatives of **21** had indicated a dynamic equilibrium between two unsymmetrical forms. But, the first direct *chemical* evidence supporting interconverting structures in the parent **21** came from clever experiments by Whitman and Carpenter (Scheme 10.1).[32]

Generation and deazetization of specifically dideuterated azo compound **22** in the presence of methyl (Z)-3-cyanoacrylate (**T**) gave mixtures of "symmetric" (**S**) and "unsymmetric" (**U**) Diels–Alder-type trapping products, whose ratio depended on the concentration of **T**.[32] Larger [**T**] gave greater proportions of S-products. The

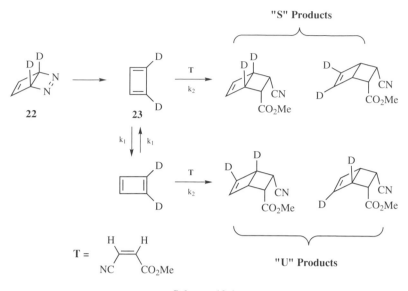

Scheme 10.1

results were interpreted according to Scheme 10.1. At the highest concentrations of **T**, initially generated **23** was trapped before significant isomerization could occur. From the concentration dependence of [**S**]/[**U**], the ratio of trapping versus isomerization rate constants k_2/k_1 could be extracted. The temperature dependence of this ratio was then measured over the range −9 to −50°C, and fitted to Eq. 10.13, where $\Delta\Delta H^{\ddagger}$ is defined as $\Delta H_1^{\ddagger}$(automerization)−$\Delta H_2^{\ddagger}$(trapping), and $\Delta\Delta S^{\ddagger}$ is defined analogously. The best values were found to be $\Delta\Delta H^{\ddagger} = 1.61$ kcal/mol and $\Delta\Delta S^{\ddagger} = 8.2$ cal/(mol K).

$$[S] / [U] = 1 + k_2[T]/k_1 = 1 + [T]\exp(\Delta\Delta H^{\ddagger} / RT - \Delta\Delta S^{\ddagger} / R) \qquad (10.13)$$

With application of reasonable values for trapping parameters $\Delta H_2^{\ddagger}$ and $\Delta S_2^{\ddagger}$, it was possible to bracket the enthalpy and entropy of activation for isomerization of cyclobutadiene. Hence, $\Delta H_1^{\ddagger}$ was estimated to fall between 1.6 and 10 kcal/mol, where the upper limit was consistent with theoretical predictions for square-planar cyclobutadiene. Most surprising, though, was the conclusion that $\Delta S^{\ddagger}$ for automerization must lie between −17 and −32 cal/(mol K), based on the $\Delta S^{\ddagger}$ values normally observed for Diels–Alder reactions as a model for $\Delta S_2^{\ddagger}$.

To explain the unexpected negative entropy value, Carpenter[33] made the dramatic proposal that bond-shift isomerization of cyclobutadiene proceeds predominantly via carbon tunneling through the barrier separating the two geometries. In support of this mechanism, he modeled the reaction as the stretching of a diatomic with two energy minima separated by 0.98 Å, and an effective mass for each "pseudoatom" of 26 atomic mass units (C_2H_2). With the simplest truncated parabola form for the barrier, and an estimated barrier of 10.8 kcal/mol based on a 1000 cm^{-1} stretching frequency, a Bell-type calculation predicted tunneling rate constants of 8.08×10^4

and 4.65×10^5 s^{-1} at -50 and $-10°$C, respectively. Comparison to classical rates of 1.01×10^2 and 4.82×10^3 s^{-1} at these temperatures, calculated for a 10.8 kcal/mol barrier and normal $\Delta S^{\ddagger}$, indicated that tunneling accounted for >97% of the reaction below 0°C. Combining the tunneling and overbarrier rate constants predicted apparent activation parameters of $\Delta H^{\ddagger} = 4.6$ kcal/mol and $\Delta S^{\ddagger} = -15$ cal/(mol K), consistent with the experimental measurements.

Subsequent calculations by other workers,[34–36] utilizing more sophisticated theoretical methods, have confirmed the notion that carbon tunneling dominates the automerization of cyclobutadiene. Interestingly, the more recent calculations suggest that the tunneling isomerization is even more rapid than Carpenter's estimates.

Spectroscopic experiments have produced compelling evidence for rapid bond-shifting in cyclobutadiene even at extremely low temperatures. Michl and co-workers[30,37,38] found that irradiation of anhydride **24** with polarized light generated aligned samples of cyclobutadiene in Ar matrices at 10 K. When polarized IR spectroscopy was used, bands assigned to in-plane and out-of-plane vibrations clearly showed opposite polarization. Under all irradiation conditions, however, polarized IR spectroscopy could detect no difference in the degree of polarization of the long-axis in-plane polarized vibrations and those polarized along the short-axis. Similar results were obtained for a variety of isotopically labeled derivatives of **24**. It was thus concluded that automerization was fast even at 10 K, at least on the order of several minutes. Moreover, attempts to generate unsymmetrically labeled cyclobutadienes, for example, **23** and **25**, always produced 50:50 mixtures of products in the IR spectra.

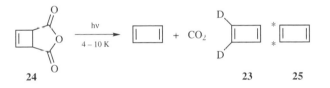

24   **23**   **25**

Dramatic ^{13}C NMR experiments confirmed the rapidity of bond-shifting of **21** at cryogenic temperatures, and further increased the minimum tunneling rate.[38] Dipolar cross-polarization solid-state ^{13}C NMR spectra on Ar matrix isolated vicinally ^{13}C labeled cyclobutadiene **25**, generated from irradiation of labeled **24**, could be modeled only by a rapidly interconverting mixture of two rectangular forms. It was concluded that automerization must be occurring with a rate constant of at least 10^3 s^{-1} at 25 K, rapid on the NMR timescale. Although the experiments could not distinguish between classical and tunneling mechanisms, it was recognized that the magnitude of calculated barriers precluded an overbarrier thermally activated process at these temperatures.

10.3.4. Tunneling in Triplet Carbene Reactions

The chemistry of carbenes is inextricably linked with issues of spin state. This is equally true for tunneling as for classical reactions. The factors that determine the electronic ground states in carbenes, and the energy splitting between different states, have been eloquently described in the three chapters of the previous volume of this series.[39–42] Hence, we will review very briefly only the most relevant features of carbene electronic state and reactivity.

In the simplest "zeroth-order" terms, carbenes have two unshared electrons occupying two orbitals, an in-plane hybridized σ-orbital and an out-of-plane p-orbital. The two lowest electronic states are a triplet (**26**), in which the two electrons have the same spin and occupy the separate orbitals, and a singlet (**27**), in which the two electrons are spin-paired in the in-plane orbital. In general, the difference in energy between carbene triplet and singlet states is rather small, and easily swayed by substituent and other structural effects. As in CH_2 itself, carbenes with only aryl substituents tend to have triplet ground states, where the carbene p-electron is delocalized into the π-system. Electron resonance spectroscopy (EPR) has been an especially informative tool in the study of triplet carbenes because of their magnetic properties. Lone-pair donating substituents, on the contrary, favor ground-state singlet carbenes, as is easily rationalized by the contribution of resonance structures such as **28**.

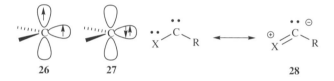

26   27   28

Quite a variety of carbene reactions have been determined to involve quantum tunneling. In general, the highly reactive carbenes, irrespective of electronic state, undergo many reactions with very low energy barriers. Moreover, a number of carbene reactions require only minimal geometry reorganization, or in other words, have narrow energy barriers that facilitate tunneling. In the following sections, we will consider some of the most prototypal tunneling reactions of triplet carbenes. Then, because their reactions are fundamentally so different, we will separately consider tunneling in singlet carbenes.

10.3.4.1. Triplet Carbene Intermolecular Hydrogen Abstraction Reactions

By the early 1970s, extensive product analyses had firmly established that triplet carbenes display reactivity in solution not unlike that of radicals.[40,42] The unpaired electrons in separate orbitals in a triplet carbene logically behave independently. Moreover, the in-plane sigma "radical" of the carbene might be expected to be particularly reactive, in analogy to other sp^2 radicals. Hence, for example, triplet carbenes add to alkenes in radical-like stepwise fashion to produce intermediate triplet diradicals, and hence cyclopropanes with loss of alkene stereochemistry. Triplet carbenes also abstract hydrogens from C–H bonds. The resulting triplet radical pairs can either recombine to give C–H "insertion" products, after a spin-flip to a singlet pair, or diffuse apart. In contrast, singlet carbenes add concertedly to alkenes, affording cyclopropanes with retained stereochemistry, as well as undergoing concerted C–H insertion.

It is thus not surprising that H-abstractions of triplet carbenes also benefit from QMT. The first inklings of nonclassical triplet carbene reactivity arose in a study by Moss and Dolling[43] in the early 1970s. In attempts to probe the triplet and singlet energetics and reactivity of phenylcarbene **29**, they investigated the temperature dependence of singlet-carbene cycloaddition versus triplet H-abstraction reactions with 2-butene. It was found that on irradiations of phenyldiazomethane in neat 2-butene, singlet carbene addition giving stereoretained cyclopropropanes (**30**) dominated product mixtures down to 143 K in solution. Surprisingly, however, a dramatic and discontinuous change was observed below the melting point of the alkene (134 K), where now C–H abstraction products (**31–33**) became major (Scheme 10.2).

This key paper was followed by a flurry of activity in this area, spanning several years.[44–47] A variety of workers reported attempts to deconvolute the temperature dependence of carbene singlet/triplet equilibria and relative reactivities from the influence of solid matrices. Invariably, in low-temperature solids, H-abstraction reactions were found to predominate over other processes. Somewhat similar results were obtained in studies of the temperature and phase dependency of the selectivity of C–H insertion reactions in alkanes. While, for example, primary versus tertiary C–H abstraction became increasingly selective as the temperature was lowered in solution, the reactions became dramatically less selective in the solid phase as temperatures were lowered further. Similar work of Tomioka and co-workers[45] explored variations of OH (singlet reaction) versus C–H (triplet reaction) carbene insertions with alcohols as a function of temperature and medium. Numerous attempts were made in these reports to explain the results based on increases in triplet carbene population

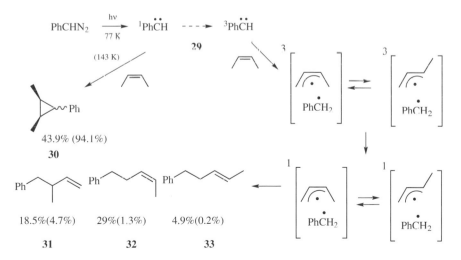

Scheme 10.2 Comparison of reactions of phenylcarbene in solid (77 K) and liquid (143 K) Z-2-butene. Percentages at 143 K are shown in parentheses.

at lower temperatures, favoring H-abstraction over other reactions. However, the origins of the striking "matrix effects" in solids were not satisfactorily explained.

It is worth noting that the unusual matrix effects were most generally associated with carbenes possessing triplet ground states. Similar investigations of phenylchlorocarbene, which possesses a singlet ground state, over a wide variety of reaction conditions, routinely failed to show similar unusual temperature or matrix dependencies.[45] Although PhCCl might not be a perfect model for other singlet aryl carbenes, these results appeared to indicate that matrices do not inherently promote singlet C–H insertion reactions.

A satisfying explanation for much of these low-temperature results came from direct spectroscopic measurements of carbene reactions. Landmarks in this area came from investigations by Platz[48] and Gaspar[49] and their co-workers, who found that decay of triplet aryl carbenes (including phenylcarbene, diphenylcarbene, and fluorenylidene) could be directly measured in low-temperature organic glasses via EPR spectroscopy. Triplet carbene signals could be detected immediately following irradiation of corresponding aryl diazomethanes, but rapidly disappeared once photolysis ceased. The decay of the carbene signals was attributed to hydrogen abstraction from solvent based on four observations: (a) C–H abstraction products were observed on thawing the glasses; (b) the triplet carbenes were longer lived in perdeuterated matrices; (c) no carbene decay was detected in perfluoroalkane matrices; and (d) decay rates were consistent with the relative hydrogen donating ability of the glass materials.[48]

$$^3\text{PhCPh} \; + \; \text{RH} \; \xrightarrow{77\,\text{K}} \; \left[\; \overset{3}{\underset{35}{\text{PhCHPh}}} \; + \; \text{R} \cdot \; \right]$$

A seriously complicating factor in kinetic analyses of the carbene reactions, however, was that first-order plots of $\ln(I/I_0)$ were distinctly curved. This signaled a typical dependence of rate on a distribution of matrix sites, where a range of "fast" reacting and "slow" reacting sites were being detected. As is often the case, plots of $\ln(I/I_0)$ versus $t^{1/2}$ or $t^{1/3}$ were found to be approximately linear. For comparison purposes, Platz made the simplifying assumption that the first ca. 20% of decay, which was found to be approximately "first-order," corresponded mainly to a single matrix site or to very similar sites.[47,48,50] Arrhenius plots of the initial decay rates at various temperatures, however, gave very unusual activation parameters. Both E_a and A values were much smaller than expected. For example, for Ph$_2$C (**34**) in toluene, $E_a = 2.1$ kcal/mol and log $A(\text{s}^{-1}) = 2.4$. On the basis of similar solution H-atom abstraction reactions, E_a values more in the range 6–12 kcal/mol and log A values of 8–11 should be expected. In fact, it was noted that the most optimistic activation parameters would predict 77 K rate constants $<10^{-6}$ M^{-1}s^{-1}, compared with experimentally observed rate constant of 6.3×10^{-4}. Therefore, in analogy to earlier work on radical H-abstraction reactions, it was proposed that the carbene reactions proceeded via QMT.

Tunneling also provided a handy explanation for the anomalous product distributions in solid matrices. At low temperatures, classical overbarrier processes such as addition to alkenes or singlet insertions into O–H bonds, for example, become negligible. H-abstraction reactions via QMT, however, do not decrease in rate as rapidly as temperature is lowered, and continue even at the lowest temperatures. Moreover, the physical constraints of the solid matrices prevent the carbenes from sampling a variety of C–H bonds as in solution. The result is that the triplet carbenes abstract the most available hydrogens through tunneling, erasing much of the selectivity seen at higher temperatures.

In later work, Platz and co-workers[51a] fit the temperature dependence of the carbene decays to Eckart-type asymmetric barriers, from which estimates of the barrier heights and widths could be extracted. Good agreement with the experimental rate versus temperature behavior was obtained; although it was noted that the data covered a rather limited temperature range. Classical barrier heights for reactions of diphenylcarbene with toluene and diethyl ether, for example, were estimated to be 7 and 5.4 kcal/mol, respectively. These values are in line with those of radical abstraction reactions,[16,52] lending credence to the "single" site model, as well as the proposed tunneling mechanism.

A major problem still remained unsolved, however. Although both tunneling and classical models predicted very large kinetic isotope effects for H versus D abstractions, for example, ca. $k_H/k_D \approx 10^4$ at 77 K for the tunneling mechanism, carbene initial decay rates were generally found to be only about 10 times faster in perhydro versus perdeutero glasses. These results suggested that very different sites were being interrogated in the two matrices, that is, "slow" hydrocarbon rates were being compared with "fast" deuterated sites. Supporting this notion, analyses of products of the matrix reactions showed expected very large isotope effects.[48,51]

It was later found that the triplet radical pair intermediates (e.g., **35**) resulting from H-abstraction in the low-temperature glasses could also be directly detected by EPR spectroscopy.[51c] Importantly, the rates of radical pair appearance matched the rates of carbene disappearance. When perdeuterated matrices were used, although the carbene EPR signals decayed in the dark as previously observed, radical pair resonances did not increase. It thus became clear that the D-abstraction was, in fact, inhibited and that the carbenes followed other pathways for reaction in deuterated media. These results spotlighted a significant problem with basing conclusions on rates of decay measured by the relatively "slow" EPR method. By the time the irradiation ceased, it was likely that most of the carbenes had already reacted, and the detected EPR signals arose from a minority of carbenes in relatively slowly reacting sites. Moreover, there was no guarantee that the same population of matrix sites was being observed at different temperatures, raising suspicions about the Arrhenius data.

Final resolution of these problems, particularly the complications from multiple matrix sites, came from investigations using spectroscopic methods with higher time resolution, viz. laser flash photolysis.[53] Short laser pulse irradiation of diazofluorene (**36**) in cold organic glasses produced the corresponding fluorenylidene (**37**), which could be detected by UV/VIS spectroscopy. Now, in contrast to the results from EPR spectroscopy, single exponential decays of the carbene could be observed in matrices with differing H-donating ability. In general, as the temperature of the glasses was lowered, the pseudo first-order rates of decay of the carbene showed two regimes. For example, in a methylcyclohexane/toluene glass, an Arrhenius plot of ln k versus $1/T$ was linear over the range of 122–100 K, with E_a = 5.5 kcal/mol and log (A/s^{-1}) = 15.8. Below 100 K, however, the rate of carbene disappearance became much less sensitive to temperature, giving, for example, E_a = 1.2 kcal/mol and log A = 6.2, over the range 100–90 K. Moreover, at low temperatures, the reaction rates were hundreds of times faster than predicted by extrapolation of the higher-temperature Arrhenius plots. It was thus suggested that at higher temperatures, the carbene was undergoing classically activated hydrogen abstraction. But at lower temperatures, where thermally activated reaction became impossible, hydrogen transfer via QMT, which was much less temperature dependent, became competitive and eventually dominant. It was noted that changes in matrix viscosity with temperature might also be playing a role.

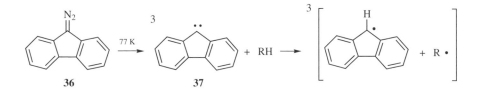

Laser flash irradiation of diazofluorene in perdeuterated matrices, in contrast, gave severely nonexponential decay of the carbene spectra.[53] Analyses of the products formed in the low-temperature matrices showed that, as with the EPR studies, the carbene was not undergoing D-abstraction. LFP of the diazo compound **36** in $CFCl_3$-CF_2BrCF_2Br glasses gave linear first-order decays, and linear Arrhenius plots, which were attributed to classical Cl and Br abstractions.

Platz and co-workers[54] also found evidence for the involvement of QMT in the hydrogen abstraction reactions of certain triplet arylcarbenes in solution at ambient temperatures. In classical activated processes, the difference in activation energies for C–H versus C–D abstractions should reflect the difference in zero point vibrational energies, with a maximum $E_a(D) - E_a(H)$ = ca. 1.2 kcal/mol.[11] Moreover, it has been shown that the preexponential terms for H versus D transfers should be comparable, with $A_H/A_D > 0.7$.[14] When product studies were used to measure relative reaction rates of several diarylcarbenes in mixtures of toluene/toluene-D_8 over the range −78 to 135°C, differences in activation energies $E_a(D) - E_a(H)$ as large as 1.88 kcal/mol, and A_H/A_D ratios as small as 0.31, were measured. It was suggested that these deviations from predictions of transition state theory indicated significant contributions of QMT to the abstraction reactions in solution. Interestingly, there was some indication that increased steric hindrance in the approach of the carbene to solvent molecules led to greater QMT importance.

More recently, evidence has been reported that triplet carbenes can abstract H from H_2 via QMT.[55] It was found that warming **38, 39**, and **40** in Ar matrices doped with ~2% H_2 to 30 K caused disappearance of the IR spectra of the carbenes, and concurrent appearance of corresponding H_2 adducts. It was suggested, based on $^3CH_2 + H_2$ calculations in the literature as well as analogy to the reactions of triplet carbenes with hydrocarbons, that these reactions involved H-abstraction to give triplet radical pairs, followed by ISC and recombination. B3LYP/6-31G** calculations suggested that, while the abstraction reactions should all be mildly exothermic, the barriers (4.4, 5.7, and 5.4 kcal/mol for reactions giving **41**, **42**, and **43**, respectively) were prohibitively high for reaction at these cryogenic temperatures. Hence, QMT provided a likely route for reaction. As expected, regardless of whether classical or tunneling mechanisms were involved, no reactions were observed between the carbenes and D_2.

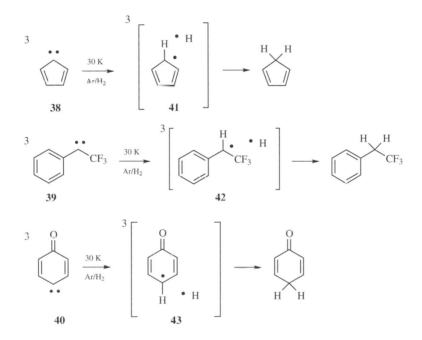

Interestingly, a bis-carbene/diradical was also found to react with H_2 at low temperatures.[55] Earlier work had shown that *p*-bis(chloromethylene)phenylene (**44**) exhibits spectroscopy and reactivity consistent with a $\sigma^2\pi^2$ quinonoid singlet diradical electronic state, rather than a bis-singlet carbene. The diradical was found to react with H_2 at 30 K, giving chlorocarbene **46**. The corresponding singlet carbene **46** could not be induced to react further with H_2 under any circumstances. It was proposed that H-abstraction gave first radical pair **45**, which then produced carbene **46** by H-atom addition. Calculations indicated that the initial H-abstraction should be exothermic by -3.6 kcal/mol, but must surmount a barrier of 6.4 kcal/mol. Hence, QMT was again implicated to facilitate the reaction. The unreactivity of singlet carbene **46** (as well as other singlet carbenes) was attributed to the requirement of direct insertion into the H–H bond, where the necessity of transfer of both hydrogens via tunneling was suggested to be less probable.[56]

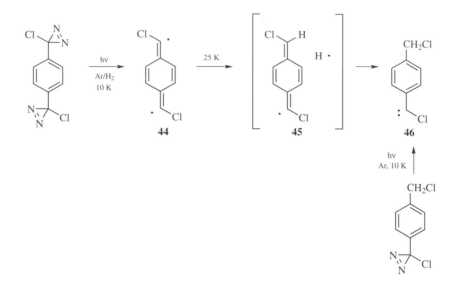

10.3.4.2. Triplet Carbene Intramolecular Reactions

Attempts to directly observe intramolecular reactions in triplet carbenes spectroscopically, borrowing from the results with intermolecular H-abstractions, have met with mixed success. Platz and co-workers[57] found that irradiation of diazo compound **47** at temperatures as low as 4.2 K gave triplet diradical **49** as the sole EPR-active product. Analogous results were obtained with the corresponding CD_3 compound. It was suggested that an especially narrow barrier might facilitate rapid H- or D-abstraction by triplet carbene **48** via tunneling. However, it was noted that it was impossible to exclude the possibilities of other mechanisms for product formation, such as H-transfer in excited diazo precursors, or by secondary irradiation of the incipiently formed carbene.

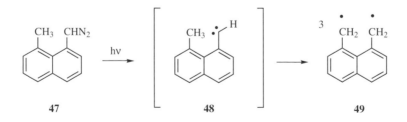

On the contrary, Chapman and McMahon[58,59] found that H-transfer in o-tolylcarbene can be directly observed at very low temperatures, through several different spectroscopic methods. Irradiation of Ar matrix isolated diazo compound **50** at 4.2 to 10 K gave triplet carbene **51**, which could be characterized by EPR, IR, and UV/VIS spectroscopy. The various spectra of **51** slowly decayed at temperatures as low as 4.2 K ($t_{1/2}$ ca. 64 h), and o-xylylene (**52**) could be observed to grow correspondingly in the IR and UV/VIS spectra. First-order plots of carbene decay were nonlinear, suggesting multiple site problems, but were reasonably linear when ln (A) for the UV/VIS or EPR absorptions were plotted versus $t^{1/2}$. Arrhenius plots over the temperature range 4.6–30 K showed two regions, with temperature insensitivity observed below 12 K. Hence, it was suggested that hydrogen migration was facilitated by QMT at the lowest temperatures. The reaction of **51** to **52** must involve a multiplicity change from triplet to singlet at some point, yet no rate enhancement was observed in Xe matrices, which might be expected to increase rates of intersystem crossing via an external heavy atom effect.[27] In the absence of any calculational estimates of the possible energy barrier for H-abstraction by the triplet carbene in this case, it is difficult to deconvolute the role of spin-flip from that of through-barrier tunneling. It was suggested that a direct rearrangement of triplet carbene to triplet **52** might be occurring, however. Platz and co-workers observed similar low temperature reactions of triplet mesitylcarbene.[60]

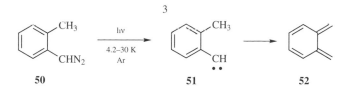

As expected, introducing a CD_3 group in **51** impeded the reaction considerably. The deuterated carbene was found to be stable up to 59 K in Xe. It was not possible to find conditions where the trideutero carbene decayed. Hence, k_H/k_D or A_H/A_D ratios, which might help support a tunneling mechanism, could not be determined.

The related [1,2]-H shift in the isomeric triplet 1-phenylethylidene (**53**) was also investigated.[58,59] The triplet carbene, generated from irradiation of the corresponding diazo compound, was characterized in low temperature inert matrices by EPR, IR, and UV/VIS spectroscopy. In this case, the carbene was stable in Ar up to the temperature limits of the matrix (36 K). Irradiation, however, readily converted the carbene to styrene.

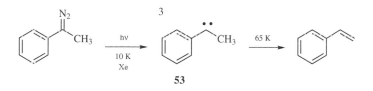

The thermal conversion of carbene **53** to styrene could also be observed in Xe matrices. Although stable at 12 K for at least 24 h, increasing amounts of carbene were observed to convert to styrene as the temperature was raised gradually to ca. 60 K. At each intermediate temperature between 12 and 60 K, "site clearing" behavior was detected, where some amounts of carbene disappeared rapidly, followed by no further decay. At around 60 K, the carbene rearranged with exponential first-order kinetics, suggesting that the Xe matrix had softened sufficiently to alleviate multiple matrix site problems. Under these conditions, technical difficulties precluded the use of EPR or UV/VIS spectroscopy to follow the carbene decay, but the process could be tracked with IR spectroscopy. Because of matrix instability above 65 K, it was not possible to measure the temperature dependence of the [1,2]-H shift, but based on Eyring absolute rate theory it was estimated that $\Delta G^{\ddagger} = 4.7$ kcal/mol. Minimal decay of the corresponding CD_3 carbene was observed in Xe from 12 to 80 K in the dark, although the carbene readily underwent D-shift photochemically.

It was suggested that H-shift occurs in carbene **53** through thermally populated S_1 carbene, which in turn rearranges through a negligible barrier on the singlet surface. Thus, the $\Delta G^{\ddagger}$ of 4.7 kcal/mol was proposed to represent a lower limit to the $T_0 - S_1$ energy gap, which would be consistent with other aryl carbenes. However, no explanation was offered for the mechanism of rearrangement of the portions of triplet carbene observed to react at lower temperatures. It seems reasonable that the lower temperature decays might represent rearrangements of carbenes in certain matrix sites that allowed tunneling of triplet carbene directly to the singlet product. The heavy-atom Xe matrix might thus help facilitate such a spin-forbidden process. Again, the absence of rearrangement of the deuterated carbene **53-d₃** would be consistent with either a classical or a tunneling reaction mechanism at these low temperatures.

10.3.5. Tunneling in Singlet Carbene Reactions

In contrast to triplet carbenes, reactions of singlet carbenes require no change in multiplicity.[39,40,42] Generally, singlet carbene reactions are very exothermic and have very low energy barriers. Moreover, only minimal geometric distortions are often required to reach transition states in singlet carbene rearrangements or, in other words, narrow energy barriers separate singlet carbenes from products. Singlet carbenes thus have considerable potential for rearrangement via QMT. However, it must be recognized that singlet carbenes display a remarkable range of reactivity depending on substitution. At one extreme, singlet methylene, CH_2, is one of the most indiscriminately reactive intermediates in organic chemistry.[42,61] But, at the other end of the spectrum, some singlet carbenes can be isolated as stable, crystalline compounds at room temperature. The reactivity and stability of singlet carbenes depend critically on the degree of π-electron donation from lone pairs on adjacent centers, with stabilization associated with resonance structures such as **28**. Powerfully, electron-donating ligands such as N often render carbenes stable at room temperature.[41] Singlet chlorocarbenes, though, are highly reactive intermediates in solution, and very rapidly (with lifetimes typically on the order of microseconds) undergo a wide variety of inter- and intramolecular reactions.[39,62] Yet, the halogen substitution moderates the reactivity of the carbenes just enough to permit spectroscopic investigation with a variety of techniques.

10.3.5.1. Hydrogen Shifts in Singlet Carbenes   An early indication that tunneling might be involved in rearrangements of ground-state singlet carbenes came from time-resolved photoacoustic calorimetry measurements of the rearrangement of methylchlorocarbene (**54**) to vinyl chloride (**55**). Measuring the heat evolution from laser-pulse irradiation of methylchlorodiazirine allowed measurement of the lifetime of carbene **54**, which was ca. 740 ns at 295 K in heptane solution.[63] Arrhenius treatment of reaction rates measured from 11.5 to 60.0°C gave E_a = 4.9 kcal/mol and log A (s^{-1}) = 9.7. It was noted that the lower than expected pre-exponential factor might arise from QMT assisted hydrogen shift.

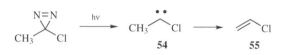

Subsequent work found greater evidence in support of tunneling.[64] Carbene **54** was generated with nanosecond laser pulses in solution, and its decay monitored by following the kinetics of ylide formation with varying amounts of pyridine dopant. The rate of rearrangement of the carbene to **55**, measured at various temperatures, ranged from 8.4 × 10^5 s^{-1} at 248 K to 30.8 × 10^5 s^{-1} at 343 K. The Arrhenius plot of the rate constants was strongly curved, becoming relatively flat at lower temperatures, where QMT was suggested to dominate. Moreover, rates of rearrangement were considerably greater, particularly at lower temperatures, than predicted by theoretical calculations of activation parameters ($\Delta H^{\ddagger}$ = 10.6 kcal/mol and $\Delta S^{\ddagger}$ = −3.2 eu at the MP4 level). On the basis of the comparison between theory[65] and experiment, it was suggested that more than 85% of the hydrogen shift occurred via tunneling at temperatures ≤298 K.

Interestingly, in a comparison of the CD$_3$ and CH$_3$ carbenes, an unusual temperature dependence of the kinetic isotope effect was observed.[64] In contrast to typical reactions, the ratio of rates of H versus D shift, k_H/k_D, actually increased as temperature was raised. In fact, k_D was measured to be larger than k_H at 248 K. It was suggested that these results required a normal temperature dependence of the isotope effect for the classical component of the reaction, but an unusual diminished isotope effect for the QMT reaction.

Theoretical calculations published concurrently appeared to support these suggestions.[65] Inclusion of tunneling corrections for the calculated rate constants was found to lower the apparent E_a and A values, consistent with the experimentally observed numbers. Moreover, the calculations appeared to indicate that the tunneling correction for D was larger than for H. Such an atypical isotope effect would explain the increase of k_H/k_D with increasing temperature.

More recent theoretical work has raised questions about these conclusions, however.[66,67] Particularly extensive calculational treatment of the rearrangement of **54** to vinyl chloride by several research groups failed to duplicate the predictions of an atypical kinetic isotope effect. These later studies indicate that tunneling effects should indeed be greater for H-shift than for the heavier D rearrangement. Consequently, the k_H/k_D ratio should actually decrease at higher temperatures. The discrepancy in predicted results was eventually traced to an error in the earlier calculations. Nevertheless, it was confirmed that considerable QMT (ca. 80%) should occur in rearrangement of the protio carbene **54** in solution at ambient temperatures. With inclusion of tunneling and solvent effects, it was possible to reproduce the experimental activation parameters satisfactorily. It was also noted that facile tunneling would explain failed attempts by other workers to observe **54** in low temperature matrices.[66]

Still unresolved, however, is why the earlier experimental results seemed to indicate that D tunnels more rapidly than H at lower temperatures in rearrangement of **54**.[64] Moreover, the calculations indicate that although tunneling is a major contributor, Arrhenius plots of the rearrangement of **54** to vinyl chloride should be approximately linear over the experimental temperature range.[66,67] It should be noted that the methods used to measure rates of decay of **54** do not reveal any information on the reaction products. It has been suggested that intermolecular reactions might have begun to intercede at lower temperatures, causing the observed Arrhenius curvature. Similarly, it seems possible that the deuterated carbene might have been reacting through pathways other than D-shift. Careful product studies would be required to confirm these possibilities.

A similar difficulty was uncovered in the related 1,2-H shift of another singlet chlorocarbene, **56**.

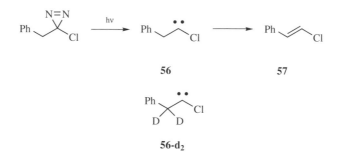

The spectroscopy of **56** has been studied extensively through laser flash photolysis.[68] The temperature dependence of decay of **56** in solution gave curved Arrhenius plots, which were interpreted as an indication of QMT, among other possibilities. However, subsequent work showed that as the temperature was lowered and rearrangement slowed, increasing amounts of reaction with starting diazirine began to compete.[69] Hence, the measured decay of carbene at lower temperatures reflected not only rearrangement, but other processes as well. When the spectroscopic measurements were conducted under higher dilution conditions, Arrhenius plots became linear down to at least −70°C. Based primarily on the linearity of these plots over that temperature range, it was concluded that QMT was unimportant under ambient conditions.[69]

Ironically, the most recent calculational treatments of the rearrangement from **56** to **57** indicate that even at room temperature tunneling does, in fact, dominate the H-shift process.[66] Phenyl substitution in **56** should lower the energy barrier for rearrangement compared with **54**, not only increasing the rate of classical rearrangement in the carbene, but also leading to greater tunneling. Calculations[66] suggest that QMT contributes at least 50% to rearrangement of **56** at ambient temperatures. It has also been determined that curvature in the predicted Arrhenius temperature dependence should be negligible over the experimentally examined temperature range. These results emphasize the perils in assuming that curved Arrhenius plots must accompany tunneling. Since tunneling is also thermally activated, it can show a temperature dependence that can parallel that of classical overbarrier reactions. Hence, significant curvature in Arrhenius plots may typically be observed only at temperatures low enough to approach the temperature independent tunneling region.

Direct evidence for QMT in the hydrogen shifts of **56** came from low temperature spectroscopic experiments.[70] It was found that irradiation of the corresponding diazirine at 10 K in an Ar matrix produced small amounts of carbene **56**, only detectable by UV spectra which matched those observed in solution by laser flash spectroscopy. The major product observed by both IR and UV spectroscopy was the rearranged styrene product **57**. The carbene UV spectrum was found to decay rapidly after irradiation of the precursor ceased. On the contrary, similar irradiation of the D_2-diazirine produced a UV spectrum of carbene **56-d$_2$** which was approximately twice as intense as that obtained for the protio compound. The deuterated carbene **56-d$_2$** could also be observed in the IR spectrum. Even at 10 K, the labeled carbene was found to decay slowly, with concurrent formation of the corresponding deuterio styrene. On warming the matrix, rearrangement rates of **56-d$_2$** increased, up to the highest temperatures permitted by the matrix (ca. 35 K). Although nonexponential decays were observed, signaling inhomogeneous matrix sites, the kinetic data were fitted to a "stretched" exponential, giving an approximate $t^{1/2}$ dependence. An Arrhenius-type plot of the logarithms of the resulting "k" values versus $1/T$ was found to be curved, and showed only a very small temperature dependence between 10 and 24 K.

These results indicate QMT in the D- (and presumably, H-) shifts in **56**. It was noted that the rate of the low temperature rearrangement was much higher than could be expected based on extrapolation from room temperature solution data. When the matrix rate constants were fitted to a standard Bell parabolic model, a "classical" barrier of 1.47 kcal/mol and log A (s^{-1}) = 6.33 were derived, although it was recognized that the temperature-dependent region was narrow and the spectroscopic data were imprecise. Corresponding measurements in solution had determined E_a = 4.8 kcal/mol for the rearrangement. It was also suggested that the increase in rate in the temperature range 24–35 K might be due to matrix softening rather than thermal activation.

Consideration of the geometries of singlet carbenes such as **54** and **56**, and the corresponding H-shift transition states, offers a clue to the facility of tunneling in these rearrangements.[66,67] For example, the ground state of **54** has one C–H bond aligned approximately perpendicular to the C–Cl bond, and clearly reflects a stabilizing hyperconjugative electron donation from the C–H bond into the carbene p-orbital. That C–H bond is slightly elongated compared to the other CH_3 bonds (1.101 Å compared with 1.095Å) and has a somewhat distorted H–C–C angle of 103.9°. Because of this hyperconjugative interaction, geometric transformation to the transition state (**58**) requires only minor movement of the hydrogen.

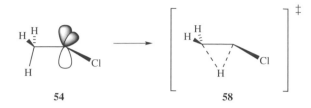

54   58

Evidence for QMT in the [1,4]-H shift of a singlet carbene has also been reported. Tomioka and co-workers[71] found that irradiation of Ar matrix isolated formyl-chlorodiazirine **59** at 13 K produced the corresponding singlet carbene **60**, along with several other products. The B3LYP calculated IR spectrum fitted best to the "anti,anti" conformer shown. On standing in the dark at 13 K, the carbene IR spectrum slowly decayed, and the H-shifted product ketene **61** grew. The decay of the carbene absorbances was linear when plotted against $t^{1/2}$, indicating typical multiple matrix site dependence, and giving a "rate constant" of $1.4 \times 10^{-4} \, \mathrm{s}^{-1/2}$.

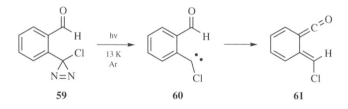

59   60   61

Calculations at the B3LYP/6-31G* level predicted an energy barrier of 8.5 kcal/mol for the [1,4]-H shift. The authors noted that this barrier would be too high to surmount at these temperatures, confirming that QMT assisted H-transfer was probable. As expected, whether or not tunneling is involved, the D-formylcarbene was stable under the same conditions. It is interesting to compare the reaction of this singlet carbene with the related rearrangement of the triplet *o*-tolyl carbene **51**. In the case of the triplet carbene, H-transfer would be expected to occur through a planar geometry, maximizing overlap of one methyl H with the in-plane sigma type unpaired electron of the carbene, thus facilitating the radical-like abstraction (and, in the absence of intersystem crossing, leading directly to triplet ππ* *o*-xylylene). However, scrutiny of the published transition state calculated for the singlet carbene **60** shows a significantly twisted geometry, suggesting a hydride-like transfer to the out-of plane p-orbital of the carbene.[71] Cursory analysis suggests that in-plane transfer of the hydrogen to the carbene in a planar geometry would, in fact, be forbidden by orbital symmetry. Simply put, carbene **60** has 8 π-electrons, but product ketene **61** has 10 π-electrons.

10.3.5.2. Singlet Carbene C–H Insertions

Although [1,2]-H shifts are formally carbene C–H insertions, these rearrangements have different orbital symmetry aspects than those of intramolecular insertions. As described above, overwhelming evidence exists that triplet carbenes undergo abstraction–recombination reactions to produce what are formally C–H insertions, often via QMT. On the contrary, despite considerable exploration, no evidence has been uncovered for tunneling in inter-molecular C–H insertions of singlet carbenes. Hence, for example, Tomioka and co-workers found that, in contrast to triplet carbenes, PhCCl and other related sin-glet ground-state carbenes exhibit little change in selectivity in CH insertions, etc., on changing from solution to solid glasses.[45] Sheridan and Zuev found,[55] similarly, that PhCCl and several other singlet carbenes could not be induced to insert into H_2 bonds under conditions where triplet carbenes react. Triplet carbenes can react in a stepwise, radical-like fashion to transfer H atoms from R–H and H–H via tunnel-ing.[39] On the contrary, singlet carbenes must undergo concerted, three-center inser-tion in the R–H or H–H sigma bonds.[40] It has been suggested[55] that one reason why tunneling is rarer for singlet carbenes is the more stringent requirement involving more atoms (and their masses) in the insertion process. It has been found, however, that the extremely reactive singlet carbene difluorovinylidine does add H_2 at cryo-genic temperatures, suggesting that tunneling might be involved in a sufficiently low-barrier insertion.[72]

Tunneling has been implicated in concerted intramolecular C–H insertions, however. *tert* Butylchlorocarbene (**62**) has been generated and studied in Ar or N_2 matrices at 11 K, produced by irradiation of the corresponding diazirine.[73] It was found that on standing in the dark at 11 K, the IR and UV/VIS spectra of **62** slowly disappeared, and the IR spectrum of C–H insertion product **63** grew concurrently. First-order plots of the decay of the IR bands of **62** were somewhat curved, but were found to be approximately linear when ln[carbene] was plotted against $t^{1/2}$. The rate of rearrangement varied from approximately 4×10^{-4} s^{-1} to 5×10^{-5} s^{-1}, the latter measured after roughly half the initial carbene signal had decayed.

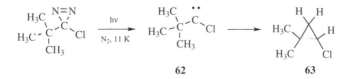

62   **63**

Several observations pointed to QMT in this [1,3]-insertion process.[73] The reac-tion rate was found to be completely insensitive to temperature up to 30 K. It was also noted that the reaction rate at 11 K was impossibly rapid compared with the literature reported solution rate of ca. 10^6 s^{-1} measured at room temperature by laser flash photolysis.[74] A "two point" Arrhenius plot of the rates at 298 and 11 K would afford $E_a = 0.5$ kcal/mol and $A = 2 \times 10^6$ s^{-1}, entirely inconsistent with activation parameters known for other alkylchlorocarbene rearrangements.[62] Moreover, such parameters would predict a 10^7 increase in rate over a temperature increase from 10 to 30 K, thus implying a very curved Arrhenius temperature dependence. Subsequent unpublished work found a calculated barrier at the B3LYP/6-311+G** level of 10 kcal/mol for the reaction from **62** to **63**.[75] Finally, the D$_9$-carbene was found to be indefinitely stable at low temperatures.

Spurred by these results, Moss and Liu sought evidence for tunneling in the related selectively deuterated *t*-butylchlorocarbene **64** in solution.[76]

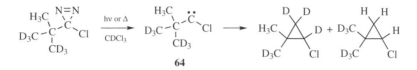

Product analysis by ^{1}H NMR indicated an isotope effect at 118°C of $k_H/k_D = 2.14$, corrected for numbers of H versus D. On lowering the temperature to −12°C, however, it was found that the isotope effect increased to 3.25. Referring to earlier experimental results on the C–H shift in methylchlorocarbene,[64,65] the authors cited the normal temperature dependence of the isotope effect as evidence against tunneling in **64**. In retrospect, however, as noted above, theoretical support for an atypical inverse temperature dependence in methylchlorocarbene has been refuted.[66,67] Hence, the involvement of tunneling in **62/64** at ambient temperatures is still an open question.

10.3.5.3. Carbon Tunneling in Singlet Carbenes

As described earlier in this chapter, reactions dominated by movement of heavier nuclei, because of smaller positional uncertainty or shorter de Broglie wavelengths, are rarer than those that involve transfers of hydrogen. However, the low energy barriers and small translations that are often needed to accomplish some fundamental rearrangements in singlet carbenes have been found to enable facile tunneling involving carbon migration.

Considerable experimental[77,78] and theoretical[66] evidence indicates that ring expansions of singlet cyclopropylhalocarbenes (**65**) have minimal contributions from tunneling. Carbene **65** has been shown to be indefinitely stable in low temperature matrices.[77,78]

On the contrary, rearrangements of cyclobutylhalocarbenes to the corresponding halocyclopentenes could be anticipated to have lower energy barriers, because of greater relief of ring strain and less hyperconjugative stabilization in the carbenes themselves. Indeed, rearrangement rates measured in solution are greater for cyclo-butylcarbenes than for the analogous cyclopropylcarbenes.[62] Hence, there might be reason to think that heavy atom tunneling in the ring expansion of singlet cyclobutyl-carbenes might be feasible. To thwart a facile H-shift that might complete,[62] however, it would be necessary to block the cyclobutyl carbon adjacent to the carbene center.

A joint experimental and theoretical effort[79,80] demonstrated that rearrange-ment of appropriately substituted singlet cyclobutylhalocarbenes can occur solely via carbon tunneling. In contrast to previously reported cyclopropylcarbene sys-tems,[77,78] irradiation of chlorodiazirine **66** at 8 K in a N_2 matrix produced no detect-able carbene **67**, but instead gave only chlorocyclopentene **68** in the IR spectrum. In contrast, irradiation of matrix isolated fluorodiazirine **69** gave first mainly the diazo isomer **70**. Subsequent irradiation with visible light produced carbene **71**. Two conformational isomers of carbene **71** could be distinguished in the IR spec-trum. One conformer was found to rearrange slowly to fluorocyclopentene **72** at 8 K in the dark. The decay followed approximately linear first-order kinetics for at least the first 20% of decomposition, with a rate constant of 4.0×10^{-6} s^{-1}. At higher temperatures, for example, 16 K, the rate of disappearance of this isomer increased, and the other conformer began to rearrange as well. In Ar matrices, both carbene conformers underwent ring expansion even at 8 K, with the fastest rate measured being ca. 4×10^{-5} s^{-1}.

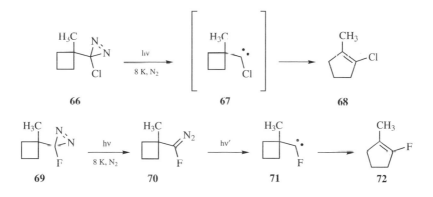

It was proposed that the more labile IR bands corresponded to the *exo*-carbene, **71-exo**, which has the methyl and fluoro substituents aligned properly for rearrange-ment. The slower decaying IR bands were assigned to *endo*-carbene, **71-endo**, which was presumed to require rotation to the *exo*-conformer before undergoing ring ex-pansion.

71-*endo*   71-*exo*

DFT calculations[79] predicted that rearrangement of chlorocarbene **67** to cyclopentene **68** is exothermic by -85.6 kcal/mol, with a barrier height of 3.1 kcal/mol. Considerable evidence indicates that, because of stronger π-donation, fluorine is more
stabilizing toward singlet carbenes than is chlorine.[62,81] Consistent with this, the ring
expansion of fluorocarbene **71** was calculated to be -78.4 kcal/mol exothermic and
to have a significantly larger barrier of 6.5 kcal/mol. It was noted that the barriers for
rearrangement of either **67** or **71** preclude the possibility of classical reaction at these
cryogenic temperatures. Moreover, the very small increases in rate of rearrangement
with increases in temperature were considered to be inconsistent with classically
thermally activated processes. Hence, it was suggested that the ring expansions of
67 and **71** involved QMT, dominated by carbon migration.

Support for tunneling in these reactions came from theoretical calculations. Direct
dynamics calculations[82] of the rate constants with canonical variational transition
state (CVT) theory, with tunneling contributions included via the small curvature
tunneling (SCT) approximation, confirmed the importance of heavy atom tunneling
in the ring expansions of **67** and **71**. The calculations predicted a low-temperature
limiting rate (T $<$ 20 K) of 1.4×10^4 s^{-1} for rearrangement of the chlorocarbene, corresponding to a half-life of only 10^{-4} s, and too fast for detection. It was recognized
that the results did not preclude direct product formation from excited diazirine. On
the contrary, fluorocarbene **71** was predicted to rearrange via tunneling at temperatures below 20 K with a limiting rate constant of 9.1×10^{-6} s^{-1}, satisfyingly close
(or, perhaps, fortuitously close) to the fastest observed experimental rates.

The calculations provided valuable insight into these reactions.[79] The actual tunneling process could be visualized as a transformation of carbene **71** to a highly twisted
configuration of the fluorocyclopentene **72**, followed by relaxation to an equilibrium
geometry; cf., Figure 10.5. The longest distance traversed by any carbon during tunneling was only 0.44 Å for the migrating methylene carbon, and the fluorine moved only
0.23 Å. At 8 K the reaction of **71** was predicted to occur virtually entirely via tunneling
from the lowest vibrational quantum levels. However, as temperature increased, tunneling from higher vibrational levels was found to increase correspondingly. Hence,
for example, the fraction of reaction via tunneling from $v = 0$ and 1 vibrational levels
was predicted to be equal at 39 K. The calculations further indicted that at higher
temperatures, nontunneling (overbarrier) processes began to contribute to the rearrangement, equaling the rate of tunneling reactions at 216 K; at 298 K, the overbarrier
component was predicted to be 2.7 times larger than the through-barrier contribution.

The calculational work offered an especially noteworthy point.[79] Even though
tunneling was calculated to play a significant role at ambient temperatures, it was
predicted that Arrhenius plots would be quite linear over a large temperature range.
Thus, these results emphasized once again that because both tunneling and classical
reactions are thermally activated, the linearity of Arrhenius temperature dependence
is no guarantee of the absence of tunneling.

Finally, two other experimental observations were addressed. First, the small increases in rate resulting from warming were suggested to be due to matrix softening
at higher temperatures, and Ar appeared to be a less-rigid matrix than N_2. Second,
the tunneling rate constant for rotation of **71-endo** to the reactive **71-exo** conformer

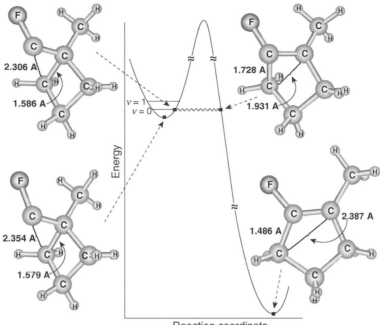

Figure 10.5. Schematic of ground-state potential curve for rearrangement of **71** to **72**. Four key structures are shown along the path. [Reproduced with permission from P. S. Zuev, R. S. Sheridan, T. V. Albu, D. G. Truhlar, D. A. Hrovat, and W. T. Borden, *Science* **2003**, *299, 867.*]

was theoretically predicted to be on the same order as that of ring expansion. The matrix surroundings were suggested to hinder rotation of the slower decaying conformer at lower temperatures.

Carbon tunneling in a second singlet chlorocarbene has also been proposed. It has proved impossible to observe noradamantylcarbene **73** spectroscopically, either by solution laser flash photolysis or with matrix isolation at low temperatures.[83] It has been suggested that the carbene rearranges too rapidly, possibly via carbon tunneling, to adamantene (**74**).

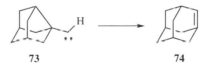

As is often the case, however, it was found that substituting Cl for H on the carbene center enabled detection of the carbene both in solution and in low-temperature matrices. Thus, LFP of diazirine **75** in solution gave carbene **76**, which could be intercepted by pyridine to give a spectroscopically observable ylide.[84] The lifetime of **76** at room temperature was found to be ca. 20 ns. This greater kinetic stability, where **76** is at least 400 times longer lived compared to estimates for parent **73** ($\tau < 50$ ps), was consistent with B3LYP calculated ring-expansion barriers of 5.3 kcal/mol for **76** versus 0.35 for **73**.[84]

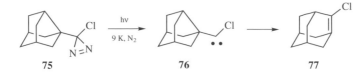

Photolysis of diazirine **75** in N_2 at 9 K produced carbene **76**, whose IR and UV/Vis spectra fit predictions by DFT calculations.[84] The IR spectra were most consistent with a carbene conformation with Cl aligned 90° to the adjacent C–CH bond. The carbene was found, by IR spectroscopy, to rearrange to chloroadamantene **77** slowly at 9 K in the dark. The rate of rearrangement was somewhat faster in Ar matrices at 9 K or at higher temperatures in N_2.

It was noted that if a conventional A value of ca. 10^{10} s^{-1} was assumed, the calculated barrier of 5.3 kcal/mol would predict a rate constant at 9 K of only $k = 1.3 \times 10^{-118}$ s^{-1} for a classical overbarrier rearrangement of **76**. The actual rate constant, however, for ring expansion of the carbene was measured to be 2.3×10^{-7} s^{-1} based on the first 10% of reaction in N_2 at 9 K. The reaction was thus attributed to carbon tunneling. Comparison of the geometry of carbene before reaction with the product geometry at the same energy for the product calculated along the intrinsic reaction pathway suggested that the longest distance traveled by any heavy atom during rearrangement was 0.44 Å for the carbene center. The Cl was estimated to require displacement of only 0.20 Å. The similarity in classical barrier height and "width" to those in the rearrangement of the 1-methylfluorocyclobutylcarbene (**71**)[79] supports the involvement of heavy atom tunneling in this case. Moreover, the minor rate accelerations observed on warming, or on switching from N_2 to Ar, were likewise attributed to changes in matrix environment.

In contrast, the larger ring adamantylchlorocarbene (**78**) has been shown to be indefinitely stable at 10 K in inert matrices.[85] Corresponding 1,2-C shift in this carbene to give **79** was calculated to have an energy barrier of 14.5 kcal/mol, higher than that of **76**.[84] In addition, similar analysis of estimated geometry changes during a tunneling pathway indicated that a larger nuclear displacement (e.g., 0.55 Å for the carbenic C) was also required. Both factors were considered to potentially contribute to the lack of tunneling in **78**.

10.3.6. Tunneling in Miscellaneous Reactive Intermediates

10.3.6.1. Tunneling in Nitrene Reactions   Nitrenes (e.g., **80**) are the nitrogen-based cousins of carbenes, likewise lacking two bonds compared with regular organic molecules. As might be expected, nitrenes share many electronic,

spectroscopic, and reactivity features with carbenes. However, although nitrenes are generally also extremely reactive species, examples of tunneling by nitrenes are very rare.

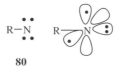

80

Nitrene chemistry has been extensively reviewed,[86] and Platz[87] has provided an excellent overview in Volume 1 of this series. It is, however, useful to highlight here several pertinent differences between carbenes and nitrenes. Carbene multiplicity is easily varied between triplet and singlet, depending on substitution. Moreover, triplet and singlet carbenes have also different electronic configurations, which together with spin differences lead to very distinct state-dependent chemistry. Nitrenes, on the contrary, generally have triplet ground states that lie substantially lower than the corresponding singlet states, at least with simple alkyl and aryl substituents. In addition, the corresponding singlet states of aryl nitrenes, for example, are open shell diradical-like species. Nevertheless, nitrenes undergo many of the same reactions as carbenes, such as C–H abstraction/insertions, 1,2-shifts, alkene additions, and so forth. Aryl nitrenes have figured prominently in photoaffinity labeling and photoresist technology,[86,87] and so there is considerable practical interest in their reactivity.

In general, aryl nitrenes abstract hydrogen from hydrocarbons more reluctantly than triplet carbenes. For example, phenyl nitrene abstracts H from toluene with a rate constant ~10^3 times lower than triplet diarylcarbenes.[87] This more moderate reactivity is apparent in experiments by Platz and Burns to directly observe intramolecular H-abstraction in a perinaphthyl nitrene.[57,88] Irradiation of azide **81** at 77 K in organic glasses produced two triplet species that could be characterized by EPR spectroscopy, nitrene **82** and biradical **83**. Both intermediates were stable at 77 K, but warming to 98 K caused rapid disappearance of the biradical signal, presumably due to ring closure; little change in the resonances of **82** were observed. Moreover, nitrene **82** was stable to further irradiation, and did not undergo photochemical H-abstraction.[88]

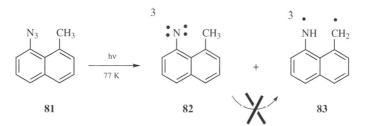

These results clearly showed that biradical **83** was not formed from ground state triplet nitrene **82**, either classically or via tunneling. The authors speculated that the biradical might have been produced from a higher energy electronically excited nitrene than was available from irradiation of **82**.[57,88] Alternatively, it is possible that the biradical might have formed directly from excited azide.

ortho-Alkyl aryl carbenes undergo facile intramolecular hydrogen shifts. As described in earlier sections, low-temperature evidence clearly indicates that [1,4]-H shifts in *o*-tolylmethylene (**51**)[58,59] and singlet chlorocarbene **60**[71] occur readily via tunneling in low temperature matrices. However, Sander and co-workers found different behavior in a similar arylnitrene.[89] Irradiation of *o*-tolylazide **84** in inert matrices at 10 K gave the triplet nitrene **85**, which could be characterized by IR spectroscopy. In contrast to the corresponding carbene, nitrene **85** was stable up to 40 K in Ar or 80 K in Xe matrices, giving no detectable H-shifted product. Subsequent irradiation led to ring expansion to an azacycloheptatetraene.

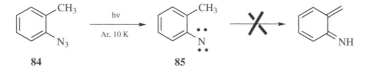

The authors[89] suggested that the lack of tunneling in the case of H-abstraction by **85** might be due to the larger T–S energy gap compared with triplet carbenes, creating a higher effective barrier corresponding to the crossing point of the triplet nitrene surface with that of the singlet product. Alternatively, it was noted that the N–H distance in nitrene **85** might be greater than the carbene–H distance in **51**, causing a prohibitively wide barrier for tunneling.

Murata et al. have reported tentative evidence for tunneling in the hydrogen shift of an alkyl arylnitrene at ambient temperatures.[90,91] Irradiation of isotopically labeled azide **86** in solution gave a mixture of C–H and C–D insertion products **88** and **89**. The ^{1}H NMR spectrum of the product mixture indicated a k_H/k_D ratio of 14.7 at 20°C, which was considered too large compared with the transition state theoretical maximum of ca. 12. Thus, QMT was proposed for the insertion reaction, which was suggested to involve H-abstraction/recombination by triplet nitrene **87**. Moreover, an Arrhenius plot of the product ratio over the range −14 to 50°C indicated $E_a(H) - E_a(D) = 1.77$, a bit larger than predicted classically. However, the authors noted that over this limited temperature range the Arrhenius plot of k_H/k_D was linear, and that $A_H/A_D = 0.63$, only slightly into the "tunneling" range. Therefore, further study was needed to confirm the importance of QMT in these reactions.

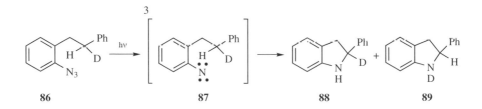

Workers have considered evidence for heavy atom tunneling in the 1,2-halo shifts of several oxynitrenes. It was observed that irradiation of nitrosylchloride (**90**) in Ar or N_2 matrices at 10 K gave small amounts of rearranged isonitrosyl compound **91** that could be detected by IR spectroscopy.[92] In the dark, the absorptions of **91** decayed, reflecting rearrangement back to **90** with rate constants of ca. 2×10^{-4} s^{-1} in Ar, and 37×10^{-4} s^{-1} in N_2. Warming the Ar sample to 25 K did not increase the rate of rearrangement, and hence it was proposed that the rearrangement involved heavy atom tunneling. Various levels of theory suggested activation barriers of ca. 8–11 kcal/mol for the 1,2-Cl shift.

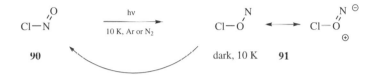

Subsequent work by Maier and co-workers[93] confirmed the low-temperature rearrangement of **91** to **90**, but raised questions about whether QMT was involved. They found similar rearrangement of Br–ON to Br–NO, with a comparable, temperature independent, rate at 10 K. Calculations indicated that for both the Cl and Br-isonitrosyl compounds, the barriers for concerted 1,2-shift were predicted to be similar to the dissociation energies to radicals. They noted several arguments against tunneling in these processes. The similarity in rates for Cl and Br shift were considered to be inconsistent with heavy atom tunneling, although it was suggested that a concerted shift might involve minimal halogen translation, and rather a simple transposition of O and N. They also pointed out that the corresponding H–ON compound was known to be stable at low temperatures, and argued that tunneling would certainly favor a 1,2-H shift. They proposed an alternative mechanism involving activationless crossing of the chloro- and bromoisonitrosyl compounds to a dissociative triplet surface instead.

10.3.6.2. Tunneling in a Cyclopropene to Triplet Carbene Rearrangement
In previous sections, tunneling in the reactions of a number of carbenes to give covalently saturated products has been described. We conclude this section with a unique example where tunneling appears to dominate the converse of this process, cleavage of a cyclopropene to a triplet carbene.

Sander and co-workers found that visible irradiation of triplet oxocyclohexadienylcarbene (**92**) in cryogenic matrices gave a new, thermally labile product which could be characterized by IR and UV/Vis spectra.[94–96] In the dark, this new product slowly isomerized back to carbene **92**. It was originally suggested that this "isomer" might correspond to a meta-stable singlet carbene, which underwent unusually slow intersystem crossing, of the order of days, back to **92**.[94] However, later theoretical work, and comparison of calculated IR spectra to experiment, showed that the new species was the ring-closed bicyclopropene **93**.[95,96]

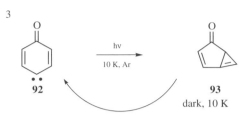

The bicyclic isomer **93** was shown to be extremely sensitive to IR radiation, which facilitated conversion back to **92**. However, when care was taken to filter most of the IR radiation during the recording of spectra, it was found that **93** ring opened with a rate in Ar at 10 K of ca. 1.4×10^{-6} s^{-1}. Reflecting dispersive kinetics, presumed to be from multiple matrix sites, decay of **93** was found to approximately follow $t^{0.7}$ "stretched exponential" dependence. Moreover, the rates of rearrangement depended on the matrix material to some extent, with approximate relative rates of Ar < N$_2$ < Kr < Xe. The derived "rate constants" were measured at various temperatures, and Arrhenius plots in all cases showed considerable curvature, with nearly flat regions at lowest temperatures, followed by rapidly increasing rates as the temperature was raised. On the basis of this typical temperature dependence, the authors proposed that the rearrangements involved heavy atom tunneling from singlet **93** to triplet **92**. Interestingly, however, the rate of rearrangement of perdeuterated **93** was found to be only marginally slower under similar conditions and matrix materials.

The Arrhenius curves were fitted to Bell-type truncated parabolic barriers,[1,2,12] arbitrarily assuming a mass of 12 for the "tunneling particle". In Ar, the curves were consistent with a reaction barrier of ca. 2.6 kcal/mol, log A (s^{-1}) $= 7$, and barrier width of 0.4 Å, consistent with expected geometry changes in the ring opening of **93**. It was suggested that the low A value reflected the necessity for singlet–triplet intersystem crossing in the rearrangement. Similarly, Xe matrices were considered to facilitate spin-flip through an external heavy atom effect. Lastly, it was suggested that the small, and under some conditions even inverse, D-effect arose from the influence of C–D versus C–H vibrations on intersystem crossing rates.[95,96]

Later calculations by others[97] at the CASPT2 level indicated that bicyclic compound **93** lies approximately 19 kcal/mol higher in energy than triplet carbene **92**. It was suggested that the predicted energetics permitted ring opening of **93** first to an open-shell singlet carbene corresponding to **92**, followed by relaxation to the triplet carbene.

10.4. CONCLUSION AND OUTLOOK

The diverse examples described in this chapter demonstrate that quantum mechanical tunneling is pervasive in both inter- and intramolecular reactions of organic reactive intermediates. This is not surprising. By their very nature, reactive intermediates routinely lie in shallow potential energy wells, protected only by low energy barriers from highly exothermic reactions. Often, only minimal geometric reorganization is required in the transformations of reactive intermediates to transition states and to subsequent products. Thus, the quantum uncertainty in geometry, or "nuclear delocalization," of reactants can easily extend through the low-lying and narrow barriers into product regions.

Tunneling is not the only deviation from classical transition state theory to result from the shallow potential energy wells for many reactive intermediates. As eloquently described by Carpenter in the first volume of this series,[98] it is becoming increasingly recognized that nonstatistical dynamic effects may play important roles in the chemistry of highly reactive species. Actual reaction trajectories may need to be considered in detail to rationalize and predict product ratios. As our understanding of these dynamic effects becomes more sophisticated, it is likely that the influences of tunneling on the different trajectories will also be important to consider.

Reactive intermediates offer obvious advantages for studying tunneling. In many instances it has been possible to strip away the distractions of thermally activated overbarrier reactions, leaving behind only QMT processes at very low temperatures. Nevertheless, as increasingly rapid spectroscopic methods are applied to the solution reactions of reactive intermediates, it is likely that more examples of tunneling will become apparent. Most commonly, organic reactive intermediates are prepared for spectroscopic interrogation through highly energetic photochemical reactions. Thus, there is ample opportunity for these species to be diverted to products through tunneling pathways as they vibrationally relax to energy minima. It will be interesting to see what role tunneling plays in the "direct" formation of products that often accompany photochemical generation of reactive intermediates. Similarly, applications of spectroscopic methods with shorter timescales should be able to uncover a wider variety of tunneling processes in matrix isolation spectroscopy at very low temperatures.

Undoubtedly, tunneling in the reactions of a greater variety of organic reactive intermediates will be revealed in the future, in both hydrogen transfers and reactions dominated by heavy atom movement. Considering the number and diversity of carbene reactions that have been shown to involve tunneling, it is surprising that there is a paucity of examples of QMT in corresponding nitrene reactions. The very few reports in this area are not completely clear cut, and have been open to interpretation. It remains to be seen why (or, perhaps, whether) tunneling is less important in nitrene reactions.

Finally, modern theoretical methods and rapidly progressing computer technology have brought remarkable calculational power to the organic chemist's desktop. With DFT methods, and inexpensive computer hardware, it is becoming relatively trivial to calculate quite reasonable geometries, electronic structures, spectra of various sorts, and transition state barriers for even quite large molecules. In contrast, calculational treatments of rates of reactions involving tunneling are more sophisticated and less common. Moreover, programs allowing reasonable predictions of tunneling rates are considerably more computationally intensive than commonly used electronic structure packages. It may be hoped that in future, as the importance of QMT in organic reactions becomes more generally recognized, it will become more routine for the practising organic chemist to include tunneling in theoretical estimations of reaction rates.

Although the potential influence of quantum uncertainty on chemical reactions was divined early in the development of quantum mechanics, appreciation of the effects of tunneling in organic reactions is still evolving. It is rapidly becoming apparent that quantum mechanical tunneling is a significant contributor to a wide variety of organic reactions ranging from biological processes to interstellar chemistry. Organic reactive intermediates have provided a particularly fruitful area for detailed examination of the factors that influence tunneling, and that role will continue.

SUGGESTED READING

R. P. Bell, *The Tunnel Effect in Chemistry*, Chapman and Hall, New York, **1980**.

E. Caldin, " Tunneling in proton-transfer reactions in solution," *Chem. Rev.* **1969**, *69*, 135.

M. Sponsler, R. Jain, F. Coms, and D. A. Dougherty, "Matrix-isolation decay kinetics of triplet cyclobutanediyls. Observation of both Arrhenius behavior and heavy-atom tunneling in C–C bond-forming reactions," *J. Am. Chem. Soc.* **1989**, *111*, 2240.

B. R. Arnold and J. Michl, "Spectroscopy of cyclobutadiene," in *Kinetics and Spectroscopy of Carbenes and Biradicals*, (Ed. M. S. Platz), Plenum Press, New York, **1990**, p. 1f.

M. S. Platz, "Atom-transfer reactions of aromatic carbenes," *Acc. Chem. Res.* **1988**, *21*, 236.

M. S. Platz, "The chemistry, kinetics, and mechanisms of triplet carbene processes in low-temperature glasses and solids," in *Kinetics and Spectroscopy of Carbenes and Biradicals*, (Ed. M. S. Platz), *Plenum Press, New York,* **1990**, pp. 143f.

P. S. Zuev, R. S. Sheridan, T. V. Albu, D. G. Truhlar, D. A. Hrovat, W. T. Borden, "Carbon tunneling from a single quantum state," *Science* **2003**, *299*, 867.

T. V. Albu, B. J. Lynch, D. G. Truhlar, A. C. Goren, D. A. Hrovat, W. T. Borden, and R. A. Moss, "Dynamics of 1, 2-hydrogen migration in carbenes and ring expansion in cyclopropylcarbenes," *J. Phys. Chem. A* **2002**, *106*, 5323.

REFERENCES

1. (a) R. P. Bell, *The Tunnel Effect in Chemistry*, Chapman and Hall, New York, **1980**. (b) R. P. Bell, *The Proton in Chemistry*, Chapman and Hall, New York, **1973**.

2. (a) E. Caldin, *Chem. Rev.* **1969**, *69*, 135. (b) T. Miyazaki, Ed., *Atom Tunneling Phenomena in Physics, Chemistry, and Biology*, Springer-Verlag, Berlin, Germany, **2003**.

3. (a) Z-X. Liang and J.P. Klinman, *Curr. Op. Struct. Biol.* **2004**, *14*, 648. (b) S. Hammes-Schiffer, *Acc. Chem. Res.* **2006**, *39*, 93. (c) A. Kohen, and J. P. Klinman, *Chem. Biol.* **1999**. *6*, R198.

4. K. Hiraoka, T. Sato, and T. Takayama, *Science* **2001**, *292*, 869.

5. S. Hammes-Schiffer, *Acc. Chem. Res.* **2001**, *34*, 273.

6. R. L. Redington, T. E. Redington, T. A. Blake, R. L. Sams, and T. J. Johnson, *J. Chem. Phys.* **2005**, *122*, 224311.

7. S. E. Wonchoba, W.-P. Hu, and D. G. Truhlar, *Phys. Rev.* **1995**, *B51*, 9985.

8. (a) B.A. Johnson, M. H. Kleinman, N. J. Turro, and M. A. Garcia-Garibay, *J. Org. Chem.* **2002**, *67*, 6944. (b) B. A. Johnson, Y. F. Hu, K. N. Houk, and M. A. Garcia-Garibay, *J. Am. Chem. Soc.* **2001**, *123*, 6941. (c) B. A. Johnson, and M. A. Garcia-Garibay, *J. Am. Chem. Soc.* **1999**, *121*, 8114. (d) A. Gamarnik, B. A. Johnson, and M. A. Garcia-Garibay, *J. Phys. Chem. A* **1998**, *102*, 5491. (e) L. M. Campos, M. V. Warrier, K. Peterfy, K. N. Houk, and M. A. Garcia-Garibay, *J Am. Chem. Soc.* **2005**, *127*, 10178.

9. (a) D. J. Miller, R. Subramanian, and W. H. Saunders, Jr., *J. Am. Chem. Soc.* **1981**, *103*, 3519. (b) M. P. Meyer, A. J. DelMonte, and D. A. Singleton, *J. Am. Chem. Soc.* **1999**, *121*, 10865.

10. (a) J. W. Moore and R. G. Pearson, *Kinetics and Mechanism*, John Wiley & Sons, Inc., New York, **1981**. (b) S. W. Benson, "*Themochemical Kinetics*," Second Edition, Wiley-Interscience, New York, **1976**.

11. See, for example, D. A. McQuarrie, *Quantum Chemistry*, University Science Books, Mill Valley, CA, **1983**.

12. Interestingly, this equation does not appear in this form in the oft-cited classic text by Bell (Ref. 1a), and refers to an earlier derivation. Bell derived a simpler expression, described in Ref. 1a, for tunneling through a parabolic barrier which affords approximately the same results. See Ref. 2a for discussion.

13. W. T. Borden, in *Reactive Intermediate Chemistry*, (Eds. R. A. Moss, M. S. Platz, and M. J. Jones, Jr.) Wiley, Hoboken, NJ, **2004**, p 961f.

14. J. Bigeleisen, *J. Phys. Chem.* **1952**, *56*, 823.

15. T. Bally, in *Reactive Intermediate Chemistry*,(Eds. R. A Moss, M. S. Platz, and M. J. Jones, Jr.), Wiley, Hoboken, NJ, **2004**, p 797f.

16. (a) E. D. Sprague and F. Williams, *J. Am. Chem. Soc.* **1971**, *93*, 767. (b) K. Takeda, J.-T. Wang. and F. Williams, *Can J. Chem.* **1974**, *52*, 2840. (c) J.-T. Wang and F. Williams, *J. Am. Chem. Soc.* **1972**, *94*, 2930. (d) A. Campion and F. Williams, *J. Am. Chem. Soc.* **1972**, *94*, 7633. (e) F. P. Sargent, M. G. Bailey, and E. M. Gardy, *Can J. Chem.* **1974**, *52*, 2171. (f) R. J. LeRoy, E. D. Sprague, and F. Williams, *J. Phys. Chem.* **1972**, *76*, 546. (g) W. Siebrand, and T. A. Wildman, *Acc. Chem. Res.* **1986**, *19*, 236.

17. M. Sponsler, R. Jain, F. Coms, and D. A. Dougherty, *J. Am. Chem. Soc.* **1989**, *111*, 2240.

18. L. R. C. Barclay, D. Griller, and K. U. Ingold, *J. Am. Chem. Soc.* **1974**, *96*, 301.

19. (a) G. Brunton, D. Griller, L. R. C. Barclay, and K. U. Ingold, *J. Am. Chem. Soc.* **1976**, *98*, 6803. (b) G. Brunton, J. A. Gray, D. Griller, L. R. C. Barclay, and K. U. Ingold, *J. Am. Chem. Soc.* **1977**, *100*, 4197.

20. B. C. Bockrath, E. W. Bittner, and T. C. Marecic, *J. Org. Chem.* **1986**, *51*, 15.

21. Platz, M. S. *Kinetics and Spectroscopy of Carbenes and Biradicals*, Plenum Press, New York, **1990**, and references therein.

22. (a) S. Buchwalter and G. L. Closs, *J. Am. Chem. Soc.* **1975**, *97*, 3857. (b) S. Buchwalter and G. L. Closs, *J. Am. Chem. Soc.* **1979**, *101*, 4688.

23. J-F. Muller, D. Muller, H. J. Dewey, and J. Michl, *J. Am. Chem. Soc.* **1978**, *100*, 1629.

24. M. Gisin, E. Rommel, J. Wirz, M. N. Burnett, and R. M. Pagni, *J. Am. Chem. Soc.* **1979**, *101*, 2216.

25. J. J. Fisher, J. H. Penn, D. Döhnert, and J. Michl, *J. Am. Chem. Soc.* **1986**, *108*, 1715.

26. J. J. Fisher and J. Michl, *J. Am. Chem. Soc.* **1987**, *109*, 583.

27. J. C. Koziar and D. O. Cowan, *Acc. Chem. Res.* **1978**, *11*, 334.

28. D. J. Cram, M. E. Tanner, and R. Thomas, *Angew. Chem. Int. Ed. Engl.* **1991**, *30*, 1024.

29. (a) T. Bally and S. Masamune, *Tetrahedron* **1980**, *36*, 343. (b) G. Maier, *Angew. Chem. Int. Ed. Engl.* **1988**, *27*, 309.

30. B. R. Arnold and J. Michl, in ref. 21, p 1.

31. G. Maier, H-O. Kalinowski, and Euler, K., *Angew. Chem. Int. Ed. Engl.*, **1982**, *21*, 693.

32. (a) D. W. Whitman and B. K. Carpenter, *J. Am. Chem. Soc.* **1980**, *102*, 4272. (b) D. W. Whitman and B. K. Carpenter, *J. Am. Chem. Soc.* **1982**, *104*, 6473.

33. B. K. Carpenter, *J. Am. Chem. Soc.* **1983**, *105*, 1700.

34. M. J. S. Dewar, K. M. Merz, Jr., and J. J. P. Stewart, *J. Am. Chem. Soc.* **1984**, *106*, 4040.

35. M-J. Huang and M. Wolfsberg, *J. Am. Chem. Soc.* **1984**, *106*, 4039.

36. R. L. Redington, *J. Chem. Phys.*, **1998**, *109*, 10781.

37. (a) B. R. Arnold and J. Michl, *J. Phys. Chem.* **1993**, *97*, 13348. (b) B. R. Arnold, J. G. Radziszewski, A. Campion, S. S. Perry, and J. Michl, *J. Am. Chem. Soc.* **1991**, *113*, 692.

38. A. M. Orendt, B. R. Arnold, J. G. Radziszewski, J. C. Facelli, K. D. Malsch, H. Strub, D. M. Grant, and J. Michl, *J. Am. Chem. Soc.* **1988**, *110*, 2648.

39. M. Jones, Jr. and R. A. Moss, in *Reactive Intermediate Chemistry* (Eds. Moss, R. A.; Platz, M. S.; Jones, M., Jr.) Wiley, Hoboken, NJ, **2004**, p. 273f.

40. G. Bertrand, in *Reactive Intermediate Chemistry* (Eds. Moss, R. A., Platz, M. S., Jones, M., Jr.), Wiley, Hoboken, NJ, **2004**, p 329f.

41. H. Tomioka, in *Reactive Intermediate Chemistry* (Eds. Moss, R. A., Platz, M. S., Jones, M., Jr.), Wiley, Hoboken, NJ, **2004**, p 375f.

42. For other useful carbene references, see the following: (a) W. Kirmse, *Carbene Chemistry*, Academic Press, New York, **1964**. (b) W. Kirmse, *Carbene Chemistry*, 2nd Edition, Academic Press, New York, **1971**. (c) M. Jones, Jr. and R. A. Moss, Eds., *Carbenes, Vol. 1*, Wiley-Interscience, New York, **1973**. (d) R. A. Moss and M. Jones, Jr., Eds., *Carbenes, Vol. 2*, Wiley-Interscience, New York, **1975**. (e) M. Regitz, Ed., *Carbene (Carbenoide) Methoden der Organische Chemie (Houben-Weyl), Vol. E19b*, Thieme, Stuttgart, **1989**. (f) U. H. Brinker, Ed., *Advances in Carbene Chemistry, Vol. 1*, JAI Press, Greenwich, CT, **1994**. (g) U. H. Brinker, Ed., *Advances in Carbene Chemistry*, Vol. 2, JAI Press, Stamford, CT, **1998**. (h) U. H. Brinker, Ed., *Advances in Carbene Chemistry*, Vol. 3. Elsevier, Amsterdam, The Netherlands, **2001**. (i) Wentrup, C. *Reactive Molecules: the Neutral Reactive Intermediates in Organic Chemistry*, Wiley, New York, **1984**. (j) H. Tomioka, in Ref. 2b, p. 147.

43. R. A. Moss and U.-H. Dolling, *J. Am. Chem. Soc.* **1971**, *93*, 954.

44. (a) R. A. Moss and M. A. Joyce, *J. Am. Chem. Soc.* **1977**, *99*, 1263 and 7399. (b) R. A. Moss and J. K. Huselton, *J. Am. Chem. Soc.* **1978**, *100*, 1324. (c) R. A. Moss and M. A. Joyce, *J. Am. Chem. Soc.* **1978**, *100*, 4475.

45. (a) H. Tomioka and Y. Izawa, *J. Am. Chem. Soc.* **1977**, *99*, 6128. (b) H. Tomioka, S. Suzuki, and Y. Izawa, *Chem. Lett.* **1982**, 863. (c) H. Tomioka, H. Okuno, and Y. Izawa, *J. Chem. Soc., Perkin Trans. 2* **1980**, 1634. (d) H. Tomioka, T. Inagaki, and Y. Izawa, *Chem. Commun.* **1976**, 1023. (e) H. Tomioka, T. Inagaki, S. Nakajura, and Y. Izawa, *J. Chem. Soc., Perkin Trans. 1* **1979**, 130.

46. (a) B. B. Wright, V. P. Senthilnathan, M. S. Platz, and C. W. McCurdy, Jr., *Tetrahedron Lett.* **1984**, *25*, 833. (b) T. G. Savino, K. Kanakarajan, and M. S. Platz, *J. Org. Chem.* **1986**, *51*, 1305. (d) B. B. Wright and M. S. Platz, *J. Am. Chem. Soc.* **1984**, *106*, 4175.

47. M. S. Platz, in Ref. 21, pp. 143f.

48. (a) V P. Senthilnathan and M. S. Platz, *J. Am. Chem. Soc.* **1980**, *102*, 7637. (b) V. P. Senthilnathan and M. S. Platz, *J. Am. Chem. Soc.* **1981**, *103*, 5503.

49. C.-T. Lin and P. P. Gaspar, *Tetrahedron Lett.* **1980**, *21*, 3553.

50. M. S. Platz, *Acc. Chem. Res.* **1988**, *21*, 236.

51. (a) M. S. Platz, V. P. Senthilnathan, B. B. Wright, and C. W. McCurdy, Jr. *J. Am. Chem. Soc.* **1982**, *104*, 6494 (b) B. B. Wright, K. Kanakarajan, and M. S. Platz, *J. Phys. Chem.* **1985**, *89*, 3574. (c) R. L. Barcus, B. B. Wright, E. Leyva, and M. S. Platz, *J. Phys. Chem.* **1987**, *91*, 6677. (d) T. G. Savino, N. Soundararajan, and M. S. Platz, *J. Phys. Chem.* **1986**, *90*, 919. (e) R. L. Barcus, M. S. Platz, and J. C. Scaiano, *J. Phys. Chem.* **1987**, *91*, 695.

52. P. A. Gray, A. A. Herod, and A. Jones, *Chem. Rev.* **1971**, *71*, 257.

53. J. Ruzicka, E. Leyva, and M. S. Platz, *J. Am. Chem. Soc.* **1992**, *114*, 897.

54. M. W. Shatter, E. Leyva, N. Soundararajan, E. Chang, D. H. S. Chang, V. Capuano, and M. S. Platz, *J. Phys. Chem.* **1991**, *95*, 7273.

55. P. S. Zuev and R. S. Sheridan, *J. Am. Chem. Soc.* **2001**, *123*, 12434.

56. (a) W. Sander, R. Hübert, E. Kraka, J. Gräfenstein, and D. Cremer, *Chem. Eur. J.* **2000**, *6*, 4567. (b) W. Sander, C. Kötting, R. Hübert, *J. Phys. Org. Chem.* **2000**, *13*, 561. The addition of H_2 to the tetrafluoro analog of **40** was noted, but possible reaction of **40** itself was not addressed.

57. (a) M. S. Platz, G. Carrol, F. Pierrat, J. Zayas, and S. Auster, *Tetrahedron* **1982**, *38*, 777. (b) M. S. Platz, *J. Am. Chem. Soc.* **1980**, *102*, 1192. (c) M. S. Platz and J. R. Burns, *J. Am. Chem. Soc.* **1979**, *101*, 4425. (d) M. S. Platz, *J. Am. Chem. Soc.* **1979**, *101*, 3398. (e) M. J. Fritz, E. L. Ramos, and M. S. Platz, *J. Org. Chem.* **1985**, *50*, 3522.

58. R. J. McMahon, and O. L. Chapman, *J. Am. Chem. Soc.* **1987**, *109*, 683.

59. O. L. Chapman, J. W. Johnson, R. J. McMahon, and P. R. West, *J. Am. Chem. Soc.* **1988**, *110*, 501. O.L. Chapman, R. J. McMahon, and P. R. West, *J. Am. Chem. Soc.* **1984**, *106*, 7973.

60. A. Admasu, M. S. Platz, A. Marcinek, J. Michalak, A. D. Gudmundsdóttir, and J. Gebicki, *J. Phys. Org. Chem.* **1997**, *10*, 207.

61. W. v. E Doering, R. G. Buttery, R. G. Laughlin, and N. Chaudhuri, *J. Am. Chem. Soc.* **1956**, *78*, 3224.

62. D. C. Merrer and R. A. Moss, in *Advances in Carbene Chemistry, Vol. 3* (Ed. U. H. Brinker) Elsevier, Amsterdam, Netherlands, **2001**, p 53.

63. (a) J. A. LaVilla and J. L. Goodman, *J. Am. Chem. Soc.* **1993**, *115*, 6877. (b) J. A. LaVilla, J. L. Goodman, *Tetrahedron Lett.* **1990**, *31*, 5109.

64. E. J. Dix, M. S. Herman, and J. L Goodman, *J. Am. Chem. Soc.* **1993**, *115*, 10424.

65. J. W. Storer and K. N. Houk, *J. Am. Chem. Soc.* **1993**, *115*, 10426.

66. T. V. Albu, B. J. Lynch, D. G. Truhlar, A. C. Goren, D. A. Hrovat, W. T. Borden, and R. A. Moss, *J. Phys. Chem. A* **2002**, *106*, 5323.

67. E. Kraka and D. Cremer, *J. Phys. Org. Chem.* **2002**, *15*, 431.

68. (a) M. T. H. Liu, *Acc. Chem. Res.* **1994**, *27*, 287. (b) M. T. H. Liu, R. Bonneau, S. Wierlacher, and W. J. Sander, *Photochem. Photobiol., A: Chem.* **1994**, *84*, 133. (c) M. T. H. Liu and R. Bonneau, *J. Am. Chem. Soc.* **1990**, *112*, 3915.

69. D. C. Merrer, R. A. Moss, M. T. H. Liu, J. T. Banks, and K. U. Ingold, *J. Org. Chem.* **1998**, *63*, 3010.

70. S. Wierlacher, W. Sander, and M. T. H. Liu, *J. Am. Chem. Soc.* **1993**, *115*, 8943.

71. N. Nakane, T. Enyo, H. Tomioka, *J. Org. Chem.* **2004**, *69*, 3538.

72. W. Sander and C. Kötting, *Chem. Eur. J.* **1999**, *5*, 24. (b) C. Kötting and W. Sander, *J. Am. Chem. Soc.* **1999**, *121*, 8891.

73. P. S. Zuev and R. S. Sheridan, *J. Am. Chem. Soc.* **1994**, *116*, 4123.

74. R. A. Moss and W. Liu, *J. Chem. Soc, Chem. Commun.* **1993**, 1597.

75. R. S. Sheridan and P. S. Zuev, *unpublished* results.

76. R. A. Moss and W. Liu, *Tetrahedron Lett.* **1996**, *37*, 279.

77. (a) G-J. Ho, K. Krogh-Jespersen, R. A. Moss, S. Shen, R. S. Sheridan, and R. Subramanian, *J. Am. Chem. Soc.* **1989**, *111*, 6875. (b) See reference 55 for a methyl substituted example.

78. G. Chu, R. A. Moss, R. R. Sauers, R. S. Sheridan, and P. S. Zuev, *Tetrahedron Lett.* **2005**, *46*, 4137.

79. P. S. Zuev, R. S. Sheridan, T. V. Albu, D. G. Truhlar, D. A. Hrovat, and W. T. Borden, *Science* **2003**, *299*, 867.

80. See also: (a) "Chemical reactions involving quantum tunneling," R. J. McMahon, *Science* **2003**, *299*, 833. (b) "Do heavy nuclei see light at the end of the tunnel?," R. Berger, *Angew. Chem. Int. Ed.* **2004**, *43*, 398.

81. R. A. Moss, S. Xue, and R. R. Sauers, *J. Am. Chem. Soc.* **1996**, *118*, 10307.

82. (a) D. G. Truhlar and B. C. Garrett, *Annu. Rev. Phys. Chem.* **1984**, *35*, 159. (b) Y. P. Liu, G. C. Lynch, T. N. Truong, D. H. Lu, D. G. Truhlar, and B. C. Garrett, *J. Am. Chem. Soc.* **1993**, *115*, 2408.

83. (a) E. L. Tae, C. Ventre, Z. Zhu, I. Likhotvorik, F. Ford, E. Tippmann, and M. S. Platz, *J. Phys. Chem. A* **2001**, *105*, 10146. (b) E. L. Tae, Z. Zhu, and M. S. Platz, *J. Phys. Chem. A* **2001**, *105*, 3803.

84. R. A. Moss, R. R. Sauers, R. S. Sheridan, J. Tian, and P. S. Zuev, *J. Am. Chem. Soc.* **2004**, *126*, 10196.

85. G. Yao, P. Rempala, C. Bashore, and R. S. Sheridan, *Tetrahedron Lett.* **1999**, *40*, 17.

86. (a) W. Lwowski, Ed., *Nitrenes*, Wiley, New York, **1970**. (b) E. F. V. Scriven, Ed., *Azides and Nitrenes*, Academic Press: New York, **1984**. (c) G. B. Schuster and M. S. Platz, *Adv. Photochem.* **1992**, *17*, 69. (d) N. P. Gritsan and M. S. Platz, *Adv. Phys. Org. Chem.* **2001**, *36*, 255.

87. M. S. Platz, in *Reactive Intermediate Chemistry*, (Eds. R. A. Moss , M. S. Platz, and M. Jones, Jr.,) Wiley: Hoboken, NJ, **2004**, p. 501f.

88. M. S. Platz, G. Carrol, F. Pierrat, J. Zayas, and S. Auster, *Tetrahedron* **1982**, *38*, 777.

89. J. Morawietz, W. Sander, and M. Traubel, *J. Org. Chem.* **1995**, *60*, 6368.

90. S. Murata, Y. Tsubone, R. Kawai, D. Eguchi, and H. Tomioka *J. Phys. Org. Chem.* **2005**, *18*, 9.

91. Also see, H. Tomioka, *Bull. Chem. Soc. Jpn.* **1998**, *71*, 1501.

92. A. Hallou, L. Schriver-Mazzuoli, A. Schriver, and P. Chaquin, *Chem. Phys.* **1998**, *237*, 251.

93. G. Maier, H. P. Reisenauer, and M. De Marco, *Chem. Eur. J.* **2000**, *6*, 800.

94. (a) W. Sander, W. Mueller, and R. Sustmann, *Angew. Chem., Int. Ed. Engl.* **1988**, *27*, 572. (b) W. Sander, A. Patyk, and G. Bucher, *J. Mol. Struct.* **1990**, *222*, 21. (c) Interestingly, this original proposal has apparently never been retracted explicitly in the literature.

95. W. Sander, G. Bucher, F. Reichel, and D. Cremer, *J. Am. Chem. Soc.* **1991**, *113*, 5311.

96. G. Bucher and W. Sander, *J. Org. Chem.* **1992**, *57*, 1346.

97. A. Sole, S. Olivella, J. M. Bofill, and J. M. Anglada, *J. Phys. Chem.* **1995**, *99*, 5934.

98. B. K Carpenter, in *Reactive Intermediate Chemistry* (Eds. R. A., Moss, M. S. Platz, and M. Jones, Jr.), Wiley, Hoboken, NJ, **2004**, p. 925f.

Reviews of Reactive Intermediate Chemistry. Edited by Matthew S. Platz, Robert A. Moss,
Maitland Jones, Jr.